Richard M. Aron, Eva A. Gallardo Gutiérrez, Miguel Martin, Dmitry Ryabogin, Ilya M. Spitkovsky, Artem Zvavitch (Eds.)

The Mathematical Legacy of Victor Lomonosov

Advances in Analysis and Geometry

Volume 2

The Mathematical Legacy of Victor Lomonosov

Operator Theory

Edited by
Richard M. Aron, Eva A. Gallardo Gutiérrez,
Miguel Martin, Dmitry Ryabogin, Ilya M. Spitkovsky, and
Artem Zvavitch

DE GRUYTER

Editors

Prof. Richard M. Aron
Kent State University
Department of Mathematical Sciences
800 E. Summit St. Kent
Kent, Ohio 44242
USA
aron@math.kent.edu

Prof. Dmitry Ryabogin
Kent State University
Department of Mathematical Sciences
800 E. Summit St. Kent
Kent, Ohio 44242
USA
ryabogin@math.kent.edu

Prof. Eva A. Gallardo Gutiérrez
Universidad Complutense de Madrid
Facultad de Matemáticas
Plaza de Ciencias 3
28040 Madrid
Spain
eva.gallardo@mat.ucm.es

Prof. Ilya M. Spitkovsky
New York University Abu Dhabi
Saadiyat Island
Division of Science
129188 Abu Dhabi
UAE
imspitkovsky@gmail.com

Prof. Miguel Martin
Universidad de Granada
Dept. de Analisis Matematico
Campus Fuentenueva S/N
18071 Granada
Spain
mmartins@ugr.es

Prof. Artem Zvavitch
Kent State University
Department of Mathematical Sciences
800 E. Summit St. Kent
Kent, Ohio 44242
USA
zvavitch@math.kent.edu

ISBN 978-3-11-065339-7
e-ISBN (PDF) 978-3-11-065675-6
e-ISBN (EPUB) 978-3-11-065346-5
ISSN 2511-0438

Library of Congress Control Number: 2020935870

Bibliographic information published by the Deutsche Nationalbibliothek
The Deutsche Nationalbibliothek lists this publication in the Deutsche Nationalbibliografie; detailed bibliographic data are available on the Internet at http://dnb.dnb.de.

Typesetting: VTeX UAB, Lithuania
Printing and binding: CPI books GmbH, Leck

www.degruyter.com

Editors' Introduction

We introduce this volume with a mixture of feelings: Fond memories for Victor Lomonosov as a colleague and friend with a marvelous sense of humor, but also a feeling of great sadness for the loss of a person possessing such mathematical brilliance. Reading—or rereading—his work, one is struck by its depth, elegance, and apparent simplicity. It is easy to be fooled by the brevity of his work: For example, there are barely 15 pages in total in his 1973 paper on the invariant subspace problem, the 1991 Israel Journal paper on the Burnside theorem in infinite dimensions, and the 2000 Israel Journal paper on the Bishop–Phelps theorem in complex Banach spaces. However, these clearly written papers, in absolutely central areas of operator theory and functional analysis, are indeed very profound and have had a significant impact on the development of analysis during the last half century.

We thank the many mathematicians who have contributed their excellent work to celebrate and memorialize Victor Lomonosov's contributions to operator theory and functional analysis. We are also grateful to De Gruyter for its support of this project.

February 2020

Richard M. Aron
Eva A. Gallardo Gutiérrez
Miguel Martín
Dmitry Ryabogin
Ilya M. Spitkovsky
Artem Zvavitch
(editors)

https://doi.org/10.1515/9783110656756-201

Per Enflo's personal thoughts about Victor Lomonosov

Early in 1973, there was excitement in the mathematical world. A young mathematician from Russia, Victor Lomonosov, had proved that operators commuting with a compact operator on a Banach space have invariant subspaces. The short, elegant proof had made 40 years of development obsolete. I was at UC Berkeley at that time. I remember that the leading operator theorists there immediately started to work hard to see if the new powerful technique could solve the entire invariant subspace problem. It seemed hopeful that this was within reach, since Lomonosov's proof even showed, that an operator commuting with an operator which commutes with a compact operator has invariant subspaces. They did not succeed, but even so it was clear that Lomonosov's new ideas had already forever changed the field of operator theory.

Who was Victor Lomonosov? It seemed that nobody at Berkeley had met him or knew anybody who had met him. Could it even be that "Lomonosov" was a made up name for the collaboration of several mathematicians, like the French "Bourbaki" ?

Fifteen years later, in the early summer of 1988, when I was sitting in my office at Kent State University, my telephone rang. The person introduced himself as Victor Lomonosov. He was visiting a friend in the US. The friend quickly took over the telephone conversation and I learned that Victor Lomonosov had plans to return and emigrate to the US in early 1989. To me, this phone call was extremely surprising, almost spooky, and at the same time very exciting. So I asked Victor to contact me immediately when he was back in the US.

When Victor came back in early 1989, I tried all regular ways to find some short term employment for him at Kent State University. But there was no money anywhere. I knew that President Michael Schwartz had the ambition to make Kent State a top level research university and I knew that he had faith in me. So I went to a reception which he was going to attend. I took him aside and I told him that Kent State had an opportunity to hire a mathematician who had just come out of the Soviet Union, and who had done some sensational, groundbreaking work. Although this work was 15 years in the past, chances were that he was still a top level mathematician. Michael Schwartz remarked that there might be political reasons for the long silence and then he asked me three questions: "Does he speak English?" I answered: "He seems very ambitious to learn to speak better." Next question: "Is he as good as you are?" Before I had commented, he continued with the third question: "Do you want him here?" So I said: "YES!" Then he said: "We will get him!" And two days later, there was money to hire Victor for 6 months. And long before the end of these 6 months, it was clear that Kent State had made a great decision. Victor got offers from several other universities, but he stayed at Kent State. And over the years he made a very important contribution

https://doi.org/10.1515/9783110656756-202

to the mathematical life of Kent State University, mainly through his own work but also through his many contacts with mathematicians from the former Soviet Union.

Not surprisingly, the cultural differences between the Soviet Union and the US sometimes showed up in the beginning of Victor's stay. Here is one example:

An American to Victor: "Oh, you do not have social security numbers in the Soviet Union. But how do you then count people?"

Victor: "One, two, three..."

I met almost daily with Victor in the beginning of his stay. We usually had lunch at Burger King, and we discussed daily life issues and mathematics. Victor worked hard to adjust to American life. Getting a driving license as quickly as possible was important. He brought his English-Russian dictionary to the written test and was allowed to use it "as long as it does not have any Ohio traffic laws in it."

And he worked hard to improve his English, by talking, reading and watching TV. In the years around 1990, perhaps in connection with the financial crisis, it had become common to blame students' failures on their professors' poor English. So, for Victor getting a tenured position at Kent State, it was clear that this could be a potential issue. But, with Victor's fast learning, nothing happened like that.

At the same time, as Victor adjusted to a life in the US, he developed his deep, ingenious, and beautiful theory of operator algebras on Banach spaces. It is based on an inequality which, when specialized to finite dimensions, gives Burnside's classical theorem. And, besides strengthening and improving many earlier results on operator algebras and invariant subspaces, it suggests the following, in some sense ultimate, conjecture for a "positive" answer to the invariant subspace problem: If A is an operator on a Banach space, then A^* has an invariant subspace. In later work, he developed further his work on extensions of Burnside's theorem.

Victor's remarkable ability to find new approaches to difficult problems led to another great triumph, when he solved the long-standing problem of whether the Bishop–Phelps' theorem holds for Banach spaces over the complex field. One version of the celebrated Bishop–Phelps' theorem from the early 1960s states that, in a Banach space over the real numbers, a closed, bounded, convex set has a support point, that is, a point where some functional attains its supremum.

In 1977, J. Bourgain proved that for Banach spaces with the Radon–Nikodym property, this holds also for Banach spaces over the complex field.

In 2000, Victor gave an example of a closed, bounded, convex set without support points in a Banach space over the complex field. And in a subsequent paper he showed that the complex version of Bishop–Phelps' theorem fails in every space which is a predual of any uniform nonself-adjoint dual operator algebra. So, his counterexample is not just an isolated "pathology," but rather it is part of a more general phenomenon.

As I already mentioned, Victor had a remarkable ability to find new approaches to difficult problems. In his own work, he was looking for new, powerful ideas. And his own papers are usually not very long. The paper "Exponential numbers of linear operators in normed spaces" started as a 4-page paper by Victor and myself. We thought that the paper had a nice idea, worth publishing. V. Gurariy joined us, and the paper grew to a 15-page paper. Then Yu. I. Lyubich joined us and it was eventually published as a 35-page paper.

Victor frequently collaborated with other mathematicians. In his joint publications—papers of high quality—there is a broad span of topics, from analysis to almost pure algebra. My discussions with him fell mostly within the area of operator theory. And for me, they were an important inspiration, both for my own efforts and for my work with doctoral students. Our discussions continued to the end of his life. I remember that, shortly before his passing, we considered the following question to which none of us had an answer: Consider a one-to-one operator T with dense range. For which T is there an operator V, similar to T, such that V and V^* have the same range? Such that V and V^* have disjoint ranges, except for $\{0\}$?

Victor's work has had a great impact. Several of his results are now classical and parts of standard courses in functional analysis. They are famous, not just among functional analysts, but in the entire mathematical world.

Östervåla
5/28 2019

Per Enflo

Photo of Victor Lomonosov

Contents

María D. Acosta and Maryam Soleimani-Mourchehkhorti

1 Bishop–Phelps–Bollobás property for positive operators between classical Banach spaces

Dedicated to the memory of Victor Lomonosov

Abstract: We prove that the class of positive operators from $L_\infty(\mu)$ to $L_1(\nu)$ has the Bishop–Phelps–Bollobás property for any positive measures μ and ν. The same result also holds for the pair (c_0, ℓ_1). We also provide an example showing that not every pair of Banach lattices satisfies the Bishop–Phelps–Bollobás property for positive operators.

Keywords: Banach space, operator, positive operator, Bishop–Phelps–Bollobás theorem, Bishop–Phelps–Bollobás property

MSC 2010: Primary 46B04, Secondary 47B99

1.1 Introduction

In 1961, Bishop and Phelps proved that for any Banach space the set of (bounded and linear) functionals attaining their norms is norm dense in the topological dual space [15]. In 1970, Bollobás gave some quantified version of that result [16]. In order to state such result, we recall the following notation. By B_X, S_X, and X^*, we denote the closed unit ball, the unit sphere, and the topological dual of a Banach space X, respectively. If X and Y are both real or both complex Banach spaces, $L(X, Y)$ denotes the space of (bounded linear) operators from X to Y, endowed with its usual operator norm.

Bishop–Phelps–Bollobás theorem (see [17, Theorem 16.1] or [19, Corollary 2.4]). *Let X be a Banach space and $0 < \varepsilon < 1$. Given $x \in B_X$ and $x^* \in S_{X^*}$ with $|1 - x^*(x)| < \frac{\varepsilon^2}{2}$, there are elements $y \in S_X$ and $y^* \in S_{X^*}$ such that $y^*(y) = 1$, $\|y - x\| < \varepsilon$ and $\|y^* - x^*\| < \varepsilon$.*

Acknowledgement: The first author was supported by projects PGC2018-093794-B-I00 (MCIU/AEI/FEDER, UE), A-FQM-484-UGR18 (FEDER, Andalucía) and FQM-185 (Junta de Andalucía/FEDER, UE). The second author was supported by a grant from IPM.
The authors kindly thank to M. Mastyło who suggested to study the property considered in this paper during a research stay in the University of Granada.

María D. Acosta, Universidad de Granada, Facultad de Ciencias, Departamento de Análisis Matemático, 18071 Granada, Spain, e-mail: dacosta@ugr.es
Maryam Soleimani-Mourchehkhorti, School of Mathematics, Institute for Research in Fundamental Sciences (IPM), P. O. Box: 19395-5746, Tehran, Iran, e-mail: m-soleimani85@ipm.ir

https://doi.org/10.1515/9783110656756-001

After a period in which a lot of attention has been devoted to extending the Bishop–Phelps theorem to operators and interesting results have been obtained about that topic (see [2]), in 2008, it was posed the problem of extending the Bishop–Phelps–Bollobás theorem for operators.

In order to state some of these extensions, it will be convenient to recall the following notion.

Definition 1.1 ([5, Definition 1.1]). Let X and Y be either real or complex Banach spaces. The pair (X, Y) is said to have the *Bishop–Phelps–Bollobás property for operators* (BPBp) if for every $0 < \varepsilon < 1$ there exists $0 < \eta(\varepsilon) < \varepsilon$ such that for every $S \in S_{L(X,Y)}$, if $x_0 \in S_X$ satisfies $\|S(x_0)\| > 1 - \eta(\varepsilon)$, then there exist an element $u_0 \in S_X$ and an operator $T \in S_{L(X,Y)}$ satisfying the following conditions:

$$\|T(u_0)\| = 1, \quad \|u_0 - x_0\| < \varepsilon \quad \text{and} \quad \|T - S\| < \varepsilon.$$

If X and Y are Banach spaces, it is known that the pair (X, Y) has the Bishop–Phelps–Bollobás property in the following cases:

- X and Y are finite-dimensional spaces [5, Proposition 2.4].
- X is any Banach space and Y has the property β (of Lindenstrauss) [5, Theorem 2.2]. The spaces c_0 and ℓ_∞ have property β.
- X is uniformly convex and Y is any Banach space ([9, Theorem 2.2] or [23, Theorem 3.1]).
- $X = \ell_1$ and Y has the approximate hyperplane series property [5, Theorem 4.1]. For instance, finite-dimensional spaces, uniformly convex spaces, $C(K)$, and $L_1(\mu)$ have the approximate hyperplane series property.
- $X = L_1(\mu)$ and Y has the Radon–Nikodým property and the approximate hyperplane series property, whenever μ is any σ-finite measure [20, Theorem 2.2] (see also [7, Theorem 2.3]).
- $X = L_1(\mu)$ and $Y = L_1(\nu)$, for any positive measures μ and ν [21, Theorem 3.1].
- $X = L_1(\mu)$ and $Y = L_\infty(\nu)$, for any positive measure μ and any localizable positive measure ν [21, Theorem 4.1] (see also [14]).
- $X = C(K)$ and $Y = C(S)$ in the real case, where K and S are compact Hausdorff topological spaces [6, Theorem 2.5].
- $X = C(K)$ and Y is a uniformly convex Banach space, in the real case [24, Theorem 2.2] (see also [22, Corollary 2.6] and [25, Theorem 5]).
- $X = C_0(L)$, for any locally compact Hausdorff topological space L, whenever Y is a $\mathbb{C}$-uniformly convex space, in the complex case [3, Theorem 2.4]. As a consequence, the pair $(C_0(L), L_p(\mu))$ has the BPBp for any positive measure μ and $1 \le p < +\infty$.
- $X = \ell_\infty^n$ and $Y = L_1(\mu)$ for any positive integer n and any positive measure μ [10, Corollary 4.5] (see also [10, Theorem 3.3], [11, Theorem 3.3] and [8, Theorem 2.9]).

- X is an Asplund space and $Y \subset C(K)$ is a uniform algebra [18, Theorem 3.6] (see also [13, Corollary 2.5]).

The paper [4] contains a survey with most of the results known about the Bishop–Phelps–Bollobás property for operators.

In this short note, we introduce a version of Bishop–Phelps–Bollobás property for positive operators between Banach lattices (see Definition 1.3). The only difference between this property and the previous one is that we assume that the operators appearing in Definition 1.1 are positive. In Section 1.2, we prove that the pair $(L_\infty(\mu), L_1(\nu))$ has the Bishop–Phelps–Bollobás property for positive operators for any positive measures μ and ν. The parallel result for $(c_0, L_1(\mu))$ is shown in Section 1.3, for any positive measure μ. As a consequence, the subset of positive operators from c_0 to ℓ_1 satisfies the Bishop–Phelps–Bollobás property. We remark that it is not known whether the pairs $(L_\infty(\mu), L_1(\nu))$ and (c_0, ℓ_1) satisfy the Bishop–Phelps–Bollobás property for operators in the real case. In both cases, the set of norm attaining operators is dense in the space of operators (see [27, Theorem B] for the first case). For the second pair, a necessary condition on the range space in order to have the Bishop–Phelps–Bollobás property for operators is known (see [10, Theorem 3.3]). We also provide an example showing that not every pair of Banach lattices satisfies the Bishop–Phelps–Bollobás property for positive operators.

1.2 Bishop–Phelps–Bollobás property for positive operators for the pair (L_∞, L_1)

We begin by recalling some notions and introducing the appropriate notion of the Bishop–Phelps–Bollobás property for positive operators. The concepts in the first definition are standard and can be found, for instance, in [1].

Definition 1.2. An *ordered vector space* is a real vector space X equipped with a vector space order, that is, an order relation $\leq$ that is compatible with the algebraic structure of X. An ordered vector space is called a *Riesz space* if every pair of vectors has a least upper bound and a greatest lower bound. A norm $\|\ \|$ on a Riesz space X is said to be a *lattice norm* whenever $|x| \leq |y|$ implies $\|x\| \leq \|y\|$. A *normed Riesz space* is a Riesz space equipped with a lattice norm. A normed Riesz space whose norm is complete is called a *Banach lattice*.

A linear mapping $T : X \to Y$ between two ordered vector spaces is called *positive* if $x \geq 0$ implies $Tx \geq 0$.

Recall that every positive linear mapping from a Banach lattice to a normed Riesz space is continuous [12, Theorem 4.3].

Definition 1.3. Let X and Y be Banach lattices. The pair (X, Y) is said to have the *Bishop–Phelps–Bollobás property for positive operators* if for every $0 < \varepsilon < 1$ there exists $0 < \eta(\varepsilon) < \varepsilon$ such that for every $S \in S_{L(X,Y)}$, such that $S \geq 0$, if $x_0 \in S_X$ satisfies $\|S(x_0)\| > 1 - \eta(\varepsilon)$, then there exist an element $u_0 \in S_X$ and a positive operator $T \in S_{L(X,Y)}$ satisfying the following conditions:

$$\|T(u_0)\| = 1, \quad \|u_0 - x_0\| < \varepsilon \quad \text{and} \quad \|T - S\| < \varepsilon.$$

Let (Ω, μ) be a measure space. We denote by $L_\infty(\mu)$ the space of real valued measurable essentially bounded functions on Ω and by $\mathbb{1}$ the constant function equal to 1 on Ω. Since an element f in $B_{L_\infty(\mu)}$ satisfies that $|f| \leq \mathbb{1}$ a. e., it is clear that a positive operator from $L_\infty(\mu)$ to any other Banach lattice satisfies the next result.

Lemma 1.4. *Let μ and ν be positive measures and T a positive operator from $L_\infty(\mu)$ to $L_1(\nu)$. Then $\|T\| = \|T(\mathbb{1})\|_1$.*

It is trivially satisfied that $\|f + g\|_1 = \|f - g\|_1$ for any positive integrable functions f and g with disjoint supports. The next result shows that if f_1 and f_2 are two positive integrable functions such that $\|f_1 - f_2\|_1$ is close to $\|f_1 + f_2\|_1$, then there are positive integrable functions g_1 and g_2 with disjoint supports and such that g_i is close to f_i for $i = 1, 2$.

Lemma 1.5. *Let (Ω, μ) be a measure space, $0 < \varepsilon < \frac{1}{5}$ and $f_1, f_2 \in L_1(\mu)$ be positive functions such that*

$$\|f_1 + f_2\|_1 \leq 1 \quad \text{and} \quad 1 - \varepsilon^2 \leq \|f_1 - f_2\|_1.$$

Then there are two positive functions g_1 and g_2 with disjoint supports in $L_1(\mu)$ and also satisfying that

$$\|g_1 + g_2\|_1 = 1 \quad \text{and} \quad \|g_i - f_i\|_1 < 7\varepsilon \quad \textit{for } i = 1, 2.$$

Proof. We define the set W given by

$$W = \{t \in \Omega : |f_1(t) - f_2(t)| \leq (1 - \varepsilon)(f_1(t) + f_2(t))\}.$$

Clearly, W is a measurable subset of Ω. We have that

$$\begin{aligned} 1 - \varepsilon^2 &\leq \|f_1 - f_2\|_1 \\ &= \int_\Omega |f_1 - f_2| \, d\mu \\ &= \int_W |f_1 - f_2| \, d\mu + \int_{\Omega \setminus W} |f_1 - f_2| \, d\mu \end{aligned}$$

$$\leq (1-\varepsilon)\int_{W}(f_1+f_2)\,d\mu+\int_{\Omega\setminus W}(f_1+f_2)\,d\mu$$
$$\leq 1-\varepsilon\int_{W}(f_1+f_2)\,d\mu.$$

As a consequence,

$$\int_{W}(f_1+f_2)\,d\mu\leq\varepsilon. \tag{1.1}$$

Now we define the sets given by

$$G_1=\Omega\setminus W\cap\{t\in\Omega: f_1(t)>f_2(t)\}\quad\text{and}\quad G_2=\Omega\setminus W\cap\{t\in\Omega: f_2(t)>f_1(t)\}.$$

Clearly, G_1 and G_2 are measurable subsets and it is satisfied that

$$(f_1-f_2)\chi_{G_1}=|f_1-f_2|\chi_{G_1}>(1-\varepsilon)(f_1+f_2)\chi_{G_1}.$$

So $f_2\chi_{G_1}\leq(2-\varepsilon)f_2\chi_{G_1}\leq\varepsilon f_1\chi_{G_1}$ and

$$\int_{G_1}f_2\,d\mu\leq\int_{G_1}\varepsilon f_1\,d\mu\leq\varepsilon. \tag{1.2}$$

By using the same argument with the function f_1, we obtain that

$$\int_{G_2}f_1\,d\mu\leq\varepsilon. \tag{1.3}$$

Since the subsets W, G_1, and G_2 are a partition of Ω, in view of (1.1), (1.2), and (1.3), we deduce that

$$\begin{aligned}\|f_1-f_1\chi_{G_1}\|_1&=\|f_1\chi_{W\cup G_2}\|_1\\&=\|f_1\chi_W\|_1+\|f_1\chi_{G_2}\|_1\\&=\int_{W}f_1\,d\mu+\int_{G_2}f_1\,d\mu\\&\leq 2\varepsilon,\end{aligned}\tag{1.4}$$

and

$$\|f_2-f_2\chi_{G_2}\|_1\leq 2\varepsilon. \tag{1.5}$$

By using that f_1 and f_2 are positive functions, we deduce that

$$
\begin{aligned}
\|f_1\chi_{G_1} + f_2\chi_{G_2}\|_1 &= \|f_1 + f_2\|_1 - \|f_1 - f_1\chi_{G_1}\|_1 - \|f_2 - f_2\chi_{G_2}\|_1 \\
&\geq \|f_1 - f_2\|_1 - 4\varepsilon \quad \text{(by (1.4) and (1.5))} \\
&\geq 1 - \varepsilon^2 - 4\varepsilon \\
&> 1 - 5\varepsilon > 0.
\end{aligned} \tag{1.6}
$$

Now we define the functions g_1 and g_2 by

$$g_i = \frac{f_i\chi_{G_i}}{\|f_1\chi_{G_1} + f_2\chi_{G_2}\|_1}, \qquad i = 1, 2.$$

It is clear that $g_i \in L_1(\mu)$ for $i = 1, 2$ and they are positive functions with disjoint supports. It is also clear that $\|g_1 + g_2\|_1 = 1$.

Since f_1 and f_2 are positive functions, we have that

$$\|f_1\chi_{G_1} + f_2\chi_{G_2}\|_1 \leq \|f_1 + f_2\|_1 \leq 1,$$

so for $i = 1, 2$ we obtain that

$$
\begin{aligned}
\|f_i\chi_{G_i}\|_1 \left| \frac{1}{\|f_1\chi_{G_1} + f_2\chi_{G_2}\|_1} - 1 \right| &= \|f_i\chi_{G_i}\|_1 \left(\frac{1}{\|f_1\chi_{G_1} + f_2\chi_{G_2}\|_1} - 1 \right) \\
&= \frac{\|f_i\chi_{G_i}\|_1 (1 - \|f_1\chi_{G_1} + f_2\chi_{G_2}\|_1)}{\|f_1\chi_{G_1} + f_2\chi_{G_2}\|_1} \\
&\leq 1 - \|f_1\chi_{G_1} + f_2\chi_{G_2}\|_1.
\end{aligned} \tag{1.7}
$$

For $i = 1, 2$, we estimate the distance from g_i to f_i as follows:

$$
\begin{aligned}
\|g_i - f_i\|_1 &= \left\| \frac{f_i\chi_{G_i}}{\|f_1\chi_{G_1} + f_2\chi_{G_2}\|_1} - f_i \right\|_1 \\
&\leq \left\| \frac{f_i\chi_{G_i}}{\|f_1\chi_{G_1} + f_2\chi_{G_2}\|_1} - f_i\chi_{G_i} \right\|_1 + \|f_i\chi_{G_i} - f_i\|_1 \\
&\leq 1 - \|f_1\chi_{G_1} + f_2\chi_{G_2}\|_1 + 2\varepsilon \quad \text{(by (1.7), (1.4) and (1.5))} \\
&< 7\varepsilon \quad \text{(by (1.6))}.
\end{aligned}
$$

□

Theorem 1.6. *For any positive measures μ and ν, the pair $(L_\infty(\mu), L_1(\nu))$ has the Bishop–Phelps–Bollobás property for positive operators.*

Moreover, in Definition 1.3, if the function f_0 where the operator S is close to attain its norm is positive, then the function f_1 where T attains its norm is also positive.

Proof. Assume that (Ω_1, μ) is a measure space. Let $0 < \varepsilon < 1$, $f_0 \in S_{L_\infty(\mu)}$, $S \in S_{L(L_\infty(\mu), L_1(\nu))}$ and assume that S is a positive operator satisfying that

$$\|S(f_0)\|_1 > 1 - \eta^2,$$

where $\eta = (\frac{\varepsilon}{58})^2$. We define the sets A, B and C given by

$$A = \{t \in \Omega_1 : -1 \le f_0(t) < -1 + \eta\}, \quad B = \{t \in \Omega_1 : 1 - \eta < f_0(t) \le 1\}$$

and

$$C = \{t \in \Omega_1 : |f_0(t)| \le 1 - \eta\}.$$

By using that S is a positive operator, we obtain that

$$\begin{aligned} 1 - \eta^2 &< \|S(f_0)\|_1 \\ &= \|S(f_0\chi_A + f_0\chi_B + f_0\chi_C)\|_1 \\ &\le \|S(f_0\chi_A)\|_1 + \|S(f_0\chi_B)\|_1 + \|S(f_0\chi_C)\|_1 \\ &\le \|S(\chi_A)\|_1 + \|S(\chi_B)\|_1 + (1-\eta)\|S(\chi_C)\|_1 \\ &\le 1 - \eta\|S(\chi_C)\|_1. \end{aligned}$$

Hence $\|S(\chi_C)\|_1 \le \eta$. By using again that S is positive, we deduce that

$$\|S(f\chi_C)\|_1 \le \|S(\chi_C)\|_1 \le \eta, \quad \forall f \in B_{L_\infty(\mu)}. \tag{1.8}$$

On the other hand, it is trivially satisfied that

$$\|f_0\chi_A + \chi_A\|_\infty \le \eta \quad \text{and} \quad \|f_0\chi_B - \chi_B\|_\infty \le \eta,$$

so

$$\|S(f_0\chi_A + \chi_A)\|_1 \le \eta \quad \text{and} \quad \|S(f_0\chi_B - \chi_B)\|_1 \le \eta. \tag{1.9}$$

By using the assumption, we obtain that

$$\begin{aligned} 1 - \eta^2 &< \|S(f_0)\|_1 \\ &\le \|S(f_0\chi_A + f_0\chi_B)\|_1 + \|S(f_0\chi_C)\|_1 \\ &\le \|S(f_0\chi_A + \chi_A)\|_1 + \|S(\chi_B - \chi_A)\|_1 + \|S(f_0\chi_B - \chi_B)\|_1 + \|S(f_0\chi_C)\|_1 \\ &\le \|S(\chi_B - \chi_A)\|_1 + 3\eta \quad \text{(by (1.9) and (1.8))}. \end{aligned}$$

As a consequence,

$$\|S(\chi_B - \chi_A)\|_1 \ge 1 - 4\eta. \tag{1.10}$$

Since $S(\chi_A)$ and $S(\chi_B)$ are positive functions and $\|S(\chi_A) + S(\chi_B)\|_1 \le 1$, we can apply Lemma 1.5 and so there are two positive functions g_1 and g_2 in $L_1(\nu)$ satisfying the following conditions:

$$\|g_1 - S(\chi_A)\|_1 < 14\sqrt{\eta} = \frac{7\varepsilon}{29}, \quad \|g_2 - S(\chi_B)\|_1 < \frac{7\varepsilon}{29},$$
$$\operatorname{supp} g_1 \cap \operatorname{supp} g_2 = \varnothing \quad \text{and} \quad \|g_1 + g_2\|_1 = 1.$$

Assume that ν is a measure on Ω_2. We obtain that

$$\|S(\chi_A)\chi_{\Omega_2 \setminus \operatorname{supp} g_1}\|_1 = \|(g_1 - S(\chi_A))\chi_{\Omega_2 \setminus \operatorname{supp} g_1}\|_1 \leq \|g_1 - S(\chi_A)\|_1 < \frac{7\varepsilon}{29} \tag{1.11}$$

and also

$$\|S(\chi_B)\chi_{\Omega_2 \setminus \operatorname{supp} g_2}\|_1 < \frac{7\varepsilon}{29}. \tag{1.12}$$

Now we define the operator $V : L_\infty(\mu) \longrightarrow L_1(\nu)$ as follows:

$$V(f) = S(f\chi_A)\chi_{\operatorname{supp} g_1} + S(f\chi_B)\chi_{\operatorname{supp} g_2} \quad (f \in L_\infty(\mu)).$$

Clearly, V is well-defined and it is a positive operator since $S \geq 0$. By applying Lemma 1.4 we have that

$$\|V\| = \|V(\mathbb{1})\|_1 = \|S(\chi_A)\chi_{\operatorname{supp} g_1} + S(\chi_B)\chi_{\operatorname{supp} g_2}\|_1 \leq \|S\| = 1.$$

Now we estimate the norm of $V - S$. If $f \in B_{L_\infty(\mu)}$, then we have that

$$\begin{aligned}
\|(V - S)(f)\|_1 &= \|S(f\chi_A)\chi_{\operatorname{supp} g_1} + S(f\chi_B)\chi_{\operatorname{supp} g_2} - S(f)\|_1 \\
&= \|S(f\chi_A)\chi_{\operatorname{supp} g_1} + S(f\chi_B)\chi_{\operatorname{supp} g_2} - S(f\chi_A) - S(f\chi_B) - S(f\chi_C)\|_1 \\
&\leq \|S(f\chi_A)\chi_{\Omega_2 \setminus \operatorname{supp} g_1}\|_1 + \|S(f\chi_B)\chi_{\Omega_2 \setminus \operatorname{supp} g_2}\|_1 + \|S(f\chi_C)\|_1 \\
&\leq \|S(\chi_A)\chi_{\Omega_2 \setminus \operatorname{supp} g_1}\|_1 + \|S(\chi_B)\chi_{\Omega_2 \setminus \operatorname{supp} g_2}\|_1 + \|S(f\chi_C)\|_1 \\
&< \frac{14\varepsilon}{29} + \eta < \frac{\varepsilon}{2} \quad \text{(by (1.11), (1.12) and (1.8)).}
\end{aligned}$$

We proved that $\|V - S\| < \frac{\varepsilon}{2}$ and so $\|V\| \geq 1 - \frac{\varepsilon}{2} > 0$. Since $f_0 \in S_{L_\infty(\mu)}$, the function f_1 given by $f_1 = \chi_B - \chi_A + f_0\chi_C \in S_{L_\infty(\mu)}$ and satisfies

$$\|f_1 - f_0\|_\infty \leq \eta < \varepsilon.$$

Since g_1 and g_2 have disjoint supports, we also have that

$$\begin{aligned}
\|V(f_1)\|_1 &= \|S(-\chi_A)\chi_{\operatorname{supp} g_1} + S(\chi_B)\chi_{\operatorname{supp} g_2}\|_1 \\
&= \|S(\chi_A)\chi_{\operatorname{supp} g_1} + S(\chi_B)\chi_{\operatorname{supp} g_2}\|_1 \\
&= \|V(\mathbb{1})\|_1 = \|V\|.
\end{aligned}$$

If we take $T = \frac{V}{\|V\|}$, the operator $T \in S_{L(L_\infty(\mu), L_1(\nu))}$, is a positive operator, attains its norm at f_1 and satisfies that

$$\|T - S\| \leq \|T - V\| + \|V - S\| = |1 - \|V\|| + \|V - S\| \leq 2\|V - S\| < \varepsilon.$$

We proved that the pair $(L_\infty(\mu), L_1(\nu))$ has the Bishop–Phelps–Bollobás property for positive operators. In case that $f_0 \geq 0$, the function f_1 also satisfies the same condition. □

1.3 A result on the Bishop–Phelps–Bollobás property for positive operators for (c_0, L_1)

Theorem 1.7. *For any positive measure μ, the pair $(c_0, L_1(\mu))$ has the Bishop–Phelps–Bollobás property for positive operators.*

Moreover, in Definition 1.3, if the element x_0 is positive, then the element u_0 where T attains its norm is also positive.

Proof. The proof of this result is similar to the proof of Theorem 1.6. In any case, we include it for the sake of completeness. Throughout this proof, we denote by $\| \ \|$ the usual norm of c_0.

Assume that Ω is the set such that (Ω, μ) is the measure space considered for $L_1(\mu)$. Let $0 < \varepsilon < 1$, $x_0 \in S_{c_0}$, $S \in S_{L(c_0, L_1(\mu))}$ and assume that S is a positive operator satisfying that

$$\|S(x_0)\|_1 > 1 - \eta^2,$$

where $\eta = (\frac{\varepsilon}{58})^2$. We define the sets A, B and C given by

$$A = \{k \in \mathbb{N} : -1 \le x_0(k) < -1 + \eta\}, \quad B = \{k \in \mathbb{N} : 1 - \eta < x_0(k) \le 1\}$$

and

$$C = \{k \in \mathbb{N} : |x_0(k)| \le 1 - \eta\}.$$

Since $x_0 \in S_{c_0}$, the sets A and B are finite and $\{A, B, C\}$ is a partition of $\mathbb{N}$.

For each positive integer n, we denote by $C_n = C \cap \{k \in \mathbb{N} : k \le n\}$, which is a finite subset of $\mathbb{N}$. By using that S is a positive operator in $S_{L(c_0, L_1(\mu))}$, we obtain that

$$\begin{aligned} 1 - \eta^2 &< \|S(x_0)\|_1 \\ &= \|S(x_0\chi_A + x_0\chi_B + x_0\chi_C)\|_1 \\ &\le \|S(x_0\chi_A)\|_1 + \|S(x_0\chi_B)\|_1 + \|S(x_0\chi_C)\|_1 \\ &\le \|S(\chi_A)\|_1 + \|S(\chi_B)\|_1 + (1 - \eta) \lim_n \{\|S(\chi_{C_n})\|_1\} \\ &\le 1 - \eta \lim_n \{\|S(\chi_{C_n})\|_1\}. \end{aligned}$$

Hence $\lim_n \{\|S(\chi_{C_n})\|_1\} \le \eta$. Since S is positive, we get that

$$\|S(x\chi_C)\|_1 \le \lim_n \{\|S(\chi_{C_n})\|_1\} \le \eta, \quad \forall x \in B_{c_0}. \tag{1.13}$$

On the other hand, it is trivially satisfied that

$$\|x_0\chi_A + \chi_A\| \le \eta \quad \text{and} \quad \|x_0\chi_B - \chi_B\| \le \eta,$$

and so

$$\|S(x_0\chi_A + \chi_A)\|_1 \leq \eta \quad \text{and} \quad \|S(x_0\chi_B - \chi_B)\|_1 \leq \eta. \tag{1.14}$$

In view of the assumption, since $\{A, B, C\}$ is a partition of $\mathbb{N}$ we obtain that

$$\begin{aligned} 1 - \eta^2 &< \|S(x_0)\|_1 \\ &\leq \|S(x_0\chi_A + x_0\chi_B)\|_1 + \|S(x_0\chi_C)\|_1 \\ &\leq \|S(x_0\chi_A + \chi_A)\|_1 + \|S(\chi_B - \chi_A)\|_1 + \|S(x_0\chi_B - \chi_B)\|_1 + \|S(x_0\chi_C)\|_1 \\ &\leq \|S(\chi_B - \chi_A)\|_1 + 3\eta \quad \text{(by (1.14) and (1.13))}. \end{aligned}$$

Hence

$$\|S(\chi_B - \chi_A)\|_1 \geq 1 - 4\eta. \tag{1.15}$$

Now we can apply Lemma 1.5 to the positive functions $S(\chi_A)$ and $S(\chi_B)$ since $\|S(\chi_A) + S(\chi_B)\|_1 \leq \|S\| = 1$. So there exist two positive functions g_1 and g_2 in $L_1(\mu)$ satisfying the following conditions:

$$\|g_1 - S(\chi_A)\|_1 < \frac{7\varepsilon}{29}, \quad \|g_2 - S(\chi_B)\|_1 < \frac{7\varepsilon}{29},$$
$$\operatorname{supp} g_1 \cap \operatorname{supp} g_2 = \varnothing \quad \text{and} \quad \|g_1 + g_2\|_1 = 1.$$

As a consequence, we have that

$$\|S(\chi_A)\chi_{\Omega\setminus \operatorname{supp} g_1}\|_1 = \|(g_1 - S(\chi_A))\chi_{\Omega\setminus \operatorname{supp} g_1}\|_1 \leq \|g_1 - S(\chi_A)\|_1 < \frac{7\varepsilon}{29} \tag{1.16}$$

and also

$$\|S(\chi_B)\chi_{\Omega\setminus \operatorname{supp} g_2}\|_1 < \frac{7\varepsilon}{29}. \tag{1.17}$$

We define the operator $U : c_0 \longrightarrow L_1(\mu)$ by

$$U(x) = S(x\chi_A)\chi_{\operatorname{supp} g_1} + S(x\chi_B)\chi_{\operatorname{supp} g_2} \quad (x \in c_0).$$

The operator U is linear, bounded, and positive. Since $U(x) = U(x\chi_{A\cup B})$ for any element $x \in c_0$ and $A \cup B$ is finite, we obtain that

$$\|U\| = \|U(\chi_{A\cup B})\|_1 = \|S(\chi_A)\chi_{\operatorname{supp} g_1} + S(\chi_B)\chi_{\operatorname{supp} g_2}\|_1 \leq \|S\| = 1.$$

Now we estimate the distance between U and S. For an element $x \in B_{c_0}$, it is satisfied

$$\begin{aligned}\|(U-S)(x)\|_1 &= \|S(x\chi_A)\chi_{\operatorname{supp} g_1} + S(x\chi_B)\chi_{\operatorname{supp} g_2} - S(x)\|_1\\ &= \|S(x\chi_A)\chi_{\operatorname{supp} g_1} + S(x\chi_B)\chi_{\operatorname{supp} g_2} - S(x\chi_A) - S(x\chi_B) - S(x\chi_C)\|_1\\ &\le \|S(x\chi_A)\chi_{\Omega\setminus \operatorname{supp} g_1}\|_1 + \|S(x\chi_B)\chi_{\Omega\setminus \operatorname{supp} g_2}\|_1 + \|S(x\chi_C)\|_1\\ &\le \|S(\chi_A)\chi_{\Omega\setminus \operatorname{supp} g_1}\|_1 + \|S(\chi_B)\chi_{\Omega\setminus \operatorname{supp} g_2}\|_1 + \|S(x\chi_C)\|_1\\ &< \frac{14\varepsilon}{29} + \eta < \frac{\varepsilon}{2} \quad \text{(by (1.16), (1.17) and (1.13)).}\end{aligned}$$

We proved that $\|U-S\| < \frac{\varepsilon}{2}$ and so $\|U\| \ge 1-\frac{\varepsilon}{2} > 0$. Since $x_0 \in S_{c_0}$, the element u_0 given by $u_0 = \chi_B - \chi_A + x_0\chi_C \in S_{c_0}$ and satisfies

$$\|u_0 - x_0\| \le \eta < \varepsilon.$$

Since g_1 and g_2 have disjoint supports, we also have that

$$\begin{aligned}\|U(u_0)\|_1 &= \|S(-\chi_A)\chi_{\operatorname{supp} g_1} + S(\chi_B)\chi_{\operatorname{supp} g_2}\|_1\\ &= \|S(\chi_A)\chi_{\operatorname{supp} g_1} + S(\chi_B)\chi_{\operatorname{supp} g_2}\|_1\\ &= \|U(\chi_{A\cup B})\|_1 = \|U\|.\end{aligned}$$

If we take $T = \frac{U}{\|U\|}$, the operator $T \in S_{L(c_0,L_1(\mu))}$, is a positive operator, attains its norm at u_0 and satisfies that

$$\|T-S\| \le \|T-U\| + \|U-S\| = \big|1-\|U\|\big| + \|U-S\| \le 2\|U-S\| < \varepsilon.$$

We proved that the pair $(c_0, L_1(\mu))$ has the Bishop–Phelps–Bollobás property for positive operators. Notice that in case that x_0 is positive, the element u_0 is also positive. □

Lastly, we provide an example showing that the property that we considered is non-trivial.

Example 1.8. Let $Y = c_0$ as a Riesz space, endowed with the norm given by

$$|||x||| = \|x\| + \left\|\left\{\frac{x_n}{2^n}\right\}\right\|_2 \quad (x \in c_0),$$

where $\|\ \|$ is the usual norm of c_0. Then the pair (c_0, Y) does not satisfy the Bishop–Phelps–Bollobás property for positive operators.

Proof. It is clear that $|||\ |||$ is a norm equivalent to the usual norm of c_0 and it is a lattice norm on Y. Also the space Y is strictly convex. So the formal identity from c_0 to Y cannot be approximated by norm attaining operators by [26, Proposition 4]. Since the formal identity is a positive operator, we are done. □

Note added in proof

The results stated in Theorems 1.6 and 1.7 have been extended to the pairs $(L_\infty(\mu), Y)$ and (c_0, Y), where Y is a uniformly monotone Banach lattice (see Arxiv-1907.08620 for more details).

Bibliography

[1] Y. A. Abramovich and C. D. Aliprantis, *An Invitation to Operator Theory*, Graduate Studies in Mathematics, **50**, American Mathematical Society, Providence, RI, 2002.

[2] M. D. Acosta, *Denseness of norm attaining mappings*, RACSAM. Rev. R. Acad. Cienc. Exactas Fís. Nat. Ser. A Mat. **100** (2006), 9–30.

[3] M. D. Acosta, *The Bishop–Phelps–Bollobás property for operators on $C(K)$*, Banach J. Math. Anal. **10** (2016), 307–319.

[4] M. D. Acosta, *On the Bishop–Phelps–Bollobás property*. Conference Function Spaces XII, 13–32, Banach Center Publ., 119, Polish Acad. Sci. Inst. Math., Warsaw, 2019.

[5] M. D. Acosta, R. M. Aron, D. García and M. Maestre, *The Bishop–Phelps–Bollobás theorem for operators*, J. Funct. Anal. **254** (2008), 2780–2799.

[6] M. D. Acosta, J. Becerra-Guerrero, Y. S. Choi, M. Ciesielski, S. K. Kim, H. J. Lee, M. L. Lourenço and M. Martín, *The Bishop–Phelps–Bollobás property for operators between spaces of continuous functions*, Nonlinear Anal. **95** (2014), 323–332.

[7] M. D. Acosta, J. Becerra-Guerrero, D. García, S. K. Kim and M. Maestre, *Bishop–Phelps–Bollobás property for certain spaces of operators*, J. Math. Anal. Appl. **414** (2014), 532–545.

[8] M. D. Acosta, J. Becerra-Guerrero, D. García, S. K. Kim and M. Maestre, *The Bishop–Phelps–Bollobás property: a finite-dimensional approach*, Publ. Res. Inst. Math. Sci. **51** (2015), 173–190.

[9] M. D. Acosta, J. Becerra-Guerrero, D. García and M. Maestre, *The Bishop–Phelps–Bollobás theorem for bilinear forms*, Trans. Am. Math. Soc. **365** (2013), 5911–5932.

[10] M. D. Acosta and J. L. Dávila, *A basis of $\mathbb{R}^n$ with good isometric properties and some applications to denseness of norm attaining operators*, J. Funct. Anal. **279** (2020), 108602.

[11] M. D. Acosta, J. L. Dávila and M. Soleimani-Mourchehkhorti, *Characterization of the Banach spaces Y satisfying that the pair (ℓ_∞^4, Y) has the Bishop–Phelps–Bollobás property for operators*, J. Math. Anal. Appl. **470** (2019), 690–715.

[12] C. D. Aliprantis and O. Burkinshaw, *Positive Operators*, Springer, Dordrecht, 2006.

[13] R. M. Aron, B. Cascales and O. Kozhushkina, *The Bishop–Phelps–Bollobás theorem and Asplund operators*, Proc. Am. Math. Soc. **139** (2011), 3553–3560.

[14] R. M. Aron, Y. S. Choi, D. García and M. Maestre, *The Bishop–Phelps–Bollobás theorem for $\mathcal{L}(L_1(\mu), L_\infty[0,1])$*, Adv. Math. **228** (2011), 617–628.

[15] E. Bishop and R. R. Phelps, *A proof that every Banach space is subreflexive*, Bull. Am. Math. Soc. **67** (1961), 97–98.

[16] B. Bollobás, *An extension to the theorem of Bishop and Phelps*, Bull. Lond. Math. Soc. **2** (1970), 181–182.

[17] F. F. Bonsall and J. Duncan, *Numerical Ranges II*, London Mathematical Society Lecture Notes Series, No. **10**, Cambridge University Press, New York-London, 1973.

[18] B. Cascales, A. J. Guirao and V. Kadets, *A Bishop–Phelps–Bollobás type theorem for uniform algebras*, Adv. Math. **240** (2013), 370–382.

[19] M. Chica, V. Kadets, M. Martín, S. Moreno-Pulido and F. Rambla-Barreno, *Bishop–Phelps–Bollobás moduli of a Banach space*, J. Math. Anal. Appl. **412** (2014), 697–719.

[20] Y. S. Choi and S. K. Kim, *The Bishop–Phelps–Bollobás theorem for operators from $L_1(\mu)$ to Banach spaces with the Radon–Nikodým property*, J. Funct. Anal. **261** (2011), 1446–1456.

[21] Y. S. Choi, S. K. Kim, H. J. Lee and M. Martín, *The Bishop–Phelps–Bollobás theorem for operators on $L_1(\mu)$*, J. Funct. Anal. **267** (2014), 214–242.

[22] S. K. Kim, *The Bishop–Phelps–Bollobás theorem for operators from c_0 to uniformly convex spaces*, Isr. J. Math. **197** (2013), 425–435.

[23] S. K. Kim and H. J. Lee, *Uniform convexity and Bishop–Phelps–Bollobás property*, Can. J. Math. **66** (2014), 373–386.
[24] S. K. Kim and H. J. Lee, *The Bishop–Phelps–Bollobás property for operators from $C(K)$ to uniformly convex spaces*, J. Math. Anal. Appl. **421** (2015), 51–58.
[25] S. K. Kim, H. J. Lee and P. K. Lin, *The Bishop–Phelps–Bollobás property for operators from $L_\infty(\mu)$ to uniformly convex Banach spaces*, J. Nonlinear Convex Anal. **17** (2016), 243–249.
[26] J. Lindenstrauss, *On operators which attain their norm*, Isr. J. Math. **1** (1963), 139–148.
[27] W. Schachermayer, *Norm attaining operators on some classical Banach spaces*, Pac. J. Math. **105** (1983), 427–438.

Asuman Güven Aksoy, Mehmet Kılıç, and Şahin Koçak

2 Isometric embeddings of finite metric trees into $(\mathbb{R}^n, d_1)$ and $(\mathbb{R}^n, d_\infty)$

Abstract: We investigate isometric embeddings of finite metric trees into $(\mathbb{R}^n, d_1)$ and $(\mathbb{R}^n, d_\infty)$. We prove that a finite metric tree can be isometrically embedded into $(\mathbb{R}^n, d_1)$ if and only if the number of its leaves is at most $2n$. We show that a finite star tree with at most 2^n leaves can be isometrically embedded into $(\mathbb{R}^n, d_\infty)$ and a finite metric tree with more than 2^n leaves cannot be isometrically embedded into $(\mathbb{R}^n, d_\infty)$. We conjecture that an arbitrary finite metric tree with at most 2^n leaves can be isometrically embedded into $(\mathbb{R}^n, d_\infty)$.

Keywords: Metric trees, embeddings, Euclidean spaces with the maximum metric

MSC 2010: Primary 54E35, Secondary 54E45, 54E50, 05C05, 47H09, 51F99

2.1 Introduction

A finite metric tree is a finite, connected, and positively weighted graph without cycles. The distance between two vertices is given by the total weight of a simple path (which is unique) between these vertices. For two in-between points of the tree, one takes the corresponding portions of the edges carrying these points into account. A leaf of a tree is a vertex with degree 1. Vertices other than the leaves are called interior vertices. A tree with a single interior vertex is called a star (or star tree). We will not allow interior vertices of degree 2 since, from the point of view of metric properties, they can be considered artificial.

Embedding new spaces into more familiar ones is a kind of innate behavior for mathematicians. For metric spaces, as the first candidate for an ambient space, the Euclidean space $(\mathbb{R}^n, d_2)$ might come into mind; but it is a complete disappointment. No metric tree (with at least three leaves) can be embedded isometrically into any $(\mathbb{R}^n, d_2)$. For a star with three leaves, this is almost obvious; and any tree with more than three leaves contains a three-star as a subspace.

There is no help in considering $(\mathbb{R}^n, d_p)$ with any $1 < p < \infty$, because these are also uniquely geodesic spaces and they do not host any tree with at least three leaves either.

Asuman Güven Aksoy, Department of Mathematics, Claremont McKenna College, 850 Columbia Avenue, Claremont, CA 91711, USA, e-mail: aaksoy@cmc.edu
Mehmet Kılıç, Şahin Koçak, Department of Mathematics, Anadolu University, 26470 Eskisehir, Turkey, e-mails: kompaktuzay@gmail.com, skocak@anadolu.edu.tr

https://doi.org/10.1515/9783110656756-002

At this point, we could take refuge in Kuratowski's embedding theorem which states that every metric space M embeds isometrically in the Banach space $L^\infty(M)$ of bounded functions on M with sup norm $\|f\|_\infty := \sup_{x\in M} |f(x)|$ where $f : M \to \mathbb{R}$ (see [5]). But a metric tree has an uncountable number of points so that our ambient space would be too huge and useless. A remedy could arise from considering the leaves of the tree only, since they determine the whole of the metric tree (as a metric space and up to isometry) as their tight span (see Theorem 8 in Dress [3]). By this approach, we get at least an isometric embedding of the metric tree into $(\mathbb{R}^n, d_\infty)$ where n is the number of leaves and d_∞ denotes the maximum metric $d_\infty(x,y) = \max_{1\le i\le n} |x_i - y_i|$.

We will show that a metric star tree can be embedded into $(\mathbb{R}^n, d_\infty)$ if and only if it has at most 2^n leaves and an arbitrary finite metric tree with more than 2^n leaves cannot be embedded into $(\mathbb{R}^n, d_\infty)$. We guess that an arbitrary finite metric tree with at most 2^n leaves can be embedded into $(\mathbb{R}^n, d_\infty)$ but we are yet unable to provide a proof for this guess.

The picture for $(\mathbb{R}^n, d_1)$ as the ambient space is more clear-cut. It was already shown by Evans ([4]) that any finite metric tree can be isometrically embedded into l_1, without explicit bounds for the dimension of the target in terms of the leaf number of the given tree. We prove by other, more geometric means that a finite metric tree can be embedded isometrically into $(\mathbb{R}^n, d_1)$ if and only if the number of its leaves is at most $2n$.

2.2 Preliminaries

For the sake of clarification, we give in the following the formal definition of a metric tree and mention some of its properties. However, in this paper we will consider a special subfamily of metric trees, namely finite connected weighted graphs without loops. Our aim is to investigate whether we can isometrically embed these finite metric trees into (R^n, d_1) or (R^n, d_∞).

Definition 2.1. Let $x, y \in M$, where (M, d) is a metric space. A *geodesic segment from x to y* (or a *metric segment*, denoted by $[x, y]$) is the image of an isometric embedding $\alpha : [a, b] \to M$ such that $\alpha(a) = x$ and $\alpha(b) = y$. A metric space is called *geodesic* if any two points can be connected by a metric segment.

A metric space (M, d) is called a *metric tree* if and only if for all $x, y, z \in M$. The following holds:

(1) there exists a unique metric segment from x to y, and
(2) $[x, z] \cap [z, y] = \{z\} \Rightarrow [x, z] \cup [z, y] = [x, y]$.

Next, we mention some useful properties of metric segments that we will use. For the proofs of these properties, we refer the reader to consult [1] and [4].

For x, y in a metric space M, write $xy = d(x, y)$. For $x, y, z \in M$, we say y is *between x and z*, denoted xyz, if and only if $xz = xy + yz$.

(1) (Transitivity of betweenness [1]) Let M be a metric space and let $a, b, c, d \in M$. If abc and acd, then abd and bcd.

(2) (Three-point property, [4, Section 3.3.1]) Let $x, y, z \in T$ (T is a complete metric tree). There exists (necessarily unique) $w \in T$ such that

$$[x,z] \cap [y,z] = [w,z] \quad \text{and} \quad [x,y] \cap [w,z] = \{w\}$$

Consequently,

$$[x,y] = [x,w] \cup [w,y], \quad [x,z] = [x,w] \cup [w,z], \quad \text{and} \quad [y,z] = [y,w] \cup [w,z].$$

Here are two examples of metric trees:

Example 2.2 (The radial metric). Define $d : \mathbb{R}^2 \times \mathbb{R}^2 \to \mathbb{R}_{\geq 0}$ by

$$d(x,y) = \begin{cases} \|x - y\| & \text{if } x = \lambda y \text{ for some } \lambda \in \mathbb{R}, \\ \|x\| + \|y\| & \text{otherwise.} \end{cases}$$

It is easy to verify that d is in fact a metric and that $(\mathbb{R}^2, d)$ is a metric tree.

Example 2.3 ("Star tree"). Fix $k \in \mathbb{N}$, and a sequence of positive numbers $(a_i)_{i=1}^k$, the *metric star tree* is defined as a union of k intervals of lengths $a_1, \dots, a_k$, emanating from a common center and equipped with the radial metric. More precisely, our tree T consists of its center o, and the points (i, t), with $1 \le i \le k$ and $0 < t \le a_i$. The distance d is defined by setting $d(o, (i, t)) = t$, and

$$d((i,t),(j,s)) = \begin{cases} |t - s| & i = j, \\ t + s & i \neq j. \end{cases}$$

Abusing the notation slightly, we often identify o with $(i, 0)$. The leaves of this metric star tree are the points (i, a_i), $i = 1, \dots, k$.

In the following, we will consider only a very special and simple type of metric trees called finite metric trees or finite simplicial metric trees. They are explained, for example, in [7], page 43 and page 73 (see also [4], Example 3.16). Topologically, they are finite trees in the sense of graph theory: There is a finite set of "vertices"; between any pair of vertices, there is at most one "edge" and the emerging graph is connected and has no loops. The degree of a vertex is defined in the graph-theoretical sense and a vertex with degree 1 is called a leaf.

In addition to this graph-theoretical structure, nonnegative weights are assigned to the edges. Moreover, to a pair of vertices a distance is assigned by considering the unique simple path (in graph-theoretical sense) connecting these vertices and adding up the weights of the edges constituting this path. This distance can be naturally extended to any pair of points of the tree and converts the graph-theoretical tree into a metric space which is a metric tree in the sense of Definition 2.1.

These special metric trees are also called finite metric trees, or finite simplicial metric trees, though they obviously contain a continuum of points as a metric space (except in the trivial case of a singleton). We will be concerned with embedding these finite metric trees into $(\mathbb{R}^n, d_1)$ or $(\mathbb{R}^n, d_\infty)$ where we will try to optimize the dimension n in dependence of the number of the leaves of the tree.

In general, metric trees are more complicated than finite simplicial metric trees. For further discussion of nonsimplicial trees and construction of metric trees related to the asymptotic geometry of hyperbolic metric spaces, we refer the reader to [2].

2.3 Embedding star trees

As motivational examples, we consider first finite metric star trees. They can be viewed as a union of k intervals of lengths $a_1, a_2, \ldots, a_k$, emanating from a common center and equipped with the radial metric, as described above (see Figure 2.1).

The following two properties give the tight embedding bounds for finite metric star trees.

Proposition 2.4. *A metric star tree (X, d) with k leaves can be isometrically embedded into $(\mathbb{R}^n, d_\infty)$ if and only if $k \leq 2^n$.*

Proof. ($\Leftarrow$) First, assume $k \leq 2^n$. Note that there are 2^n extreme points on the unit ball of the space $(\mathbb{R}^n, d_\infty)$, which are $\epsilon \in \{-1, 1\}^n$. We will denote these extreme points by

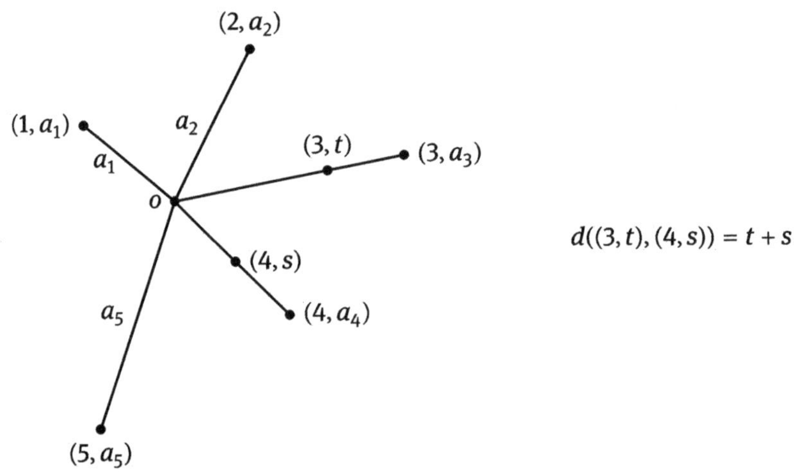

Figure 2.1: A metric star tree.

$v_1, v_2, \ldots, v_{2^n}$ and define the map $f : X \to (\mathbb{R}^n, d_\infty)$ by $f(o) = O \in \mathbb{R}^n$ and $f((i,t)) = t \cdot v_i$. The map f is then an isometric embedding from X to $(\mathbb{R}^n, d_\infty)$:

$$d_\infty(f(i,t), f(o)) = d_\infty(t \cdot v_i, O) = t = d((i,t), o),$$
$$d_\infty(f(i,t), f(i,s)) = d_\infty(t \cdot v_i, s \cdot v_i) = |t - s| = d((i,t),(i,s)),$$
$$d_\infty(f(i,t), f(j,s)) = d_\infty(t \cdot v_i, s \cdot v_j) = t + s = d((i,t),(j,s)), \quad \text{when } i \neq j.$$

(⇒) Assume that $k > 2^n$ but there exists an isometric embedding f from X to $(\mathbb{R}^n, d_\infty)$. We can assume that the embedding takes the center of X to the origin of $\mathbb{R}^n$ because translation is an isometry. Then we can find two points (i, a_i) and (j, a_j) in X with $i \neq j$ such that all corresponding coordinates of their images have the same sign. If their images are $f(i, a_i) = A_i = (A_i^1, A_i^2, \ldots, A_i^n)$ and $f(j, a_j) = A_j = (A_j^1, A_j^2, \ldots, A_j^n)$, then $|A_i^m| \le a_i$ and $|A_j^m| \le a_j$ for all $m = 1, 2, \ldots, n$ since $d_\infty(A_i, O) = d((i, a_i), o) = a_i$ and $d_\infty(A_j, O) = d((j, a_j), o) = a_j$. Hence, we obtain

$$d_\infty(A_i, A_j) = \max_{m=1}^{n}\{|A_i^m - A_j^m|\} < a_i + a_j = d((i, a_i), (j, a_j)),$$

which contradicts the assumption that f is an isometry. □

Proposition 2.5. *A metric star tree X with k leaves can be embedded isometrically into $(\mathbb{R}^n, d_1)$ if and only if $k \le 2n$.*

Proof. (⇐) First, assume $k \le 2n$. Note that there are $2n$ extreme point on the unit ball of the space $(\mathbb{R}^n, d_1)$, which are $e_i = (\delta_{i,j})_{j=1}^n$, where $\delta_{i,j}$ is the Kronecker delta. We will denote these extreme points in $\mathbb{R}^n$ by $E_1, E_2, \ldots, E_{2n}$ and define the map $f : X \to (\mathbb{R}^n, d_1)$ by $f(o) = O$ and $f((i,t)) = t \cdot E_i$. We will show that f is an isometric embedding from X to $(\mathbb{R}^n, d_1)$:

$$d_1(f(i,t), f(o)) = d_1(t \cdot E_i, O) = t = d((i,t), o),$$
$$d_1(f(i,t), f(i,s)) = d_1(t \cdot E_i, s \cdot E_i) = |t - s| = d((i,t),(i,s)),$$
$$d_1(f(i,t), f(j,s)) = d_1(t \cdot E_i, s \cdot E_j) = t + s = d((i,t),(j,s)), \quad \text{where } i \neq j.$$

(⇒) Assume that $k > 2n$ but there exists an isometric embedding f from X to $(\mathbb{R}^n, d_1)$. We can assume that the embedding takes the center of X to the origin because translation is an isometry. Then we can find two points (i, a_i) and (j, a_j) in X with $i \neq j$ such that at least one common coordinate of their images are nonzero and have the same sign. If their images are $f(i, a_i) = A_i = (A_i^1, A_i^2, \ldots, A_i^n)$ and $f(j, a_j) = A_j = (A_j^1, A_j^2, \ldots, A_j^n)$, then we obtain

$$\begin{aligned} d((i, a_i), (j, a_j)) &= d((i, a_i), o) + d(o, (j, a_j)) = d_1(A_i, 0) + d_1(0, A_j) \\ &= |A_i^1| + |A_i^2| + \cdots + |A_i^n| + |A_j^1| + |A_j^2| + \cdots + |A_j^n| \\ &> |A_i^1 - A_j^1| + |A_i^2 - A_j^2| + \cdots + |A_i^n - A_j^n| \\ &= d_1(A_i, A_j), \end{aligned}$$

which is a contradiction. □

2.4 Embedding arbitrary metric trees

For later use, we recall some metric preliminaries.

Definition 2.6. For $p = (p_1, p_2, \dots, p_n) \in (\mathbb{R}^n, d_\infty)$, we define

$$S_i^+(p) = \{q = (q_1, q_2, \dots, q_n) \in \mathbb{R}^n \mid d_\infty(p,q) = q_i - p_i\},$$
$$S_i^-(p) = \{q = (q_1, q_2, \dots, q_n) \in \mathbb{R}^n \mid d_\infty(p,q) = p_i - q_i\}$$

for $i = 1, 2, \dots, n$ and call them the sectors at the point p as shown in the following Figure 2.2 and Figure 2.3. Notice that if q belongs to $S_i^\varepsilon(p)$, $S_i^\varepsilon(q) \subseteq S_i^\varepsilon(p)$ holds, where $\varepsilon \in \{+,-\}$.

The following theorem gives a characterization of geodesics in $(\mathbb{R}^n, d_\infty)$.

Proposition 2.7 (Theorem 2.2 of [6]). *Let $p = (p_1, p_2, \dots, p_n)$, $q = (q_1, q_2, \dots, q_n) \in \mathbb{R}^n$ be two points, $q \in S_i^\varepsilon(p)$ and $\alpha : [0, d(p,q)] \to \mathbb{R}^n$ be a path such that $\alpha(0) = p$ and*

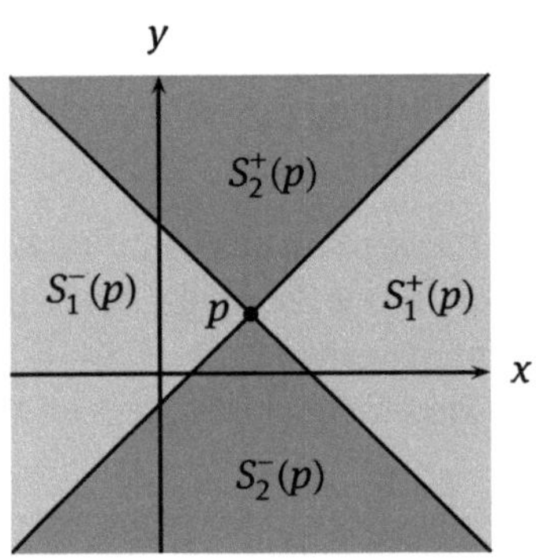

Figure 2.2: Sectors of a point p in $(\mathbb{R}^2, d_\infty)$.

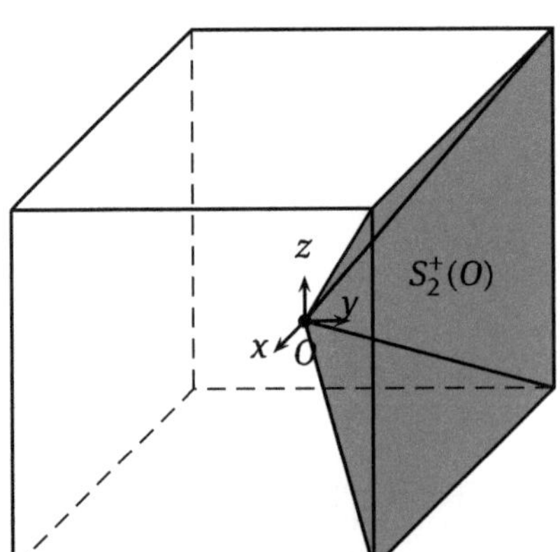

Figure 2.3: The sector $S_2^+(O)$ of the origin in $(\mathbb{R}^3, d_\infty)$.

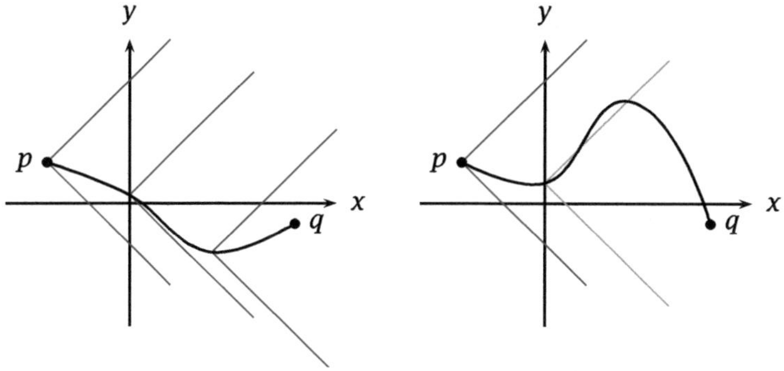

Figure 2.4: Two paths between p and q in $\mathbb{R}^2_\infty$ one of which (on the left) is a geodesic but the other is not.

$\alpha(d(p,q)) = q$. Then α is a geodesic in $\mathbb{R}^n_\infty$ if and only if $\alpha(t') \in S_i^\varepsilon(\alpha(t))$ for all $t, t' \in [0, d(p,q)]$ such that $t < t'$. (See Figure 2.4.)

We will also need the following "shortening lemma."

Lemma 2.8. *Let $\alpha : [0,b] \to (\mathbb{R}^n, d_\infty)$ be a geodesic and $0 < c < d < b$. Then $\tilde{\alpha} : [0, b-d+c] \to (\mathbb{R}^n, d_\infty)$,*

$$\tilde{\alpha}(t) = \begin{cases} \alpha(t) & \text{when } t \in [0,c] \\ \alpha(t-c+d) - \alpha(d) + \alpha(c) & \text{when } t \in [c, b-d+c] \end{cases}$$

is a geodesic.

Proof. Assume that $\alpha(b) \in S_i^\varepsilon(\alpha(0))$ for some $i \in \{1,2,\dots,n\}$ and $\varepsilon \in \{+,-\}$. Since α is a geodesic, for all $t, t' \in [0,b]$ such that $t < t'$ we get $\alpha(t') \in S_i^\varepsilon(\alpha(t))$. Given $t, t' \in [0, b-d+c]$ with $t < t'$. We consider three possibilities: if $t, t' \in [0,c]$, then $\tilde{\alpha}(t') \in S_i^\varepsilon(\tilde{\alpha}(t))$ because $\tilde{\alpha} = \alpha$ on $[0,c]$. If $t, t' \in [c, b-d+c]$, then $\tilde{\alpha}(t') \in S_i^\varepsilon(\tilde{\alpha}(t))$ because $\alpha(t'-c+d) \in S_i^\varepsilon(\alpha(t-c+d))$ and this implies $\alpha(t'-c+d)-\alpha(d)+\alpha(c) \in S_i^\varepsilon(\alpha(t-c+d)-\alpha(d)+\alpha(c))$. If $t \in [0,c]$ and $t' \in [c, b-d+c]$, we know that $\tilde{\alpha}(t') \in S_i^\varepsilon(\tilde{\alpha}(c))$ and $\tilde{\alpha}(c) \in S_i^\varepsilon(\tilde{\alpha}(t))$. Since $\tilde{\alpha}(c) \in S_i^\varepsilon(\tilde{\alpha}(t))$, $S_i^\varepsilon(\tilde{\alpha}(c)) \subseteq S_i^\varepsilon(\tilde{\alpha}(t))$; hence, we get $\tilde{\alpha}(t') \in S_i^\varepsilon(\tilde{\alpha}(t))$. Thus, the previous proposition implies that $\tilde{\alpha}$ is a geodesic. □

Proposition 2.9. *A finite metric tree with more than 2^n leaves can not be embedded isometrically into $(\mathbb{R}^n, d_\infty)$.*

Proof. Let us assume that X has k leaves, $k > 2^n$ and $f : X \to (\mathbb{R}^n, d_\infty)$ be an isometric embedding. Denote the leaves $a_1, a_2, \dots, a_k$, and their images under f by $A_1, A_2, \dots, A_k$, that is, $f(a_i) = A_i$. Let us denote the vertex points on X by b_i for $i = 1,2,\dots,k$ such that there exists an edge between a_i and b_i and assume $f(b_i) = B_i$. Note that the vector A_iB_i can not be equal to $t \cdot (A_jB_j)$ for $i \neq j$ and $t > 0$. Because if we assume $A_iB_i = t \cdot (A_jB_j)$

or equivalently $B_i - A_i = t \cdot (B_j - A_j)$, we get

$$\begin{aligned}\|A_i - A_j\| &= \|(A_i - B_i) + (B_i - B_j) + (B_j - A_j)\| \\ &= \|(1-t)(B_j - A_j) + (B_i - B_j)\| \\ &\le |1-t| \cdot \|B_j - A_j\| + \|B_i - B_j\|\end{aligned}$$

On the other hand, since the geodesic from a_i to a_j passes through b_i and b_j respectively, we have

$$\begin{aligned}\|A_i - A_j\| &= \|A_i - B_i\| + \|B_i - B_j\| + \|B_j - A_j\| \\ &= t\|B_j - A_j\| + \|B_i - B_j\| + \|B_j - A_j\| \\ &= (1+t)\|B_j - A_j\| + \|B_i - B_j\|\end{aligned}$$

and this contradicts previous inequality.

Now consider the set $\{t \cdot (A_i - B_i) \mid 0 \le t \le 1, i = 1, 2, \ldots, k\}$. According to Lemma 2.8, this set is a star tree with $k > 2^n$ leaves. But this contradicts the Proposition 2.4. □

Proposition 2.10. *A finite metric tree with more than 2n leaves cannot be embedded isometrically into* $(\mathbb{R}^n, d_1)$.

Proof of this claim is very similar to the proof above. In fact, let us assume that X has k leaves with $k > 2n$ and $f : X \to (\mathbb{R}^n, d_1)$ be an isometric embedding. Denote those leaves by $a_1, a_2, \ldots, a_k$, and their images by $A_1, A_2, \ldots, A_k$, that is, $f(a_i) = A_i$. Let us denote the vertices on X by b_i for $i = 1, 2, \ldots, k$ such that there is an edge between a_i and b_i and assume $f(b_i) = B_i$. Now consider the set $\{t \cdot (A_i - B_i) \mid 0 \le t \le 1, i = 1, 2, \ldots, k\}$. This set is a star tree with $k > 2n$ leaves but this contradicts the Proposition 2.5.

Theorem 2.11. *Let (X, d) be a metric tree. If X contains at most 2n leaves, it can be embedded isometrically into* $(\mathbb{R}^n, d_1)$.

Proof. We will prove this theorem by induction on n. If $n = 1$, the statement is obviously true. Assume that we can embed isomerically any metric tree which has $2n$ leaves into $(\mathbb{R}^n, d_1)$. Let X be any metric tree which has $2(n+1)$ leaves. We will show that X can be embedded isometrically into $(\mathbb{R}^{n+1}, d_1)$. We can choose two leaves in X such that after deleting them with the adjacent edges (and discarding possibly emerging vertices with degree 2), the rest of the tree has $2n$ leaves (see Figure 2.5). Let us call these leaves as A_0 and A_1 and their adjacent edges as B_0A_0 and B_1A_1. According to our assumption, there is an isometric embedding f from rest of the tree $Y = X - ((B_0A_0] \cup (B_1A_1])$ to $(\mathbb{R}^n, d_1)$. Define the map $F : X \to (\mathbb{R}^{n+1}, d_1)$,

$$F(x) = \begin{cases} (f(x), 0) & \text{when } x \in Y \\ (f(B_0), -d(x, B_0)) & \text{when } x \in [B_0A_0] \\ (f(B_1), d(x, B_1)) & \text{when } Mx \in [B_1A_1] \end{cases}$$

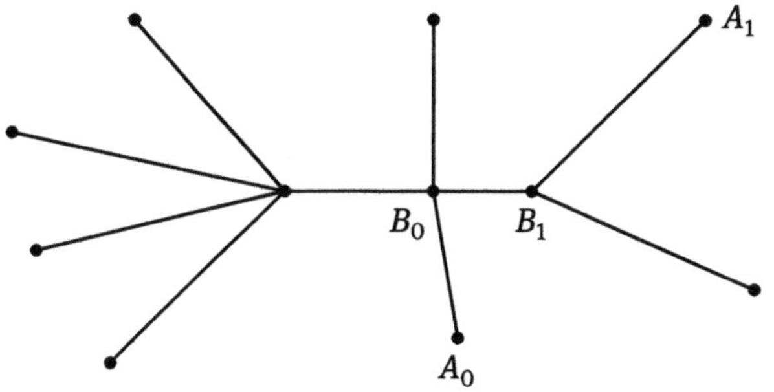

Figure 2.5: If we delete $(B_0A_0]$ and $(B_1A_1]$ and discarding the vertex B_1, we get a tree which has two fewer leaves.

We will show that F is an isometric embedding. Let $x, y \in X$ be two arbitrary points. Since f is an isometric embedding, if $x, y \in Y$, we get

$$d_1(F(x), F(y)) = d_1(f(x), f(y)) = d(x, y).$$

If $x \in (B_0A_0]$ and $y \in Y$,

$$\begin{aligned} d_1(F(x), F(y)) &= d_1(f(B_0), f(y)) + d(x, B_0) \\ &= d(B_0, y) + d(x, B_0) \\ &= d(x, y). \end{aligned}$$

If $x \in (B_1A_1]$ and $y \in Y$,

$$\begin{aligned} d_1(F(x), F(y)) &= d_1(f(B_1), f(y)) + d(x, B_1) \\ &= d(B_1, y) + d(x, B_1) \\ &= d(x, y). \end{aligned}$$

If $x \in (B_0A_0]$ and $y \in (B_1A_1]$,

$$\begin{aligned} d_1(F(x), F(y)) &= d_1(f(B_0), f(B_1)) + d(x, B_0) + d(y, B_1) \\ &= d(x, B_0) + d(B_0, B_1) + d(B_1, y) \\ &= d(x, y). \end{aligned}$$

□

Bibliography

[1] L. M. Blumenthal, *Theory and Applications of Distance Geometry*, Oxford University Press, London, 1953.

[2] M. Bridson and A. Haefliger, *Metric Spaces of Nonpositive Curvature*, Grundlehren der Mathematischen Wissenschaften, vol. 319, Springer-Verlag, Berlin, 1999.

[3] A. Dress, Trees, tight extensions of metric spaces, and the cohomological dimension of certain groups: A note on combinatorial properties of metric spaces, Adv. Math. **53** (1984), 321–402.

[4] S. Evans, *Probability and Real Trees*, Springer, Berlin, 2008.

[5] J. Heinonen, *Geometric Embeddings of Metric Spaces*, 2003.

[6] M. Kılıç, *Isometry classes of planes in* $(\mathbb{R}^3, d_\infty)$, Hacet. J. Math. Stat. **48** (4) (2019), DOI: 10.15672/HJMS.2018.571.

[7] A. Papadopoulos, *Metric Spaces, Convexity and Nonpositive Curvature* (2nd edition), European Math. Soc., 2014.

Chafiq Benhida and Raúl E. Curto

3 Iterates of the spherical Aluthge transform of 2-variable weighted shifts

To the memory of Professor Victor Lomonosov

Abstract: Let $\mathbf{T} \equiv (T_1, T_2)$ be a commuting pair of Hilbert space operators, and let $P := \sqrt{T_1^* T_1 + T_2^* T_2}$ be the positive factor in the (joint) polar decomposition of $\mathbf{T}$; that is, $T_i = V_i P$ $(i = 1, 2)$. The spherical Aluthge transform of $\mathbf{T}$ is the (necessarily commuting) pair $\Delta_{\text{sph}}(\mathbf{T}) := (\sqrt{P} V_1 \sqrt{P}, \sqrt{P} V_2 \sqrt{P})$. In this paper, we focus on the asymptotic behavior of the sequence $\{\Delta_{\text{sph}}^{(n)}(\mathbf{T})\}_{n \geq 1}$ as $n \to \infty$, where $\Delta_{\text{sph}}^{(1)}(\mathbf{T}) := \Delta_{\text{sph}}(\mathbf{T})$ and $\Delta_{\text{sph}}^{(n+1)}(\mathbf{T}) := \Delta_{\text{sph}}(\Delta_{\text{sph}}^{(n)}(\mathbf{T}))$ $(n \geq 1)$. In those cases when the limit exists, the limit pair is a fixed point for the spherical Aluthge transform, that is, a spherically quasinormal pair. For a suitable class of 2-variable weighted shifts, we establish the convergence of the sequence of iterates in the weak operator topology.

Keywords: Spherical Aluthge transform, iterates, polar decomposition, spherical isometry

MSC 2010: Primary 47B37, 47B20, 47A13, 37E25, Secondary 47A57, 44A60

3.1 Introduction

The Aluthge transform of a bounded operator T acting on a Hilbert space $\mathcal{H}$ was introduced by A. Aluthge in ([1]). If $T \equiv V|T|$ is the canonical polar decomposition of T, the Aluthge transform $\Delta(T)$ is given as $\Delta(T) := \sqrt{|T|} V \sqrt{|T|}$. One of Aluthge's motivations was to use this transform in the study of p-hyponormal and log-hyponormal operators. Roughly speaking, the idea was to convert an operator, T, into another operator, $\Delta(T)$, which shares with the first one many structural and spectral properties, but which is closer to being a normal operator. Over the last two decades, substantial and signifi-

Acknowledgement: The first named author was partially supported by Labex CEMPI (ANR-11-LABX-0007-01). The second named author was partially supported by NSF Grant DMS-1302666 and by Labex CEMPI (ANR-11-LABX-0007-01).
The authors are deeply grateful to the referee for a careful reading of the paper and for several suggestions and edits that helped improve the presentation. Some of the calculations in this paper were made with the software tool *Mathematica* [34].

Chafiq Benhida, UFR de Mathématiques, Université des Sciences et Technologies de Lille, F-59655, Villeneuve d'Ascq Cedex, France, e-mail: Chafiq.Benhida@univ-lille.fr
Raúl E. Curto, Department of Mathematics, University of Iowa, Iowa City, Iowa 52242, USA, e-mail: raul-curto@uiowa.edu

https://doi.org/10.1515/9783110656756-003

cant results about $\Delta(T)$, and how it relates to T, have been obtained by a long list of mathematicians who devoted considerable attention to this topic (see, for instance, [2], [9], [20], [25–27], [28–30]). Aluthge transforms have been generalized to the case of powers of $|T|$ different from $\frac{1}{2}$ ([4, 7]) and to the case of commuting pairs of operators ([17], [18]).

This generalization, called the *spherical Aluthge transform* of $\mathbf{T}$, is the (necessarily commuting) pair $\Delta_{\mathrm{sph}}(\mathbf{T}) := (\sqrt{P}V_1\sqrt{P}, \sqrt{P}V_2\sqrt{P})$, where $P := \sqrt{T_1^*T_1 + T_2^*T_2}$ is the positive factor in the (joint) polar decomposition of $\mathbf{T}$ and (V_1, V_2) is the joint partial isometry. In this paper, we study the asymptotic behavior of the iterates of the spherical Aluthge transform of $\mathbf{T}$; that is, the behavior as $n \to \infty$ of the sequence of commuting pairs given by $\Delta_{\mathrm{sph}}^{(1)}(\mathbf{T}) := \Delta_{\mathrm{sph}}(\mathbf{T})$ and $\Delta_{\mathrm{sph}}^{(n+1)}(\mathbf{T}) := \Delta_{\mathrm{sph}}(\Delta_{\mathrm{sph}}^{(n)}(\mathbf{T}))$ $(n \geq 1)$. We do this for a class of 2-variable weighted shifts obtained as finite-rank perturbations of spherical isometries. In those cases when the limit exists, the limit pair is a fixed point for the spherical Aluthge transform; that is, a spherically quasinormal pair. For this class of 2-variable weighted shifts, we will establish the convergence of the sequence of iterates in the weak operator topology; see the details in Section 3.4.

3.2 Notation and preliminaries

3.2.1 The classical Aluthge transform

Let $\mathcal{H}$ denote a (complex, separable) Hilbert space, and let $\mathcal{B}(\mathcal{H})$ denote the C^*-algebra of bounded linear operators on $\mathcal{H}$. For $T \in \mathcal{B}(\mathcal{H})$, let $T \equiv V|T|$ be the canonical polar decomposition of T; that is, $|T| := (T^*T)^{\frac{1}{2}}$, V is a partial isometry, and $\ker V = \ker |T| = \ker T$. The Aluthge transform of T is the operator

$$\Delta(T) := |T|^{\frac{1}{2}} V |T|^{\frac{1}{2}}.$$

The Aluthge transform has been extensively studied, in terms of algebraic, structural, and spectral properties. We list below a brief sample of the results obtained over the last several years.

(i) T is a fixed point of Δ (i. e., $\Delta(T) = T$) if and only if T is quasinormal, that is, T commutes with $|T|$.

(ii) (A. Aluthge [1]) Let $0 < p < \frac{1}{2}$ and assume that T is p-hyponormal. Then $\Delta(T)$ is $(p + \frac{1}{2})$-hyponormal.

(iii) In [25], I. B. Jung, E. Ko, and C. Pearcy showed that T and $\Delta(T)$ share many spectral properties; in particular, $\sigma(\Delta(T)) = \sigma(T)$.

(iv) In [25, Corollary 1.16], I. B. Jung, E. Ko, and C. Pearcy proved that if $\Delta(T)$ has a nontrivial invariant subspace, then so does T; and if T has dense range, then the above implication becomes an equivalence [25, Theorem 1.15].

(v) M. H. Kim and E. Ko ([28]), and F. Kimura ([29]) proved that T has property (β) if and only if $\Delta(T)$ has property (β).
(vi) In [2], T. Ando established that for all $\lambda \notin \sigma(T)$, one has $\|(T-\lambda)^{-1}\| \geq \|(\Delta(T)-\lambda)^{-1}\|$.
(vii) G. Exner proved in [21, Example 2.11] that the subnormality of T is not preserved under the Aluthge transform.
(viii) Subsequently, S. H. Lee, W. Y. Lee, and J. Yoon ([30]) showed that for $k \geq 2$, the Aluthge transform, when acting on weighted shifts, does not preserve k-hyponormality.

3.2.2 Iterates of the Aluthge transform

The iterates of the Aluthge transform are given by

$$\Delta^{(1)}(T) := \Delta(T)$$

and

$$\Delta^{(n+1)}(T) := \Delta(\Delta^{(n)}(T)) \ (n \geq 1).$$

It is easy to verify that the Aluthge transform of a weighted shift W_ω is again a weighted shift; see Subsection 3.2.5. Concretely, the weights of $\Delta(W_\omega)$ are

$$\sqrt{\omega_0\omega_1}, \sqrt{\omega_1\omega_2}, \sqrt{\omega_2\omega_3}, \sqrt{\omega_3\omega_4}, \ldots.$$

If we let

$$W_{\sqrt{\omega}} := \text{shift}\ (\sqrt{\omega_0}, \sqrt{\omega_1}, \sqrt{\omega_2}, \ldots),$$

then $\Delta(W_\omega)$ is the Schur product of $W_{\sqrt{\omega}}$ and its restriction to the closed linear span $\bigvee\{e_1, e_2, \ldots\}$. Thus, a sufficient condition for the subnormality of $\Delta(W_\omega)$ is the subnormality of $W_{\sqrt{\omega}}$. (For more on this connection, see [15].)

Next, observe that

$$\Delta^{(2)}(W_\omega) = \text{shift}\Big(\sqrt{\sqrt{\omega_0\omega_1}\sqrt{\omega_1\omega_2}}, \sqrt{\sqrt{\omega_1\omega_2}\sqrt{\omega_2\omega_3}}, \ldots\Big),$$
$$\Delta^{(3)}(W_\omega) = \text{shift}\big((\omega_0\omega_1^3\omega_2^3\omega_3)^{\frac{1}{8}}, (\omega_1\omega_2^3\omega_3^3\omega_4)^{\frac{1}{8}}, \ldots\big),$$

and

$$\Delta^{(4)}(W_\omega) = \text{shift}\big((\omega_0\omega_1^4\omega_2^6\omega_3^4\omega_4)^{\frac{1}{16}}, (\omega_1\omega_2^4\omega_3^6\omega_4^4\omega_5)^{\frac{1}{16}}, \ldots\big).$$

Thus, if we let $\omega^{(n)}$ denote the weight sequence of $\Delta^{(n)}(W_\omega)$, we have

$$\omega_k^{(n+1)} = \sqrt{\omega_k^{(n)}\omega_{k+1}^{(n)}},$$

and an induction argument shows that

$$\omega_k^{(n)} = \left(\prod_{j=0}^{n} \omega_{k+j}^{\binom{n}{j}} \right)^{\frac{1}{2^n}}. \tag{3.1}$$

The study of the limiting behavior of the iterates of the Aluthge transform has received considerable attention. Below is a list of some major results in this direction.

(i) In [25], I. B. Jung, E. Ko, and C. Pearcy conjectured that for every bounded operator T the sequence $\{\Delta^{(n)}(T)\}$ converges in norm to a quasinormal operator.

(ii) In [2, Theorem], T. Ando proved that the conjecture is true for 2×2 matrices.

(iii) In [3], J. Antezana, E. Pujals, and D. Stojanoff proved the conjecture to be true for $\dim \mathcal{H} < \infty$; see also [4].

(iv) In 2003, J. Thompson (as communicated in [26, Example 5.5]) found an example of an operator for which the sequence converges to 0 in the strong operator topology (SOT), but it does not converge in norm.

(v) In 2001, M. Yanagida found an example of a unilateral weighted shift for which the sequence of iterates does not converge in the weak operator topology (WOT) (cf. [31, p. 2, lines 15 and 16]).

(vi) In [9], M. Chō and W. Y. Lee proved that for any $0 < a < b$ there exists a unilateral weighted shift W_ω such that the sequence $\{\omega_0^{(n)}\}_{n \geq 0}$ clusters at both a and b.

Possibly the most definitive results about the convergence of the iterates of the classical Aluthge transform were obtained by K. Rion in [31].

Proposition 3.1 ([31, Proposition 1]). *The WOT and SOT convergences of $\{T_{\omega^{(n)}}\}$ are equivalent to the pointwise convergence of the sequence $\{\omega^{(n)}\}_n$, given by* (3.1).

Theorem 3.2 ([31, Theorem 7]). *Assume ω is bounded below. Then the set $\mathcal{S}$ of SOT subsequential limits of $\{\Delta^{(n)}(T_\omega)\}$ is nonempty. Moreover, $\mathcal{S}$ is a closed interval of quasinormal shifts; that is, $\mathcal{S} = [a, b]U_+$ for some $a, b > 0$.*

3.2.3 The spherical Aluthge transform

We first recall the definition of the spherical Aluthge transform (introduced in [17] and [18]). Given a commuting pair $\mathbf{T} \equiv (T_1, T_2)$ of operators acting on $\mathcal{H}$, let $P := (T_1^* T_1 + T_2^* T_2)^{\frac{1}{2}}$. Clearly, $\ker P = \ker T_1 \bigcap \ker T_2$. For $x \in \ker P$, let $V_i x := 0$ $(i = 1, 2)$; for $y \in \operatorname{Ran} P$, say $y = Px$, let $V_i y := T_i x$ $(i = 1, 2)$. It is easy to see that V_1 and V_2 are well-defined, and extend continuously to $\overline{\operatorname{Ran} P}$. We then have

$$\begin{pmatrix} T_1 \\ T_2 \end{pmatrix} = \begin{pmatrix} V_1 P \\ V_2 P \end{pmatrix} = \begin{pmatrix} V_1 \\ V_2 \end{pmatrix} P, \tag{3.2}$$

as operators from $\mathcal{H}$ to $\mathcal{H} \oplus \mathcal{H}$. Moreover, this is the canonical polar decomposition of $\begin{pmatrix} T_1 \\ T_2 \end{pmatrix}$. It follows that $\begin{pmatrix} V_1 \\ V_2 \end{pmatrix}$ is a partial isometry from $(\ker P)^\perp$ onto $\overline{\operatorname{Ran}\begin{pmatrix} T_1 \\ T_2 \end{pmatrix}}$.

The spherical Aluthge transform of $\mathbf{T}$ is $\Delta_{\text{sph}}(\mathbf{T}) \equiv (\widehat{T_1}, \widehat{T_2})$, where

$$\widehat{T_i} := P^{\frac{1}{2}} V_i P^{\frac{1}{2}} \quad (i = 1, 2).$$

The spherical Aluthge transform was introduced in [17]; its general theory was developed in [18]. One of the basic results follows.

Lemma 3.3 (cf. [18]). $\Delta_{\text{sph}}(\mathbf{T})$ *is commutative.*

The equality of the spectra of an operator and its Aluthge transform (mentioned in Subsection 3.2.1) can be extended to commuting pairs $\mathbf{T}$ (cf. [8]). That is, one can use a bit of homological algebra applied to the appropriate Koszul complexes to prove directly that for a commuting pair $\mathbf{T} \equiv (T_1, T_2)$

$$\sigma_T(\Delta_{\text{sph}}(\mathbf{T})) = \sigma_T(\mathbf{T}), \tag{3.3}$$

where $\sigma_T(\mathbf{T})$ is the Taylor spectrum of $\mathbf{T}$. (For more information on the notion of Taylor spectrum and related results, the reader is referred to [11], [12], [33].)

Moreover, if $\mathbf{T} \equiv (T_1, T_2)$ is Taylor invertible and we represent it as a column matrix, then one can see that P is also invertible, and in this case,

$$\Delta_{\text{sph}}(\mathbf{T}) = (P^{\frac{1}{2}} \oplus P^{\frac{1}{2}}) \mathbf{T} P^{-\frac{1}{2}}.$$

3.2.4 Spherically quasinormal pairs

It is well known that the fixed points of the classical Aluthge transform are the quasinormal operators, that is, those operators $T = V|T|$ such that V and $|T|$ commute (equivalently, T and $|T|$ commute). For the spherical Aluthge transform, the fixed commuting pairs are the so-called spherically quasinormal pairs, which we now define. First, we need some terminology.

Following A. Athavale-S. Podder ([6]) and J. Gleason ([23]), we say that

(i) $\mathbf{T}$ is *matricially quasinormal* if T_i commutes with $T_j^* T_k$ for all $i, j, k = 1, 2$;

(ii) $\mathbf{T}$ is *(jointly) quasinormal* if T_i commutes with $T_j^* T_j$ for all $i, j = 1, 2$; and

(iii) $\mathbf{T}$ is *spherically quasinormal* if T_i commutes with

$$P := T_1^* T_1 + T_2^* T_2,$$

for $i = 1, 2$. Also, recall that $\mathbf{T}$ is said to be *normal* if $T_1 T_2 = T_2 T_1$ and T_i is normal $(i = 1, 2)$.

It follows that

$$\begin{aligned}\text{normal} &\implies \text{matricially quasinormal} \implies \text{(jointly) quasinormal} \\ &\implies \text{spherically quasinormal} \implies \text{subnormal ([6, Proposition 2.1])} \\ &\implies k\text{-hyponormal} \implies \text{hyponormal.} \qquad (3.4)\end{aligned}$$

On the other hand, results of R. E. Curto, S. H. Lee, and J. Yoon (cf. [16]), and of J. Gleason ([23]) show that the reverse implications in (3.4) do not necessarily hold.

In [19, Theorem 2.2], R. E. Curto and J. Yoon showed that the spherically quasinormal commuting pairs are precisely the fixed points of the spherical Aluthge transform; moreover, it follows from the results in [18, Section 2] that if $\mathbf{T}$ is spherically quasinormal then (V_1, V_2) is a commuting pair. In [18], it was also shown that every spherically quasinormal 2-variable weighted shift is a positive multiple of a spherical isometry (see Theorem 3.8). In order to state this result, we need a brief discussion of unilateral and 2-variable weighted shifts, which follows.

3.2.5 Unilateral weighted shifts

For $\omega \equiv \{\omega_n\}_{n=0}^{\infty}$, a bounded sequence of positive real numbers (called *weights*), let $W_\omega \equiv \text{shift}(\omega_0, \omega_1, \ldots) : \ell^2(\mathbb{Z}_+) \to \ell^2(\mathbb{Z}_+)$ be the associated unilateral weighted shift, defined by $W_\omega e_n := \omega_n e_{n+1}$ (all $n \geq 0$), where $\{e_n\}_{n=0}^{\infty}$ is the canonical orthonormal basis in $\ell^2(\mathbb{Z}_+)$. The *moments* of $\omega \equiv \{\omega_n\}_{n=0}^{\infty}$ are given as

$$\gamma_k \equiv \gamma_k(W_\omega) := \begin{cases} 1, & \text{if } k = 0 \\ \omega_0^2 \cdots \omega_{k-1}^2, & \text{if } k > 0. \end{cases} \qquad (3.5)$$

The (unweighted) unilateral shift is $U_+ := \text{shift}(1, 1, 1, \ldots)$, and for $0 < a < 1$ we let $S_a := \text{shift}(a, 1, 1, \ldots)$.

We now recall a well-known characterization of subnormality for unilateral weighted shifts, due to C. Berger (cf. [10, III.8.16]) and independently established by Gellar and Wallen ([22]): W_ω is subnormal if and only if there exists a probability measure σ supported in $[0, \|W_\omega\|^2]$ (called the *Berger measure* of W_ω) such that

$$\gamma_k(W_\omega) = \omega_0^2 \cdot \cdots \cdot \omega_{k-1}^2 = \int t^k \, d\sigma(t) \quad (k \geq 1).$$

Observe that U_+ and S_a are subnormal, with Berger measures δ_1 and $(1 - a^2)\delta_0 + a^2\delta_1$, respectively, where δ_p denotes the point-mass probability measure with support the singleton set $\{p\}$. On the other hand, the Berger measure of the Bergman shift B_+ (acting on $A^2(\mathbb{D})$, and with weights $\omega_n := \sqrt{\frac{n+1}{n+2}}$ $(n \geq 0)$) is the Lebesgue measure on the interval $[0, 1]$.

3.2.6 2-variable weighted shifts

Consider now two double-indexed positive bounded sequences $\alpha_{\mathbf{k}}, \beta_{\mathbf{k}} \in \ell^\infty(\mathbb{Z}_+^2)$, $\mathbf{k} \equiv (k_1, k_2) \in \mathbb{Z}_+^2$ and let $\ell^2(\mathbb{Z}_+^2)$ be the Hilbert space of square-summable complex sequences indexed by $\mathbb{Z}_+^2$. (Recall that $\ell^2(\mathbb{Z}_+^2)$ is canonically isometrically isomorphic to $\ell^2(\mathbb{Z}_+) \bigotimes \ell^2(\mathbb{Z}_+)$.) We define the 2-variable weighted shift $\mathbf{T} \equiv (T_1, T_2) = W_{(\alpha,\beta)}$ by

$$T_1 e_{\mathbf{k}} := \alpha_{\mathbf{k}} e_{\mathbf{k}+\boldsymbol{\varepsilon}_1} \quad \text{and} \quad T_2 e_{\mathbf{k}} := \beta_{\mathbf{k}} e_{\mathbf{k}+\boldsymbol{\varepsilon}_2}, \tag{3.6}$$

where $\boldsymbol{\varepsilon}_1 := (1, 0)$ and $\boldsymbol{\varepsilon}_2 := (0, 1)$. Clearly,

$$T_1 T_2 = T_2 T_1 \Longleftrightarrow \beta_{\mathbf{k}+\boldsymbol{\varepsilon}_1} \alpha_{\mathbf{k}} = \alpha_{\mathbf{k}+\boldsymbol{\varepsilon}_2} \beta_{\mathbf{k}} \quad (\text{all } \mathbf{k} \in \mathbb{Z}_+^2). \tag{3.7}$$

Moreover, for $\mathbf{k} \in \mathbb{Z}_+^2$ we have

$$T_1^* e_{0,k_2} = 0 \quad \text{and} \quad T_1^* e_{\mathbf{k}} = \alpha_{\mathbf{k}-\boldsymbol{\varepsilon}_1} e_{\mathbf{k}-\boldsymbol{\varepsilon}_1} \quad (k_1 \geq 1); \tag{3.8}$$

$$T_2^* e_{k_1,0} = 0 \quad \text{and} \quad T_2^* e_{\mathbf{k}} := \beta_{\mathbf{k}-\boldsymbol{\varepsilon}_2} e_{\mathbf{k}-\boldsymbol{\varepsilon}_2} \quad (k_2 \geq 1). \tag{3.9}$$

In an entirely similar way, one can define multivariable weighted shifts. The weight diagram of a generic 2-variable weighted shift is shown in Figure 3.1.

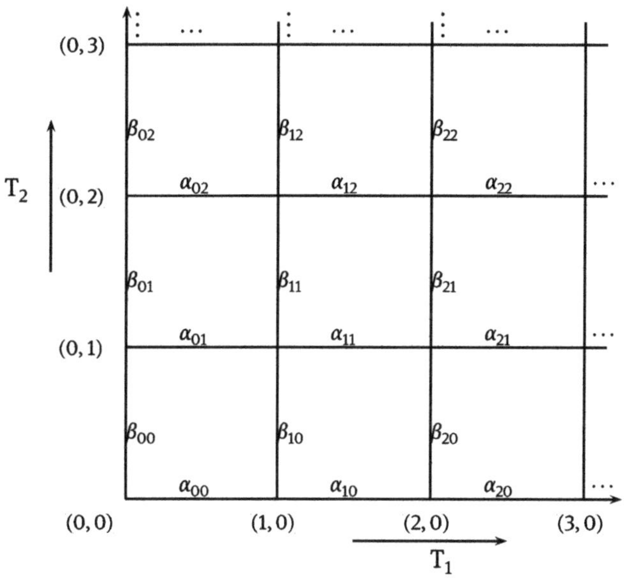

Figure 3.1: Weight diagram of a generic 2-variable weighted shift.

When all weights are equal to 1, we obtain the so-called Helton–Howe shift; that is, the shift that corresponds to the pair of multiplications by the coordinate functions in the

Hardy space $H^2(\mathbb{T}\times\mathbb{T})$ of the 2-torus, with respect to normalized arclength measure on each unit circle $\mathbb{T}$ (cf. [23]). This shift can also be represented as $(U_+\otimes I, I\otimes U_+)$, acting on $\ell^2(\mathbb{Z}_+)\otimes\ell^2(\mathbb{Z}_+)$.

We now recall the definition of *moments* for a commuting 2-variable weighted shift $\mathbf{T}\equiv(T_1,T_2)=W_{(\alpha,\beta)}$. Given $\mathbf{k}\equiv(k_1,k_2)\in\mathbb{Z}_+^2$, the moment of $\mathbf{T}\equiv(T_1,T_2)=W_{(\alpha,\beta)}$ of order $\mathbf{k}$ is

$$\gamma_{\mathbf{k}}\equiv\gamma_{\mathbf{k}}(W_{(\alpha,\beta)}):=\begin{cases}1, & \text{if } k_1=0 \text{ and } k_2=0\\ \alpha_{(0,0)}^2\cdots\alpha_{(k_1-1,0)}^2, & \text{if } k_1\ge 1 \text{ and } k_2=0\\ \beta_{(0,0)}^2\cdots\beta_{(0,k_2-1)}^2, & \text{if } k_1=0 \text{ and } k_2\ge 1\\ \alpha_{(0,0)}^2\cdots\alpha_{(k_1-1,0)}^2\beta_{(k_1,0)}^2\cdots\beta_{(k_1,k_2-1)}^2, & \text{if } k_1\ge 1 \text{ and } k_2\ge 1.\end{cases} \tag{3.10}$$

We remark that, due to the commutativity condition (3.7), $\gamma_{\mathbf{k}}$ can be computed using any nondecreasing path from $(0,0)$ to (k_1,k_2).

To detect hyponormality, there is a simple criterion.

Theorem 3.4 ([13], Six-point test). *Let $\mathbf{T}\equiv(T_1,T_2)$ be a 2-variable weighted shift, with weight sequences α and β. Then*

$$\mathbf{T}\ \textit{is hyponormal}\ \Leftrightarrow\begin{pmatrix}\alpha_{\mathbf{k}+\varepsilon_1}^2-\alpha_{\mathbf{k}}^2 & \alpha_{\mathbf{k}+\varepsilon_2}\beta_{\mathbf{k}+\varepsilon_1}-\alpha_{\mathbf{k}}\beta_{\mathbf{k}}\\ \alpha_{\mathbf{k}+\varepsilon_2}\beta_{\mathbf{k}+\varepsilon_1}-\alpha_{\mathbf{k}}\beta_{\mathbf{k}} & \beta_{\mathbf{k}+\varepsilon_2}^2-\beta_{\mathbf{k}}^2\end{pmatrix}\ge 0\quad(\textit{all}\ \mathbf{k}\in\mathbb{Z}_+^2).$$

A straightforward generalization of the above mentioned Berger–Gellar–Wallen result was proved in [24]. That is, a commuting pair $\mathbf{T}\equiv(T_1,T_2)$ admits a commuting normal extension if and only if there is a probability measure μ (which we call the Berger measure of $\mathbf{T}$) defined on the 2-dimensional rectangle $R=[0,a_1]\times[0,a_2]$ (where $a_i:=\|T_i\|^2$) such that

$$W_{\boldsymbol{\alpha}}\text{ is subnormal}\Leftrightarrow\gamma_{\mathbf{k}}=\int t_1^{k_1}t_2^{k_2}\,d\mu(t_1,t_2)\quad(\text{all }\mathbf{k}\in\mathbb{Z}_+^2).$$

Thus, the study of subnormality for multivariable weighted shifts is intimately connected to multivariable real moment problems.

3.3 Spherically quasinormal 2-variable weighted shifts

In this section, we present a characterization of spherical quasinormality for 2-variable weighted shifts, which was announced in [17] and proved in [18] and [19]. Before we state it, we list some simple facts about quasinormality for 2-variable weighted shifts.

Remark 3.5 (cf. [16]). We first observe that no 2-variable weighted shift can be matricially hyponormal, as a simple calculation shows. Also, a 2-variable weighted shift

$\mathbf{T} \equiv (T_1, T_2) = W_{(\alpha,\beta)}$ is (jointly) quasinormal if and only if $\alpha_{(k_1,k_2)} = \alpha_{(0,0)}$ and $\beta_{(k_1,k_2)} = \beta_{(0,0)}$ for all $k_1, k_2 \geq 0$. This can be seen via a simple application of (3.7) and (3.8). As a result, up to a scalar multiple in each component, a quasinormal 2-variable weighted shift is identical to the so-called Helton–Howe shift. This fact is entirely consistent with the one-variable result: a unilateral weighted shift W_ω is quasinormal if and only if $W_\omega = cU_+$ for some $c > 0$.

The following result describes the weight diagram of $\Delta_{\mathrm{sph}}(\mathbf{T}) \equiv (\widehat{T_1}, \widehat{T_2})$.

Proposition 3.6 ([18]). *Let $\mathbf{T} \equiv (T_1, T_2) = W_{(\alpha,\beta)}$ be a 2-variable weighted shift. Then*

$$\widehat{T}_1 e_{\mathbf{k}} = \alpha_{\mathbf{k}} \frac{(\alpha^2_{\mathbf{k}+\epsilon_1} + \beta^2_{\mathbf{k}+\epsilon_1})^{1/4}}{(\alpha^2_{\mathbf{k}} + \beta^2_{\mathbf{k}})^{1/4}} e_{\mathbf{k}+\epsilon_1} \tag{3.11}$$

$$\widehat{T}_2 e_{\mathbf{k}} = \beta_{\mathbf{k}} \frac{(\alpha^2_{\mathbf{k}+\epsilon_2} + \beta^2_{\mathbf{k}+\epsilon_2})^{1/4}}{(\alpha^2_{\mathbf{k}} + \beta^2_{\mathbf{k}})^{1/4}} e_{\mathbf{k}+\epsilon_2} \tag{3.12}$$

for all $\mathbf{k} \in \mathbb{Z}^2_+$.

We now recall the class of spherically isometric commuting pairs of operators (cf. [5], [6], [23]). A commuting pair $\mathbf{T} \equiv (T_1, T_2)$ is a spherical isometry if $T_1^* T_1 + T_2^* T_2 = I$.

Lemma 3.7 ([5, Proposition 2]). *Any spherical isometry is subnormal.*

Theorem 3.8 ([16, Theorem 3.1]; cf. [18, Lemma 10.3]). *For a commuting 2-variable weighted shift $W_{(\alpha,\beta)} = (T_1, T_2)$, the following statements are equivalent:*

(i) *$W_{(\alpha,\beta)} \equiv (T_1, T_2)$ is a spherically quasinormal 2-variable weighted shift;*
(ii) *(algebraic condition) $T_1^* T_1 + T_2^* T_2 = C \cdot I$, for some positive constant C;*
(iii) *(weight condition) for all $\mathbf{k} \equiv (k_1, k_2) \in \mathbb{Z}^2_+$, $\alpha^2_{(k_1,k_2)} + \beta^2_{(k_1,k_2)} = C$, for some positive constant $C > 0$;*
(iv) *(moment condition) for all $\mathbf{k} \equiv (k_1, k_2) \in \mathbb{Z}^2_+$, $\gamma_{\mathbf{k}+\epsilon_1} + \gamma_{\mathbf{k}+\epsilon_2} = C\gamma_{\mathbf{k}}$, for some positive constant $C > 0$.*

3.3.1 Construction of spherically quasinormal 2-variable weighted shifts

As observed in [18], within the class of 2-variable weighted shifts there is a simple description of spherical isometries, in terms of the weight sequences $\alpha \equiv \{\alpha_{(k_1,k_2)}\}$ and $\beta \equiv \{\beta_{(k_1,k_2)}\}$. Indeed, since spherical isometries are (jointly) subnormal, we know that the unilateral weighted shift associated with the 0th row in the weight diagram must be subnormal. Thus, without loss of generality, we can always assume that the 0th row corresponds to a subnormal unilateral weighted shift, and denote its weights by $\{\alpha_{(k,0)}\}_{k=0,1,2,\ldots}$. Also, in view of Theorem 3.8 we can assume that $c = 1$. Using the identity,

$$\alpha_{\mathbf{k}}^2 + \beta_{\mathbf{k}}^2 = 1 \quad (\mathbf{k} \in \mathbb{Z}_+^2) \tag{3.13}$$

and the above mentioned 0th row, we can compute $\beta_{(k,0)} := \sqrt{1 - \alpha_{k,0}^2}$ for $k = 0, 1, 2, \ldots$. With these new values at our disposal, we can use the commutativity property (3.7) to generate the values of α in the first row; that is,

$$\alpha_{(k,1)} := \alpha_{(k,0)}\beta_{(k+1,0)}/\beta_{(k,0)}.$$

We can now repeat the algorithm, and calculate the weights $\beta_{(k,1)}$ for $k = 0, 1, 2, \ldots$, again using the identity (3.13). This in turn leads to the α weights for the second row, and so on. For more on this construction, the reader is referred to [19]. In particular, it is worth noting that the construction may stall if the sequence $\{\alpha_{(k,0)}\}_{k \geq 0}$ is not strictly increasing.

Proposition 3.9 ([14, Proposition 12.14]). *Let*

$$\alpha_{(0,0)} := \sqrt{p}, \quad \alpha_{(1,0)} := \sqrt{q}, \quad \alpha_{(2,0)} := \sqrt{r} \quad \text{and} \quad \alpha_{(3,0)} := \sqrt{r},$$

and assume that $0 < p < q < r < 1$. *Then the algorithm described in this section fails at some stage. As a consequence, there does not exist a spherical isometry interpolating these initial data.*

Remark 3.10. In Proposition 3.9, the reader may have noticed that the 0th row is not subnormal; for, it is well known that, up to a constant, the only subnormal unilateral weighted shifts with two equal weights are U_+ and S_a ([32, Theorem 6]). Thus, save for these two special (trivial) cases, assuming subnormality of the 0th row will automatically guarantee that $\alpha_{(k,0)}$ is strictly increasing; therefore, in the sequel we will always assume that the 0th row is subnormal.

3.4 Iterates of the spherical Aluthge transform

For notational convenience, in this section we will switch from pairs (T_1, T_2) to pairs (S, T). Given a 2-variable weighted shift $(S, T) \equiv W_{\alpha,\beta}$, recall that the spherical Aluthge transform is given by

$$(\Delta_{\text{sph}}(S, T))_1 e_{\mathbf{k}} = \alpha_{\mathbf{k}} \frac{(\alpha_{\mathbf{k}+\epsilon_1}^2 + \beta_{\mathbf{k}+\epsilon_1}^2)^{1/4}}{(\alpha_{\mathbf{k}}^2 + \beta_{\mathbf{k}}^2)^{1/4}} e_{\mathbf{k}+\epsilon_1}$$

and

$$(\Delta_{\text{sph}}(S, T))_2 e_{\mathbf{k}} = \beta_{\mathbf{k}} \frac{(\alpha_{\mathbf{k}+\epsilon_2}^2 + \beta_{\mathbf{k}+\epsilon_2}^2)^{1/4}}{(\alpha_{\mathbf{k}}^2 + \beta_{\mathbf{k}}^2)^{1/4}} e_{\mathbf{k}+\epsilon_2}$$

for all $\mathbf{k} \in \mathbb{Z}_+^2$.

Thus, it is clear that each of the iterates of $\Delta_{\text{sph}}(S,T)$ is a 2-variable weighted shift. We now define, recursively, two weight sequences, $S_n(i,j)$ and $T_n(i,j)$ using the horizontal and vertical components of the iterates of the spherical Aluthge transform. For $n=0$, we let $S_0(i,j) := \alpha_{(i,j)}$ and $T_0(i,j) := \beta_{(i,j)}$. For $n>0$, $S_n(i,j)$ and $T_n(i,j)$ are the weights of the horizontal and vertical actions of $\Delta_{\text{sph}}^{(n)}(S,T)$. This easily leads to the following expressions:

$$(\Delta_{\text{sph}}(S_n,T_n))_1 e_{(i,j)} = S_{n+1}(i,j)e_{(i,j)}$$
$$(\Delta_{\text{sph}}(S_n,T_n))_2 e_{(i,j)} = T_{n+1}(i,j)e_{(i,j)}.$$

It follows that

$$S_{n+1}e_{(0,0)} = S_n(0,0)\frac{(S_n(1,0)^2+T_n(1,0)^2)^{1/4}}{(S_n(0,0)^2+T_n(0,0)^2)^{1/4}}e_{(1,0)}$$

and

$$T_{n+1}e_{(0,0)} = T_n(0,0)\frac{(S_n(0,1)^2+T_n(0,1)^2)^{1/4}}{(S_n(0,0)^2+T_n(0,0)^2)^{1/4}}e_{(0,1)}.$$

As in the 1 variable case [31], one observes that the asymptotic behavior anywhere impacts the asymptotic behavior at the origin $(0,0)$; as a result, and without loss of generality, we can focus attention on the recursively defined sequences

$$S_{n+1}(0,0) = S_n(0,0)\frac{(S_n(1,0)^2+T_n(1,0)^2)^{1/4}}{(S_n(0,0)^2+T_n(0,0)^2)^{1/4}} \tag{3.14}$$

and

$$T_{n+1}(0,0) = T_n(0,0)\frac{(S_n(0,1)^2+T_n(0,1)^2)^{1/4}}{(S_n(0,0)^2+T_n(0,0)^2)^{1/4}}. \tag{3.15}$$

We will now restrict attention to finite rank perturbations of spherical isometries. We will prove that the iterates of Δ_{sph} converge in the WOT to a spherical isometry. The proof entails consideration of cases of increasing complexity. First, we need some notation.

Let $\mathbf{k} \in \mathbb{Z}_+^2$ and let $\mathcal{L}_{\mathbf{k}} := \bigvee\{e_{\mathbf{k}+\mathbf{p}} : \mathbf{p} \in \mathbb{Z}_+^2\}$; that is, $\mathcal{L}_{\mathbf{k}}$ is the closed subspace generated by the orthonormal canonical basis vectors in the quadrant determined by the lattice point $\mathbf{k}$. Equivalently, $\mathcal{L}_{\mathbf{k}} = \mathbf{k} + \mathbb{Z}_+^2$.

Remark 3.11.
(i) Observe that if $\mathcal{M}_{k_2}$ represents the range of T^{k_2} and if $\mathcal{N}_{k_1}$ represents the range of S^{k_1}, then $\mathcal{L}_{\mathbf{k}} = \mathcal{M}_{k_2} \bigcap \mathcal{N}_{k_1}$. Also, for $k_1 = k_2 = 1$, the space $\mathcal{L}_{(1,1)}$ is the *core* of the 2-variable weighted shift (cf. [18, paragraph immediately following Lemma 3.4]).
(ii) It is easy to show that all iterates of the spherical Aluthge transform leave the subspaces $\mathcal{L}_{\mathbf{k}}$ invariant.

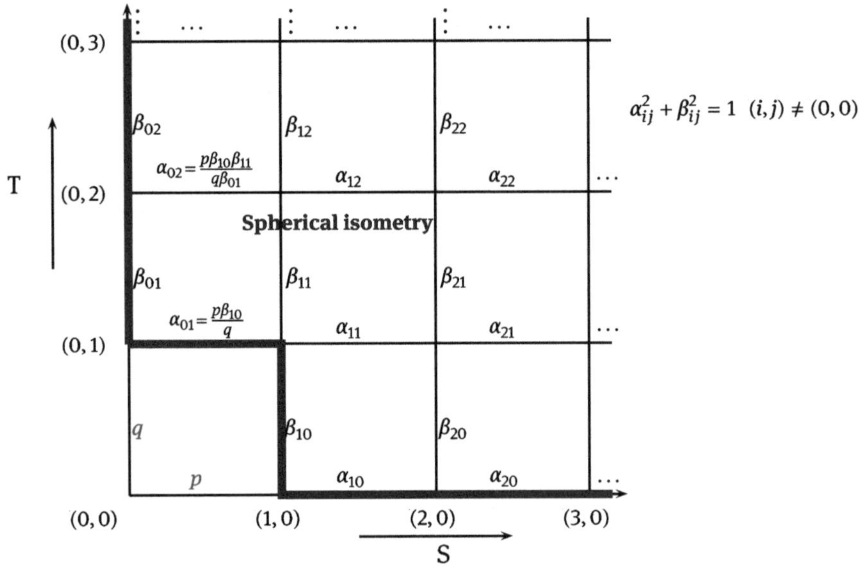

Figure 3.2: Weight diagram for the 2-variable weighted shift in Theorem 3.12.

Theorem 3.12 (Case 1: 1-cell perturbation). *Consider the 2-variable weighted shift given by the weight diagram in Figure 3.2. Then the iterates of the spherical Aluthge transform of (S, T) converge in the WOT to a spherical isometry.*

Proof. Since the spherical Aluthge transform leaves invariant the subspace where (S, T) is a spherical isometry (i. e., the subspace $\mathcal{M}_1 \wedge \mathcal{N}_1$), it is enough to focus attention on the asymptotic behavior of the iterates at the origin. It is not hard to see that $\Delta_{\mathrm{sph}}(S, T)$ has the same structure, and the same is true of $\Delta^2_{\mathrm{sph}}(S, T)$, $\Delta^3_{\mathrm{sph}}(S, T)$, ...

Thus, for this special case, the asymptotic behavior of the spherical Aluthge iterates is controlled by the pair

$$\begin{cases} p_n := S_n(0,0) \\ q_n := T_n(0,0). \end{cases}$$

Observe that

$$\begin{cases} p_1 = p(p^2 + q^2)^{-1/4} \\ q_1 = q(p^2 + q^2)^{-1/4}, \end{cases}$$

$$\begin{cases} p_2 = p(p^2 + q^2)^{-3/8} \\ q_2 = q(p^2 + q^2)^{-3/8} \end{cases}$$

and, in general, for $n > 2$,

$$\begin{cases} p_n = p(p^2 + q^2)^{-\sum_{k=2}^{n+1}(\frac{1}{2})^k} \\ q_n = q(p^2 + q^2)^{-\sum_{k=2}^{n+1}(\frac{1}{2})^k}. \end{cases}$$

From this, it readily follows that, in the limit, we obtain

$$\begin{cases} p_\infty = p(p^2+q^2)^{-\frac{1}{2}} \\ q_\infty = q(p^2+q^2)^{-\frac{1}{2}}. \end{cases}$$

Since

$$p_\infty^2 + q_\infty^2 = 1,$$

we see that the sequence of iterates does converge to a spherical isometry. □

Remark 3.13. For future use, we record the following identity involving p_n and q_n in the Proof of Theorem 3.12:

$$p_n^2 + q_n^2 = (p^2+q^2)^{2^{-n}}. \tag{3.16}$$

Theorem 3.14 (Case 2: 2-cell perturbation). *Consider the 2-variable weighted shift given by the weight diagram in Figure 3.3. Then the iterates of the spherical Aluthge transform of (S, T) converge in the WOT to a spherical isometry.*

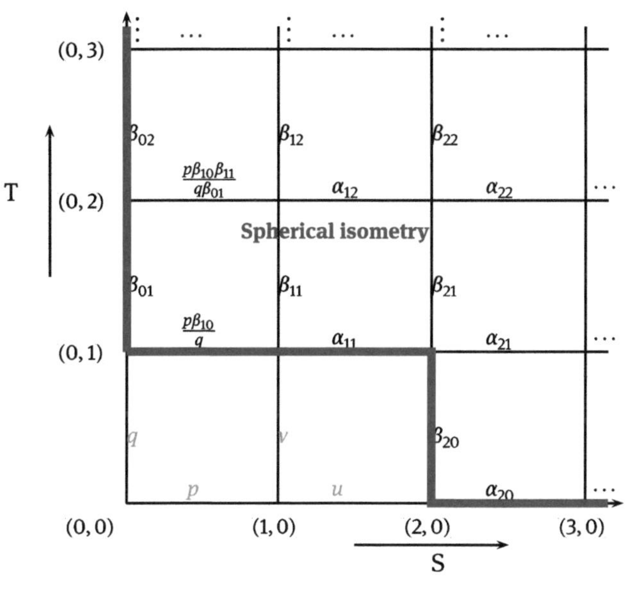

Figure 3.3: Weight diagram for the 2-variable weighted shift in Theorem 3.14.

Proof. Observe first that the restriction of (S, T) to the invariant subspace $\mathcal{L}_{(1,0)}$ is a 2-variable weighted shift satisfying the conditions in Theorem 3.12, with the parameters u and v taking the place of p and q. In particular, we know from (3.16) that

$$u_n^2 + v_n^2 = (u^2+v^2)^{2^{-n}}. \tag{3.17}$$

Moreover, by (3.14) the values of $S_{n+2}(0,0)$ are determined by the values of $S_{n+1}(0,0)$, $T_{n+1}(0,0)$, $S_{n+1}(1,0)$ and $T_{n+1}(1,0)$, and the last two values follow the pattern for the weights in Theorem 3.12, since the lattice point $(1,0)$ is in the subspace $\mathcal{L}_{(1,0)}$. We now observe that

$$S_{n+1}(0,0)^2 + T_{n+1}(0,0)^2 = \frac{S_n(0,0)^2\sqrt{u_n^2+v_n^2} + T_n(0,0)^2}{\sqrt{S_n(0,0)^2 + T_n(0,0)^2}}.$$

(Notice that $T_n(0,0)^2$ appears without another factor in the numerator because $T_{n+1}(0,0)$ uses information about the lattice points $(0,0)$ and $(0,1)$, and of course the restriction of (S,T) to $\mathcal{L}_{(0,1)}$ is a spherical isometry.)

It follows that both the expressions for $S_{n+1}(0,0)^2+T_{n+1}(0,0)^2$ and $u_{n+1}^2+v_{n+1}^2$, which are needed for $S_{n+2}(0,0)$ and $T_{n+2}(0,0)$, depend directly on the quantity $u_n^2+v_n^2$, whose asymptotic behavior is given by (3.17). It is now not hard to check that $S_{n+2}(0,0)^2 + T_{n+2}(0,0)^2$ converges to 1 as $n \to \infty$. At the same time, the reader will notice that convergence does not easily follow from the convergence of the sequence $\{u_n^2+v_n^2\}$, but the concrete asymptotic pattern in (3.17) is important; that is, one has a sequence of the form $c^{2^{-n}}$, where c is a positive constant. □

Theorem 3.15 (Case 3: 3-cell perturbation). *Consider the 2-variable weighted shift given by the weight diagram in Figure 3.4. Then the iterates of the spherical Aluthge transform of (S,T) converge in the WOT to a spherical isometry.*

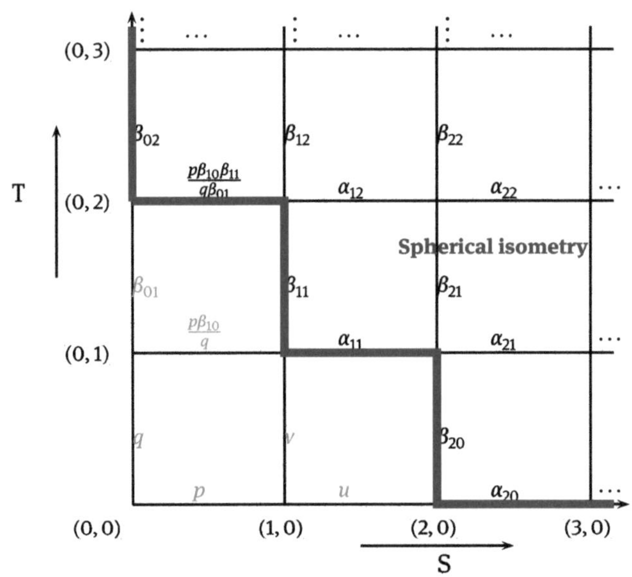

Figure 3.4: Weight diagram for the 2-variable weighted shift in Theorem 3.15.

Proof. Observe that the restriction of (S, T) to the invariant subspace $\mathcal{L}_{(0,1)}$ satisfies the hypotheses in Theorem 3.12. Using this information, one now needs to imitate the proof of Theorem 3.14 to reach the desired conclusion. □

Theorem 3.16 (Case 4: multicell perturbation). *Consider the 2-variable weighted shift given by the weight diagram in Figure 3.5. Then the iterates of the spherical Aluthge transform of (S, T) converge in the WOT to a spherical isometry.*

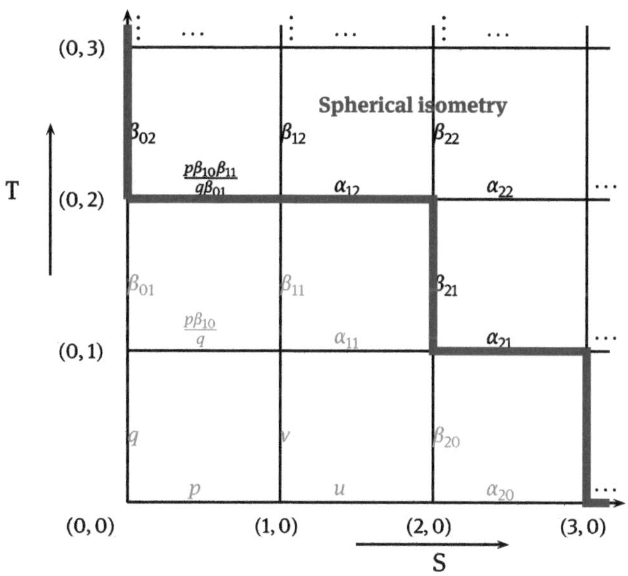

Figure 3.5: Weight diagram for the 2-variable weighted shift in Theorem 3.16.

Proof. As the reader will surely anticipate, this case reduces to the previous cases, through a series of steps. For instance, the restriction of (S, T) to $\mathcal{L}_{(1,0)}$ fits Case 3, and once this information is incorporated, Case 4 becomes similar to Case 2. □

We conclude this section with two open questions, which we plan to discuss in a separate paper.

Question 3.17. *Let (T_1, T_2) be a commuting pair of operators on a finite dimensional Hilbert space. Does the sequence of iterates $\Delta^n_{\text{sph}}(T_1, T_2)$ converge in the norm?*

Remark 3.18. One very special case of Theorem 3.12 has to do with taking the Helton–Howe shift and altering only the weights $\alpha_{(0,0)}$ and $\beta_{(0,0)}$. By commutativity, we must have $x := \alpha^2_{(0,0)} = \beta^2_{(0,0)}$. Call this new shift (S_x, T_x). (Strictly speaking, (S_x, T_x) does not satisfy the hypotheses of Theorem 3.12, since the Helton–Howe shift (S_1, T_1) is not a spherical isometry, but $(\frac{1}{\sqrt{2}}S_x, \frac{1}{\sqrt{2}}T_x)$ is.) One can then prove that the Berger measure of (S_x, T_x) is $(1-x)\delta_{(0,0)} + x\delta_{(1,1)}$. When we take the spherical Aluthge transform, the atoms remain unchanged, but the densities become $1-\sqrt{x}$ and $\sqrt{x}$, respectively. As we keep iterating, the square root becomes fourth root, eighth root, etc., so the Berger measure of the nth iterate is given by $(1-\sqrt[2^n]{x})\delta_{(0,0)} + \sqrt[2^n]{x}\delta_{(1,1)}$. As the number of iterates grow, this expression converges to 1, so in the limit we get only $\delta_{(1,1)}$, that is, the Berger measure of the Helton–Howe shift.

Remark 3.19. The reader must have surely noticed that in Theorem 3.12 the parameters p and q determine the asymptotic behavior of the iterates. On the other hand, due to the commutativity of (S, T) those parameters are directly related, that is, $q\alpha_{01} = p\beta_{10}$; in other words, q depends on p and the data encapsulated by the spherical isometry $(S, T) \restriction_{\mathcal{L}_{(1,0)} \vee \mathcal{L}_{(0,1)}}$. That is, the asymptotic behavior in that case depends on one degree of freedom, given by, for instance, p. In Theorem 3.14, the number of degrees of freedom is two (think about the parameters p and u as being free), while in Theorem 3.15 the number of degrees of freedom is three. We leave it to the reader to determine the number of degrees of freedom in Theorem 3.16 and in more general cases.

Question 3.20. *What is the asymptotic behavior of the iterates of the spherical Aluthge transform of 2-variable weighted shifts with finitely atomic Berger measures?*

Bibliography

[1] A. Aluthge, *On p-hyponormal operators for* $0 < p < 1$, Integral Equ. Oper. Theory **13** (1990), 307–315.

[2] T. Ando, *Aluthge transforms and the convex hull of the spectrum of a Hilbert space operator*, in *Recent advances in operator theory and its applications*, Oper. Theory, Adv. Appl. **160** (2005), 21–39.

[3] J. Antezana, E. Pujals and D. Stojanoff, *Convergence of iterated Aluthge transform sequence for diagonalizable matrices*, Adv. Math. **216** (2007), 255–278.

[4] J. Antezana, E. Pujals and D. Stojanoff, *Iterated Aluthge transform: A brief survey*, Rev. Unión Mat. Argent. **49** (2008), 29–41.

[5] A. Athavale, *On the intertwining of joint isometries*, J. Oper. Theory **23** (1990), 339–350.

[6] A. Athavale and S. Podder, *On the reflexivity of certain operator tuples*, Acta Math. Sci. (Szeged) **81** (2015), 285–291.

[7] C. Benhida, *Mind Duggal Transforms*, Filomat **33** (2019), 5863–5870.

[8] C. Benhida, R. E. Curto, S. H. Lee and J. Yoon, *Joint spectra of spherical Aluthge transforms of commuting n-tuples of Hilbert space operators*, C. R. Math. Acad. Sci. Paris **357** (2019), 799–802.

[9] M. Chō, I. B. Jung and W. Y. Lee, *On Aluthge transforms of p-hyponormal operators*, Integral Equ. Oper. Theory **53** (2005), 321–329.

[10] J. Conway, *The Theory of Subnormal Operators*, Mathematical Surveys and Monographs, **36**, Amer. Math. Soc., Providence, 1991.

[11] R. Curto, *On the connectedness of invertible n-tuple*, Indiana Univ. Math. J. **29** (1980), 393–406.

[12] R. Curto, Applications of several complex variables to multi-parameter spectral theory, in *Surveys of Recent Results in Operator Theory, vol. II*, J. B. Conway and B. B. Morrel (eds.), Longman Publishing Co., London, 1988, pp. 25–90.
[13] R. E. Curto, *Joint hyponormality: A bridge between hyponormality and subnormality*, Proc. Symposia Pure Math. **51** (1990), Part II, 69–91.
[14] R. E. Curto, Two-variable weighted shifts in multivariable operator theory, in *Handbook of Analytic Operator Theory*, K. Zhu (ed.), CRC Press, 2019, pp. 17–63.
[15] R. E. Curto and G. Exner, *Berger measure for some transformations of subnormal weighted shifts*, Integral Equ. Oper. Theory **84** (2016), 429–450.
[16] R. E. Curto, S. H. Lee and J. Yoon, *Quasinormality of powers of commuting pairs of bounded operators*, preprint, 2019.
[17] R. E. Curto and J. Yoon, *Toral and spherical Aluthge transforms for* 2*-variable weighted shifts*, C. R. Acad. Sci. Paris **354** (2016), 1200–1204.
[18] R. E. Curto and J. Yoon, *Aluthge transforms of* 2*-variable weighted shifts*, Integral Equ. Oper. Theory (2018) **90**, 52.
[19] R. E. Curto and J. Yoon, Spherically quasinormal pairs of commuting operators, *Analysis of Operators on Function Spaces (The Serguei Shimorin Memorial Volume)*, A. Aleman, H. Hedenmalm, D. Khavinson and M. Putinar (eds.), *Trends in Mathematics*, Birkhäuser Verlag, 2019, pp. 213–237.
[20] K. Dykema and H. Schultz, *Brown measure and iterates of the Aluthge transform for some operators arising from measurable actions*, Trans. Am. Math. Soc. **361** (2009), 6583–6593.
[21] G. R. Exner, *Aluthge transforms and n-contractivity of weighted shifts*, J. Oper. Theory **61** (2009), 419–438.
[22] R. Gellar and L. J. Wallen, *Subnormal weighted shifts and the Halmos–Bram criterion*, Proc. Jpn. Acad. **46** (1970), 375–378.
[23] J. Gleason, *Quasinormality of Toeplitz tuples with analytic symbols*, Houst. J. Math. **32** (2006), 293–298.
[24] N. P. Jewell and A. R. Lubin, *Commuting weighted shifts and analytic function theory in several variables*, J. Oper. Theory **1** (1979), 207–223.
[25] I. B. Jung, E. Ko, and C. Pearcy, *Aluthge transform of operators*, Integral Equ. Oper. Theory **37** (2000), 437–448.
[26] I. B. Jung, E. Ko and C. Pearcy, *Spectral pictures of Aluthge transforms of operators*, Integral Equ. Oper. Theory **40** (2001), 52–60.
[27] I. B. Jung, E. Ko and C. Pearcy, *The iterated Aluthge transform of an operator*, Integral Equ. Oper. Theory **45** (2003), 375–387.
[28] M. K. Kim and E. Ko, *Some connections between an operator and its Aluthge transform*, Glasg. Math. J. **47** (2005), 167–175.
[29] F. Kimura, *Analysis of non-normal operators via Aluthge transformation*, Integral Equ. Oper. Theory **50** (2004), 375–384.
[30] S. H. Lee, W. Y. Lee and J. Yoon, *Subnormality of Aluthge transform of weighted shifts*, Integral Equ. Oper. Theory **72** (2012), 241–251.
[31] K. Rion, *Convergence properties of the Aluthge sequence of weighted shifts*, Houst. J. Math. **42** (2016), 1217–1226.
[32] J. Stampfli, *Which weighted shifts are subnormal*, Pac. J. Math. **17** (1966), 367–379.
[33] J. L. Taylor, *A joint spectrum for several commuting operators*, J. Funct. Anal. **6** (1970), 172–191.
[34] Wolfram Research Inc., *Mathematica*, Version 11.3, Wolfram Research Inc., Champaign, IL, 2018.

Jesús M. F. Castillo

4 The freewheeling twisting of Hilbert spaces

Abstract: Everybody knows what a Hilbert space is. A twisted Hilbert space instead is a Banach space X admitting a subspace Y isomorphic to a Hilbert space such that the corresponding quotient Z/Y is also isomorphic to a Hilbert space. The first non-trivial example was obtained by Enflo–Lindenstrauss–Pisier but the central object for us is the Kalton–Peck Z_2 space. This paper is to explain why twisted Hilbert spaces are important in Banach space theory, what is known and what is not known about them and which problems the construction of a theory of twisted Hilbert spaces must tackle.

Keywords: Twisted Hilbert spaces

MSC 2010: 46B20, 46B70, 46E30, 46M18

4.1 Hilbert spaces revisited

A Hilbert space is a complete normed space whose norm $\|\cdot\|$ comes induced by an inner product $(\cdot,\cdot)$ in the form $\|x\| = (x,x)^{1/2}$. The orthogonal projection (which should not be linear, but it is) provides a contractive projection onto every closed subspace. Every Hilbert space is isometric to some $\ell_2(\Gamma)$ but this, as we attempt to explain throughout this paper, is not the end of the game. On the isomorphic level, which is much more interesting for us, a Hilbert space is a Banach space such that every closed subspace is complemented; and thus, locally speaking, a Hilbert space is a Banach space with the property that there exists a constant $C > 0$ such that every finite dimensional subspace is C-complemented. This is contained in the classical proof of Lindenstrauss and Tzafriri that a Banach space is a Hilbert space if and only if every closed subspace is complemented [58]. See also [52] for a quantitative improvement.

A good starting point for the line of research explained in these notes is the Eilenberg–McLane program [33]; see also [18] for a more detailed exposition of the program in Banach spaces. In it, the authors establish as a foundational line that only categories and functors are objects of study in mathematics. This bluntly said, means that something like "a Banach space" does not exist. To understand this assertion, let

Acknowledgement: This research was supported by Project MTM2016-76958-C2-1-P and Project IB16056 de la Junta de Extremadura.
650 thanks must be returned to an anonymous referee for helping to improve the fun of writing this paper.

Jesús M. F. Castillo, Instituto de Matemáticas, Universidad de Extremadura, Avda. de Elvas s/n, 06011 Badajoz, Spain, e-mail: castillo@unex.es

https://doi.org/10.1515/9783110656756-004

us focus on Hibert spaces and be nitpicking: which Hilbert space is one considering: $\ell_2, L_2(0,1)$, $L_2(\mathbb{R})$, the Schatten class S_2, the Hardy space H_2 ...? The fact that all of them are isometric is just an outstanding theorem. So, better let us consider them as different spaces (which, I now, mean for the time being that we are attempting to make a theory not "up to isometries" but yes "up to isomorphisms"; oh well...) The key observation here is that those spaces do never come alone: each of them is part of a family; that of ℓ_p or L_p or S_p or H_p spaces, for p ranging from 1 to ∞ (if one restricts the attention to Banach spaces) or ranging from 0 to ∞ if quasi-Banach spaces are allowed as well. This family is the functor. And the Eilenberg–McLane claim is then that one needs to understand the family *first* to *then* understand the space. With a different bias, Cabello says it quite clearly in [8] "*Most decent Banach and quasi-Banach spaces carry natural module structures over some familiar Banach algebra.*" Which somehow can be read as: Banach spaces not carrying a natural module structure over some familiar Banach algebra are ... expendable. So, since a necessary ingredient in the definition of a functor is the category where it acts, it is necessary to define the category on which we will consider "the" Hilbert space.

(Complex) interpolation theory is a natural place where the Eilenberg–MacLane program is subtly verified. By the virtues of classical Riesz interpolation theorem, when a linear operator $\ell_\infty \longrightarrow \ell_\infty$ also acts continuously from ℓ_1 to ℓ_1 it automatically acts continuously from ℓ_p to ℓ_p (and with an explicit estimate of its norm, just in case one prefers to think finite-dimensionally). This is sometimes abbreviated by saying that ℓ_p spaces form an interpolation scale. As analogously, L_p, S_p, or H_p spaces do. But the smoking gun in this crime scene is that it is not however true that an operator sending, say, L_1 to L_1 and L_∞ to L_∞ also sends ℓ_2 to ℓ_2 (whatever that means): it just sends L_2 to L_2.

There is a way to say that: those Hilbert spaces live in different categories, even if, as mere Banach spaces, all of them are isometric. The underlying category is determined by the underlying algebra that endows them the module structure: ℓ_2 is an ℓ_∞-Banach module, L_2 is an L_∞-Banach module, S_2 is an $\mathfrak{L}(\ell_2, \ell_2)$-Banach module, ...

4.2 Twisted Hilbert spaces

A twisted Hilbert space is a Banach space X admitting a subspace Y isomorphic to a Hilbert space such that the corresponding quotient Z/Y is also isomorphic to a Hilbert space. In homological terms, it is the middle term in an exact sequence

$$0 \longrightarrow H \longrightarrow X \longrightarrow H' \longrightarrow 0$$

in which both H, H' are Hilbert spaces. There is no loss of generality (for the moment) in assuming that $H = H'$, and so we will do in what follows. Recall that an exact sequence $0 \to Y \to X \to Z \to 0$ of Banach spaces and linear continuous operators is a diagram in which the kernel of each arrow coincides with the image of the preceding one. That, by the open mapping theorem, means that Y is isomorphic to a subspace of X and X/Y is isomorphic to Z. The study of twisted sums was fueled by the attempt to solve what is known as Palais problem (although Palais rejects this assignation [60]): does there exist a twisted Hilbert space that is not (isomorphic to) a Hilbert space? That is what we will call a nontrivial twisted Hilbert space. The existence of such objects was first proved by Enflo, Lindenstrauss, and Pisier [34], but the construction which is of paramount importance to our purposes is that of the Kalton–Peck space Z_2 presented in [55]. In part, because while the ELP space is, to some extent, an existence result, the space Z_2 is actually constructed. How is Z_2 obtained? In [42], Kalton showed that exact sequences $0 \to Y \to X \to Z \to 0$ of (quasi-) Banach spaces are in correspondence (the perfect example of what is called a natural transformation, in categorical terms [37]) with certain nonlinear maps $F : Z \to Y$, called quasi-linear maps. So, twisted Hilbert spaces appear generated by quasi-linear maps $F : H \longrightarrow H$. To indicate that, it is customary to call such space $H \oplus_F H$. The way in which $H \oplus_F H$ is constructed out of F is simple: it is the product vector space $H \times H$ endowed with the quasi-norm

$$\|(y, x)\|_F = \|y - Fx\|_H + \|x\|_H.$$

The quasi-linearity properties of F make $\|(\cdot,\cdot)\|_F$ a quasi-norm; and a deep theorem of Kalton [43] shows that the convex hull of the closed unit ball of $\|(\cdot,\cdot)\|_F$ is actually the unit ball of an equivalent norm. So $H \oplus_F H$ is actually a Banach space (with no recognizable norm at hand, however). As we already know for certain, $H \oplus_F H$ is a Hilbert space if and only if H (the left H) is complemented. Which, in homological terms is said: the exact sequence splits. How does the quasi-linear map F tests whether $H \oplus_F H$ is a Hilbert space or not? This way: if there exists a linear map $L : H \to H$ so that $\|F - L\| < +\infty$ (i. e., $F - L$ is a bounded map). In the highway of Banach space theory, this is the exit toward Ulam's stability results, a topic we will not pursue here, but that the interested reader can peruse in [11]. So, quasi-linear maps $F : H \to H$ that cannot be approximated by linear maps will be called nontrivial maps. And thus the question is: how does one construct a nontrivial quasi-linear map on a Hilbert space?

In [55], Kalton and Peck presented an explicit construction of quasi-linear maps on Banach spaces with unconditional basis; in particular, in separable Hilbert spaces,

$$\mathsf{KP}_\phi(x) = x\phi\left(\log \frac{\|x\|}{|x|}\right)$$

where $\phi : \mathbb{R} \to \mathbb{R}$ is a Lipschitz map $\mathbb{R} \to \mathbb{R}$. These maps are usually called *Kalton–Peck maps*. The choice of the function $\phi(t) = t$ is especially rewarding and we simply write KP for that map. This method was refined by Kalton [45, 47] and extended to Köthe function spaces. With the same (verbatim) definition. Recall that a Köthe function space $\mathcal{K}$ over a σ-finite measure space (Σ, μ) is a linear subspace of $L_0(\Sigma, \mu)$, the vector space of all measurable functions, endowed with a quasi-norm (or a norm) such that whenever $|f| \leq g$ and $g \in \mathcal{K}$, then $f \in \mathcal{K}$ and $\|f\| \leq \|g\|$ and so that for every finite measure subset $A \subset \Sigma$ the characteristic function 1_A belongs to $\mathcal{K}$.

The space $\ell_2 \oplus_{\mathsf{KP}} \ell_2$ is called Z_2, the standard Kalton–Peck space. If the reader is prompt to believe that $\mathsf{KP}(x) = x \log \frac{\|x\|}{|x|}$ is a quasi-linear map, and our word on this, Z_2 is a Banach space. Is a twisted Hilbert space since one has a natural exact sequence

$$0 \longrightarrow \ell_2 \longrightarrow Z_2 \longrightarrow \ell_2 \longrightarrow 0$$

in which the inclusion is the operator $x \longrightarrow (x, 0)$ and the quotient map the operator $(y, x) \longrightarrow x$. It is not so simple to check that Z_2 is not a Hilbert space; namely, that KP cannot be approximated by a linear map. Everything depends on the more or less obscure wizardry one has to make to be certain that, since $\mathsf{KP}(e_n) = 0$, the only linear map that can be close to KP is one having the form $L(e_n) = \lambda_n e_n$ for a certain bounded sequence (λ_n): The idea is averaging an arbitrary linear map L that is at finite distance from Ω to get a new linear map L' at the same distance from Ω as L and such that $L'(\varepsilon x) = \varepsilon L'(x)$ for every $\varepsilon \in \{-1, +1\}^{\mathbb{N}}$. It is straightforward that a linear map with that property must have the form $L'(x) = (\lambda_n x_n)$ for some sequence λ. See [13] or else [21] for details. Done that, the rest is simple: $\|\mathsf{KP}(\sum^N e_n) - L(\sum^N e_n)\| \leq M\sqrt{N}$ is mandatory for some $M < +\infty$: but $\|L(\sum^N e_n)\| = \|\sum \lambda_n e_n\| \leq C\sqrt{N}$ since (λ_n) is bounded while $\|\mathsf{KP}(\sum^N e_n)\| \sim N\sqrt{N}$ so no bound M is possible.

The nature of the Enflo–Lindenstrauss–Pisier example ELP is, however, very different. In our terms, what they did was to construct quasi-linear maps $g_n : \ell_2^n \to \ell_2^{2^n}$ increasingly (in n) far from linear maps. With these, they constructed the (finite dimensional) spaces $\ell_2^{2^n} \oplus_{g_n} \ell_2^n$ to then form the space $\ell_2(\ell_2^{2^n} \oplus_{g_n} \ell_2^n)$.

4.3 Complex interpolation and twisted Hilbert spaces

We essentially need a few elements of complex interpolation , which will be presented via the Kalton–Montgomery analytic families approach [54]. Let U be an open subset of the complex plane conformally equivalent to the open unit disc. The closure of U will be denoted $\overline{U}$ and its boundary ∂U. Let Σ be suitable a complex *ambient* Banach space.

Definition 4.1. A *Kalton space of analytic functions on* U is a Banach space $\mathcal{F} \equiv (\mathcal{F}(U, \Sigma), \|\cdot\|_{\mathcal{F}})$ of analytic functions $f : U \to \Sigma$ satisfying the following conditions:

(a) For each $z \in U$, the evaluation map $\delta_z : \mathcal{F} \to \Sigma$ is bounded.
(b) If $\varphi : U \to \mathbb{D}$ is a conformal equivalence and $f : U \to \Sigma$ is an analytic map, then $f \in \mathcal{F}$ if and only if $\varphi \cdot f \in \mathcal{F}$, and in this case $\|\varphi \cdot f\|_{\mathcal{F}} = \|f\|_{\mathcal{F}}$.

This definition appears formalized in [54] with the name of admissible space of analytic functions, though previous papers of Kalton already contain several forerunners, with different names. The Kalton space can be viewed as a generalization of the Calderon space in classical complex interpolation schema. Given a Kalton space $\mathscr{F}$, for each $z \in U$ we define

$$X_z = \{x \in \Sigma : x = f(z) \text{ for some } f \in \mathscr{F}\}$$

with the norm $\|x\| = \inf\{\|f\|_{\mathscr{F}} : x = f(z)\}$ so that X_z is isometric to $\mathscr{F}/\ker \delta_z$. The family $(X_z)_{z\in U}$ is called an *analytic family* of Banach spaces on U. A function $f_{x,z} \in \mathscr{F}$ such that $f_{x,z}(z) = x$ and $\|f_{x,z}\|_{\mathscr{F}} \le c\|x\|_z$ is called a *c-extremal* (for x at z). The map $\Omega_z : X_z \to \Sigma$ defined as $\Omega_z(x) = f_x'(z)$ it is usually called the derivative (of the analytic family) at z, or simply a derivation or differential.

The key point here is that a derivation Ω_z acts as a quasi-linear map on X_z, in the sense that it generates the quasi-Banach space

$$d_{\Omega_z} X_z = \{(w, x) \in \Sigma \times X : w - \Omega_z x \in X_z\}$$

(endowed with the quasi-norm $\|(w,x)\| = \|w - \Omega_z x\| + \|x\|$) which is a twisted sum space in the sense that there exists an exact sequence

$$0 \longrightarrow X_z \longrightarrow d_{\Omega_z} X_z \longrightarrow X_z \longrightarrow 0.$$

In other words or, better, in the same words we were using at the Introduction, whenever X is a Banach space that "appears" in a complex interpolation scale, that is, X is, up to renorming, a space X_z obtained from a Kalton space of analytic functions; there is a natural twisting of X_z. If the exact sequence above is or not trivial (i. e., if Ω_z is or not trivial) has to be checked on a case-by-case basis.

One example is in order to round the square: consider the interpolation couple (ℓ_1, ℓ_∞) on the unit strip $\mathbb{S} = \{z : 0 \le \operatorname{Re} z \le 1\}$ and set $z = 1/2$. This means that our Calderon space this time is going to be the continuous bounded ℓ_∞-valued functions on $\mathbb{S}$, holomorphic on the interior of $\mathbb{S}$ and such that $f(it) \in \ell_\infty$ and $f(1 + it) \in \ell_1$. The Riesz–Thorin theorem [4] yields the interpolation space $(\ell_1, \ell_\infty)_{1/2} = \ell_2$, and thus the preceding ideas automatically produce a twisted Hilbert space, the one generated by the derivative $\Omega_{1/2}$, whatever it is. But we can actually identify this map. Given positive norm one $x \in \ell_2$, an extremal can be given by $f_x(z) = x^{2z}$, since we simplify a bit just picking 0 and 1 to represent points *it* and $1 + it$—$f_x(1) = x^2 \in \ell_1$ and $f_x(0) = 1 \in \ell_\infty$. Hence, differentiation yields $x \to 2x \log|x|$, understood as the map that associated to the sequence x the sequence $(2x(n) \log|x(n)|)_n$. But this is (up to the 2 factor and for norm one vectors) the Kalton–Peck map and, therefore, Z_2 is isomorphic to $d_{\Omega_{1/2}} X_{1/2}$.

And since ideas are more powerful that realizations, by the same token, any Banach space X such that $(X, X^*)_{1/2}$ is a Hilbert space generates a derivation $\Omega_{1/2}$ and then a twisted Hilbert space $d_{\Omega_{1/2}}(X, X^*)_{1/2}$. And, per finire in bellezza, one quite easily encounters the situation "$(X, X^*)_{1/2}$ is a Hilbert space" [62, around Theorem 3.1];

see also [70] and [21, Proposition 6.1]. More precisely, let $\overline{X}^*$ denote the antidual of X (namely, X^* under the multiplication $\lambda x^* = \overline{\lambda} x^*$). Then we have the following.

Proposition 4.2. *Let X be a Banach space with a monotone shrinking basis. Then $(X, \overline{X}^*)_{1/2} = \ell_2$ with equality of norms.*

We are thus one epsilon away from saying: any suitable Banach space X generates a twisted Hilbert space ∂X. A serious warning must be done, though: "suitable" is an oversimplification. Indeed, (complex) interpolation for couples requires: a couple (X_0, X_1); and linear continuous embeddings $X_0 \to \Sigma$ and $X_1 \to \Sigma$ into an ambient Hausdorff topological vector space (usually a Banach space). Each of those ingredients has effect in the final result. For instance, it is obvious that interpolation between ℓ_1 and ℓ_1 yields ℓ_1. But, one is inadvertently assuming that the inclusions are into $\Sigma = \ell_\infty$ and are the obvious ones. Compare with the following result of Garling and Montgomery-Smith [36].

Theorem 4.3. *There is a Banach couple (A_0, A_1) such that A_0 and A_1 are isometric to ℓ_1, and for $(A_0, A_1)_\theta$ contains a complemented copy of c_0.*

And yes, that is because the embeddings are an essential part of the interpolation process. Less spectacularly said: pick $e : \ell_1 \to \ell_\infty$, the embedding into even positions, and $o : \ell_1 \to \ell_\infty$ the embedding into odd positions. Now $(\ell_1, \ell_1)_\theta = 0$.

All that said, yes, summing up: any suitable Banach space X generates a twisted Hilbert space ∂X.

4.4 Scary monsters

We pass then to test the twisted Hilbert space $\partial X = d_{\Omega_{1/2}}(X, X^*)_{1/2}$ that appears for natural choices of X. Among them, we rank:

- L_p-spaces
- Lorentz spaces
- Orlicz spaces
- Schreier's space
- Tsirelson's space
- James's space
- ℓ_2-amalgams of previous examples

Let us fix and simplify the notation: since we will just consider the twisted Hilbert space $d_{\Omega_{1/2}}(X, X^*)_{1/2}$ generated by complex interpolation at 1/2 when $(X, X^*)_{1/2}$ is a Hilbert space, we will call this derived space ∂X and, when necessary, the associated derivation map will be called Ω_X. But when it is not necessary we will just call it Ω. Keep in mind that no matter what X has been chosen, ∂X is a twisted Hilbert space.

This notation makes clear that the properties of the derived space ∂X depend on properties of X.

4.4.1 L_p-spaces

What has been said above for the couple (ℓ_1, ℓ_∞) can be repeated verbatim for other couples $(L_1(\mu), L_\infty(\mu))$, the associated derivation being still $\Omega x = x \log \frac{\|x\|}{|x|}$ and the twisted Hilbert space ∂L_2. The properties of the derived space can be different (*are* different in fact) depending on whether the underlying measure is atomic or not. For instance, the associated exact sequence

$$0 \longrightarrow L_2(\mu) \longrightarrow \partial L_2(\mu) \longrightarrow L_2(\mu) \longrightarrow 0$$

has strictly singular quotient map only when $L_2(\mu)$ is a sequence space. Recall that an operator is said to be strictly singular when it is not an isomorphism on any infinite dimensional subspace. In complete analogy, a quasi-linear map $\Omega : Z \to Y$ is said to be singular if its restriction to every infinite dimensional closed subspace is never trivial. Of course, one can prove that a quasi-linear map is singular if and only if the quotient operator in any associated exact sequence is strictly singular. Analogously, a derivation Ω will be called singular if the associated exact sequence it induces is singular.

4.4.2 Lorentz spaces

Lorentz $L_{p,q}(\mu)$ spaces can be considered *more general* versions of L_p-spaces; in particular, they are generated by real interpolation out of the couple (L_∞, L_1), while complex interpolation just produces the L_p-spaces. Complex interpolation between two Lorentz spaces produces Lorentz spaces in the obvious way $(L_{p_0,q_0}, L_{p_1,q_1})_\theta = L_{p,q}$ with $p^{-1} = (1-\theta)p_0^{-1} + \theta p_1^{-1}$ and $q^{-1} = (1-\theta)q_0^{-1} + \theta q_1^{-1}$. The associated derivation has been obtained in [10] as the map

$$\Omega_\theta(x) = q\left(\frac{1}{q_1} - \frac{1}{q_0}\right)\mathsf{KP}(x) + \left(\frac{q}{p}\left(\frac{1}{q_0} - \frac{1}{q_1}\right) - \left(\frac{1}{p_0} - \frac{1}{p_1}\right)\right)\kappa(x).$$

Here, $\mathsf{KP}(\cdot)$ is the Kalton–Peck map above and $\kappa(\cdot)$ is the so-called Kalton map [42] given by $\kappa(x) = x\, r_x$ where r_x is the rank function $r_x(t) = m\{s : |x(s)| > |x(t)| \text{ or } |x(s)| = |x(t)| \text{ and } s \le t\}$. The derivative above is strictly singular when $q_0 \neq q_1$ [10, Example 1 and Proposition 2].

4.4.3 Orlicz spaces

Still more general versions of Lorentz spaces are the Orlicz function spaces over a measure space (Σ, μ). Recall that an *N-function* is a map $\varphi : [0,\infty) \to [0,\infty)$ which is strictly increasing, continuous, $\varphi(0) = 0$, $\varphi(t)/t \to 0$ as $t \to 0$, and $\varphi(t)/t \to \infty$ as $t \to \infty$. An N-function φ satisfies the Δ_2*-property* if there exists a number $C > 0$ such that $\varphi(2t) \leq C\varphi(t)$ for all $t \geq 0$. For $1 < p < \infty$, $\varphi(t) = t^p$ is N-function satisfying the Δ_2-property. When an N-function φ satisfies the Δ_2-property, the *Orlicz space* $L_\varphi(\mu)$ is given by $L_\varphi(\mu) = \{f \in L_0(\mu) : \varphi(|f|) \in L_1(\mu)\}$ endowed with the norm $\|f\| = \inf\{r > 0 : \int \varphi(|f|/r)d\mu \leq 1\}$. Given two N-functions φ_0 and φ_1 satisfying the Δ_2-property, and $0 < \theta < 1$ then $\varphi^{-1} = (\varphi_0^{-1})^{1-\theta}(\varphi_1^{-1})^\theta$ defines an N-function φ satisfying the Δ_2-property, and $(L_{\varphi_0}(\mu), L_{\varphi_1}(\mu))_\theta = L_\varphi(\mu)$. If $t = \varphi_0^{-1}(t) \cdot \varphi_1^{-1}(t)$ then $(L_{\varphi_0}(\mu), L_{\varphi_1}(\mu))_{1/2} = L_2(\mu)$ with associated derivation defined for $0 \leq f \in L_2(\mu)$, $\|f\|_2 = 1$ by

$$\Omega(f) = f \log \frac{\varphi_1^{-1}(f^2)}{\varphi_0^{-1}(f^2)} = 2f \log \frac{\varphi_1^{-1}(f^2)}{f}.$$

Note that once Ω has been defined for normalized $0 \leq f \in X$, we define $\Omega(0) = 0$ and $\Omega(g) = g \cdot \Omega(|g|/\|g\|)$ for $0 \neq g \in X$.

4.4.4 Schreier's space

Schreier's space $\mathscr{S}$ [65] (see also [22, 24] for a related exposition) is arguably the first *exotic* Banach space (in this case, for being the first space without the weak Banach–Saks property; namely, weakly null sequences do not have convergent averages). Its construction follows the rather general schema of fixing a compact family $\mathcal{F} \subset \{0,1\}^{\mathbb{N}}$ of finite subsets of $\mathbb{N}$ to then define the space $\mathscr{S}_\mathcal{F}$ as the completion of the space of finitely supported sequences with respect to the norm

$$\|x\|_\mathcal{F} = \sup_{A\in\mathcal{F}} \|1_A x\|_1.$$

For instance, if $\mathcal{F}(n)$ is the family of subsets having n elements then $\mathscr{S}_{\mathcal{F}(n)}$ is (a renorming of) c_0. Thus, to obtain something new one has to allow finite sets of arbitrarily large size. Schreier's space is obtained choosing as $\mathcal{F}$ the family of *admissible* sets, namely, those verifying $|A| \leq \min A$. The Schreier space, simply called from now on $\mathscr{S}$, is a Banach space with unconditional basis. Nobody knows for sure which is the derived space $\partial\mathscr{S}$. Since $\|\sum^N e_n\|_\mathscr{S} = \frac{n}{2}$ and $\|\sum^N e_n\|_{\mathscr{S}^*} = \log n$ we are certain that the associated exact sequence at 1/2, namely,

$$0 \longrightarrow \ell_2 \longrightarrow \partial\mathscr{S} \longrightarrow \ell_2 \longrightarrow 0$$

is not trivial. In [21], it was introduced the parameter for a Köthe space $\mathcal{K}$:

$$M_\mathcal{K}(n) = \sup\{\|x_1 + \cdots + x_n\| : x_1, \ldots, x_k \text{ disjoint in the unit ball of } \mathcal{K}\}.$$

However, since $M_S(n) = n = M_{S^*}(n)$ we cannot directly use the technique of [21] to decide whether it is singular. Many different spaces $\mathscr{S}_{\mathcal{F}}$ have appeared in the literature (Schachermayer's space [64], Leung's space [56], Lunatic spaces [17], all the higher order Schreier hierarchies spaces of Alspach and Argyros [1], [25], ...) and used in several different contexts. The study of the derived spaces $\partial\mathscr{S}_{\mathcal{F}}$ has not yet started.

4.4.5 Baernstein spaces

Baernstein $\mathscr{B}_p$ [3] (see also [16]) introduced the next precursors of Tsirelson space as follows:

$$\|x\|_{\mathscr{B}_p} = \sup\big\|(\|1_{E_k}x\|_1)\big\|_p$$

where the supremum is taken on the finite sequences of consecutive admissible sets $E_1 < E_2 < \cdots < E_n$. Baernstein spaces are reflexive for $1 < p < +\infty$, have unconditional basis and still, like $\mathscr{S}$, fail the Banach–Saks property. It seems interesting to determine the properties $\partial\mathscr{B}_p$.

4.4.6 Tsirelson's spaces

Next station: Tsirelson's space $\mathscr{T}$ [69, 16]; the first reflexive space without copies of any ℓ_p. Its norm is too complex to describe it here. Moreover, it is still uncharted map what occurs with $\partial\mathscr{T}$. Now observe that $M_{\mathscr{T}}(n) = n$ while $M_{\mathscr{T}^*}(n) \sim \log_2(n)$. Let us then show that the differential $\Omega_{1/2}$ corresponding to the pair $(\mathscr{T}, \mathscr{T}^*)$ is singular: indeed, [21, Proposition 5.1] states that if $\Omega_{1/2}$ is not disjointly singular then there exists a subspace $W \subset (\mathscr{T}, \mathscr{T}^*)_{1/2}$ spanned by disjointly supported vectors and a constant K such that

$$\left|\log\frac{M_0(n)}{M_1(n)}\right| M_W(n) \le K M_0(n)^{1/2} M_1(n)^{1/2}.$$

Since $(\mathscr{T}, \mathscr{T}^*)_{1/2}$ is a Hilbert space $M_W(n) = \sqrt{n}$ for every subspace, which therefore yields

$$\left|\log\frac{n}{\log n}\right| \sqrt{n} \le K\sqrt{n}(\log n)^{1/2}$$

namely

$$\log n \le K(\log n)^{1/2} + \log\log n$$

which is impossible. Disjoint singularity is equivalent to singularity on ℓ_2, and thus, the associated exact sequence

$$0 \longrightarrow \ell_2 \longrightarrow \partial\mathscr{T} \longrightarrow \ell_2 \longrightarrow 0$$

has strictly singular quotient map.

This makes even more surprising that Suárez [67] has been able to show that replacing $\mathscr{T}$ by its 2-convexification, say $\mathscr{T}_2$ (please see the next entry) the space $\partial\mathscr{T}_2$ is actually a weak Hilbert space, providing in this way the first twisted Hilbert weak Hilbert space. The analysis of $\partial\mathscr{T}_2$ contains further surprises: the space does not contain Z_2 and is not contained in either Z_2 or ELP. This moreover opens the door to study spaces such as $\partial\mathscr{S}_2$, where $\mathscr{S}_2$ is the 2-convexification of Schreier space. Further variations are waiting in line: Casazza and Nielsen proved in [15] that the symmetric convexified Tsirelson space $\mathscr{T}_{c2}$ (has symmetric basis, of course and) is of weak cotype 2 but not of cotype 2. The space $\partial\mathscr{T}_{c2}$ is claiming a second look.

4.4.7 *p*-convexifications

Given a Banach space λ with a 1-unconditional basis (e_n) and given $1 \le p < +\infty$, its p-convexification λ_p is defined as the completion of the space of finitely supported sequences endowed with the norm

$$\left\|\sum_{n=1}^{\infty} \lambda_n e_n\right\|_{\lambda_p} = \left\|\sum_{n=1}^{\infty} |\lambda_n|^p e_n\right\|_{\lambda}^{1/p}.$$

According to [21, Proposition 3.6], $\lambda_p = (\lambda, \ell_\infty)_{1/p}$ with recognizable associated derivation (see below). In particular, the 2-convexification of λ is $\lambda_2 = (\lambda, \ell_\infty)_{1/2}$ with recognizable derivation. A different thing is to estimate the derivation associated to $(\lambda_2, \lambda_2^*)_{1/2} = \ell_2$ and, therefore, which one is the derived space $\partial\lambda_2$.

4.4.8 James's space

James space $\mathscr{J}$ [38] is a nonreflexive separable space isometric to its bidual. It is a case worth consideration because it is off-limits: no unconditional basis or Köthe structure are present. It is still true that $(\mathscr{J}, \mathscr{J}^*)_{1/2} = \ell_2$ and therefore the question of determining $\partial\mathscr{J}$ is of paramount importance.

4.4.9 ℓ_2-amalgams

Calderón's paper [14] contains a rather general interpolation result for vector sums. Let Λ be a Köthe space defined on a measure space M. Given a Banach space X, one can form the vector valued space Banach $\Lambda(X)$ of measurable functions $f : M \to X$ such that the function $\widehat{f}(\cdot) = \|f(\cdot)\|_X : M \to \mathbb{R}$ given by $t \to \|f(t)\|_X$ is in Λ, endowed with the norm $\|\|f(\cdot)\|_X\|_\Lambda$. Precisely, Calderon's paper contains the interpolation formula

$$(\lambda_0(X), \lambda_1(X^*))_\theta = \lambda_0^{1-\theta}\lambda_1^\theta(X, X^*)_\theta$$

valid for an interpolation couple (λ_0, λ_1) of Köthe spaces over the same measure space with λ_0 reflexive. The associated derivation has been calculated in [26]. In the case more interesting for us in these notes, we get the following.

Theorem 4.4. *Let λ be a reflexive Köthe space with associated derivative ω at $(\lambda, \lambda^*)_{1/2}$ and let X be a reflexive Banach space with associated derivation Ω at $(X, X^*)_{1/2}$. Then $(\lambda(X), \lambda^*(X^*))_{1/2} = (\lambda, \lambda^*)_{1/2}(X, X^*)_{1/2}$ with associated derivation Φ defined on the dense subspace of simple functions $f = \sum_{n=1}^N a_n 1_{A_n}$ as*

$$\Phi(f) = \omega(\widehat{f}(\cdot)) \sum_{n=1}^{N} \frac{a_n}{\|a_n\|} 1_{A_n} + \sum_{n=1}^{N} \Omega(a_n) 1_{A_n}.$$

This result allows one to obtain a great variety of new twisted Hilbert spaces. A couple of examples are in order:

- Pick the sequence of finite dimensional couples $(\ell_{p_n}^{k_n}, \ell_{p_n^*}^{k_n})$ and their ℓ_2 amalgams $\ell_2(\ell_{p_n}^{k_n})$ and $\ell_2(\ell_{p_n^*}^{k_n})$. The derivative at $(\ell_2(\ell_{p_n}^{k_n}), \ell_2(\ell_{p_n^*}^{k_n}))_{1/2} = \ell_2(\ell_2^{k_n}) = \ell_2$ is

$$\Omega\left(\left(\sum_{j=1}^{k_n} x_j^n u_j^n\right)_n\right) = \left(\frac{2}{p_n} - \frac{2}{p_n^*}\right) \sum_{j=1}^{k_n} x_j^n \log \frac{|x_j^n|}{\|\sum_{j=1}^{k_n} x_j^n u_j^n\|_2} u_j^n.$$

- A particularly interesting choice is, according to [40] and [61, p. 21], when $\lim p_n = 2$ and $k_n \to \infty$ adequately chosen, in which case the space $\ell_2(\ell_{p_n}^{k_n})$ is asymptotically Hilbert and non-Hilbert. It is likely that the derived space $\partial\ell_2(\ell_{p_n}^{k_n})$ is asymptotically Hilbert, too.

4.5 Perspectives

After the previous reasonably thick list of examples of twisted Hilbert spaces, time is ripe to tackle the classification of twisted Hilbert spaces. What is the meaning of "classification" here? A good step would be to identify properties that twisted Hilbert spaces may or may not enjoy, including the identification of properties that all twisted Hilbert spaces must have.

Obvious properties shared by all twisted Hilbert spaces are: superreflexivity and, in general, all 3-space properties that Hilbert spaces enjoy (see [23]). Among these, the property of being near-Hilbert: that is, to be of type p for all $p < 2$ and cotype q for all $q > 2$. Obvious properties that no twisted Hilbert space can have are those implying that copies of ℓ_2 are complemented, such as Maurey extension property. In between, one encounters either properties of Hilbert spaces twisted Hilbert might enjoy (i.e„ that most known examples do enjoy, but for which it is not known if all twisted Hilbert spaces do), say:

1. Is every twisted Hilbert space isomorphic to its dual?
2. Is every twisted Hilbert space isomorphic to its square?
3. Is every twisted Hilbert space isomorphic to its hyperplanes?
4. Do twisted Hilbert spaces admit complex structures?

or properties that Hilbert spaces cannot enjoy but that maybe some twisted Hilbert space could:
1. Do there exist nonergodic (nontrivial) twisted Hilbert spaces?
2. Do there exist a twisted Hilbert space that is not isomorphic to its hyperplanes?
3. Do there exist nonprime twisted Hilbert spaces?

A couple of words about these properties: It has been conjectured that Z_2 could be the first natural Banach space not isomorphic to its hyperplanes [41, 44]. Connected to this: Do hyperplanes of Z_2 admit a complex structure? Indeed, if one could prove that hyperplanes of Z_2 do not admit complex structures then it is clear that they cannot be isomorphic to Z_2 since there are obvious complex structures on Z_2. This approach was undertaken in [20]. It is also unknown whether Z_2 is prime [41, 44]; namely, if it contains infinite dimensional complemented subspaces other than Z_2. What is known [44] is that complemented subspaces of Z_2 isomorphic to their square are Z_2. Ergodicity has been invented by Ferenczi and Rosendal [35], where they conjectured that every Banach non-Hilbert space is ergodic. Avoiding a few technicalities, a Banach space is ergodic when it contains a continuum of subspaces so that the relation of isomorphism between them mimicries that of $\{-1,1\}^{\mathbb{N}}$: $a \le b$ if and only if a and b are equal up to a finite number of elements. Ergodicity was brought to twisted Hilbert space affairs by Cuellar [31] who showed that every nonergodic Banach space must be near Hilbert. Which are the perfect near Hilbert spaces? Yes, twisted Hilbert spaces. So Cuellar [31] raised the still open question of whether every nontrivial twisted Hilbert space must be ergodic. Since Anisca [2] proved that weak Hilbert spaces are ergodic, Suárez's example [67] of a weak Hilbert twisted Hilbert space shows that the Ferenczi–Rosendal conjecture is still alive. An illustrative example of where the difficulties lie: the question of whether the Kalton–Peck Z_2 space is ergodic is open. And the problem is not only the order condition, is that we just effectively know ... three nonisomorphic subspaces of Z_2. A Banach space is Happy [39, 40] if all its closed subspaces have the approximation property (Hereditary APProximation propertY). All known examples of Happy spaces are asymptotically Hilbertian and Szankowski [68] established that Happy spaces are near Hilbert. Suárez example [67] above is weak Hilbert and therefore Happy. What can be said about Happy twisted Hilbert spaces?

And, finally, properties a nontrivial twisted Hilbert cannot enjoy because they oblige it to be Hilbert, say:
1. A twisted Hilbert space with unconditional basis is a Hilbert space.
2. A twisted Hilbert space with either type 2 or cotype 2 is a Hilbert space.

All derived twisted Hilbert spaces ∂X associated to a Köthe space are isomorphic to their duals. This is one of those things that, as David Yost says, is very well known for all people who knows it. In particular, see [7, Corollary 4]. But it is not known if all twisted Hilbert spaces are isomorphic to their duals. There are twisted Hilbert spaces that do not contain complemented copies of ℓ_2, like Z_2; and other that contain them, like ELP. More yet, the Kalton–Peck sequence

$$0 \longrightarrow \ell_2 \longrightarrow Z_2 \longrightarrow \ell_2 \longrightarrow 0$$

has strictly singular quotient map. On the other hand, the exact sequence

$$0 \longrightarrow \ell_2 \longrightarrow \mathrm{ELP} \xrightarrow{\pi} \ell_2 \longrightarrow 0$$

has the opposite behavior: every subspace of ELP having infinite dimensional image by π contains an infinite dimensional subspace on which π is an isomorphism. This is due to the fact that ELP, like every space having the form $\ell_2(F_n)$ for finite dimensional F_n, is reflexive and has property W_2: every weakly null sequence contains a weakly 2-summable subsequence. It is then clear that Z_2 does not enjoy property W_2.

Following [28], an exact sequence $0 \to Y \to Z \to X \to 0$ of separable Banach spaces is said to be $\mathcal{C}$-trivial if every operator $Y \to C[0,1]$ can be extended to an operator $Z \to C[0,1]$. The Kalton–Peck sequence $0 \to \ell_2 \to Z_2 \to \ell_2 \to 0$ is $\mathcal{C}$-trivial [51], and it is known that non-$\mathcal{C}$-trivial sequences $0 \to \ell_2 \to E \to \ell_2 \to 0$ exist [53]. Maybe these properties can be used in the classification of twisted Hilbert spaces. Two Banach space variations worth consideration are: Let us say that a separable Banach space X is *solid* when every operator $X \to C[0,1]$ can be extended to an operator $C[0,1] \to C[0,1]$ through any embedding $X \to C[0,1]$ (this is Kalton's notion of $\mathcal{C}$-extensible space [53, 51]); and X will be called *liquid* if every operator $Y \to C[0,1]$ from every subspace $Y \subset X$ can be extended to an operator $X \to C[0,1]$. For instance, ℓ_2 is liquid but not solid [53] while ℓ_1 is solid but not liquid [51] and c_0 is liquid [57] and solid [66]. Do there exist solid twisted Hilbert spaces?

Thus, the properties of twisted Hilbert spaces that could be (so far) used for their classification are:
- Property W_2
- Ω is strictly singular
- Ω is strictly co-singular
- Containing (not containing) complemented copies of ℓ_2
- Containing (not containing) (complemented) copies of Z_2
- Containing (not containing) (complemented) copies of an ELP space; that is, a space $\ell_2(F_n)$ for finite dimensional F_n
- To be weak-Hilbert
- To be asymptotically Hilbert
- To be liquid
- To be solid

Observe that we know so far four nonisomorphic twisted Hilbert spaces: ELP, Z_2, $\partial\mathscr{S}_2$, $\partial\mathscr{T}_2$. If we move however in the direction of considering complex spaces, Kalton [48] creates a variations $Z_2(\alpha)$ of Z_2 for complex number α and proves that $Z_2(\alpha)$ and $Z_2(\beta)$ are not isomorphic to $\alpha \neq \beta$. Thus, there is a continuum of nonisomorphic twisted Hilbert spaces, all of them providing strictly singular exact sequences

$$0 \longrightarrow \ell_2 \longrightarrow Z_2(\alpha) \longrightarrow \ell_2 \longrightarrow 0.$$

The particular case of weak Hilbert spaces is specially interesting: We do not know if there exists a continuum of nonisomorphic weak Hilbert twisted Hilbert spaces. A try could be to consider, for different δ, the variations $\mathscr{T}_2(\delta)$ of Tsirelson space and its 2-convexification, which are totally incomparable for different δ. What about the derived spaces $\partial\mathscr{T}_2(\delta)$ for $0 < \delta \leq 1/2$? Are they mutually nonisomorphic?

A different approach is possible: it would require first to prove that whenever wH is a weak Hilbert space then also ∂wH is weak Hilbert (something we do not know); then, find an argument showing that iterated derivations (see [6]) $\partial wH, \partial^2 wH, \partial^3 wH, \ldots$ are (first) possible and (then) nonisomorphic.

4.6 Intermezzo. All derivations on Köthe spaces are Kalton–Peck maps

Let us show and explain now how and why all derivations on Köthe spaces are in a sense Kalton–Peck's derivations. First of all, observe that most of the previous examples of spaces yielding $(X, X^*)_{1/2} = \ell_2$ are actually Köthe spaces. There is a reason for that, and is declared by Kalton and Montgomery-Smith in [54, p. 1151] *One of the drawbacks of the complex method is that in general it seems relatively difficult to calculate complex interpolation spaces. There is one exception to this rule, which is the case when one has a pair of Banach lattices.* Why the case of Köthe spaces is different? Because there complex interpolation becomes plain and simple factorization in the sense we explain now.

Given two Köthe spaces on the same base space, we define the space

$$YZ = \{f \in L_0 : f = yz : y \in Y, z \in Z\}$$

endowed with the quasi-norm $\|x\| = \inf \|y\|_Y \|z\|_Z$ where the infimum is taken over all factorizations as above. Recall that the p-convexification, $1 \leq p < +\infty$ of a Köthe function space X is defined to be the space $X_p = X^{1/p} = \{f \in L_0 : |f|^p \in X\}$ endowed with the norm $\|x\|_p = \|\, |x|^p \,\|^{1/p}$. This immediately yields

$$X_p = (X, \ell_\infty)_{1/p}$$

since $x = |x|(\operatorname{sign} x)$, and thus $x^{1/p} = x^{1/p}(\pm 1)$; that is, $X_p = X^{1/p}\ell_\infty^{1/p^*}$. This suggests that $X_p = (X, \ell_\infty)_{1/p}$. Indeed, this is it: the Lozanovskii decomposition formula [59] allows to show (see [54, Theorem 4.6]) that, in general,

$$X_\theta = (X_0, X_1)_\theta = X_0^{1-\theta} X_1^\theta$$

with

$$\|x\|_\theta = \inf\{\|y\|_0^{1-\theta}\|z\|_1^\theta : y \in X_0, z \in X_1, |x| = |y|^{1-\theta}|z|^\theta\},$$

where the infimum is taken over all factorizations of x.

Now, by homogeneity we may always assume that $\|y\|_0 = \|z\|_1$ for y, z in this infimum. When $\|y\|_0, \|z\|_1 \leq K\|x\|_\theta$, we shall say that $|x| = |y|^{1-\theta}|z|^\theta$ is a K-*optimal decomposition* for x. The value $\Omega_\theta(x)$ of the derivative Ω_θ at x we know can be calculated by derivation of a K-extremal f_x of the Kalton space at x. In the factorization language, this corresponds to: obtain a K-optimal decomposition $a_0(x)$, $a_1(x)$ for x. Since $\|x\|_\theta = \|a_0(x)\|_0 = \|a_1(x)\|_1$, set $f_x \in \mathcal{H}$ given by $f_x(z) = |a_0(x)|^{1-z}|a_1(x)|^z$ for positive x and $f_x = (\operatorname{sgn} x) f_{|x|}$ for general x, to obtain

$$\Omega_\theta(x) = (f_x)'(\theta) = \big|a_0(x)\big|^{1-\theta}\big|a_1(x)\big|^\theta \log\frac{|a_1(x)|}{|a_0(x)|}x = x\log\frac{|a_1(x)|}{|a_0(x)|}.$$

It is in this way that all derivations in the Köthe space ambient, are Kalton–Peck-like maps, as we claimed above. Unfortunately, this does not mean that we know "who" is the derivation Ω generated by a concrete Köthe space X, even if this does not prevent one to know things about the twisted Hilbert space ∂X. For example, many relevant properties of the twisted Hilbert space $\partial\mathscr{T}_2$ generated by the 2-convexification of Tsirelson's space are studied although there is no direct knowledge of the corresponding derivation. The problem is the very limited information we have about the Lozanovskii factorization $\ell_1 = \mathcal{K}\mathcal{K}^*$ out of the case of rearrangement invariant Köthe spaces. Schreier, Tsirelson, etc. (as well as their p-convexifications) are strongly non-symmetric.

Perhaps a rawer manifestation of the difference between "what-we-know-about spaces" and "what-we-know-about-derivations" is Watbled's theorem 4.2: the proof provides no hint about the associated derivation (which, likely, it is not a "Kalton–Peck" derivation).

4.7 Stranger things

Derivation is not the only way to obtain twisted Hilbert spaces (perhaps). We can invoke homology, too. Consider the simplest case of twisted Hilbert spaces: those obtained via complex interpolation as

$$0 \longrightarrow \ell_2 \longrightarrow \partial X \longrightarrow \ell_2 \longrightarrow 0$$

out from a Banach space X with unconditional basis (most of the examples so far are in this category). We will call the (vector space) set of these spaces $\mathrm{Ext}_\infty(\ell_2)$. It can be shown that they are in good correspondence with twisted sum spaces E,

$$0 \longrightarrow \mathbb{R} \longrightarrow E \longrightarrow \ell_1 \longrightarrow 0$$

whose (vector space) space we will call $\mathrm{Ext}(\ell_1, \mathbb{R})$. I imagine the reader's surprise finding out that quasi-Banach nonlocally convex spaces play their role here. But they do: it is impossible for a Banach space to contain an uncomplemented copy of $\mathbb{R}$, but not for a quasi-Banach space, say $0 \to \mathbb{R} \to L_p \to L_p/\mathbb{R} \to 0$ for $0 < p < 1$; recall that $L_p, 0 < p < 1$ does not admit any single nonzero linear continuous functional. There is a simple way to describe the correspondence

$$\mathrm{Ext}_\infty(\ell_2) \to \mathrm{Ext}\,(\ell_1, \mathbb{R}).$$

Pick an element in $\mathrm{Ext}_\infty(\ell_2)$, form the tensor product

$$0 \longrightarrow \ell_2 \otimes \ell_2 \longrightarrow \partial X \otimes \ell_2 \longrightarrow \ell_2 \otimes \ell_2 \longrightarrow 0$$

in the category of ℓ_∞-Banach modules. In that category, $\ell_2 \otimes \ell_2 = \ell_1$, so one has a sequence

$$0 \longrightarrow \ell_1 \longrightarrow \partial X \otimes \ell_2 \longrightarrow \ell_1 \longrightarrow 0$$

which will necessarily be described by a quasi-linear map Ω. Let $\Sigma : \ell_1 \to \mathbb{R}$ be the sum operator, and form the exact sequence which has associated quasi-linear map $\Sigma\Omega$.

The correspondence in the other direction can be roughly described as: once you get a quasi-linear map $\omega : \ell_1 \to \mathbb{R}$, it induces a map $\beta : \ell_2 \times \ell_2 \to \mathbb{R}$ given by $\beta(x, y) = \omega(x \cdot y)$, where $x \cdot y$ is the pointwise product. This map is biquasi-linear, in the sense that both $\beta(x, \cdot)$ and $\beta(\cdot, y)$ are quasi-linear maps. So, in the same way that bilinear maps $\ell_2 \times \ell_2 \to \mathbb{R}$ correspond to linear maps $\ell_2 \to \ell_2^*$, it can be shown (after some additional work) that biquasi-linear maps $\ell_2 \times \ell_2 \to \mathbb{R}$ correspond to quasi-linear maps $\ell_2 \to \ell_2^*$. These provide exact sequences $0 \to \ell_2^* \to \diamond \to \ell_2 \to 0$, and this is it.

A test question here is to clarify the elements in $\mathrm{Ext}(\ell_1, \mathbb{R})$ that correspond to the derivatives we have already obtained in ℓ_2. For our purposes here, it is the other way around the one we are thinking of: construct "interesting" elements of $\mathrm{Ext}(\ell_1, \mathbb{R})$ and then shift them to $\mathrm{Ext}_\infty(\ell_2, \ell_2)$. The hunt for stranger derivations however goes now as follows: observe that the Kalton–Peck map corresponds to the simplest quasi-linear map $\ell_1 \to \mathbb{R}$, the Ribe's map [63] given by

$$\mathscr{R}(x) = \sum_n x_n \log \frac{\|x_n\|}{|x|_1}$$

(which is "the scalar version of KP"). To give an idea of why $\mathscr{R}$ is "simple," observe that, being true that $\mathscr{R}$ is nontrivial, it is one subspace away from being trivial. Precisely, every infinite dimensional subspace of ℓ_1 contains a further infinite dimensional subspace where $\mathscr{R}$ is trivial. See [27, Lemma 2] for details and generalizations. Following this line of thinking, it is clear that the title of weirdest centralizer on ℓ_2 should be awarded to the one corresponding to the weirdest quasi-linear map $\ell_1 \to \mathbb{R}$. And, who is this? Unquestionably, the title goes to Kalton's strictly singular quasi-linear map $\ell_1 \to \mathbb{R}$ [49]: there is a quasi-linear map $\mathscr{K} : \ell_1 \to \mathbb{R}$ which is not trivial on any infinite dimensional subspace of ℓ_1. To see how strange is this, observe that if $0 \to \mathbb{R} \to K \to \ell_1 \to 0$ is the exact sequence $\mathscr{K}$ defines, the space K enjoys the outrageous property of making every infinite dimensional closed subspace of K to contain a certain prefixed point. Nothing is known about the corresponding twisted Hilbert space.

Looking back the history of twisted Hilbert spaces, one observes that the appearance of the ELP space was more an existence theorem than an example, in the sense that "non-Hilbert twisted Hilbert spaces exist because finite dimensional twisted Hilbert spaces increasingly far from being Hilbert exist." This local approach to twisted sums was developed in [12]. However, when the space Z_2 rushed in the idea spread that twisted Hilbert spaces could be weird: it is not only that the Kalton–Peck sequence is singular, it is not only that Z_2 does not contain complemented copies of ℓ_2 ... is that Z_2 is actually harsher: every operator $Z_2 \to X$ is either strictly singular or an isomorphism on some complemented copy of Z_2 [44]. By the time of [55], Z_2 was considered an extremal twisted Hilbert space because of the fact that its finite dimensional pieces $0 \to \ell_2^n \to Z_2^n \to \ell_2^n \to 0$, who necessarily verify $d(Z_2^n, \ell_2^{2n}) \to \infty$, in fact, verify that $d(Z_2^n, \ell_2^{2n}) \sim \log n$ and this is the maximum speed for twisted Hilbert spaces. In other words, if E_n are $2n$-dimensional twisted Hilbert spaces then $d(E_n, \ell_2^{2n}) \le c \log n$ for some $c > 0$. In this sense, Z_2 is an extremal twisted Hilbert space.

That is ok, but ... what about weird twisted Hilbert spaces?

We know by now a few additional things, and some are concealed in [67]: the existence of a weak Hilbert twisted Hilbert space $\partial\mathscr{T}_2$ is a major achievement that somehow qualifies as the weirdest twisted Hilbert space and, therefore, both its structure and the derivation that it defines deserves scrutiny. The analysis of its structure is [67]: apart from being weak Hilbert, $\partial\mathscr{T}_2$ is basically incomparable with either Kalton–Peck or Enflo–Lindenstrauss–Pisier spaces: it is not a subspace or quotient from them. In saying this, what is important for us is what the associated derivation Ω actually does. We know that it cannot be "faster" than the Kalton–Peck's map, so what? The answer (well, an answer) is already in the paper: what Ω does to make $\partial\mathscr{T}_2$ weak Hilbert is not to grow fast, is to grow slow. Actually, $d(\partial\mathscr{T}_2{}^n, \ell_2^{2n})$ grows slower that any iteration $\log\log\cdots\log n$. Precisely, [67, Corollary 1]: *There is $C > 0$ such that for all finite-dimensional subspaces $E \subset \partial\mathscr{T}_2$ and all $m = 1, 2, \dots$ one has*

$$d(E, \ell_2^{\dim E}) \le C^m \log_m (\dim E).$$

Thus, the land to chart is that of slowly growing derivations: How slowly can a nontrivial derivation grow?

Once the speed race is abandoned, another line to be considered is that of symmetry. Given a Köthe space λ constructed on a base measure space S, a measure-preserving $\sigma : S \to S$ and $x \in \lambda$, the meaning of the element $x \circ \sigma$ is clear. Now, a quasi-linear map on λ is said to be symmetric when $\|(\Omega x) \circ \sigma - \Omega(x \circ \sigma)\| \leq C\|x\|$. Symmetric quasi-linear maps are somehow at the pinnacle of evolution. The Kalton–Peck map is symmetric, in fact, given any permutation σ of the integers, $(\Omega x) \circ \sigma = \Omega(x \circ \sigma)$. So, the look for weird derivations should go in the quest for highly nonsymmetric ones. A way to consider this is: each reasonable quasi-linear map (see next section) on a Köthe space has associated a group of symmetries (a group of measure preserving maps of the base space). The Kalton–Peck's map, being symmetric, has the biggest possible group of symmetries. The hunt is now for maps with small group of symmetries.

4.8 Lark's modules in aspic: the centralizer issue

Back to the starting point and the Eilenberg–McLane program, we have already begun to grasp the idea that there is a Hilbert space in each category, and that the standard categories we have been (aware or unawarely) working with are ℓ_∞-modules and L_∞-modules. Kalton observed [45] that the quasi-linear map detects the module structure in the following way. Let A be either ℓ_∞ or L_∞. An *A-centralizer* defined on a Banach A-module X is a homogeneous map $\Omega : X \to L_0$ such that $\Omega(ax) - a\Omega(x) \in X$ for all $a \in A$ and, moreover,

$$\|\Omega(ax) - a\Omega(x)\| \leq C\|a\|\|x\|$$

for some constant C. Not so simple as it seems but A-centralizers, like derivations, act as quasi-linear maps. Consequently, induce exact sequences

$$0 \longrightarrow X \longrightarrow \partial_\Omega X \longrightarrow X \longrightarrow 0.$$

These sequences are not only exact sequences in the category of Banach spaces but in the category of A-modules (see [45, 5]). Observe that Banach sequence spaces which are ℓ_∞-modules are not necessarily Köthe spaces: in fact, every twisted Hilbert space $\ell_2 \oplus_\Omega \ell_2$ is an ℓ_∞-module when Ω is an ℓ_∞-centralizer while it cannot have unconditional basis: any twisted Hilbert space with unconditional basis is a Hilbert space [50]. Or else, the Kalton–Peck space Z_2 is not even a Köthe space. Thus, *these*, and not other more classical, are the right categories where Hilbert and twisted Hilbert spaces live, what distinguishes them, and mark the way in which we must, accordingly, distinguish them.

And of course that the Kalton–Peck map KP on $L_2(\Sigma, m)$ is an $L_\infty(\Sigma, m)$-centralizer and derivations are centralizers in the corresponding module structure.

So, there is time for summarizing and moving outside.

As we have already said, each base measure space (S, m) provides the corresponding category of $L_\infty(S, m)$-modules, whose most prestigious examples are the Köthe spaces on (S, m). The fact that, say, ℓ_∞ and L_∞ are isomorphic does not mean that ℓ_∞-modules and L_∞-modules are the same, because they are not. So, saying ℓ_2 means working in the category of ℓ_∞-modules and the twisted Hilbert that emerge via ℓ_∞-centralizers are again ℓ_∞-modules, no longer Köthe spaces, no longer Banach spaces with unconditional basis. While saying $L_2(S, m)$ means working in the category of $L_\infty(S, m)$-modules and the twisted Hilbert that emerge via $L_\infty(S, m)$-centralizers are again $L_\infty(S, m)$-modules, no longer Köthe spaces.

But there are other Hilbert spaces and other A-module structures in the real world. The Schatten S_2 class is the perfect example. Given a Hilbert space H, the Schatten class [32] $S_p(H)$ is the vector space

$$S_p(H) = \{\tau \in \mathfrak{L}(H) : (a_n(\tau)) \in \ell_p\}$$

endowed with the obvious quasi-norm $\|\tau\|_p = \|(a_n(\tau))\|_p$. Here, $(a_n(\tau))$ is the sequence of the singular numbers of the operator τ. To make it simple, compact self-adjoint operators κ on complex Hilbert spaces admit a representation $\kappa = \sum \lambda_n(\tau)\varphi_n \otimes \varphi_n$ for some real sequence $(\lambda_n(\tau)) \in c_0$ and some orthonormal sequence (φ_n). The sequence $(\lambda_n(\tau))$ is that of eigenvalues of τ. From that, one can obtain that compact nonnecessarily self-adjoint operators admit a representation $\kappa = \sum a_n(\tau)\psi_n \otimes \varphi_n$ for some real sequence $(\lambda_n) \in c_0$ and some orthonormal sequences (ψ_n), (φ_n). The numbers $a_n(\tau)$ are no longer eigenvalues of τ, but the square roots of the eigenvalues of $\tau^*\tau$. These are called the singular numbers of τ. In the case $p = 2$, the quasi-norm above becomes a norm and $S_2(H)$ becomes a Hilbert space.

This Hilbert space, as isometric to ℓ_2 as any, carries its own module structure: that of $\mathfrak{L}(H)$-module. The vanishing point of the theory on display here is that, for the same "abstract" reasons we have tried to explain, as soon as

$$\Omega(\iota) = \tau \log |\tau|$$

makes sense, and it defines an $\mathfrak{L}(H)$-centralizer on $S_2(H)$ that, accordingly, generates its own twisted Hilbert space ♣:

$$0 \longrightarrow S_2(H) \longrightarrow ♣ \longrightarrow S_2(H) \longrightarrow 0.$$

It is not easy to check out by brute force that $\Omega(x) = x \log |x|$ is a $\mathfrak{L}(H)$-centralizer on $S_2(H)$, but this is what Kalton does in [46]. There are simpler ways, and knowing the interpolation formula

$$(S_1(H), S_\infty(H))_{1/2} = S_2(H)$$

is one of them, once the general theory as been incorporated.

The story does not end here since the construction moved much further: it reached the domain of noncommutative L_p-spaces [5], even if intrinsic technicalities somehow blurred the panorama; and in the imaginative company of Correa [29] it cleanly set in the domain of operator spaces, up to the point that Correa shows that there is a nontrivial exact sequence

$$0 \longrightarrow \ell_2 \longrightarrow \ell_2 \longrightarrow \ell_2 \longrightarrow 0$$

in the category of operator spaces.

Beyond that is the fact that the Kalton–Peck map $\Omega x = x \log |x|$ has a stubborn determination to be a centralizer on any category in which one can barely make a sense of it. Consequently, each of those categories has, in addition to its own Hilbert space, its own Z_2 space, which is always a nontrivial twisted Hilbert space. See the noncommutative Z_2 space [9] or the operator space Z_2 in [29].

4.9 Coda. All centralizers on Köthe spaces are derivations

Kalton did much more than identifying the Kalton–Peck map as the derivation corresponding to the scale of ℓ_p-spaces. He developed [45, 47] a deep theory connecting derivations, twisted sums, and complex interpolation scales in the specific case of Köthe function spaces. A warning is in order: complex interpolation occurs, by its nature, in complex Banach spaces, while one usually treats real Banach spaces. The ditch between them we can usually wade without too much splatter. In this context, it make sense to define a centralizer Ω as *real* if it sends real functions to real functions.

Theorem 4.5.

1. *Given a complex interpolation pair (X_0, X_1) of Köthe function spaces and a point $0 < \theta < 1$, the derivation Ω_θ is an L_∞-centralizer on the space X_θ.*
2. *For every real L_∞-centralizer Ω on a separable superreflexive Köthe function space X, there is a number $\varepsilon > 0$ and an interpolation pair (X_0, X_1) of Köthe function spaces so that $X = X_\theta$ for some $0 < \theta < 1$ and $\varepsilon\Omega - \Omega_\theta : X_\theta \to X_\theta$ is a bounded map.*
3. *The spaces X_0, X_1 are uniquely determined up to equivalent renorming.*
4. *The induced centralizer Ω_θ is bounded as a map $X_\theta \to X_\theta$ for some θ if and only if $X_0 = X_1$, up to an equivalent renorming.*

This outstanding theorem not only says that any given (real) centralizer is actually a derivation obtained from an interpolation couple; it is even able to precisely point to which spaces have been chosen as ends. The first of those statements means that by giving a centralizer on a Köthe space $\mathcal{K}$ one is pointing to a precise interpolation couple (X_0, X_1) so that $\mathcal{K} = (X_0, X_1)_\theta$ and Ω is the derivation obtained from that scale at

θ. In the case of centralizers on Hilbert spaces, this means that by giving Ω on $L_2(\Sigma, m)$ Kalton's theorem points at some single Köthe space X such that Ω is (up to a bounded factor) the associated derivation of the scale (X, X^*) at 1/2. The second part of the statement is equally surprising: inside that chosen scale one can even specify which is the space X in the scale to be chosen. One example is in order, the Kalton–Peck map corresponds to the scale of ℓ_p spaces: $-2x \log|x|$ corresponds to $(\ell_1, \ell_\infty)_{1/2}$, but in general $(\frac{1}{p} - \frac{1}{p^*})x \log|x|$ corresponds to $(\ell_p, \ell_{p^*})_{1/2}$.

Kalton [47] did not stop there: If the centralizer Ω is not real, then there exist three Köthe spaces so that Ω is, up to a bounded map, the derivation at 0 corresponding to the family formed by three Köthe spaces equidistributed in the unit sphere. See [19] for further developments. Correa [30] exhibits an example of a (complex) centralizer in ℓ_2 obtained from a family of three Orlicz spaces that cannot be obtained as a derivation of two Köthe spaces.

Bibliography

[1] D. Alspach, S. Argyros, *Complexity of Weakly Null Sequences*, Dissertationes Math. **CCCXXI**, 1993.

[2] R. Anisca, *The ergodicity of weak Hilbert spaces*, Proc. Am. Math. Soc. **138** (2010), 1405–1413.

[3] A. Baernstein, *On reflexivity and summability*, Stud. Math. **42** (1972), 91–94.

[4] J. Bergh, J. Löfström, *Interpolation Spaces. An Introduction*, Springer-Verlag, 1976.

[5] F. Cabello, J. M. F. Castillo, S. Goldstein, Jesús Suárez, *Twisting noncommutative L_p-spaces*, Adv. Math. **294** (2016), 454–488.

[6] F. Cabello, J. M. F. Castillo, N. J. Kalton, *Complex interpolation and twisted twisted Hilbert spaces*, Pac. J. Math. **276** (2015), 287–307.

[7] F. Cabello Sánchez, *Nonlinear centralizers in homology*, Math. Ann. **358** (2014), 779–798.

[8] F. Cabello Sánchez, *Pointwise tensor products of function spaces*, J. Math. Anal. Appl. **418** (2014), 317–335.

[9] F. Cabello Sánchez, *The noncommutative Kalton–Peck spaces*, J. Noncommut. Geom. **11** (2017), 1395–1412.

[10] F. Cabello Sánchez, *Factorization in Lorentz spaces, with an application to centralizers*, J. Math. Anal. Appl. **446** (2017), 1372–1392.

[11] F. Cabello Sánchez, J. M. F. Castillo, *Banach Space Techniques Underpinning a Theory for Nearly Additive Mappings*, Dissertationes Math., **404**, 2002.

[12] F. Cabello Sánchez and J. M. F. Castillo, *Uniform boundedness and twisted sums of Banach spaces*, Houst. J. Math. **30** (2004), 523–536.

[13] F. Cabello Sánchez, J. M. F. Castillo, J. Suárez, *On strictly singular nonlinear centralizers*, Nonlinear Anal. **75** (2012), 3313–3321.

[14] A. Calderón, *Intermediate spaces and interpolation, the complex method*, Stud. Math. **24** (1964), 113–190.

[15] P. G. Casazza, N. J. Nielsen, *A Banach space with a symmetric basis which is of weak cotype 2 but not of cotype 2*, Stud. Math. **157** (2003), 1–16.

[16] P. G. Casazza, T. J. Shura, *Tsirelson's Space*. Lecture Notes in Math., **1363**, Springer-Verlag, 1989.

[17] J. M. F. Castillo, *A variation on Schreier's space*, Riv. Mat. Univ. Parma **2** (1993), 319–324.
[18] J. M. F. Castillo, *The hitchhiker guide to categorical Banach space theory, Part I*, Extr. Math. **25** (2010), 103–149.
[19] J. M. F. Castillo, W. H. G. Correa, V. Ferenczi, M. González, *On the stability of the differential process generated by complex interpolations*, J. Inst. Mat. Jussieu, to appear. arXiv:1712.09647v3.
[20] J. M. F. Castillo, W. Cuellar, V. Ferenczi, Y. Moreno, *Complex structures on twisted Hilbert spaces*, Isr. J. Math. **222** (2017), 787–814.
[21] J. M. F. Castillo, V. Ferenczi, M. Gonzalez, *Singular exact sequences generated by complex interpolation*, Trans. Am. Math. Soc. **369** (2016), 4671–4708.
[22] J. M. F. Castillo, M. González, *An approach to Schreier's space*, Extr. Math. **6** (2–3) (1991), 166–169.
[23] J. M. F. Castillo, M. González, *Three-space Problems in Banach Space Theory*, Springer Lecture Notes in Math., **1667**, 1997.
[24] J. M. F. Castillo, M. González, F. Sánchez, *M-ideals of Schreier type and the Dunford–Pettis property*, in: Non-Associative algebra and its applications, Santos González (ed.), Mathematics and its Applications, **303**, Kluwer Acad. Press, 1994, pp. 80–85.
[25] J. M. F. Castillo, M. González and F.Sánchez, *Oscillation of weakly null and Banach–Saks sequences*, Bull. U.M.I. **11-A** (1997), 685–695.
[26] J. M. F. Castillo, D. Morales, J. Suárez, *Derivation of vector valued complex interpolation scales*, J. Math. Anal. Appl. **468** (2018), 461–472.
[27] J. M. F. Castillo, Y. Moreno, *Strictly singular quasi-linear maps*, Nonlinear Anal. **49** (2002), 897–904.
[28] J. M. F. Castillo, J. Suárez, *Extension of operators into Lindenstrauss spaces*, Isr. J. Math. **169** (2009), 1–27.
[29] W. H. G. Correa, *Twisting operator spaces*, Trans. Am. Math. Soc. **370** (2018), 8921–8957.
[30] W. H. G. Correa, *Complex interpolation of families of Orlicz spaces and the derivation process*, Israel J. Math., to appear. arXiv:1906.01013v2.
[31] W. Cuellar-Carera, *Non-ergodic Banach spaces are near-Hilbert*, Trans. Am. Math. Soc. **370** (2018), 8691–8707.
[32] J. Diestel, H. Jarchow and A. Tonge, *Absolutely Summing Operators*, Cambridge Studies in Advanced Math., **43**, Cambridge University Press, 1995.
[33] S. Eilenberg and S. MacLane, *General theory of natural equivalences*, Trans. Am. Math. Soc. **58** (1945), 231–294.
[34] P. Enflo, J. Lindenstrauss and G. Pisier, *On the "three-space" problem for Hilbert spaces*, Math. Scand. **36** (1975), 199–210.
[35] V. Ferenczi, Ch. Rosendal, *Ergodic Banach spaces*, Adv. Math. **195** (2005), 259–282.
[36] D. J. H. Garling, S. J. Montgomery-Smith, *Complemented subspaces of spaces obtained by interpolation*, J. Lond. Math. Soc. (2) **44** (3) (1991), 503–513.
[37] E. Hilton and K. Stammbach, *A Course in Homological Algebra*, GTM, **4**, Springer-Verlag.
[38] R. C. James, *A non-reflexive Banach space isometric with its second conjugate space*, Proc. Natl. Acad. Sci. USA **37**, (1951), 174–177.
[39] B. Johnson, A. Szankowski, *Hereditary approximation property*, Ann. Math. **176** (2012), 1987–2001.
[40] W. B. Johnson, *Banach spaces all of whose subspaces have the approximation property*, in: Seminaire d'Analyse Fonctionnelle, 79/80, Ecole Polytechnique. Palaiseau. Exp. n° 16. Cf. also, Special Topics of Applied Mathematics. Functional Analysis, Numerical Analysis and Optimization. Proceedings Bonn 1979, edited by J. Frehse, D. Pallaschke, and V. Trottenberg, North Holland, 1980, pp. 15–26.

[41] W. B. Johnson, J. Lindenstrauss and G. Schechtman. *On the relation between several notions of unconditional structure*, Isr. J. Math. **37** (1980), 120–129.

[42] N. J. Kalton, *The three-space problem for locally bounded F-spaces*, Compos. Math. **37** (1978), 243–276.

[43] N. J. Kalton, *Convexity, type and the three-space problem*, Stud. Math. **69** (1981), 247–287.

[44] N. J. Kalton, *The space Z_2 viewed as a symplectic Banach space*, in: Proc. Research Workshop on Banach Space Theory (1981), Univ. of Iowa, 1982, 97–111.

[45] N. J. Kalton, *Nonlinear commutators in interpolation theory*, Mem. Amer. Math. Soc., **385**, 1988.

[46] N. J. Kalton, *Trace-class operators and commutators*, J. Funct. Anal. **86** (1989), 41–74.

[47] N. J. Kalton, *Differentials of complex interpolation processes for Köthe function spaces*, Trans. Am. Math. Soc. **333** (1992), 479–529.

[48] N. J. Kalton, *An elementary example of a Banach space not isomorphic to its complex conjugate*, Can. Math. Bull. **38** (1995), 218–222.

[49] N. J. Kalton, *The basic sequence problem*, Stud. Math. **116** (1995), 167–187.

[50] N. J. Kalton, *Twisted Hilbert spaces and unconditional structure*, J. Inst. Math. Jussieu **2** (2003), 401–408.

[51] N. J. Kalton, *Extension of linear operators and Lipschitz maps into $\mathcal{C}(K)$-spaces*, New York J. Math. **13** (2007), 317–381.

[52] N. J. Kalton, *The complemented subspace problem revisited*, Stud. Math. **188** (2008), 223–257.

[53] N. J. Kalton, *Automorphisms of $\mathcal{C}(K)$ spaces and extension of linear operators*, Ill. J. Math. **52** (2008), 279–317.

[54] N. J. Kalton, S. Montgomery-Smith, *Interpolation of Banach spaces*, Chapter 36 in: Handbook of the Geometry of Banach Spaces, W. B. Johnson and J. Lindenstrauss (eds.), pp. 1131–1175.

[55] N. J. Kalton and N. T. Peck, *Twisted sums of sequence spaces and the three space problem*, Trans. Am. Math. Soc. **255** (1979), 1–30.

[56] D. Leung, *Uniform convergence of operators and Grothendieck spaces with the Dunford–Pettis property*, Math. Z. **197** (1988), 21–32.

[57] J. Lindenstrauss and A. Pełczyński, *Contributions to the theory of the classical Banach spaces*, J. Funct. Anal. **8** (1971), 225–249.

[58] J. Lindenstrauss, L. Tzafriri, *On the complemented subspaces problem*, Isr. J. Math. **9** (1971), 263–269.

[59] G. Ya. Lozanovskii, *On some Banach lattices*, Sib. Math. J. **10** (1969), 584–599.

[60] A. Pietsch, *History of Banach Spaces and Linear Operators*, Birkhauser, 2007.

[61] G. Pisier, *The Volume of Convex Bodies and Banach Space Geometry*, Cambridge Tracts in Mathematics, **94**, Cambridge University Press.

[62] G. Pisier, H. Xu, *Non-commutative Lp-spaces*, Chapter 34 in Handbook in the Geometry of Banach Spaces, vol. 2, W. B. Johnson and J. Lindenstrauss (eds.), Elsevier, 2003, pp. 1459–1518.

[63] M. Ribe, *Examples for the nonlocally convex three space problem*, Proc. Am. Math. Soc. **73** (1979), 351–355.

[64] W. Schachermayer, *The Banach-Saks property is not L_2-hereditary*, Isr. J. Math. **40** (1981), 340–344.

[65] J. Schreier, *Ein Gegenbeispiel zur Theorie der Schwachen Konvergenz*, Stud. Math. **2**, (1930), 58–62.

[66] A. Sobczyk, *On the extension of linear transformations*, Trans. Am. Math. Soc. **55** (1944), 153–169.

[67] J. Suárez de la Fuente, *A weak Hilbert space that is a twisted Hilbert space*, J. Inst. Math. Jussieu, DOI: 10.1017/S1474748018000221.

[68] A. Szankowski, *Subspaces without the approximation property*, Isr. J. Math. **30** (1978), 123–129.
[69] B. Tsirel'son, *Not every Banach space contains an imbedding of l_p or c_0*, Funct. Anal. Appl. **8** (1974), 138–141.
[70] F. Watbled, *Complex interpolation of a Banach space with its dual*, Math. Scand. **87** (2000), 200–210.

Lixin Cheng, Mikio Kato, and Wen Zhang

5 A survey of ball-covering property of Banach spaces

To perpetuate the memory of Victor Lomonosov

Abstract: In this survey paper, we present a brief review of the research area of ball-covering property of Banach spaces; and some new results about this topic are included.

Keywords: Ball-covering property, geometric and topological property, Banach space

MSC 2010: 46B20, 46B04

5.1 Introduction

There are many topics studying representation of subsets of a Banach space by balls. For example, the Mazur intersection property, the sphere packing problem, measure of noncompactness, and the ball topology. (See, for instance, [1, 2, 11, 17, 19, 27, 29, 36, 40].) In this survey, we focus on the recent development of the study of **ball-covering property of Banach spaces** [4], which is also called **spheres covering by balls** [13].

The letter X will always be a Banach space and X^* its dual. We use B_X to denote the closed unit ball, and S_X, the unit sphere of X. We denote by $B(x,r)$ (resp., $\bar{B}(x,r)$) the open (resp., closed) ball centered at x with radius r. For a Banach space X, its unit sphere S_X can certainly be covered by a family $\mathcal{B}$ of open (or, closed) balls of X off the origin. In particular, if X is separable (resp., $\dim X < \infty$), then $\mathcal{B}$ can be chosen consisting of countably (resp., finitely) many balls with arbitrarily small radii. These facts raise the following questions naturally.

Problem 5.1. *For what (nonseparable) Banach spaces X, do there exist coverings $\mathcal{B}$ consisting of countably many balls of X off the origin?*

Acknowledgement: Support by National Natural Science Foundation of China, No. 11731010, and in part by JSPS KAKENHI Grant No. 16K05207.
The authors would like to thank the referee for the comments and advice which made this article better and the Functional Analysis group at Xiamen University for their very helpful conversations and suggestions.

Lixin Cheng, Wen Zhang, School of Mathematical Science, Xiamen University, Xiamen 361005, P. R. China, e-mails: lxcheng@xmu.edu.cn, wenzhang@xmu.edu.cn
Mikio Kato, Faculty of Engineering, Kyushu Institute of Technology, Kitakyushu 804-8550, Japan, e-mail: katom@shinshu-u.ac.jp

https://doi.org/10.1515/9783110656756-005

A Banach spaces X admitting a unit sphere covering $\mathcal{B}$ consisting of countably many balls off the origin is said to have the **ball-covering property** (or BCP for short). Since every open ball B off the origin is the union of a sequence of closed balls contained in B, and since every closed ball off the origin is contained in an open ball off the origin, for an infinite dimensional Banach space, we can blur the distinction between "open" and "closed" balls in a ball-covering $\mathcal{B}$ of S_X.

The following example shows that a nonseparable space may have the ball-covering property.

Example 5.2. *Suppose that $X = \ell_\infty$, and $e_n = \chi_{\{n\}} \in \ell_\infty$ for $n = 1, 2, \ldots$. Then for every $0 < \delta < 1$ and $n \in \mathbb{N}$, let $x_n = (1+\delta)e_n$. Then $\mathcal{B} = \{\bar{B}(\pm x_n, 1) : n \in \mathbb{N}\}$ is a ball-covering of S_{ℓ_∞}.*

Problem 5.3. *If $\dim(X) = n \in \mathbb{N}$, what is the least number $\kappa = \min \mathcal{B}^\sharp$ for all ball-coverings $\mathcal{B}$ of X? (where $\mathcal{B}^\sharp$ denotes the cardinality of $\mathcal{B}$).*

For a convex function f defined on a Banach space X, its subdifferential mapping $\partial f : X \to 2^{X^*}$ is defined for $x \in X$ by

$$\partial f(x) = \{x^* \in X^* : f(y) - f(x) \geq \langle x^*, y - x\rangle, \text{ for all } y \in X\}. \tag{5.1}$$

If f is continuous, then for each $x \in X$, $\partial f(x)$ is a nonempty w^*-compact convex set (see, for instance, [30]). We say a (single-valued) mapping $\varphi : X \to X^*$ is a selection of the (set-valued) subdifferential mapping $\partial f : X \to 2^{X^*}$ if $\varphi(x) \in \partial f(x)$ for all $x \in X$. Please note that a continuous convex function $f : X \to \mathbb{R}$ is Gâteaux differentiable at x if and only if $\partial f(x)$ is a singleton. In this case, we have $\partial f(x) = \{df(x)\}$, where $df(x) \in X^*$ denotes the Gâteaux derivative of f at x.

The following theorem [9] is a key characterization in study of the ball-covering property.

Theorem 5.4 ([9]). *Suppose that X is a Banach space, and $\{z_n\} \subset S_X$. Then $\mathcal{B} \equiv \{B(x_n, r_n)\}$ forms a ball-covering of X for some $x_n \in \mathbb{R}^+ z_n$ with $\|x_n\| \geqslant r_n$ for all $n \in \mathbb{N}$ if and only if for every selection φ for the subdifferential mapping $\partial\|\cdot\|$ of the norm $\|\cdot\|$, $\{\varphi(x_n)\}$ positively separates points of X, that is, $\sup_{n\in\mathbb{N}}\langle\varphi(x_n), x\rangle > 0$ for every $x \neq 0$ in X.*

Proof. Sufficiency. Without loss of generality, we can assume that $\|z_n\| = 1$ and $r_n = \|x_n\|$ for all $n \in \mathbb{N}$. Fix a selection ψ for the subdifferential mapping $\partial\|\cdot\|$. Suppose, to the contrary, that there exists $y \in S_X$ such that $y \notin B(kz_n, k)$ for all $k, n \in \mathbb{N}$; that is,

$$\|kz_n - y\| \geq k, \quad \text{for all } k, n \in \mathbb{N}. \tag{5.2}$$

Equivalently,

$$\frac{\|z_n - \frac{1}{k}y\| - \|z_n\|}{\frac{1}{k}} \geq 0, \quad \text{for all } k, n \in \mathbb{N}. \tag{5.3}$$

Note that $\|\psi(x)\| = 1$ and $\langle \psi(x), x\rangle = \|x\|$ for all $0 \neq x \in X$. Then it follows from (5.3) that

$$\left\langle \psi\left(z_n - \frac{1}{k}y\right), -y \right\rangle \geq \frac{\|z_n - \frac{1}{k}y\| - \|z_n\|}{\frac{1}{k}} \geq 0, \quad \text{for all } k, n \in \mathbb{N}. \tag{5.4}$$

Therefore,

$$\left\langle \psi\left(z_n - \frac{1}{k}y\right), y \right\rangle \leq 0, \quad \text{for all } k, n \in \mathbb{N}. \tag{5.5}$$

For each fixed $n \in \mathbb{N}$, let z_n^* be a w^*-cluster point of $(\psi(z_n - \frac{1}{k}y))_{k=1}^{\infty}$. Then $z_n^* \in \partial\|z_n\|$ and with $\langle z_n^*, y\rangle \leq 0$. Now, we define a new selection for $\partial\|\cdot\|$ by

$$\varphi(x) = \begin{cases} \psi(x), & x \neq z_n, \\ z_n^*, & x = z_n. \end{cases}$$

Thus, $\{\varphi(z_n)\}$ does not positively separate points of X.

Necessity. Let $\mathcal{B} = \{B(x_n, r_n)\}$ be a ball-covering of S_X. Then for each $y \in S_X$, there is $m \in \mathbb{N}$ such that $y \in B(x_m, \|x_m\|)$. Consequently,

$$\frac{\|x_m - ty\| - \|x_m\|}{t} \leq \|x_m - y\| - \|x_m\| \equiv -\delta, \quad \text{for some } \delta > 0 \text{ and for all } 0 < t \leq 1. \tag{5.6}$$

Since

$$\max_{x^* \in \partial\|x_m\|} \langle x^*, -y\rangle = \lim_{t \to 0^+} \frac{\|x_m - ty\| - \|x_m\|}{t} \leq -\delta,$$
$$\min_{x^* \in \partial\|x_m\|} \langle x^*, y\rangle \geq \delta. \tag{5.7}$$

Clearly, (5.7) is equivalent to that for every selection φ of the subdifferential mapping $\partial\|\cdot\|$, $\langle \varphi(x_m), y\rangle \geq \delta > 0$. □

As an example of application of Theorem 5.4, we will show the following result.

Example 5.5. *Let X be the n-dimensional Euclidean space $\mathbb{R}^n$. Then the least number κ of balls in a ball-covering of S_X is $n + 1$.*

Proof. Note that the norm $\|\cdot\|$ is everywhere differentiable in $X \setminus \{0\}$, and with $d\|x\| = x/\|x\|$ for all $x \neq 0$. We first show that $\kappa \geq n + 1$. Suppose that $\mathcal{B} = \{B(x_j, r_j)\}_{j=1}^m$ with $\kappa = m$ is a minimal ball-covering of S_X. Then by Theorem 5.4, $\{\varphi_j\}_{j=1}^m$ positively separates points of X, where $\varphi_j \equiv d\|x_j\| = x_j/\|x_j\|$ for all $1 \leq j \leq m$. Consequently, $m \geq n + 1$.

On the other hand, let $(e_j)_{j=1}^n$ be the standard unit vector basis of X, and let $e_0 = -\frac{1}{\sqrt{n}}\sum_{j=1}^n e_j$. Then $\{e_j\}_{j=0}^n$ positively separates points of X. By Theorem 5.4 again, S_X admits a ball-covering of $n + 1$ balls. Hence, $\kappa \leq n + 1$. □

5.2 Minimal ball-coverings of finite dimensional spaces

For a Banach space X, we denote by $\mathscr{B}(X)$ the collection of all ball-coverings $\mathcal{B}$ of the sphere S_X. If $\mathcal{B}_0 \in \mathscr{B}(X)$ satisfies $\mathcal{B}_0^\sharp = \min\{\mathcal{B}^\sharp : \mathcal{B} \in \mathscr{B}(X)\}$, that is, the cardinality $\mathcal{B}_0^\sharp$ of $\mathcal{B}_0$ is the minimum in $\{\mathcal{B}^\sharp : \mathcal{B} \in \mathscr{B}(X)\}$, then we say that $\mathcal{B}_0$ is a minimal ball-covering. We also use $\mathcal{B}_{\min}$ to denote a minimal ball-covering of S_X, and $\kappa \equiv \kappa(X) = \mathcal{B}_{\min}^\sharp$, the smallest cardinality of all ball-coverings.

5.2.1 κ of n-dimensional spaces

For ℓ_p spaces of n-dimension ($n > 1$), we have

Theorem 5.6. *Let $X = \ell_p^n$. Then:*
(i) [4] $\kappa = n + 1$, *if* $1 < p < \infty$;
(ii) [18] $\kappa = 4$ *when* $n = 2$ *and* $\kappa = n + 1$ *when* $n > 2$, *if* $p = 1$;
(iii) [4], $\kappa = 2n$, *if* $p = \infty$.

Remark 5.7. Theorem 5.6(ii) was partly due to Zhifang Hu and Xin Zhao in an unpublished paper [18]. Since all exposed points of the dual unit ball $B_{\ell_\infty^n}$ of ℓ_1^n are just $\sum_{j=1}^n \pm e_j$, they use Theorem 5.4 to show that every minimal subset of the 2^n point set $\{\sum_{j=1}^n \pm e_j\}$ positively separating points of ℓ_1^n contains just $n + 1$ elements when $n > 2$.

Theorem 5.8 ([4]). *Suppose X is an n dimensional space. Then:*
(i) $2n \geq \kappa \geq n + 1$;
(ii) $\kappa = n + 1$, *if X is smooth and*
(iii) $\kappa = 2n$ *if and only if X is isometric to* ℓ_∞^n.

Theorem 5.9 ([6]). *Let $n, k \in \mathbb{N}$ with $n + 1 \leq k \leq 2n$. Then there is an n-dimensional space X such that $\kappa(X) = k$.*

5.2.2 On optimal radii of minimal ball-coverings

For a ball-covering $\mathcal{B} = \{B(x_i, r_i)\}_{i\in I}$ of S_X, the number $r_\mathcal{B} \equiv \sup_{i\in I} r_i$ is called the radius of $\mathcal{B}$. We use $\gamma_{\inf}$ to denote the exact lower bound of the radius set $\{r_\mathcal{B} : \mathcal{B}$ is minimal$\}$ of minimal ball-coverings.

Lemma 5.10 ([24]). *Let Ω be a n-simplex with vertex-set $\{x_i\}_{i=0}^n$ in the Euclidean space $\mathbb{R}^n$ ($n \geqslant 2$) with $0 \in \operatorname{int}(\Omega)$, S be the circumhypersphere of Ω, and r be the radius of S, and let $\rho_{ij} = \|x_i - x_j\|$. Similarly, let Ω_j be the n-simplex $\operatorname{co}[\{x_i\}_{i\neq j} \cup \{0\}]$, S_j be the circumhypersphere of Ω_j. r_j be the radius of S_j and d_j be the distance from 0 to M_j =*

$\mathrm{co}(x_i)_{i\neq j}$. *Then*

$$r^2 = -\frac{1}{2}\frac{D_0(x_0, x_1, \ldots, x_n)}{D(x_0, x_1, \ldots, x_n)},$$

where

$$D_0 \equiv D_0(x_0, x_1, \ldots, x_n) = \begin{vmatrix} 0 & \rho_{01}^2 & \rho_{02}^2 & \cdots & \rho_{0n}^2 \\ \rho_{10}^2 & 0 & \rho_{12}^2 & \cdots & \rho_{1n}^2 \\ \cdots & & \cdots & & \cdots \\ \rho_{n0}^2 & \rho_{n1}^2 & \rho_{n2}^2 & \cdots & 0 \end{vmatrix},$$

and

$$D \equiv D(x_0, x_1, \ldots, x_n) = \begin{vmatrix} 0 & 1 & 1 & \cdots & 1 \\ 1 & 0 & \rho_{01}^2 & \cdots & \rho_{0n}^2 \\ \cdots & & \cdots & & \cdots \\ 1 & \rho_{n0}^2 & \rho_{n1}^2 & \cdots & 0 \end{vmatrix},$$

$r_j = \frac{r^2}{2d_j}$ $(j = 0, 1, \ldots, n)$, $r^{n+1} \leqslant (\frac{2}{n})^{n+1} \prod_{j=0}^{n} r_j$. *In particular,* $\rho \equiv \rho_{ij} = \sqrt{\frac{2(n+1)}{n}}r$ $(0 \leqslant i, j \leqslant n)$ *whenever* Ω *is a regular n-simplex.*

The following result is due to Xiaojing Zhang [39].

Theorem 5.11 ([39]). *Let* X *be the Euclidean space of* n *dimension with* $n \geqslant 2$, *and* $\{x_i\}_{i=0}^n$ *be the vertex-set of an inscribed regular simplex of sphere* $\frac{n}{2}S_X$. *Then* $S_X \subset \bigcup_{i=0}^n \bar{B}(x_i, \frac{n}{2})$.

Proof. Let $y_i = \frac{2}{n}\{x_i\}_{i=0}^n$. Then $\{y_i\}_{i=0}^n$ is the vertex-set of inscribed regular simplex Ω of S_X. It follows from Lemma 5.10

$$\rho = \rho_{ij} = \|y_i - y_j\| = \sqrt{\frac{2(n+1)}{n}}, \quad 0 \leqslant i \neq j \leqslant n.$$

Since every $\Omega_i = \mathrm{co}[\{y_l\}_{l\neq j} \cup \{0\}]$ has a circumhypersphere which is the sphere of $B(rz_i, r)$ with $z_i \in S_X$,

$$r^2 = -\frac{1}{2}\frac{D_0(y_0, y_1, \ldots, y_n)}{D(y_0, y_1, \ldots, y_n)},$$

where

$$D_0 \equiv D_0(0, y_1, \ldots, y_n) = -n(-\rho^2)^{n-1},$$

and

$$D \equiv D(0, y_1, \ldots, y_n) = (-\rho^2)^{n-1}[2n - (n-1)\rho^2],$$

we obtain $r = \sqrt{\frac{-D_0}{2D}} = \frac{n}{2}$. Let $M_j = \mathrm{co}(y_i)_{i\neq j}$ and $d_j = \mathrm{dist}(0, M_j)$, then $r = \frac{1}{2d_j}$, $d_j = \frac{1}{n}$.

Now we consider $\Omega' \equiv \mathrm{co}(z_i)_{i=0}^n$, which is an inscribed regular n-simplex with $M_j' = \mathrm{co}(z_i)_{i\neq j}$, $\Omega_j' = \mathrm{co}(z_i)_{i\neq j} \cup (0)$ and $d_j' = \mathrm{dist}(0, M_j)$ $(= d = \frac{1}{n})$, $0 \leqslant i \leqslant n$.

Now, we will show that $S_X \subset \bigcup_{i=0}^n \bar{B}(\frac{n}{2}z_i, \frac{n}{2})$. Suppose, to the contrary, there exists $x \in S_X \setminus \bigcup_{i=0}^n \bar{B}(\frac{n}{2}z_i, \frac{n}{2})$. Then $\|x - \frac{1}{n}z_i\| > \frac{1}{n}$ for all $0 \leqslant i \leqslant n$. This implies that $\langle x, z_i\rangle < \frac{1}{n}$, $i = 0, 1, \dots, n$. Therefore,

$$\frac{1}{n} \leqslant \max_{y\in\Omega'}\langle x, y\rangle = \max_{y\in \mathrm{co}(z_i)_{i=0}^n}\langle x, y\rangle = \max_{0\leqslant i\leqslant n}\langle x, z_i\rangle < \frac{1}{n},$$

this is a contradiction! Hence, $S_X \subset \bigcup_{i=0}^n \bar{B}(\frac{n}{2}z_i, \frac{n}{2})$. Since Ω is isometric to Ω', $S_X \subset \bigcup_{i=0}^n \bar{B}(x_i, \frac{n}{2})$. □

Corollary 5.12. *Let X be the Euclidean space of n dimension with $n \geqslant 2$. Then $\gamma_{\inf} \leq \frac{n}{2}$.*

In 2008, Guochen Lin and Xisheng Shen [20] considered the minimum radius problem of ball-coverings $\mathcal{B}$ with the cardinality $\mathcal{B}^\sharp = m$ of the n dimensional Euclidean space by the neural network method. They gave a new computing formula for the minimum radius $\gamma_{\inf}$. As a consequence, they showed the following result lemma.

Lemma 5.13 ([16]). *Let $\Omega = \mathrm{co}\{x_i\}_{i=1}^{n+1}$ be a n-simplex of the n dimensional Euclidean space, R be the radius of its circumhypersphere, $P \in \mathrm{int}\,\Omega$ (the interior of Ω) and d_i be the distance from P to $\Omega_i = \mathrm{co}\{x_j : 1 \leqslant j \leqslant n+1, j \neq i\}$. Then $\prod_{i=1}^{n+1} d_i \leqslant (\frac{R}{n})^{n+1}$, and the equality holds if and only if Ω is a regular simplex.*

Theorem 5.14 ([20]). *Let X be the Euclidean space of n dimension with $n \geqslant 2$. Then $\gamma_{\inf} = \frac{n}{2}$.*

Sketch of the proof. It suffices to prove

$$\text{(i)} \qquad \max_{\{x_i\}_{i=1}^{n+1}\subset S_{\mathbb{R}^n}} \min_{\|x\|=1} \max_{1\leqslant i\leqslant n+1} \langle x_i, x\rangle \leq \frac{1}{n}$$

for every sequence $\{x_i\}_{i=1}^{n+1} \subset S_{\mathbb{R}^n}$ with $0 \in \mathrm{int}\,\mathrm{co}\{x_i\}_{i=1}^{n+1}$, and (ii) the maximum is attained if and only if $\mathrm{co}\{x_i\}_{i=1}^{n+1}$ is a regular simplex.

Let $\Omega = \mathrm{co}\{x_i\}_{i=1}^{n+1}$, and $d_i = d(0, \Omega_i)$. Then by Lemma 5.13, we have $d_{i_0} \equiv \min_{1\leqslant i\leqslant n+1} d_i \leqslant \frac{1}{n}$. Let w be the best approximate point of 0 in Ω_{i_0} and $x = \frac{w}{\|w\|}$, we have $\langle x_{i_0}, w\rangle \leqslant 0$ and $\max_{1\leqslant i\leqslant n+1}\langle x_i, x\rangle = d_{i_0} \leqslant \frac{1}{n}$. Then $d_{i_0} < \frac{1}{n}$ when Ω is not regular and $d_{i_0} = \frac{1}{n}$ when Ω is regular. Now, choose $\{B(\frac{n}{2}x_i, \frac{n}{2})\}_{i=1}^{n+1}$ and we have $\gamma_{\inf} = \frac{n}{2}$. □

Remark 5.15. Recently, Zeyu Chen (in a private communication) further proved that for the n dimensional Euclidean space $\mathbb{R}^n$, the exact lower bound $\gamma_{\inf}$ cannot be attained.

Remark 5.16. It follows from Theorem 5.6(iii) and Example 5.2, $\gamma_{\inf}(\ell_\infty^n) = 1$ when $n > 2$. But apart from this and Euclidean spaces, we know little about $\gamma_{\inf}$ for a general finite dimensional space X. We do not even know $\gamma_{\inf}$ of ℓ_p^n with $1 \leq p \neq 2 < \infty$.

Problem 5.17. *For* $X = \ell_p^n$, $\gamma_{\inf} = ?$

Remark 5.18. For ball-covering of finite dimensional normed spaces, [15], [22], [23], and [35] also contain some interesting results. For example, L. Lin, F. Zhang, and M. Zhang [23] showed every n-dimensional normed space with a minimal ball-covering of $2n-1$ balls contains an $(n-1)$-dimensional subspace isometric to $(\mathbb{R}^{n-1}, \|\ \|_\infty)$. A ball-covering $\mathcal{B}$ of S_X is said to be symmetric provided $B \in \mathcal{B}$ implies $-B \in \mathcal{B}$. About symmetric ball-coverings, it was shown by Cheng and Fu [15] that for every n-dimensional normed space X, each minimal symmetric ball-covering contains exactly $2n$ balls. They also showed that if $X = \mathbb{R}^n$, the Euclidean space of n-dimension, the least bound of radii of minimal symmetric ball-coverings is just $\frac{\sqrt{n}}{2}$, and it can never be attained. The BCP was also discussed in hereditarily indecomposable spaces [21].

5.3 Separability and BCP

We know that a Banach space X is separable if and only if for every $\varepsilon > 0$ there is a ball-covering $\mathcal{B}$ consisting of countably many balls with radius $r(\mathcal{B}) < \varepsilon$. Therefore, every separable Banach space admits the BCP. If X is not separable, then for every $\varepsilon > 0$ there exists an uncountable net $\{x_\iota\} \subset S_X$ such that

$$\|x_\xi - x_\eta\| > 1 - \varepsilon \quad \text{for all } \xi \neq \eta,$$

and this implies that if X has a ball-covering $\mathcal{B}$ of countable balls with $r(\mathcal{B}) < 1/2$, then it is separable. On the other hand, by the separation theorem of convex sets, we observe that a Banach space X admitting the BCP entails that its dual X^* is w^*-separable. Thus, for a reflexive Banach space X, it admits the BCP if and only if it is separable. In this section, we focus the following two questions.

Question 5.19. Whether there is a number $r_0 > 0$ satisfying
(i) every Banach space X with a countable ball-covering $\mathcal{B}$ with $r(\mathcal{B}) < r_0$ is separable;
(ii) for all $r > r_0$ there is a nonseparable X admitting a countable ball covering $\mathcal{B}$ such that $r(\mathcal{B}) < r$.

Question 5.20. If X^* is w^*-separable, can we renorm X such that X has the BCP with respect to the new norm?

For a collection $\mathcal{B}$ of balls, if the radius of each ball in $\mathcal{B}$ is less or equal to r, we also denote it by $\mathcal{B}(r)$. The following theorem, incorporating Example 5.2 gives an affirmative answer to Question 5.19 by $r_0 = 1$.

Theorem 5.21 ([4]). *Suppose* $0 < r < 1$. *If* X *admits a countable ball-covering* $\mathcal{B}$ *with* $r(\mathcal{B}) \leq r$, *then* X *is separable.*

Proof. Let $\mathcal{B} = \{B(x_n, r_n)\}_{n=1}^{\infty}$ be a ball-covering of S_X with $0 < r_n \le r$ for all $n \in \mathbb{N}$. Let $\eta = \frac{1-r}{2}$, and $r_\eta = r + \eta$. Then $\mathcal{B}_1(r_\eta) \equiv \{B(x_n, r_n + \eta)\}_{n=1}^{\infty}$ covers $B_X \setminus (1-\eta)B_X$. Consequently, there is a collection $\mathcal{B}(r_\eta)$ of countably many balls whose union contains B_X. Since each ball in $\mathcal{B}(r_\eta)$ can be covered by a sequence of balls with radii less or equal to r_η^2, B_X can be covered by countable collection $\mathcal{B}_2(r_\eta^2)$ of balls with radii at most r_η^2. Inductively, for every $n \in \mathbb{N}$ there is a countable collection $\mathcal{B}_n(r_\eta^n)$ of balls with radii at most r_η^n. Since $r_\eta^n \to 0$ (as $n \to \infty$), we obtain that B_X can be covered by a sequence of balls with arbitrarily small radii, which implies that X is separable. □

Theorem 5.22. *Suppose that X is a (Gâteaux) smooth Banach space. Then it has the BCP if and only if X^* is w^*-separable.*

Proof. It suffices to show sufficiency. Since X is a Gâteaux smooth, every norm-attaining functional $x^* \in S_{X^*}$ is a w^*-exposed point of B_{X^*}, that is, there exists $x \in S_X$ such that $d\|x\| = x^*$ (see, for instance, [30]). By the Bishop–Phelps theorem [3], norm-attaining functionals are dense in X^*. Since X^* is w^*-separable, there exists a sequence $\{z_n^*\} \subset S_{X^*}$, which positively separates points of X. Consequently, there is a sequence $\{y_n^*\}$ of w^*-exposed points of B_{X^*} which positively separates points of X. Let $\{y_n\} \subset S_X$ be such that $d\|y_n\| = y_n^*$ for all $n \in \mathbb{N}$. By Theorem 5.4, there is a ball-covering $\mathcal{B} = \{B(x_n, r_n)\}$ with $x_n \in \mathbb{R}^+ y_n$ for all $n \in \mathbb{N}$. □

The answer to Problem 5.20 is also affirmative. The following theorem is due to [9], where a generalized Pełczyński lemma for bi-orthogonal systems is used.

Theorem 5.23 ([9]). *Suppose that X is a Banach space with a w^*-separable dual. Then for every $\varepsilon > 0$ there exists an equivalent norm $|\cdot|$ on X such that:*
(i) $(1+\varepsilon)^{-1}\|x\| \le |x| \le (1+\varepsilon)\|x\|$ *for all $x \in X$;*
(ii) *X has the ball-covering property with respect to $|\cdot|$.*

In 2009, V. P. Fonf and C. Zanco [13] showed independently that a Banach space X with a w^*-separable dual can be renormed such that X admits a countable ball-covering consisting of uniformly bounded balls.

Theorem 5.24 ([13]). *For a Banach space X, the following assertions are equivalent:*
(i) *X is w^*-separable.*
(ii) *For any $\varepsilon > 0$ there are a $(1+\varepsilon)$-equivalent norm on X and $R > 0$, such that with respect to the new norm, there is a countable ball-covering $\mathcal{B}$ with $r(\mathcal{B}) \le R$.*

Cheng, Kadets, Wang, and Zhang [7] further proved a sharp quantitative version of the Fonf and Zanco renorming theorem above.

Theorem 5.25 ([7]). *Let X be a Banach space with a w^*-separable dual. Then for every $0 < \varepsilon < \frac{1}{3}$ there is an equivalent norm $|\cdot|$ satisfying $\|\cdot\| \le |\cdot| \le (1+\varepsilon)\|\cdot\|$ such that there is a countable ball-covering $\mathcal{B}(r(\varepsilon))$ with the radius $r(\varepsilon) \le \frac{1+\varepsilon}{\varepsilon}$.*

They also showed in [7] that for sufficiently small $0 < \varepsilon < \frac{1}{3}$ the Banach space $L[0,1]$ admits a $(1+\varepsilon)$-equivalent norm $|\cdot|$ such that with respect to the new norm, there is a countable ball-covering with its radius $r(\varepsilon)$ being arbitrarily close to $\frac{1}{\varepsilon}$, that is, $\varepsilon r(\varepsilon) \to 1$ as $\varepsilon \to 0$. But $\frac{1}{\varepsilon}$ can never be attained.

5.4 Some geometric and topological properties characterized by BCP

In this section, we will see that some geometric and topological properties of Banach spaces can be described by ball-coverings of its finite dimensional subspaces. For a ball-covering $\mathcal{B}$ of a Banach space X, we say that it is α-off the origin for some $\alpha > 0$ if $\inf_{b\in B} \|b\| \geq \alpha$ for every $B \in \mathcal{B}$.

A Banach space Y is said to be finitely representable in a Banach space X, if for every $\varepsilon > 0$ and for every subspace $Y_0 \subset Y$ of finite dimension, there exist a subspace X_0 of X and a linear isomorphism $T : Y_0 \to X_0$ such that $\|T\|\|T^{-1}\| < 1+\varepsilon$. X is said to be superreflexive if every Banach space Y is reflexive whenever it is finitely representable in X.

Theorem 5.26 ([6]). *Suppose that X is a Banach space. Then it is superreflexive if and only if there exists an equivalent norm on X such that (with respect to the new norm) there are positive-valued functions $f, g : N \mapsto R^+$ such that for every $n \in N$ and every n dimensional subspace Y, there is a minimal ball-covering $\mathcal{B}$ of Y satisfying:*
(i) $\mathcal{B}^{\#} = n+1$;
(ii) $r(\mathcal{B}) \leq f(n)$;
(iii) *$\mathcal{B}$ is $g(n)$-off the origin.*

Theorem 5.27. *Let X be a Banach space satisfying that its norm is densely Gâteaux differentiable. Then X has BCP if and only if there is a sequence $\{x_n^*\}$ of w^*-exposed points of B_{X^*} which is positively separates points of X.*

Proof. Sufficiency. Note that for every $x_m^* \in \{x_n^*\}$ there exists $x_m \in S_X$ such that $d\|x_m\| = x_m^*$, and note $\partial\|x_m\| = \{d\|x_m\|\}$. We are done by Theorem 5.4.

Necessity. Suppose that $\mathcal{B} = \{B(x_n, r_n)\}_{n=1}^{\infty}$ is a ball-covering of S_X with $\|x_n\| \geq r_n$ for all $n \in \mathbb{N}$. Since smooth points of X (i. e., Gâteaux differentiability points of the norm) are dense in X, for each x_n there exists a sequence $(x_{nj}) \subset X$ such that $x_{nj} \to x_n$. Clearly, $B(x_n, r_n) \subset \bigcup_{j=1}^{\infty} B(x_{nj}, \|x_{nj}\|)$. This entails that $\mathcal{B}_1 \equiv \{B(x_{ij}, \|x_{ij}\|)\}_{i,j\in\mathbb{N}}$ is again a ball-covering of S_X, and each ball in $\mathcal{B}_1$ is centered at a smooth point of X. Let $\varphi_{ij} = d\|x_{ij}\|$. Then, again by Theorem 5.4, $\{\varphi_{ij}\}_{ij,\in\mathbb{N}}$ positively separates points of X. □

Recall that a Banach space X is said to be a Gâteaux differentiability space if every continuous convex function defined on X is densely Gâteaux differentiable. Separable

Banach spaces and Gâteaux smoothable Banach spaces are Gâteaux differentiability spaces (see, for instance, [30]).

Corollary 5.28. *Suppose that X is a Gâteaux differentiability space. Then it admits BCP if and only if there exists a sequence of w^*-exposed points of B_{X^*} which positively separates points of X.*

A Banach space X is said to be (Gâteaux) smooth provided for each $x(\neq 0) \in X$ there exists $x^* \in X^*$ such that

$$\lim_{t\to 0}\frac{\|x+ty\|-\|x\|}{t} = \langle x^*, y\rangle \quad \text{for all } y \in X.$$

X is called uniformly smooth provided $\frac{\rho(\tau)}{\tau} \to 0$, as $\tau \to 0^+$, where ρ is defined for $\tau \in (0,\infty)$ by

$$\rho(\tau) = \sup\left\{\frac{\|x+\tau y\| + \|x-\tau y\| - 2}{2} : x, y \in S_X\right\}.$$

Smooth and uniformly smooth Banach spaces can also be characterized by behavior of ball-coverings of their finite dimensional subspaces.

Theorem 5.29 ([10]). *A Banach space X is Gâteaux smooth if and only if for every $n \in \mathbb{N}$, for $\{x_j^*\}_{j=0}^n \subset S_{X^*}$ and $\{x_j\}_{j=0}^n \subset S_X$ satisfying $\langle x_j^*, x_j\rangle = 1$, and for every subspace $Y \supset \{x_j\}_{j=0}^n$ positively separated by $\{x_j^*\}_{j=0}^n$, there is a ball-covering $\{B(y_j, r_j)\}_{j=0}^n$ of S_Y, with $y_j \in R^+x_j$, $0 \le j \le n$.*

A bounded set $A \subset X^*$ is said to be an α-norming set for some $1 \ge \alpha > 0$ if

$$\alpha\|x\| \le \sigma_A(x) \equiv \sup_{x\in A}\langle x^*, x\rangle \le \alpha^{-1}\|x\| \quad x \in X.$$

A 1-norming set is said to be a norming set.

Theorem 5.30 ([10]). *A Banach space X is uniformly smooth if and only if for every $n \in \mathbb{N}$, and for $\{x_j^*\}_{j=0}^n \subset S_{X^*}$ and $\{x_j\}_{j=0}^n \subset S_X$ satisfying $\langle x_j^*, x_j\rangle = 1$, and for every subspace $Y \supset \{x_j\}_{j=0}^n$ such that $\{x_j^*\}_{j=0}^n$ is α-norming Y, there is a ball-covering $\{B(y_j, \|y_j\| - \alpha/2)\}_{j=0}^n$ of Y, with $y_j \in R^+x_j$, $0 \le j \le n$.*

A Banach space X is said to be uniformly nonsquare provided there exists $\varepsilon > 0$ so that $\min\|x \pm y\| \le 1 - \varepsilon$ for every pair $x, y \in S_X$, which is equivalent to that ℓ_1^2 is not representable in X. The following result is due to [8].

Theorem 5.31. *A Banach space X is uniformly nonsquare if and only if there exist two positive numbers α and β such that for every two-dimensional subspace U of X there is a ball-covering $\mathcal{B}_U$ of S_U satisfying*

$$\mathcal{B}_U^\sharp = 3, \quad \mathcal{B}_U \text{ is } \alpha\text{-off the origin; and } r(\mathcal{B}_U) \le \beta.$$

We say that a Banach space is a universal finite representability space (UFRS) if every Banach space is finitely representable in it. For example, c_0, and the (reflexive) space $(\sum_{n\geq 1}\bigoplus l_n^n)_2$ of ℓ_2-sum of the n dimensional spaces ℓ_n^n are universal finite representability spaces.

Theorem 5.32 ([38]). *A Banach space X is not a UFRS if and only if there exist $n \geq 2$ and $\alpha, \beta > 0$ satisfying that for every n-dimensional subspace Y of X, there is a minimal ball-covering $\mathcal{B}$ of S_Y such that:*
(i) $\mathcal{B}^\# \leq 2n-1$;
(ii) $r(\mathcal{B}) \leq \beta$;
(iii) $\mathcal{B}$ is α-off the origin.

The concept of B-convex was first introduced by Anatole Beck in 1962 to characterize the existence of strong law of large numbers for random variables which take values in Banach spaces.

A Banach space X is said to be *B-convex* if there exists $n \in \mathbb{N}$ and $\varepsilon > 0$ such that for any $\{x_j\}_{j=1}^n \subset S_X$, we have

$$\min\left\{\left\|\sum_{j=1}^{n}\theta_j x_j\right\| : \theta_j \in \{\pm 1\}, 1 \leq j \leq n\right\} < n-\varepsilon,$$

or equivalently, ℓ_1^n is not finitely representable in X.

Theorem 5.33 ([38]). *A Banach space X is B-convex if and only if there exist $n \in \mathbb{N}$ and $\alpha, \beta > 0$ such that for every n dimensional subspace Y of X, there is a minimal ball-covering $\mathcal{B}$ of X^*/Y^0 ($Y^0 = \{x^* \in X^* : \langle x^*, x\rangle = 0$ for all $x \in Y\}$) satisfying:*
(i) $\mathcal{B}^\# \leq 2n-1$;
(ii) $r(\mathcal{B}) \leq \beta$; and
(iii) $\mathcal{B}$ is α-off the origin.

The notion of topology b_X for a Banach space X was introduced by G. Godefroy and N. J. Kalton [17] for discussing the question of uniqueness of a compact Hausdorff “consistent” topology and relations to the unique predual property. A point $x_0 \in X$ has a b_X-base of neighborhoods of the form

$$V = X \setminus \bigcup_{j=1}^{n} \bar{B}(x_j, r_j),$$

where $n \in \mathbb{N}$ and $x_1, x_2, \ldots, x_n \in X$ with $\|x_0 - x_j\| > r_j$. The ball topology has the following property [17, p. 197]:
(1) For fixed $y \in X$, the map $x \to x + y$ is b_X-continuous;
(2) For fixed $\lambda > 0$ the map $x \to \lambda x$ is b_X-continuous;
(3) The map $x \to -x$ is b_X-continuous.

We say that a point x_0 in a topological space is a G_δ provided there is a sequence $\{V_n\}$ of open neighborhoods of x_0 such that $\{x_0\} = \bigcap_{n\in\mathbb{N}} V_n$.

Theorem 5.34 ([10]). *A Banach space X endowed with the ball topology b_X satisfies that each point of X is a b_X-G_δ-point if and only if X admits the BCP. Therefore, every b_X-compact set in a Banach space with the BCP is b_X-sequentially compact.*

Remark 5.35. Since the BCP is not invariant under linear isomorphisms (see the next section), the b_X-G_δ-property of points in a Banach space X is not topologically invariant.

B. Wang [37] showed that if a Banach space X with a w^*-separable dual admits the Mazur intersection property, then it has the BCP.

5.5 On stability of the BCP

5.5.1 Negative results

We have already known that the BCP of Banach spaces may have many interesting geometric and topological properties. However, the BCP of a Banach space is not invariant under linear isomorphisms, and it is not inherited by its closed subspaces.

Let $X = \ell_\infty$. Then the norm of the quotient space ℓ_∞/c_0 is

$$\|x\|_Q \equiv \limsup_n |x(n)|, \quad \text{for all } x = (x(n)) \in \ell_\infty.$$

For fixed $0 < \lambda \leq 1$, let

$$\|x\|_\lambda = \lambda\|x\| + (1-\lambda)\|x\|_Q, \quad \text{for all } x \in \ell_\infty.$$

Then $\|\cdot\|_\lambda$ is an equivalent norm on ℓ_∞.

Theorem 5.36 ([5]).
(i) $(\ell_\infty, \|\cdot\|_\lambda)$ *has the BCP if and only if* $1 \geq \lambda > \frac{1}{2}$;
(ii) *The quotient space* ℓ_∞/c_0 *does not have the BCP.*

Theorem 5.37 ([7]).
(i) $\ell_1[0,1]$ *does not have the BCP;*
(ii) ℓ_∞ *has the BCP;*
(iii) $\ell_1[0,1]$ *is linearly isometric to a subspace of* ℓ_∞.

5.5.2 Positive results

Recently, Zhenghua Luo and Bentuo Zheng [26] showed that for a sequence of Banach spaces, the BCP is inherited by their ℓ_p-sum for $1 \le p \le \infty$. In the following, we will present a different and generalized approach to the interesting result.

Let $s = \{x = (x(n))_{n=1}^{\infty}, x(n) \in \mathbb{R}\}$ be a Banach sequential space, and $e_n = \delta_{m,n} = 1$, if $m = n$, $= 0$, if $m \ne n$. The "coefficients" e_n^* is defined for $x = (x(n)) \in s$ by $\langle e_n^*, x\rangle = x(n)$, $n = 1, 2, \ldots$. We say that s is monotone provided $\|e_n\| = 1 = \|e_n^*\|$ for all $n \in \mathbb{N}$, and for all sequences $(a_n)_{n=1}^{\infty}$ with $a_n \in \mathbb{R}$,

$$\left\|\sum_{j=1}^{n} a_j e_j\right\| \le \left\|\sum_{j=1}^{n+1} a_j e_j\right\|, \quad \text{for all } n \in \mathbb{N};$$

A monotone s is called strictly monotone provided for all finite sequences $a = (a_j)_{j=1}^{n}$, $b = (b_j)_{j=1}^{n}$ with $a_n, b_n \in \mathbb{R}^+$, $b_j \ge a_j$ $(1 \le j \le n)$, $b \ne a$ imply

$$\left\|\sum_{j=1}^{n} a_j e_j\right\| < \left\|\sum_{j=1}^{n} b_j e_j\right\|.$$

Clearly, every Banach space with a 1-unconditional basis is linearly isometric to a monotone sequential space. s is said to be coordinate smooth if e_n, $n \in \mathbb{N}$ are Gâteaux differentiability points of the norm of s. For example, ℓ_p $(1 < p \le \infty)$ and c_0 are monotone and coordinate smooth, and ℓ_p, $1 \le p < \infty$ are strictly monotone.

For a sequence (X_n) of Banach spaces, the s-sum of (X_n) is defined by

$$X \equiv \Big(\bigoplus_{n\in\mathbb{N}} X_n\Big)_s = \{x = (x_1, x_2, \ldots, x_n, \ldots) : x_n \in X_n, n \in \mathbb{N}\}$$

endowed with the norm $\|x\| = \|(\|x_1\|, \|x_2\|, \ldots)\|_s$.

Theorem 5.38. *Let s be a monotone and coordinate smooth Banach sequential space, and $\{X_n\}$ be a sequence of Banach spaces. Then the sphere S_X of the sum $X \equiv (\bigoplus_{n\in\mathbb{N}} X_n)_s$ admits a ball-covering $\mathcal{B} = \{B(x_{mn}, r_{mn}) : m, n \in \mathbb{N}\}$ with $\{x_{mn}\}_{n=1}^{\infty} \in X_m$ for all $m \in \mathbb{N}$ if and only if every X_n admits the BCP.*

Proof. Sufficiency. Suppose that $X_m = (X_m, \|\cdot\|_m)$, $m \in \mathbb{N}$ have the BCP. Then for each fixed $m \in \mathbb{N}$, there is a ball-covering $\mathcal{B}_m = \{B(x_{mn,r_{mn}})\}_{n=1}^{\infty}$ of $S_m \equiv S_{X_m}$. By Theorem 5.4, this is equivalent to that for each selection φ_m of the subdifferential mapping $\partial\|\cdot\|_m$, $\{\varphi_m(x_{mn})\}$ positively separates points of X_m. Since $\pm e_m$ are smooth points of s, $\partial\|e_m x_{mn}\|_X = e_m \partial\|x_{mn}\|_m$, where $X \ni e_m x_{mn} = x_{mm}$, if $n = m$; $= 0$, if $n \ne m$. This entails that for each selection φ of the subdifferential mapping $\partial\|\cdot\|_X$, the double sequence $\{\varphi(x_{mn})\}$ positively separates points of X. We complete the proof of sufficiency again by Theorem 5.4.

Necessity. Suppose that X has a ball-covering $\mathcal{B} = \{B(x_{mn}, r_{mn}) : m, n \in \mathbb{N}\}$ with $\{x_{mn}\}_{n=1}^{\infty} \in X_m$ for all $m \in \mathbb{N}$. Then for each fixed $m \in \mathbb{N}$, $\mathcal{B}_m = \{B(x_{mn}, r_{mn}) \cap X_m : n \in \mathbb{N}\}$ is again a ball-covering of X_m. □

Corollary 5.39 ([26]). *Suppose that $s \in \{\ell_p, c_0 : 1 < p \leq \infty\}$, and that $\{X_n\}$ is a sequence of Banach spaces. Then the sphere S_X of the sum $X \equiv (\bigoplus_{n\in\mathbb{N}} X_n)_s$ admits a ball-covering $\mathcal{B} = \{B(x_{mn}, r_{mn}) : m, n \in \mathbb{N}\}$ with $\{x_{mn}\}_{n=1}^{\infty} \in X_m$ for all $m \in \mathbb{N}$ if and only if every X_n admits the BCP.*

Theorem 5.40. *Let s be a separable strictly monotone Banach sequential space, and $\{X_n\}$ be a sequence of Banach spaces. Then the sphere S_X of the sum $X \equiv (\bigoplus_{n\in\mathbb{N}} X_n)_s$ has the BCP if and only if every X_n has the BCP.*

Proof. Sufficiency. Since s is separable, it is a Gâteaux differentiability space. By Corollary 5.28, there exist two sequences $\{s_n\} \subset S_s$ and $\{s_n^*\} \subset S_{s^*}$ with $d\|s_n\| = s_n^*$ for all $n \in \mathbb{N}$ so that $\{s_n^*\}$ positively separates points of s. Since $X_n, n = 1, 2, \ldots$ have the BCP, by Theorem 5.4, for each fixed $m \in \mathbb{N}$, there is a sequence $\{x_{mn}\}_{n=1}^{\infty} \subset S_{(X_m, \|\cdot\|_m)}$ such that for each selection φ_m for the subdifferential mapping $\partial\|\cdot\|_m$ of the norm $\|\cdot\|_m$, $\{\varphi_m(x_{mn})\}$ positively separates points of X_m. Put $s_n = (s_n(j))$ and $d\|s_n\|(= s_n^*) = (s_n^*(j))$. Then

$$1 = \langle s_n^*, s_n \rangle = \sum_{j=1}^{\infty} s_n^*(j) \cdot s_n(j) = \sum_{j=1}^{\infty} |s_n^*(j) \cdot s_n(j)|.$$

Now, let

$$x_n = s_n \cdot (x_{mn}) = \left(\bigoplus_{m=1}^{\infty} s_n(m) x_{mn} \right)_s \in X.$$

Then $\|x_n\| = \|s_n\| = 1$ for all $n \in \mathbb{N}$, and

$$\partial\|x_n\| = \left\{ \left(\bigoplus_{m=1}^{\infty} s_n^*(m) x_{mn}^* \right)_{s^*} : x_{mn}^* \in \partial\|x_{mn}\|, m, n \in \mathbb{N} \right\}. \tag{5.8}$$

This entails that $\partial\|x_n\| = (\bigoplus_{m=1}^{\infty} s_n^*(m) \partial\|x_{mn}\|)_{s^*}$. Or, equivalently, for each selection φ of $\partial\|\cdot\|_X$, there exists a sequence of selections φ_m for $\partial\|\cdot\|_m$ such that $\varphi(x_n) = (\bigoplus_{m=1}^{\infty} s_n^*(m) \varphi(x_{mn}))_{s^*}$. Clearly, $\{\varphi(x_n)\}$ positively separates points of X.

Necessity. Suppose, to the contrary, that X_m does not have the BCP for some $m \in X$. Then for any sequence $(x_{mn}) \subset S_m \equiv S_{X_m}$ there is a selection φ_m for $\partial\|\cdot\|_m$ such that $\{\varphi_m(x_{mn})\}$ does not positively separate points of X_m. By (5.8), for any sequence $\{x_n\} \subset S_X$, there is a selection φ for $\|\cdot\|$ such that $\{\varphi(x_n)\}$ does not positively separate points of X. □

Corollary 5.41 ([26]). *Suppose that $s = \ell_1$, and that $\{X_n\}$ is a sequence of Banach spaces. Then the sphere S_X of the sum $X \equiv (\bigoplus_{n\in\mathbb{N}} X_n)_s$ admits the BCP if and only if every X_n admits the BCP.*

Theorem 5.42 ([26]). *Let (Ω, Σ, μ) be a finite measure space, and X be a normed space. Then $L_p(\mu, X)$ $(1 \le p < +\infty)$ has the BCP if and only if X has the BCP.*

Nevertheless, $L_\infty[0,1]$ fails the BCP.

Remark 5.43. S. Shang and Y. Cui showed some new properties of BCP in [31], [32], [33], and [34]. Especially, S. Shang [31] first considered the heredity of the BCP of the ℓ_p-sum $X \bigoplus_{\ell_p} Y$ $(1 \le p \le +\infty)$. Making use of Corollary 5.28, he showed that if X and Y are Gâteaux differentiability spaces, then $X \bigoplus_{\ell_p} Y$ admits the BCP if and only if both X and Y have the BCP. In [25], by a direct construction of a countable ball-covering, Z. Luo, J. Luo, and B. Wang further proved that the assumption of "Gâteaux differentiability space" in the Shang's result can be dropped.

5.6 Final remarks

So far, we do not know whether there is a nonseparable Banach space which guarantees the invariance of the BCP under linear isomorphisms. Though we have known that a Banach space admits the BCP if and only if X^* is w^* separable in the renorming sense. Therefore, our questions are classified into two categories: One is toward the positive direction, and the other is toward the negative one.

5.6.1 Questions toward positive direction

We have already known that the BCP is not invariant under linear isomorphism, and the BCP of a Banach space is not inherited by its subspaces (Theorems 5.36 and 5.37). However, all known counterexamples are not Asplund spaces, even not Gâteaux differentiability spaces. Thus, the following questions are natural.

Problem 5.44. *For what Banach spaces X (except the trivial case that X are reflexive), is the BCP of X invariant under linear isomorphisms?*

Problem 5.45. *Is the BCP of X invariant if X is an Asplund space, or more general, a Gâteaux differentiability space?*

Problem 5.46. *For what classes of Banach spaces X is the BCP invariant under quotient mappings?*

Problem 5.47. *For what Banach spaces X is the BCP of X heritable by its closed subspaces?*

For a Banach space X, if we use (Ω, ρ) to denote all continuous seminorms p on X endowed with the metric ρ defined for $p_1, p_2 \in \Omega$ by

$$\rho(p_1, p_2) = \sup_{x \in B_X} |p_1(x) - p_2(x)|,$$

then Ω is a complete metric space, and w^* separability of X^* implies that equivalent norms on X with the BCP are dense in Ω.

Problem 5.48. *For a nonseparable Banach X with a w^* separable dual, do equivalent norms on X with the BCP form a dense G_δ subset of Ω?*

5.6.2 A question toward negative direction

Since $\ell_1^*[0,1] = \ell_\infty[0,1]$ is w^* separable and does not has the BCP, and since $\ell_1[0,1]$ has the Radon–Nikodým property (RNP), not all Banach spaces with w^* separable duals and with the RNP are contained in the "invariant" class of Banach spaces for the BCP.

Problem 5.49. *If a Banach space with the BCP satisfies that the BCP is invariant under linear isomorphisms, must X be separable?*

Finally, we should mention here that P. L. Papini [28] collected and discussed some results concerning different coverings for the unit ball or the unit sphere of Banach spaces including ball coverings off origin. Generally, balls which cover L^p were discussed in [12], slices and balls which cover unit sphere of certain Banach spaces were discussed in [14].

In this paper, Theorems 5.38 and 5.40 are new results and other proofs are based on original proofs which were showed in the references.

Bibliography

[1] R. R. Akhmerov, M. I. Kamenskii, A. S. Potapov, A. E. Rodkina and B. N. Sadovskii, *Measures of Noncompactness and Condensing Operators*, Birkhäuser, Basel, 1992.

[2] J. Appell, *Recent Trends in Nonlinear Analysis*, Birkhäuser, Basel, 2000.

[3] E. Bishop and R. R. Phelps, *A proof of every Banach space is subreflexive*, Bull. Am. Math. Soc. **67** (1) (1961), 97–98.

[4] L. Cheng, *Ball-covering property of Banach spaces*, Isr. J. Math. **156** (2006), 111–123.

[5] L. Cheng, Q. Cheng and X. Liu, *Ball-covering property of Banach spaces is not preserved under linear isomorphisms*, Sci. China Ser. A **51** (2008), 143–147.

[6] L. Cheng, Q. Cheng and H. Shi, *Minimal ball-covering in Banach spaces and their application*, Stud. Math. **192** (1) (2009), 15–27.

[7] L. Cheng, V. Kadets, B. Wang and W. Zhang, *A note on ball-covering property of Banach spaces*, J. Math. Anal. Appl. **371** (2010), 249–253.

[8] L. Cheng, Z. Luo, X. Liu and W. Zhang, *Several remarks on ball-coverings of normed spaces*, Acta Math. Sin. **26** (9) (2010), 1667–1672.

[9] L. Cheng, H. Shi and W. Zhang, *Every Banach spaces with a w^*-separable dual has a $1+\varepsilon$-equivalent norm with the ball covering property*, Sci. China Ser. A, Math. **52** (2009), 1869–1874.

[10] L. Cheng, B. Wang, W. Zhang and Y. Zhou, *Some geometric and topological properties of Banach spaces via ball coverings*, J. Math. Anal. Appl. **377** (2) (2011), 874–880.

[11] P. Enflo, *Banach spaces which can be given an equivalent uniformly convex norm*, Isr. J. Math. **13** (1972), 281–288.

[12] V. P. Fonf, M. Levin and C. Zanco, *Covering L^p spaces by balls*, J. Geom. Anal. **24** (4) (2014), 1891–1897.

[13] V. P. Fonf and C. Zanco, *Covering spheres of Banach spaces by balls*, Math. Ann. **344** (2009), 939–945.

[14] V. P. Fonf and C. Zanco, *Covering the unit sphere of certain Banach spaces by sequences of slices and balls*, Can. Math. Bull.. **57** (1) (2014), 42–50.

[15] R. Fu and L. Cheng, *Ball-coverings property of Banach spaces* (Chinese), J. Math. Study **39** (1) (2006), 39–43.

[16] L. Gerber, *The orthocentric simplex as an extreme simplex*, Pac. J. Math. **56** (1975), 97–111.

[17] G. Godefroy and N. J. Kalton, *The ball topology and its applications*, in: Contemp. Math., **85**, Amer. Math. Soc., 1989, pp. 195–237.

[18] Z. Hu and X. Zhao, *On minimal ball-coverings of ℓ_1^n*, preprint.

[19] R. C. James, *Uniformly non-square Banach spaces*, Ann. Math. **80** (1964), 542–550.

[20] G. Lin and X. Shen, *A neural network method for the minimum radius problem of ball coverings* (Chinese), Xiamen Daxue Xuebao **47** (6) (2008), 797–800.

[21] L. Lin, *Relations between hereditarily indecomposable spaces and spaces with the ball-covering property* (Chinese), J. East China Norm. Univ. Natur. Sci. Ed. **3** (2008), 8–11.

[22] L. Lin, *Some properties of n-dimensional normed spaces with a minimal ball-covering of $n+1$ balls* (Chinese), J. Huabei Coal Ind. Teach. College **3** (2008), 18–22.

[23] L. Lin, F. Zhang and M. Zhang, *Every n-dimensional space with a minimal ball-covering of $2n-1$ balls contains an $(n-1)$-dimensional subspace isometric to $(\mathbb{R}^{n-1}, \|\,\|_\infty)$* (Chinese), J. Math. Study **41** (4) (2008), 407–415.

[24] G. Liu, *On radius of a circumhypersphere of n-simplexes in E^n* (Chinese), J. Soochow Univ. **6** (1) (1990), 1–5.

[25] Z. Luo, J. Liu and B. Wang, *A remark on the ball-covering property of product spaces*, Filomat **31** (12) (2017), 3905–3908.

[26] Z. Luo and B. Zheng, *Stability of ball-covering property*, Stud. Math. **250** (1) (2020), 19–34.

[27] S. Mazur, *Über schwache Konvergenz in den Raümen (L_p)*, Stud. Math. **4** (1933), 128–133.

[28] P. L. Papini, *Covering the sphere and the ball in Banach spaces*, Commun. Appl. Anal. **13** (2009), 579–586.

[29] A. Pełczýnski, *All separable Banach spaces admit for every $\varepsilon > 0$ fundamental and total biorthogonal sequences bounded by $1+\varepsilon$*, Stud. Math. **55** (1976), 295–304.

[30] R. R. Phelps, *Convex Functions, Monotone Operators and Differentiability*, Lect. Notes in Math., **1364**, Springer-Verlag, 1989.

[31] S. Shang, *Differentiability and ball-covering property in Banach spaces*, J. Math. Anal. Appl. **434** (1) (2016), 182–190.

[32] S. Shang and Y. Cui, *Ball-covering property in uniformly non-l_3^1 Banach spaces and application*, Abstr. Appl. Anal. **1** (2013), 1–7.

[33] S. Shang and Y. Cui, *Locally 2-uniform convexity and ball-covering property in Banach space*, Banach J. Math. Anal. **9** (1) (2015), 42–53.

[34] S. Shang and Y. Cui, *Dentable point and ball-covering property in Banach spaces*, J. Convex Anal. **25** (3) (2018), 1045–1058.

[35] H. Shi and X. Zhang, *Minimal ball covering of the unit sphere in R^n* (Chinese), Xiamen Daxue Xuebao **45** (5)(2006), 621–623.
[36] J. Vanderwerff, *Mazur intersection properties for compact and weakly convex sets*, Can. Math. Bull. **41** (1998), 225–230.
[37] B. Wang, *The ball-covering property of two spaces with w^*-separable duality* (Chinese), J. Math. Study **43** (4) (2010), 393–396.
[38] W. Zhang, *Characterizations of universal finite representability and B-convexity of Banach spaces via ball coverings*, Acta Math. Sin. **28** (7) (2012), 1369–1374.
[39] X. Zhang, *Radius of a minimal ball-covering in the space R^n* (Chinese), J. Math. Study **40** (1) (2007), 109–113.
[40] V. Zizler, *Renorming concerning Mazur's intersection property of balls for weakly compact convex sets*, Math. Ann. **276** (1986), 61–66.

Andrea Colesanti and Galyna Livshyts

6 A note on the quantitative local version of the log-Brunn–Minkowski inequality

In memory of Victor Lomonosov

Abstract: We show that there exists an $\epsilon(n) > 0$, depending only on the dimension n, so that for any symmetric convex body K in the $\epsilon(n)$-neighborhood of B_2^n (in the C^2 metric), the log-Brunn–Minkowski inequality

$$|\lambda K +_0 (1-\lambda)L| \geq |K|^{\lambda}|L|^{1-\lambda}$$

holds. The proof is based on the previous results from [7], as well as an additional "third derivative" argument, which allows us to establish a uniform neighborhood. As a consequence, we conclude that the uniform cone volume measure determines a symmetric convex body uniquely, provided that it is in a fixed neighborhood of any ball.

Keywords: Convex bodies, log-concave, Brunn–Minkowski, cone-measure

MSC 2010: Primary 52

6.1 Introduction

The Brunn–Minkowski theory is the study of convexity properties of measures, in particular, of the Lebesgue measure. The classical Brunn–Minkowski inequality in its classical formulation asserts that for every pair of convex bodies K and L and for every $\lambda \in [0,1]$,

$$|\lambda K + (1-\lambda)L|^{1/n} \geq \lambda|K|^{1/n} + (1-\lambda)|L|^{1/n}, \tag{6.1}$$

Acknowledgement: The authors would like to thank Alina Stancu and Mohammad Ivaki for precious conversations on the subject of this paper. The first author is supported by the G. N. A. M. P. A., project *Problemi sovradeterminati e questioni di ottimizzazione di forma*. The second author is supported by the NSF CAREER DMS-1753260. The work was partially supported by the National Science Foundation under Grant No. DMS-1440140 while the second author was in residence at the Mathematical Sciences Research Institute in Berkeley, California, during the Fall 2017 semester. The authors are very grateful to the anonymous referee for the helpful comments.

Andrea Colesanti, Dipartimento di Matematica e Informatica "U. Dini", Università degli Studi di Firenze, Firenze, Italy, e-mail: colesant@math.unifi.it
Galyna Livshyts, School of Mathematics, Georgia Institute of Technology, Atlanta, Georgia, USA, e-mail: glivshyts6@math.gatech.edu

https://doi.org/10.1515/9783110656756-006

where $|\cdot|$ stands for the volume (Lebesgue measure). Equality conditions are completely characterized: equality holds if and only if K and L are either contained in parallel hyperplanes, or they are homothetic. A standard homogeneity argument shows that (6.1) is equivalent to the (a priori, stronger) inequality:

$$|\lambda K + (1-\lambda)L| \geq |K|^{\lambda}|L|^{1-\lambda}, \tag{6.2}$$

which again holds for every pair of convex bodies K and L and every $\lambda \in [0,1]$. We refer, for example, to the extensive survey by Gardner [10] on the subject, or to the book of Schneider [17].

Recently, a number of questions have emerged in regards to possible improvements of the Brunn–Minkowski inequality in the presence of symmetry, and other structural assumptions. Many different variants of the Brunn–Minkowski inequality have been studied, in particular in the context of the L_p-Brunn–Minkowski theory, where the standard Minkowski addition is replaced by the p-addition, where p varies in $\mathbb{R}$ (see [17, Chapter 9]). While the case $p \geq 1$ is completely understood, recently the attention focused on the values $p < 1$, and the case $p = 0$, to which the present paper is devoted. Define the geometric average of two convex bodies K and L, with parameter λ, as

$$\lambda K +_0 (1-\lambda)L = \{x \in \mathbb{R}^n : \langle x,u\rangle \leq h_K^{\lambda}(u)h_L^{1-\lambda}(u),\ \forall u \in \mathbb{S}^{n-1}\},$$

where h_K and h_L are the support functions of K and L, respectively. Recall the definition

$$h_K(x) = \max_{y\in K}\langle x,y\rangle.$$

The log-Brunn–Minkowski conjecture (see Boroczky, Lutwak, Yang, and Zhang [1]) states that

$$|\lambda K +_0 (1-\lambda)L| \geq |K|^{\lambda}|L|^{1-\lambda} \tag{6.3}$$

for every pair of symmetric convex sets K and L. Important applications and motivations for this conjecture can be found in [2], [3]. In particular, it was shown by Saraglou [16] that this conjecture is stronger than the famous B-conjecture (see [8]). Note that the straightforward inclusion

$$\lambda K +_0 (1-\lambda)L \subset \lambda K + (1-\lambda)L$$

shows that (6.3) is stronger than the classical Brunn–Minkowski inequality.

It is not difficult to see that the condition of symmetry in (6.3) is necessary: for instance, it may fail for a pair of intervals in dimension 1 if one of them is not centered. Böröczky, Lutwak, Yang, and Zhang [1] showed that this conjecture holds for $n = 2$. Saraglou [16] proved that the conjecture is true when the sets K and L are unconditional (i. e., they are symmetric with respect to every coordinate hyperplane).

Rotem [15] showed that log-Brunn–Minkowski conjecture holds for complex convex bodies.

Recently, local versions of the log-Brunn–Minkowski inequality have been considered. In the paper [7], the authors prove the following fact. *Let $R > 0$ and let ϕ be a smooth function on the unit sphere; then there exists $a > 0$ such that if the support functions of K and L are $Re^{\epsilon_1\phi}$, $Re^{\epsilon_2\phi}$, respectively, with $0 \leq \epsilon_1, \epsilon_2 < a$, then* (6.3) *holds.* In this note, we improve the previous result. Below B_2^n denotes the unit ball in $\mathbb{R}^n$ and $\mathbb{S}^{n-1}$ stands for the unit sphere. By $\|\cdot\|_{C^2(\mathbb{S}^{n-1})}$, we denote the C^2-norm on the sphere (see Section 6.2 for details).

Theorem 6.1. *Let $R > 0$ and $n \geq 2$. There exists $\epsilon(n) > 0$ such that for every symmetric convex C^2-smooth body K in $\mathbb{R}^n$ such that $\|h_K - R\|_{C^2(\mathbb{S}^{n-1})} \leq \epsilon(n)R$, where h_K is the support function of K, we have*

$$|\lambda K +_0 (1-\lambda)RB_2^n| \geq |K|^\lambda |RB_2^n|^{1-\lambda} \quad \forall \lambda \in [0,1]. \tag{6.4}$$

Moreover, the equality holds if and only if K is a ball centered at the origin.

We remark, that a result from [7] is also contained in the recent paper by Kolesnikov and Milman [13] (see Theorem 1.2), and the main result of the present paper (Theorem 6.1) was later included in the work of Chen, Huang, Li, and Liu [4] as Theorem 1.9, where the argument was based on the results from [13] and the continuity method.

We shall use the notation s_K for the surface area measure of K on the sphere (i. e., the push-forward of the Hausdorff measure on the boundary of K to the sphere under the Gauss map; see Section 6.2 for more details). The cone volume measure of a convex set K is the measure on the sphere, defined as

$$c_K(\Omega) = \frac{1}{n} \int_\Omega h_K(u)\, ds_K(u).$$

It was conjectured by Lutwak [14] that the cone volume measure determines a symmetric smooth convex body uniquely. A partial case of this conjecture when the cone volume measure is uniform was posed earlier by Firey [9].

As a corollary of our main result, we deduce a local uniqueness result.

Corollary 1. *Let $n \geq 2$ and let $R > 0$ be a constant. There exists $\epsilon = \epsilon(n) > 0$, which depends only on the dimension, such that, given a symmetric C^2-smooth convex body K satisfying*

$$\|R - h_K\|_{C^2(\mathbb{S}^{n-1})} \leq \epsilon(n)R,$$

and $dc_K(u) = R^n du$, one has that K coincides with the Euclidean ball of radius R.

6.2 Preliminaries

We work in the Euclidean n-dimensional space $\mathbb{R}^n$. The unit ball shall be denoted by B_2^n and the unit sphere by $\mathbb{S}^{n-1}$. The Lebesgue volume of a measurable set $A \subset \mathbb{R}^n$ is denoted by $|A|$.

We say that a convex body K is of class $C^{2,+}$ if ∂K is of class C^2 and the Gauss curvature is strictly positive at every $x \in \partial K$. In particular, if K is $C^{2,+}$ then it admits unique outer unit normal $\nu_K(x)$ at every boundary point x. Recall that the Gauss map $\nu_K : \partial K \to \mathbb{S}^{n-1}$ is the map assigning the collection of unit normals to each point of ∂K.

We recall that an orthonormal frame on the sphere is a map which associates to every $x \in \mathbb{S}^{n-1}$ an orthonormal basis of the tangent space to $\mathbb{S}^{n-1}$ at x. Let $\psi \in C^2(\mathbb{S}^{n-1})$; we denote by $\psi_i(u)$ and $\psi_{ij}(u)$, $i, j \in \{1, \dots, n-1\}$, the first and second covariant derivatives of ψ at $u \in \mathbb{S}^{n-1}$, with respect to a fixed local orthonormal frame on an open subset of $\mathbb{S}^{n-1}$. We define the matrix

$$Q(\psi; u) = (q_{ij})_{i,j=1,\dots,n-1} = (\psi_{ij}(u) + \psi(u)\delta_{ij})_{i,j=1,\dots,n-1}, \tag{6.5}$$

where the δ_{ij}'s are the usual Kronecker symbols. On an occasion, instead of $Q(\psi; u)$ we write $Q(\psi)$. Note that $Q(\psi; u)$ is symmetric by standard properties of covariant derivatives. In what follows, we shall often consider ψ to be a support function of a convex body K. In this case, $Q(\psi)$ is called *curvature matrix* of K; this name comes from the fact that $\det(Q(\psi))$ is the density of the curvature measure s_K and, therefore,

$$|K| = \frac{1}{n} \int_{\mathbb{S}^{n-1}} h_K(u) \det Q(h_K, u) du.$$

(See, for instance, Koldobsky [12] for the proof.) We recall here a fact that will be frequently used in the paper (a proof can be deduced, for instance, from [17, Section 2.5]).

Proposition 6.2. *Let $K \in \mathcal{K}^n$ and let h be its support function. Then K is of class $C^{2,+}$ if and only if $h \in C^2(\mathbb{S}^{n-1})$ and*

$$Q(h; u) > 0, \quad \forall\, u \in \mathbb{S}^{n-1}.$$

In view of the previous result, we say that a function h defined on $\mathbb{S}^{n-1}$ is of class $C^{2,+}(\mathbb{S}^{n-1})$ if $h \in C^{2,+}(\mathbb{S}^{n-1})$ and $Q(h; u)$ is positive definite for every $x \in \mathbb{S}^{n-1}$. For $g \in C^2(\mathbb{S}^{n-1})$, we set

$$\|g\|_{C^2(\mathbb{S}^{n-1})} = \|g\|_{L^\infty(\mathbb{S}^{n-1})} + \|\nabla_S g\|_{L^\infty(\mathbb{S}^{n-1})} + \sum_{i,j}^{n-1} \|g_{ij}\|_{L^\infty(\mathbb{S}^{n-1})},$$

where $\nabla_S g$ denotes the spherical gradient of g (i. e., the vector having first covariant derivatives as components). We also set

$$\|g\|^2_{L^2(\mathbb{S}^{n-1})} = \int_{\mathbb{S}^{n-1}} g^2(u) du, \quad \|\nabla_S g\|^2_{L^2(\mathbb{S}^{n-1})} = \int_{\mathbb{S}^{n-1}} |\nabla_S g(u)|^2 du.$$

6.2.1 Cofactor matrices

For a natural number N, denote by $\mathrm{Sym}(N)$ the space of $N \times N$ symmetric matrices. Given $A \in \mathrm{Sym}(N)$ we denote by a_{jk} its jkth entry and write $A = (a_{jk})$. For $j, k = 1, \ldots, N$, we set

$$c_{jk}(A) = \frac{\partial \det}{\partial a_{jk}}(A). \tag{6.6}$$

The matrix $C(A) = (c_{jk}(A))$ is called the cofactor matrix of A. We also set, for $j, k, r, s = 1\ldots, N$,

$$c_{jk,rs}(A) = \frac{\partial^2 \det}{\partial a_{jk} \partial a_{rs}}(A). \tag{6.7}$$

Recall that

$$\det(A) = \frac{1}{N!} \sum \delta\begin{pmatrix} j_1, \ldots, j_N \\ k_1, \ldots, k_N \end{pmatrix} a_{j_1 k_1} \cdots a_{j_N k_N}, \tag{6.8}$$

where the sum is taken over all possible indices $j_s, k_s \in \{1, \ldots, N\}$ (for $s = 1, \ldots, N$) and the Kronecker symbol

$$\delta\begin{pmatrix} j_1, \ldots, j_N \\ k_1, \ldots, k_N \end{pmatrix}$$

equals 1 (resp., -1) when $j_1, \ldots, j_N$ are distinct and $(k_1, \ldots, k_N)$ is an even (resp., odd) permutation of $(j_1, \ldots, j_N)$; otherwise it is 0. Using (6.8), along with (6.6) and (6.7), we derive for every $j, k, r, s \in \{1, \ldots, N\}$:

$$\begin{aligned} c_{jk}(A) &= \frac{1}{(N-1)!} \sum \delta\begin{pmatrix} j, j_1, \ldots, j_{N-1} \\ k, k_1, \ldots, k_{N-1} \end{pmatrix} a_{j_1 k_1} \cdots a_{j_{N-1} k_{N-1}}, \\ c_{jk,rs}(A) &= \frac{1}{(N-2)!} \sum \delta\begin{pmatrix} r, j, j_1, \ldots, j_{N-2} \\ s, k, k_1, \ldots, k_{N-2} \end{pmatrix} a_{j_1 k_1} \cdots a_{j_{N-2} k_{N-2}}. \end{aligned} \tag{6.9}$$

Remark 6.3. If $A \in \mathrm{Sym}(N)$ is invertible, then by (6.9),

$$(c_{jk}(A)) = \det(A)\, A^{-1}.$$

In particular, if $A = I_N$ (the identity matrix of order N), then $(c_{jk})(I_N) = I_N$.

Remark 6.4. Observe that, by (6.9), for every $A = (a_{jk}) \in \mathrm{Sym}(N)$,

$$\sum_{j,k=1}^{N} c_{jk}(A) a_{jk} = N \det(A).$$

Remark 6.5. Let $A = (a_{ij}) \in \mathrm{Sym}(N)$ and let $M > 0$ be such that

$$|a_{jk}| \le M, \quad \forall j, k = 1\ldots, N.$$

Then there exists some constant $c = c(N)$ (*i. e.*, depending only on N) such that, for every $j, k, r, s = 1, \dots, N$,

$$|c_{jk}(A)| \leq c(N)\, M^{N-1}, \quad |c_{jk,rs}(A)| \leq c(N)\, M^{N-2}.$$

Note that if $g \equiv c$ on $\mathbb{S}^{n-1}$ then $Q(g; u) = cI_{n-1}$ for every $u \in \mathbb{S}^{n-1}$.

6.2.2 The Cheng–Yau lemma and an extension

Let $h \in C^3(\mathbb{S}^{n-1})$. Consider the cofactor matrix $y \to C[Q(h; y)]$. This is a matrix of functions on $\mathbb{S}^{n-1}$. The lemma of Cheng and Yau ([5]) asserts that each column of this matrix is divergence-free.

Lemma 6.6 (Cheng–Yau). *Let $h \in C^3(\mathbb{S}^{n-1})$. Then, for every index $j \in \{1, \dots, n-1\}$ and for every $y \in \mathbb{S}^{n-1}$,*

$$\sum_{i=1}^{n-1} (c_{ij}[Q(h; y)])_i = 0,$$

where the subscript i denotes the derivative with respect to the ith element of an orthonormal frame on $\mathbb{S}^{n-1}$.

For simplicity of notation, we shall often write $C(h)$, $c_{ij}(h)$ and $c_{ij,kl}(h)$ in place of $C[Q(h)]$, $c_{ij}[Q(h)]$ and $c_{ij,kl}[Q(h)]$, respectively. As a corollary of the previous result, we have the following integration by parts formula. If $h, \psi, \phi \in C^2(\mathbb{S}^{n-1})$, then

$$\int_{\mathbb{S}^{n-1}} \phi\, c_{ij}(h)(\psi_{ij} + \psi\, \delta_{ij}) dy = \int_{\mathbb{S}^{n-1}} \psi\, c_{ij}(h)(\phi_{ij} + \phi\, \delta_{ij}) dy. \tag{6.10}$$

Note that we adopt the summation convention over repeated indices. The lemma of Cheng and Yau admits the following extension (see Lemma 2.3 in [6]).

Lemma 6.7. *Let $h, \psi \in C^3(\mathbb{S}^{n-1})$. Then, for every $k \in \{1, \dots, n-1\}$ and for every $y \in \mathbb{S}^{n-1}$*

$$\sum_{l=1}^{n-1} (c_{ij,kl}[Q(h; y)](\psi_{ij} + \psi\delta_{ij}))_l = 0.$$

Correspondingly, we have for every $h, \psi, \varphi, \phi \in C^2(\mathbb{S}^{n-1})$,

$$\begin{aligned} &\int_{\mathbb{S}^{n-1}} \psi\, c_{ij,kl}(h)(\varphi_{ij} + \varphi\delta_{ij})(\phi_{kl} + \phi\, \delta_{kl}) dy \\ &= \int_{\mathbb{S}^{n-1}} \phi\, c_{ij,kl}(h)(\varphi_{ij} + \varphi\delta_{ij})(\psi_{kl} + \psi\, \delta_{kl}) dy. \end{aligned} \tag{6.11}$$

6.2.3 A Poincaré inequality for even functions on the sphere

Here, we use some basic facts from the theory of spherical harmonics, which can be found, for instance, in [11, 12] or in [17, Appendix]. We denote by Δ_σ the spherical Laplace operator (or Laplace–Beltrami operator), on $\mathbb{S}^{n-1}$. The first eigenvalue of Δ_σ is 0, and the corresponding eigenspace is formed by constant functions. The second eigenvalue of Δ_σ is $n-1$, and the corresponding eigenspace is formed by the restrictions of linear functions of $\mathbb{R}^n$ to $\mathbb{S}^{n-1}$. The third eigenvalue is $2n$, which implies, in particular, that for any **even** function $\psi \in C^2(\mathbb{S}^{n-1})$ such that

$$\int_{\mathbb{S}^{n-1}} \psi du = 0,$$

one has

$$\int_{\mathbb{S}^{n-1}} \psi^2 du \le \frac{1}{2n} \int_{\mathbb{S}^{n-1}} |\nabla_s \psi|^2 du. \tag{6.12}$$

6.3 Computations of derivatives

Let $\psi \in C^2(\mathbb{S}^{n-1})$, and let $s > 0$. We consider the function $h_s(u) = e^{s\psi(u)}$. We will denote derivatives with respect to the parameter s by a dot, for example,

$$\dot{h}_s(u) = \frac{d}{ds} h(u), \quad \ddot{h}_s(u) = \frac{d^2}{ds^2} h(u), \ \ldots$$

Note that

$$\dot{h}_s = \psi h_s, \quad \ddot{h}_s = \psi^2 h_s, \quad \dddot{h}_s = \psi^3 h_s. \tag{6.13}$$

Remark 6.8. As we may interchange the order of derivatives, for every $j, k = 1, \ldots, n-1$ we have

$$q_{jk}(\dot{h}) = \dot{q}_{jk}(h),$$

and thus

$$\dot{Q}(h) = Q(\dot{h}).$$

Similar equalities hold for successive derivatives in s.

Consider the volume function

$$f(s) = \frac{1}{n} \int_{\mathbb{S}^{n-1}} h_s(u) \det(Q(h_s; u)) du. \tag{6.14}$$

If h_s is the support function of a convex body K_s (as it will be in the sequel), f represents the volume of K_s.

Remark 6.9. The entries of $Q(h_s)$ are continuous functions of the second derivatives of h_s and $Q(h_0) > 0$. Hence there exists $\eta_0 > 0$ such that if $\psi \in C^2(\mathbb{S}^{n-1})$ is such that $\|\psi\|_{C^2(\mathbb{S}^{n-1})} \le \eta_0$, then

$$Q(e^{s\psi}; u) > 0 \quad \forall\, u \in \mathbb{S}^{n-1},\ \forall\, s \in [-2, 2]. \tag{6.15}$$

We shall use notation

$$\mathcal{U} = \{\psi \in C^2(\mathbb{S}^{n-1}) : \|\psi\|_{C^2(\mathbb{S}^{n-1})} \le \eta_0\}.$$

Note that if $\psi \in \mathcal{U}$ then $f > 0$ in $[-2, 2]$. Moreover, in the case $h_0 \equiv 1$ we have $Q(h_0) = I_{n-1}$, and

$$f(0) = \frac{1}{n}|\mathbb{S}^{n-1}|. \tag{6.16}$$

Lemma 6.10. *In the notation introduced above, we have, for every s:*

$$f'(s) = \int_{\mathbb{S}^{n-1}} \psi h_s \det(Q(h_s))du; \tag{6.17}$$

$$f''(s) = \int_{\mathbb{S}^{n-1}} [\psi^2 h_s \det(Q(h_s)) + \psi h_s c_{jk}(h_s) q_{jk}(\psi h_s)]du; \tag{6.18}$$

$$\begin{aligned} f'''(s) = &\int_{\mathbb{S}^{n-1}} h_s[\psi^3 \det(Q(h_s)) + 2\psi^2 c_{jk}(h_s) q_{jk}(\psi h_s)]du \\ &+ \int_{\mathbb{S}^{n-1}} h_s\{\psi[c_{jk,rs}(h_s) q_{jk}(\psi h_s) q_{rs}(\psi h_s) + c_{jk}(h_s) q_{jk}(\psi^2 h_s)]\}du. \end{aligned} \tag{6.19}$$

Proof. For brevity, we write h instead of h_s. We differentiate the function f in s:

$$\begin{aligned} f'(s) &= \frac{1}{n}\int_{\mathbb{S}^{n-1}} [\dot{h}\det(Q(h)) + h c_{jk}(h)\dot{q}_{jk}(h)]dy \\ &= \frac{1}{n}\int_{\mathbb{S}^{n-1}} [\dot{h}\det(Q(h)) + h c_{jk}(h) q_{jk}(\dot{h})]dy \\ &= \frac{1}{n}\int_{\mathbb{S}^{n-1}} [\dot{h}\det(Q(h)) + \dot{h} c_{jk}(h) q_{jk}(h)]dy \\ &= \int_{\mathbb{S}^{n-1}} \dot{h}\det(Q(h))dy. \end{aligned}$$

Above we have used Remarks 6.4 and 6.8, and the integration by parts formula (6.10).

Passing to the second derivative, we get:

$$\begin{aligned} f''(s) &= \int_{\mathbb{S}^{n-1}} [\ddot{h}\det(Q(h)) + \dot{h} c_{jk}(h)\dot{q}_{jk}(h)]du \\ &= \int_{\mathbb{S}^{n-1}} [\ddot{h}\det(Q(h)) + \dot{h} c_{jk}(h) q_{jk}(\dot{h})]du. \end{aligned}$$

Finally,

$$f'''(s) = \int_{\mathbb{S}^{n-1}} [\dddot{h}\det(Q(h)) + 2\ddot{h}c_{jk}(h)q_{jk}(\dot{h})]du$$
$$+ \int_{\mathbb{S}^{n-1}} \{\dot{h}[c_{jk,rs}(h)q_{jk}(\dot{h})q_{rs}(\dot{h}) + c_{jk}q_{jk}(\ddot{h})]\}du.$$

Equalities (6.17), (6.18), and (6.19) follow from (6.13). □

The next corollary has appeared in [7].

Corollary 6.11. *In the notation introduced before we have*

$$f'(0) = \int_{\mathbb{S}^{n-1}} \psi du; \tag{6.20}$$

$$f''(0) = \int_{\mathbb{S}^{n-1}} [n\psi^2 - |\nabla_s\psi|^2]du. \tag{6.21}$$

Proof. Equality (6.20) follows immediately from (6.17). Moreover, plugging $s = 0$ in (6.18), and using the facts

$$c_{jk}(h_0) = \delta_{jk} \quad \text{and} \quad q_{jk}(\psi) = (\psi_{jk} + \psi\delta_{kj})$$

for every $j, k = 1, \ldots, n-1$, we get

$$f''(0) = \int_{\mathbb{S}^{n-1}} [n\psi^2 + \psi\Delta_s\psi]du.$$

By the divergence theorem on $\mathbb{S}^{n-1}$, we deduce (6.21). □

Lemma 6.12. *For every $\rho > 0$, there exists $\eta > 0$, such that if $\psi \in \mathcal{U}$ is an even function and it verifies:*

–
$$\int_{\mathbb{S}^{n-1}} \psi du = 0;$$

–
$$\|\psi\|_{C^2(\mathbb{S}^{n-1})} \leq \eta;$$

then

$$|(\log f)'''(s)| \leq \rho\|\nabla_s\psi\|^2_{L^2(\mathbb{S}^{n-1})}, \quad \forall s \in [-2, 2],$$

where f is defined as in (6.14) and $h_s = e^{s\psi}$.

Proof. We have

$$(\log f)'''(s) = \frac{f'''(s)}{f(s)} - 3\frac{f'(s)f''(s)}{f^2(s)} + 2\frac{(f')^3(s)}{f^3(s)}.$$

We first fix $\eta_1 > 0$ such that $\|\psi\|_{C^2(\mathbb{S}^{n-1})} \leq \eta_1$ implies

$$f(s) \geq \frac{1}{4n}|\mathbb{S}^{n-1}| = \frac{1}{4}f(0), \quad \forall s \in [-2,2].$$

Hence

$$|(\log f)'''(s)| \leq C_0(|f'''(s)| + |f'(s)f''(s)| + |(f')^3(s)|) = C_0(T_1 + T_2 + T_3),$$

for some constant $C_0 = C_0(n)$. Throughout this proof, we will denote by C a generic positive constant dependent on the dimension n and η_1.

Bound on the term T_3
There exists C such that

$$\|h_s\|_{C^2(\mathbb{S}^{n-1})} = \|e^{s\psi}\|_{C^2(\mathbb{S}^{n-1})} \leq C,$$

for every $s \in [-2,2]$ and for every $\psi \in \mathcal{U}$. Therefore,

$$h_s(u)\det(Q(h_s;u)) \leq C, \quad \forall \psi \in \mathcal{U}.$$

Consequently, by Lemma 6.10, we may write two types of estimates

$$|f'(s)| \leq C\|\psi\|_{C^2(\mathbb{S}^{n-1})}, \quad |f'(s)| \leq C\|\psi\|_{L_2(\mathbb{S}^{n-1})}.$$

By (6.12), there exists $\eta' > 0$ such that

$$|(f')^3(s)| \leq \frac{\rho}{3C_0}\|\nabla_s\psi\|_{L^2(\mathbb{S}^{n-1})}, \tag{6.22}$$

if ψ verifies $\|\psi\|^2_{C^2(\mathbb{S}^{n-1})} \leq \eta'$.

Bound of the term T_2
By Lemma 6.10, (6.12), and the integration by parts formula (6.10), we have

$$\begin{aligned}
|f''(s)| &\leq C\|\psi\|^2_{L^2(\mathbb{S}^{n-1})} + \left|\int_{\mathbb{S}^{n-1}} \psi h_s c_{jk}(h_s)(\psi h_s\delta_{jk} + (\psi h_s)_{jk})du\right| \\
&\leq C\|\psi\|^2_{L^2(\mathbb{S}^{n-1})} + \left|\int_{\mathbb{S}^{n-1}} c_{jk}(h_s)(\psi h_s)_j(\psi h_s)_k du\right| \\
&\leq C\|\psi\|^2_{L^2(\mathbb{S}^{n-1})} + C\|\nabla_s\psi\|^2_{L^2(\mathbb{S}^{n-1})} \\
&\leq C\|\nabla_s\psi\|^2_{L^2(\mathbb{S}^{n-1})}
\end{aligned}$$

(note that the first term was bounded using the argument as for the previous part of this proof). Hence we have the bound (6.22) for T_2 as well.

Bound of the term T_1

Equality (6.19) provides an expression of $f'''(s)$ as the sum of four terms. Each of them can be treated as in the previous two cases, with the exception of

$$\left| \int_{\mathbb{S}^{n-1}} \psi h_s c_{jk,rs}(h_s) q_{rs}(\psi h_s) q_{jk}(\psi h_s) du \right|.$$

We estimate it as follows:

$$\begin{aligned}
&\left| \int_{\mathbb{S}^{n-1}} \psi h_s c_{jk,rs}(h_s) q_{rs}(\psi h_s) q_{jk}(\psi h_s) du \right| \\
&\quad \le \left| \int_{\mathbb{S}^{n-1}} \psi^2 h_s^2 c_{jk,rs}(h_s) q_{rs}(\psi h_s) \delta_{jk} du \right| + \left| \int_{\mathbb{S}^{n-1}} \psi h_s c_{jk,rs}(h_s) q_{rs}(\psi h_s)(\psi h_s)_{jk} du \right| \\
&\quad \le C\|\psi\|_{C^2(\mathbb{S}^{n-1})} \|\psi\|^2_{L_2(\mathbb{S}^{n-1})} + \left| \int_{\mathbb{S}^{n-1}} c_{jk,rs}(h_s) q_{rs}(\psi h_s)(\psi h_s)_j(\psi h_s)_k du \right| \\
&\quad \le C\|\psi\|_{C^2(\mathbb{S}^{n-1})} \|\psi\|^2_{L_2(\mathbb{S}^{n-1})} + C\|\psi\|_{C^2(\mathbb{S}^{n-1})} \|\nabla_s \psi\|^2_{L_2(\mathbb{S}^{n-1})} \\
&\quad \le C\|\psi\|_{C^2(\mathbb{S}^{n-1})} \|\nabla_s \psi\|^2_{L_2(\mathbb{S}^{n-1})}.
\end{aligned}$$

We deduce that the upper bound (6.22) can be established for T_1. This concludes the proof. □

Lemma 6.13. *Let f be defined by (6.14). There exists $\eta > 0$ such that for every even $\psi \in \mathcal{U}$ so that $\|\psi\|_{C^2(\mathbb{S}^{n-1})} \le \eta$, the function $\log(f(s))$, is concave in $[-2,2]$. Moreover, it is strictly concave in this interval unless ψ is constant.*

Proof. We first assume that

$$\int_{\mathbb{S}^{n-1}} \psi du = 0. \tag{6.23}$$

For every $s \in [-2,2]$, there exists $\bar{s}$ between 0 and s such that

$$(\log f)''(s) = (\log f)''(0) + s(\log f)'''(\bar{s}) = \frac{f(0)f''(0) - f'(0)^2}{f(0)^2} + s(\log f)'''(\bar{s}).$$

It is shown in Lemma 6.12 that, for an arbitrary $\rho > 0$ there exists $\eta > 0$ such that if $\|\psi\|_{C^2(\mathbb{S}^{n-1})} \le \eta$ then

$$(\log f)'''(s) \le \rho\|\nabla_s \psi\|^2_{L^2(\mathbb{S}^{n-1})}, \quad \forall s \in [-2,2].$$

Using the last inequality along with Corollary 6.11 and (6.23), we have

$$(\log f)'''(s) \le \frac{n}{|\mathbb{S}^{n-1}|} \left[\int_{\mathbb{S}^{n-1}} (n\psi^2(u) - |\nabla_s \psi(u)|^2) du \right] + 2\rho\|\nabla_s \psi\|^2_{L^2}.$$

By (6.12), we deduce

$$(\log f)''(s) \leq \|\nabla_s \psi\|^2_{L^2(\mathbb{S}^{n-1})}\left(2\rho - \frac{1}{2|\mathbb{S}^{n-1}|}\right),$$

which is negative as long as

$$\rho < \frac{1}{4|\mathbb{S}^{n-1}|},$$

and, with this choice, strictly negative unless ψ is a constant function.

Next, we drop the assumption (6.23). For $\psi \in C^2(\mathbb{S}^{n-1})$, let

$$m_\psi = \frac{1}{|\mathbb{S}^{n-1}|}\int_{\mathbb{S}^{n-1}} \psi du, \quad \text{and} \quad \bar{\psi} = \psi - m_\psi.$$

Clearly, $\bar{\psi} \in C^2(\mathbb{S}^{n-1})$ and $\bar{\psi}$ verifies condition (6.23). Moreover,

$$\|\bar{\psi}\|_{C^2(\mathbb{S}^{n-1})} \leq \|\psi\|_{C^2(\mathbb{S}^{n-1})} + |m_\psi| \leq 2\|\psi\|_{C^2(\mathbb{S}^{n-1})}.$$

Consequently, $\bar{\psi} \in \mathcal{U}$ if $\|\psi\|_{C^2(\mathbb{S}^{n-1})} \leq \eta_0/2$. We also have

$$\bar{h}_s := e^{s\bar{\psi}} = e^{s(\psi - m_\psi)} = e^{-sm_\psi} h_s.$$

Hence

$$Q(\bar{h}_s) = e^{-sm_\psi} Q(h_s).$$

Consider

$$\bar{f}(s) := \frac{1}{n}\int_{\mathbb{S}^{n-1}} \bar{h}_s \det(Q(\bar{h}_s))du = e^{-nsm_\psi} f(s).$$

We observe that $\log(\bar{f})$ and $\log(f)$ differ by a linear term and convexity (resp., strict convexity) of f is equivalent to convexity (resp. strict convexity) of $\bar{f}$. On the other hand, by the first part of this proof $\log(\bar{f})$ is concave as long as $\|\bar{\psi}\|_{C^2(\mathbb{S}^{n-1})}$ is sufficiently small, and this condition is verified when, in turn, $\|\psi\|_{C^2(\mathbb{S}^{n-1})}$ is sufficiently small. The proof is concluded. □

6.4 Proofs

Proof of Theorem 6.1. We assume $R = 1$; the general case can be deduced by a scaling argument.

We first suppose that $\|h-1\|_{C^2(\mathbb{S}^{n-1})} \le 1/4$. This implies that $h > 0$ on $\mathbb{S}^{n-1}$ and, therefore, we may write h in the form $h = e^{\psi}$, where $\psi = \log(h) \in C^2(\mathbb{S}^{n-1})$.

We select $\epsilon_0 > 0$ such that $\|h-1\|_{C^2(\mathbb{S}^{n-1})} \le \epsilon_0$ implies $\|\psi\|_{C(\mathbb{S}^{n-1})} \le \eta_0$, that is, $\psi \in \mathcal{U}$ (see Remark 6.9). As a consequence of Proposition 6.2, $h_s = e^{s\psi}$ is the support function of a $C^{2,+}$ convex body, for every $s \in [-2,2]$. In particular, for every $\lambda \in [0,1]$, the function $e^{\lambda\psi}$ is the support function of

$$\lambda K + 0(1-\lambda)B_2^n.$$

There exists $\epsilon > 0$ such that $\|h-1\|_{C^2(\mathbb{S}^{n-1})} \le \epsilon$ implies $\|\psi\|_{C^2(\mathbb{S}^{n-1})} \le \eta$, where $\eta > 0$ is the quantity indicated in Lemma 6.13. By the conclusion of Lemma 6.13, the function $f(\lambda) = |\lambda K + 0(1-\lambda)B_2^n|$ is log-concave, and hence (6.4) follows. The equality case follows from the fact that the log-concavity of f is strict unless ψ is a constant function, which corresponds to the case when K is a ball. □

Below we shall sketch the proof of the corollary; we shall follow essentially the same scheme as in [1].

Sketch of the proof of Corollary 1. First, by integrating the condition $dc_K(u) = R^n du$ over the sphere, we get $|K| = |RB_2^n|$. Theorem 6.1 implies (see [1]):

$$\int_{\mathbb{S}^{n-1}} \log \frac{R}{h_K} dc_K(u) \ge \log \frac{|RB_2^n|}{|K|} = 0,$$

or, equivalently,

$$\int_{\mathbb{S}^{n-1}} \log R dc_K(u) \ge \int_{\mathbb{S}^{n-1}} \log h_K dc_K(u). \tag{6.24}$$

Using the fact that $dc_K(u) = R^n du$ once again, and then applying Theorem 6.1 again, we see that the right-hand side of the above is equal to

$$\int_{\mathbb{S}^{n-1}} \log h_K d_{RB_2^n}(u) \ge \int_{\mathbb{S}^{n-1}} \log R d_{RB_2^n}(u). \tag{6.25}$$

Note that the above is equal to

$$\int_{\mathbb{S}^{n-1}} \log R dc_K(u).$$

We have obtained a chain of inequalities starting and ending with the same expression, and hence equality must hold in all the inequalities. Therefore, K is a Euclidean ball. Since, in addition, $|K| = |RB_2^n|$, we see that $K = RB_2^n$, which completes the proof. □

Remark 6.14. As was mentioned earlier, the inequality (6.24) is called the log-Minkowski inequality; when one of the bodies is a ball, (6.24) in fact follows as a corollary of the well-known Blaschke–Santalo inequality (see, e. g., Schnieder [17]). However, for the proof we need here both directions, the inequality (6.24) as well as the inequality (6.25), and such a result cannot be obtained by completely elementary methods.

We also emphasize that in the proof we crucially need the estimate on the third derivative of our functional, which does not depend on the function ψ: if we did not have such an estimate, we could not conclude the existence of a neighborhood in the C^2-metric, and hence could not talk about the uniqueness result.

Bibliography

[1] K. J. Böröczky, E. Lutwak, D. Yang, G. Zhang, *The log-Brunn–Minkowski inequality*, Adv. Math. **231** (2012), 1974–1997.

[2] K. J. Böröczky, E. Lutwak, D. Yang, G. Zhang, *The logarithmic Minkowski problem*, J. Am. Math. Soc. **26** (2013), 831–852.

[3] K. J. Böröczky, E. Lutwak, D. Yang, G. Zhang, *Affine images of isotropic measures*, J. Differ. Geom. **99** (2015), 407–442.

[4] S. Chen, Y. Huang, Q. Li, J. Liu, *L_p-Brunn–Minkowski inequality for $p \in (1 - \frac{c}{n^{\frac{3}{2}}}, 1)$*, preprint.

[5] S. T. Cheng, S. T. Yau, *On the regularity of solutions of the n-dimensional Minkowski problem*, Commun. Pure Appl. Math. **29** (1976), 495–516.

[6] A. Colesanti, D. Hug, E. Saorín-Gómez, Monotonicity and concavity of integral functionals, Commun. Contemp. Math. **19** (2017), 1–26.

[7] A. Colesanti, G. Livshyts, A. Marsiglietti, *On the stability of Brunn–Minkowski type inequalities*, J. Funct. Anal. **273** (2017), 1120–1139.

[8] D. Cordero-Erausquin, M. Fradelizi, B. Maurey, *The (B) conjecture for the Gaussian measure of dilates of symmetric convex sets and related problems*, J. Funct. Anal. **214** (2004), 410–427.

[9] W. J. Firey, *Shapes of worn stones*, Mathematika **21** (1974), 1–11.

[10] R. Gardner, *The Brunn–Minkowski inequality*, Bull. Am. Math. Soc. **39** (2002), 355–405.

[11] H. Groemer, *Geometric Applications of Fourier Series and Spherical Harmonics*, Cambridge University Press, Cambridge, 1996.

[12] A. Koldobsky, *Fourier Analysis in Convex Geometry*, Math. Surveys and Monographs, AMS, Providence RI, 2005.

[13] A. V. Kolesnikov, E. Milman, *Local L_p-Brunn–Minkowski inequalities for $p < 1$*, preprint.

[14] E. Lutwak, *The Brunn–Minkowski–Firey theory I: Mixed volumes and the Minkowski problem*, J. Differ. Geom. **38** (1) (1993), 131–150.

[15] L. Rotem, *A letter: The log-Brunn–Minkowski inequality for complex bodies*, http://www.tau.ac.il/~liranro1/papers/complexletter.pdf.

[16] C. Saraglou, *Remarks on the conjectured log-Brunn–Minkowski inequality*, Geom. Dedic. **177** (2015), 353–365.

[17] R. Schneider, *Convex Bodies: the Brunn–Minkowski Theory*, 2nd expanded edition, Encyclopedia of Mathematics and its Applications, Cambridge University Press, Cambridge, 2013.

James E. Corbin and Peter Kuchment

7 Spectra of "fattened" open book structures

Dedicated to the memory of the great mathematician and friend, Victor Lomonosov

Abstract: We establish convergence of spectra of Neumann Laplacian in a thin neighborhood of a branching 2D structure in 3D to the spectrum of an appropriately defined operator on the structure itself. This operator is a 2D analog of the well known, by now, quantum graphs. As in the latter case, such considerations are triggered by various physics and engineering applications.

Keywords: Open book structure, Neumann Laplacian, thin structure, spectrum

MSC 2010: 35P99, 58J05, 58J90, 58Z05

7.1 Introduction

We consider a compact subvariety M of $\mathbb{R}^3$ that locally (in a neighborhood of any point) looks like either a smooth submanifold or an "open book" with smooth two-dimensional "pages" meeting transversely along a common smooth one-dimensional "binding";[1] see Figure 7.1. Clearly, any compact smooth submanifold of $\mathbb{R}^3$ (with or without a boundary) qualifies as an open book structure with a single page. Another example of such structure is shown in Figure 7.2.

A "fattened" version M_ϵ of M is an (appropriately defined) ϵ-neighborhood of M, which we call a "fattened open book structure."

Consider now the Laplace operator $-\Delta$ on the domain M_ϵ with Neumann boundary conditions (**"Neumann Laplacian"**), which we denote A_ϵ. As a (nonnegative) elliptic operator on a compact manifold, it has discrete finite multiplicity spectrum $\lambda_n^\epsilon := \lambda_n(A^\epsilon)$ with the only accumulation point at infinity. The result formulated in this work is that when $\epsilon \to 0$, each eigenvalue λ_n^ϵ converges to the corresponding eigenvalue λ_n of an operator A on M, which acts as $-\Delta_M$ (2D Laplace–Beltrami) on each 2D stratum (**page**) of M, with appropriate junction conditions along 1D strata (**bindings**).

1 We do not provide here the general definition of what is called Whitney stratification (see, e. g., [1, 21, 29, 42]), resorting to a simple description through local models.

Acknowledgement: The work of both authors has been partially supported by the NSF DMS-1517938 grant.

James E. Corbin, Peter Kuchment, Department of Mathematics, Texas A& M University, College Station, TX, USA, e-mails: jxc8004@math.tamu.edu, kuchment@math.tamu.edu

https://doi.org/10.1515/9783110656756-007

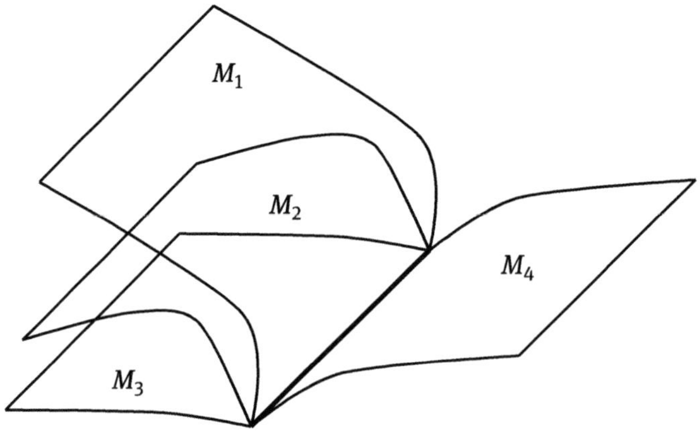

Figure 7.1: An open book structure with **"pages"** M_k meeting at a **"binding."**

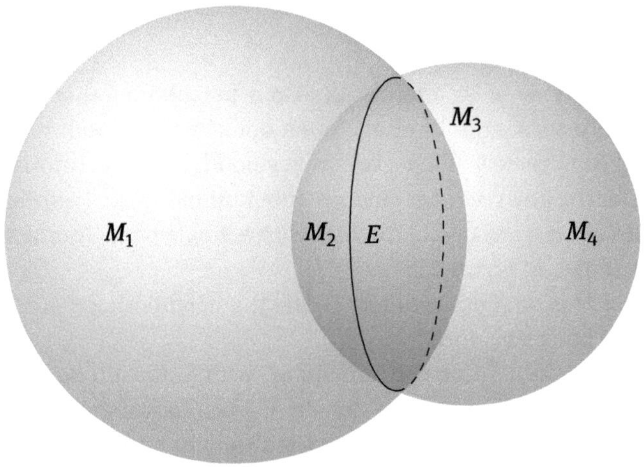

Figure 7.2: A transversal intersection of two spheres yields an open book structure with four pages and a circular binding. The requirement of absence of zero-dimensional strata prohibits adding a third sphere with a generic triple intersection. Tangential contacts of spheres are also disallowed.

Similar results have been obtained previously for the case of fattened graphs (see [27, 36], as well as books [2, 32] and references therein), that is, M being one-dimensional.

The case of a *smooth* submanifold $M \subset \mathbb{R}^3$ is not that hard and has been studied well under a variety of constraints set near M (e. g., [2, 19, 22, 24]). Having singularities along strata of lower dimensions significantly complicates considerations, even in the quantum graph case [4–6, 8, 22, 24, 25, 27, 28, 36, 40].

Our considerations are driven by the similar types of applications (see, e. g., [2, 9, 11–18, 23, 23, 24, 35–39]), as in the graph situation.

Section 7.2 contains the descriptions of the main objects: open book structures and their fattened versions, the Neumann Laplacian A, etc. The next Section 7.3 contains formulation of the result. The proof is reduced to constructing two families of “averaging” and “extension” operators. This construction is even more technical than in the quantum graph case and will be provided in another, much longer text. The last Section 7.4 contains the final remarks and discussions.

In this article, the results are obtained under the following restrictions: the width of the fattened domain shrinks “with the same speed” around all strata; no “corners” (0D strata) are present; the pages intersect transversely at the bindings. Some of them will be removed in a further work.

7.2 The main notions

7.2.1 Open book structures

Simply put, an open book structure[2] M is connected and consists of finitely many connected, compact smooth submanifolds (with or without boundary) of $\mathbb{R}^3$ (**strata**) of dimensions two and one, such that they only intersect along their boundaries and each stratum’s boundary is the union of some lower dimensional strata [21]. We also assume that the strata intersect at their boundaries transversely. In other words, locally M looks either as a smooth surface, or an “open book” with pages meeting at a nonzero angle at a “binding.” Up to a diffeomorphism, a neighborhood of the binding is shown in Figure 7.3.

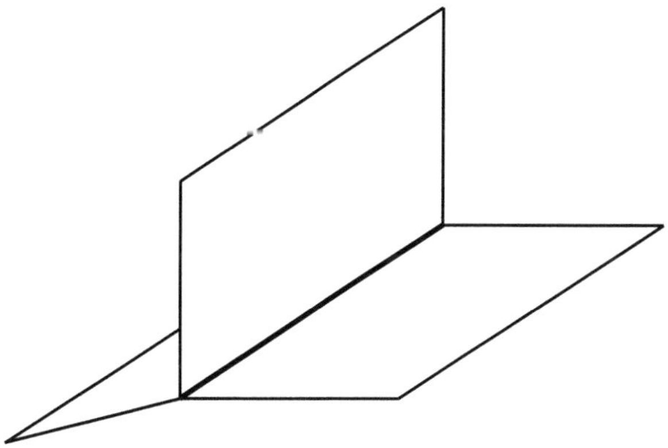

Figure 7.3: A local model of a binding neighborhood.

2 One can find open book structures in a somewhat more general setting being discussed in algebraic topology literature, for example, in [33, 43].

7.2.2 The fattened structure

We can now define the **fattened open book structure** M_ϵ.

Let us remark first of all that there exists $\epsilon_0 > 0$ so small that for any two points x_1, x_2 on the same page of M, the closed intervals of radius ϵ_0 normal to M at these points do not intersect. This ensures that the $\epsilon < \epsilon_0$-fattened neighborhoods do not form a "connecting bridge" between two points that are otherwise far away from each other along M. We will assume that in all our considerations $\epsilon < \epsilon_0$, which is not a restriction, since we will be interested in the limit $\epsilon \to 0$.

We denote the ball of radius r about x as $B(x, r)$.

Definition 7.1. Let M denote an open book structure in $\mathbb{R}^3$ and $\epsilon_0 > 0$, as defined above. We define for any $\epsilon < \epsilon_0$ the corresponding fattened domain M_ϵ as follows:

$$M_\epsilon := \bigcup_{x \in M} B(x, \epsilon). \tag{7.1}$$

7.2.3 Quadratic forms and operators

We adopt the standard notation for Sobolev spaces (see, e. g., [30]). Thus, $H^1(\Omega)$ denotes the space of square integrable with respect to the Lebesgue measure functions on a domain $\Omega \subset \mathbb{R}^n$ with square integrable first-order weak derivatives.

Definition 7.2. Let Q_ϵ be the closed nonnegative quadratic form with domain $H^1(M_\epsilon)$, given by

$$Q_\epsilon(u) = \int_{M_\epsilon} |\nabla u|^2 \, dM_\epsilon. \tag{7.2}$$

We also refer to $Q_\epsilon(u)$ as the **energy** of u.

This form is associated with a unique self-adjoint operator A_ϵ in $L_2(M_\epsilon)$. The following statement is standard (see, e. g., [7, 30]).

Proposition 7.3. *The form Q_ϵ corresponds to the **Neumann Laplacian** $A_\epsilon = -\Delta$ on M_ϵ with its domain consisting of functions in $H^2(M_\epsilon)$ whose normal derivatives at the boundary ∂M_ϵ vanish.*

Its spectrum $\sigma(A_\epsilon)$ is discrete and nonnegative.

Moving now to the limit structure M, we equip it with the surface measure dM induced from $\mathbb{R}^3$.

Definition 7.4. Let Q be the closed, nonnegative quadratic form (**energy**) on $L_2(M)$ given by

$$Q(u) = \sum_k \int_{M_k} |\nabla_{M_k} u|^2 \, dM \tag{7.3}$$

with domain $\mathcal{G}^1$ consisting of functions u for whose the energy $Q(u)$ is finite and that are continuous across the bindings between pages M_k and $M_{k'}$:

$$u|_{\partial M_k \cap E_m} = u|_{\partial M_{k'} \cap E_m}. \tag{7.4}$$

Here, ∇_{M_k} is the gradient along M_k and restrictions in (7.4) to the binding E_m coincide as elements of $H^{1/2}(E_m)$.

Unlike the fattened graph case, by the Sobolev embedding theorem [7] the restriction to the binding is not continuous as an operator from $\mathcal{G}^1$ to $C(E_m)$, it only maps to $H^{1/2}(E_m)$. This distinction significantly complicates the analysis of fattened stratified surfaces in comparison with fattened graphs.

Proposition 7.5. *The operator A associated with the quadratic form Q acts on each M_k as*

$$Au := -\Delta_{M_k} u, \tag{7.5}$$

with the domain $\mathcal{G}^2$ consisting of functions on M such that the following conditions are satisfied:

–

$$\|u\|^2_{L_2(M)} + \|Au\|^2_{L_2(M)} < \infty, \tag{7.6}$$

– *continuity across common bindings E_m of pairs of pages M_k, $M_{k'}$:*

$$u|_{\partial M_k \cap E_m} = u|_{\partial M_{k'} \cap E_m}, \tag{7.7}$$

– ***Kirchhoff condition** at the bindings:*

$$\sum_{k:\partial M_k \supset E_m} D_{\nu_k} u(E_m) = 0, \tag{7.8}$$

where $-\Delta_{M_k}$ is the Laplace–Beltrami operator on M_k and D_{ν_k} denotes the normal derivative to ∂M_k along M_k.

The spectrum of A is discrete and nonnegative.

The proof is simple, standard, and similar to the graph case. We thus omit it.

7.3 The main result

Definition 7.6. We denote the ordered in nondecreasing order eigenvalues of A as $\{\lambda_n\}_{n\in\mathbb{N}}$, and those of A_ϵ as $\{\lambda_n^\epsilon\}_{n\in\mathbb{N}}$.

For a real number Λ not in the spectrum of A_ϵ, we denote by $\mathcal{P}_\Lambda^\epsilon$ the spectral projector of A_ϵ in $L_2(M_\epsilon)$ onto the spectral subspace corresponding to the half-line $\{\lambda \in \mathbb{R} \mid \lambda < \Lambda\}$.

Similarly, $\mathcal{P}_\Lambda$ denotes the analogous spectral projector for A. We then denote the corresponding (finite dimensional) spectral subspaces as $\mathcal{P}_\Lambda^\epsilon L_2(M_\epsilon)$ and $\mathcal{P}_\Lambda L_2(M)$ for M_ϵ and M, respectively.

We now introduce two families of operators needed for the proof of the main result.

Definition 7.7. A family of linear operators J_ϵ from $H^1(M_\epsilon)$ to $\mathcal{G}^1$ is called **averaging operators** if for any $\Lambda \notin \sigma(A_\epsilon)$ there is an ϵ_0 such that for all $\epsilon \in (0, \epsilon_0]$ the following conditions are satisfied:

- For $u \in \mathcal{P}_\Lambda^\epsilon L_2(M_\epsilon)$, J_ϵ is "nearly an isometry" from $L_2(M_\epsilon)$ to $L_2(M)$ with an $o(1)$ error, that is,

$$\left|\|u\|^2_{L_2(M_\epsilon)} - \|J_\epsilon u\|^2_{L_2(M)}\right| \leq o(1)\|u\|^2_{H^1(M_\epsilon)} \tag{7.9}$$

  where $o(1)$ is uniform with respect to u.
- For $u \in \mathcal{P}_\Lambda^\epsilon L_2(M_\epsilon)$, J_ϵ asymptotically "does not increase the energy," that is,

$$Q(J_\epsilon u) - Q_\epsilon(u) \leq o(1)Q_\epsilon(u) \tag{7.10}$$

  where $o(1)$ is uniform with respect to u.

Definition 7.8. A family of linear operators K_ϵ from $\mathcal{G}^1$ to $H^1(M_\epsilon)$ is called **extension operators** if for any $\Lambda \notin \sigma(A)$ there is an ϵ_0 such that for all $\epsilon \in (0, \epsilon_0]$ the following conditions are satisfied:

- For $u \in \mathcal{P}_\Lambda L_2(M)$, K_ϵ is "nearly an isometry" from $L_2(M)$ to $L_2(M_\epsilon)$ with $o(1)$ error, that is,

$$\left|\|u\|^2_{L_2(M)} - \|K_\epsilon u\|^2_{L_2(M_\epsilon)}\right| \leq o(1)\|u\|^2_{\mathcal{G}^1} \tag{7.11}$$

  where $o(1)$ is uniform with respect to u.
- For $u \in \mathcal{P}_\Lambda L_2(M)$, K_ϵ asymptotically "does not increase" the energy, that is,

$$Q_\epsilon(K_\epsilon u) - Q(u) \leq o(1)Q(u) \tag{7.12}$$

  where $o(1)$ is uniform with respect to u.

Existence of such averaging and extension operators is known to be sufficient for spectral convergence of A_ϵ to A (see [32]). For the sake of completeness, we formulate and prove this in our situation.

Theorem 7.9. *Let M be an open book structure and its fattened partner $\{M_\epsilon\}_{\epsilon\in(0,\epsilon_0]}$ as defined before. Let A and A_ϵ be the operators on M and M_ϵ as in Definitions 7.3 and 7.5.*

Suppose there exist averaging operators $\{J_\epsilon\}_{\epsilon\in(0,\epsilon_0]}$ and extension operators $\{K_\epsilon\}_{\epsilon\in(0,\epsilon_0]}$ as stated in Definitions 7.7 and 7.8.

Then, for any n

$$\lambda_n(A_\epsilon) \underset{\epsilon\to 0}{\to} \lambda_n(A).$$

We start with the following standard (see, e. g., [34]) min-max characterization of the spectrum.

Proposition 7.10. *Let B be a self-adjoint nonnegative operator with discrete spectrum of finite multiplicity and $\lambda_n(B)$ be its eigenvalues listed in nondecreasing order. Let also q be its quadratic form with the domain D. Then*

$$\lambda_n(B) = \min_{W\subset D}\ \max_{x\in W\setminus\{0\}} \frac{q(x,x)}{(x,x)}, \tag{7.13}$$

where the minimum is taken over all n-dimensional subspaces W in the quadratic form domain D.

Proof of Theorem 7.9. Proof now employs Proposition 7.10 and the averaging and extension operators J, K to “replant” the test spaces W in (7.13) between the domains of the quadratic forms Q and Q_ϵ.

Let us first notice that due to the definition of these operators (the near-isometry property), for any fixed finite-dimensional space W in the corresponding quadratic form domain, for sufficiently small ϵ the operators are injective on W. Since we are only interested in the limit $\epsilon \to 0$, we will assume below that ϵ is sufficiently small for these operators to preserve the dimension of W. Thus, taking also into account the inequalities (7.9)–(7.12), one concludes that on any fixed finite dimensional subspace W one has the following estimates of Rayleigh ratios:

$$\frac{Q(J_\epsilon u)}{\|J_\epsilon u\|^2_{L_2(M)}} \le (1+o(1))\frac{Q_\epsilon(u)}{\|u\|^2_{L_2(M_\epsilon)}} \tag{7.14}$$

$$\frac{Q_\epsilon(K_\epsilon u)}{\|K_\epsilon u\|^2_{L_2(M_\epsilon)}} \le (1+o(1))\frac{Q(u)}{\|u\|^2_{L_2(M)}}. \tag{7.15}$$

Let now $W_n \subset \mathcal{G}^1$ and $W_n^\epsilon \subset H^1(M_\epsilon)$ be n, such that

$$\lambda_n = \max_{x\in W_n\setminus\{0\}} \frac{Q(x,x)}{(x,x)}, \tag{7.16}$$

and

$$\lambda_n^\epsilon = \max_{x\in W_n^\epsilon\setminus\{0\}} \frac{Q_\epsilon(x,x)}{(x,x)}. \tag{7.17}$$

Due to the min-max description and inequalities (7.14) and (7.15), one gets

$$\lambda_n \leq \sup_{u\in J_\epsilon(W_n^\epsilon)} \frac{Q(J_\epsilon u)}{\|J_\epsilon u\|^2_{L_2(M)}} \leq (1+o(1))\lambda_n^\epsilon, \tag{7.18}$$

and

$$\lambda_n^\epsilon \leq \sup_{u\in K_\epsilon(W_n)} \frac{Q_\epsilon(K_\epsilon u)}{\|K_\epsilon u\|^2_{L_2(M_\epsilon)}} \leq (1+o(1))\lambda_n. \tag{7.19}$$

Thus, $\lambda_n - \lambda_n^\epsilon = o(1)$, which proves the theorem. □

The long technical task, to be addressed elsewhere, consists in proving the following statement.

Theorem 7.11. *Let M be an open book structure and its fattened partner $\{M_\epsilon\}_{\epsilon\in(0,\epsilon_0]}$ as defined before. Let A and A_ϵ be operators on M and M_ϵ as in Definitions 7.3 and 7.5. There exist averaging operators $\{J_\epsilon\}_{\epsilon\in(0,\epsilon_0]}$ and extension operators $\{K_\epsilon\}_{\epsilon\in(0,\epsilon_0]}$ as stated in Definitions 7.7 and 7.8.*

This leads to the main result of this text.

Theorem 7.12. *Let M be an open book structure and its fattened partner $\{M_\epsilon\}_{\epsilon\in(0,\epsilon_0]}$. Let A and A_ϵ be operators on M and M_ϵ as in Definitions 7.3 and 7.5.*

Then, for any n

$$\lambda_n(A_\epsilon) \underset{\epsilon\to 0}{\to} \lambda_n(A).$$

7.4 Conclusions and final remarks

- As the quantum graph case teaches [28, 32], allowing the volumes of the fattened bindings to shrink when $\epsilon \to 0$ slower than those of fattened pages, is expected to lead to interesting phase transitions in the limiting behavior. This is indeed the case, as it will be shown in yet another publication (see also [3]).
- It is more practical to allow presence of zero-dimensional strata (corners). The analysis and results get more complex, as we hope to show in yet another work, with more types of phase transitions.
- Resolvent convergence, rather than weaker local convergence of the spectra, as done in [32] in the graph case, would be desirable and probably achievable.
- One can allow some less restrictive geometries of the fattened domains.
- The case of Dirichlet Laplacian is expected to be significantly different in terms of results and much harder to study, as one can conclude from the graph case considerations [22].

Bibliography

[1] V. I. Arnold, S. M. Gusein-Zade and A. N. Varchenko, *Singularities of Differentiable Maps. Volume 1. Classification of Critical Points, Caustics and Wave Fronts*, Birkhäuser/Springer, New York, 2012.

[2] G. Berkolaiko and P. Kuchment, *Introduction to Quantum Graphs*, AMS 2013.

[3] J. Corbin, *Convergence of Neumann Laplacian on open book structures*, PhD Thesis, Texas A&M University 2019.

[4] G. Dell'Antonio, *Dynamics on quantum graphs as constrained systems*. Rep. Math. Phys. **59** (3) (2007), 267–279.

[5] G. Dell'Antonio and A. Michelangeli, Dynamics on a graph as the limit of the dynamics on a "fat graph", in *Mathematical Technology of Networks*, 49–64, Springer Proc. Math. Stat., **128**, Springer, Cham, 2015.

[6] G. Dell'Antonio, L. Tenuta, *Quantum graphs as holonomic constraints*. J. Math. Phys. **47** (7) (2006), 072102, 21 pp.

[7] D. E. Edmunds and W. Evans, *Spectral Theory and Differential Operators*, Oxford Science Publ., Claredon Press, Oxford, 1990.

[8] W. D. Evans and D. J. Harris, *Fractals, trees and the Neumann Laplacian*, Math. Ann. **296** (1993), 493–527.

[9] P. Exner, J. Keating, P. Kuchment, T. Sunada and A. Teplyaev (eds.), *Analysis on Graphs and its Applications*, Proc. Symp. Pure Math., AMS, 2008.

[10] P. Exner and O. Post, *Convergence of spectra of graph-like thin manifolds*, J. Geom. Phys. **54** (1) (2005), 77–115.

[11] P. Exner, P. Seba, Electrons in semiconductor microstructures: a challenge to operator theorists, in *Proceedings of the Workshop on Schrödinger Operators, Standard and Nonstandard (Dubna 1988)*, 79–100, World Scientific, Singapore, 1989.

[12] A. Figotin and P. Kuchment, *Band-gap structure of the spectrum of periodic and acoustic media. I. Scalar model*, SIAM J. Appl. Math. **56** (1) (1996), 68–88.

[13] A. Figotin and P. Kuchment, *Band-gap structure of the spectrum of periodic and acoustic media. II. 2D photonic crystals*, SIAM J. Appl. Math. **56** (1996), 1561–1620.

[14] A. Figotin and P. Kuchment, 2D photonic crystals with cubic structure: asymptotic analysis, in G. Papanicolaou (ed.), *Wave Propagation in Complex Media*, 23–30, IMA Volumes in Math. and Appl., **96**, 1997.

[15] A. Figotin and P. Kuchment, *Spectral properties of classical waves in high contrast periodic media*, SIAM J. Appl. Math. **58** (2) (1998), 683–702.

[16] M. Freidlin, *Markov Processes and Differential Equations: Asymptotic Problems*, Lectures in Mathematics ETH Zürich, Birkhäuser Verlag, Basel, 1996.

[17] M. Freidlin and A. Wentzell, *Diffusion processes on graphs and the averaging principle*, Ann. Probab., **21** (4) (1993), 2215–2245.

[18] M. Freidlin and A. Wentzell, *Diffusion processes on an open book and the averaging principle*, Stoch. Process. Appl. **113** (2004), 101–126.

[19] R. Froese and I. Herbst, *Realizing holonomic constraints in classical and quantum mechanics*, Commun. Math. Phys. **220** (3) (2001), 489–535.

[20] N. Gerasimenko and B. Pavlov, *Scattering problems on non-compact graphs*, Theor. Math. Phys. **75** (1988), 230–240.

[21] M. Goresky, *Stratified Morse Theory*, Springer Verlag 1988.

[22] D. Grieser, Thin tubes in mathematical physics, global analysis and spectral geometry, in P. Exner, J. Keating, P. Kuchment, T. Sunada and A. Teplyaev (eds.), *Analysis on Graphs and its Applications*, 565–593, Proc. Symp. Pure Math., AMS, 2008.

[23] P. Kuchment, The Mathematics of Photonics Crystals, in G. Bao, L. Cowsar, and W. Masters (eds.), *Mathematical Modeling in Optical Science*, SIAM, 2001.
[24] P. Kuchment, *Graph models for waves in thin structures*, Waves Random Media **12** (4) (2002), R1–R24.
[25] P. Kuchment, Differential and pseudo-differential operators on graphs as models of mesoscopic systems, in *Analysis and applications—ISAAC 2001 (Berlin)*, 7–30, Int. Soc. Anal. Appl. Comput., Kluwer Acad. Publ., Dordrecht, 2003.
[26] P. Kuchment and L. Kunyansky, *Spectral properties of high contrast band-gap materials and operators on graphs*, Exp. Math. **8** (1) (1999), 1–28.
[27] P. Kuchment and H. Zeng, *Convergence of spectra of mesoscopic systems collapsing onto a graph*, J. Math. Anal. Appl. **258** (2001), 671–700.
[28] P. Kuchment, and H. Zeng, Asymptotics of spectra of Neumann Laplacians in thin domains, *Advances in differential equations and mathematical physics (Birmingham, AL, 2002)*, 199–213, Contemp. Math., **327**, Amer. Math. Soc., Providence, RI, 2003.
[29] Y. C. Lu, *Singularity Theory and an Introduction to Catastrophe Theory*, Universitext, Springer-Verlag, New York, Berlin, 1980.
[30] V. Maz'ja, *Sobolev Spaces*, Springer-Verlag, Berlin, 1985.
[31] V. Maz'ya and S. Poborchi, *Differential Functions on Bad Domains*, World Scientific, New Jersey, 1997.
[32] O. Post, *Spectral Analysis on Graph-like Spaces*, Springer-Verlag, Berlin, 2012.
[33] A. Ranicki, *High-dimensional Knot Theory. Algebraic Surgery in Codimension 2*, Springer Verlag, Berlin, 1998.
[34] M. Reed and B. Simon, *Methods of Modern Mathematical Physics I: Functional Analysis*, Academic Press, San Diego, 1980.
[35] J. Rubinstein and M. Schatzman, *Spectral and variational problems on multiconnected strips*, C. R. Acad. Sci. Paris Ser. I Math. **325** (4) (1997), 377–382.
[36] J. Rubinstein and M. Schatzman, *Asymptotics for thin superconducting rings*, J. Math. Pures Appl. **77** (8) (1998), 801–820.
[37] J. Rubinstein and M. Schatzman, *On Multiply Connected Mesoscopic Superconducting Structures*, Sémin. Théor. Spectr. Géom., **15**, Univ. Grenoble I, Saint-Martin-d'Hères, 1998, pages 207–220.
[38] J. Rubinstein and M. Schatzman, *Variational problems on multiply connected thin strips I: basic estimates and convergence of the Laplacian spectrum*, Arch. Ration. Mech. Anal. (2001), 160–271.
[39] K. Ruedenberg and C. W. Scherr, *Free-electron network model for conjugated systems. I. Theory*, J. Chem. Phys. **21** (9) (1953), 1565–1581.
[40] Y. Saito, *The limiting equation of the Neumann Laplacians on shrinking domains, Elec.* J. Differ. Equ. **2000** (31) (2000), 1–25.
[41] M. Schatzman, *On the eigenvalues of the Laplace operator on a thin set with Neumann boundary conditions*, Appl. Anal. **61** (1996), 293–306.
[42] H. Whitney, *Collected Papers, vol. II*, Birkhäuser Boston, Inc., Boston, MA, 1992.
[43] H. E. Winkelnkemper, *Manifolds as Open Books*, Bull. Amer. Math. Soc. **79** (1973), 45–51.

Sheldon Dantas, Sun Kwang Kim, Han Ju Lee, and Martin Mazzitelli

8 On some local Bishop–Phelps–Bollobás properties

Dedicated to the memory of Professor Victor Lomonosov

Abstract: We continue a line of study initiated in [12, 16] about some *local versions* of Bishop–Phelps–Bollobás-type properties for bounded linear operators. We introduce and focus our attention on two of these local properties, which we call $\mathbf{L}_{p,o}$ and $\mathbf{L}_{o,p}$, and we explore the relation between them and some geometric properties of the underlying spaces, such as spaces having strict convexity, local uniform rotundity, and property β of Lindenstrauss. At the end of the paper, we present a diagram comparing all the existing Bishop–Phelps–Bollobás type properties with each other. Some open questions are left throughout the article.

Keywords: Banach space, norm attaining operators, Bishop–Phelps–Bollobás property

MSC 2010: Primary 46B04, Secondary 46B07, 46B20

8.1 Introduction

One of the main results in the theory of norm attaining functions defined on Banach spaces was proved by Errett Bishop and Robert R. Phelps in [7]. They showed that the

Acknowledgement: The second author is the corresponding author. The first author was supported by the project OPVVV CAAS CZ.02.1.01/0.0/0.0/16_019/0000778 and by the Estonian Research Council grant PRG877. The second author was partially supported by Basic Science Research Program through the National Research Foundation of Korea (NRF) funded by the Ministry of Education, Science, and Technology (NRF-2017R1C1B1002928). The third author was supported by the Research program through the National Research Foundation of Korea funded by the Ministry of Education, Science, and Technology (NRF-2016R1D1A1B03934771). The fourth author was partially supported by CONICET PIP 11220130100329CO.

Sheldon Dantas, Department of Mathematics, Faculty of Electrical Engineering, Czech Technical University in Prague, Technická 2, 166 27 Prague 6, Czech Republic, e-mail: gildashe@fel.cvut.cz, https://orcid.org/0000-0001-8117-3760
Sun Kwang Kim, Department of Mathematics, Chungbuk National University, 1 Chungdae-ro, Seowon-Gu, Cheongju, Chungbuk 28644, Republic of Korea, e-mail: skk@chungbuk.ac.kr, https://orcid.org/0000-0002-9402-2002
Han Ju Lee, Department of Mathematics Education, Dongguk University – Seoul, 04620 (Seoul), Republic of Korea, e-mail: hanjulee@dgu.ac.kr, https://orcid.org/0000-0001-9523-2987
Martin Mazzitelli, Instituto Balseiro, CNEA-Universidad Nacional de Cuyo, CONICET, San Carlos de Bariloche, Argentina, e-mail: martin.mazzitelli@crub.uncoma.edu.ar

https://doi.org/10.1515/9783110656756-008

set of all functionals which attain the maximum on a nonempty, closed, bounded, convex subset S of a real Banach space X is norm dense in the dual space X^*. On the other hand, Victor Lomonosov gave in [19] an example which shows that this statement cannot be extended to the complex case by constructing a closed bounded convex subset of some Banach space with no support points. Here, we are interested to study this result when S is the closed unit ball, which simply says that the set of all norm attaining functionals defined on a real or complex Banach space X is dense in X^* (see also [6]). We will refer this last statement as the Bishop–Phelps theorem. Joram Lindenstrauss was the first mathematician who considered the vector valued case of the Bishop–Phelps theorem (see [18]). He produced a counterexample which proves that this theorem is no longer valid for bounded linear operators in general. Nevertheless, he gave some necessary conditions to get a Bishop–Phelps-type theorem for this class of functions. For instance, if the domain X is a reflexive Banach space, then it is true that the set of all norm attaining operators from X into any Banach space Y is dense in the set of all operators from X into Y. After Lindenstrauss, a lot of attention has been paid on this topic. We refer to the survey paper [1] and the references therein for more information about denseness of norm attaining functions in various directions.

In [8], Béla Bollobás proved a stronger version of the Bishop–Phelps theorem, in such a way that whenever a norm-one functional x^* almost attains its norm at some norm-one point x, it is possible to find a new norm-one functional y^* and a new norm-one point y such that y^* attains its norm at y, y is close to x, and y^* is close to x^*. Since the norm of a functional is defined as a supremum and we can always take some point such that a given functional almost attains its norm, Bollobás result says that in the Bishop–Phelps theorem one can control the distances between the involved points and functionals. This result is known nowadays as the Bishop–Phelps–Bollobás theorem. Motivated by Lindenstrauss work, in 2008, María Acosta, Richard Aron, Domingo García, and Manuel Maestre initiated the study of the Bishop–Phelps–Bollobás theorem in the vector-valued case (see [3]). They found conditions on Banach spaces X and Y in order to get a Bishop–Phelps–Bollobás-type theorem for operators from X into Y. For instance, they characterized those spaces Y such that the Bishop–Phelps–Bollobás theorem holds for operators from ℓ_1 into Y. After more than 10 years of [3], there is a huge literature about this topic and we refer the reader to [2, 4, 5, 10, 11, 17] and the references therein for further information. Many different variants of the Bishop–Phelps–Bollobás theorem were introduced during the last years. For some of them, we refer the recent papers [12–15]. Our aim is to study local versions of these properties, as in [16]. Before we explain exactly what this means, let us introduce some notation and necessary preliminaries.

We work on Banach spaces over the field $\mathbb{K}$, which can be the real or complex numbers. We denote by S_X, B_X, and X^* the unit sphere, the closed unit ball, and the topological dual of X, respectively. The symbol $\mathcal{L}(X, Y)$ stands for the set of all bounded linear operators from X into Y and we say that $T \in \mathcal{L}(X, Y)$ attains its norm (or it is norm attaining) if there is $x_0 \in S_X$ such that

$$\|T\| = \sup_{x \in S_X} \|T(x)\| = \|T(x_0)\|.$$

Following [3], we say that a pair of Banach spaces (X, Y) satisfies the Bishop–Phelps–Bollobás property (**BPBp**, for short) if given $\varepsilon > 0$, there is $\eta(\varepsilon) > 0$ such that whenever $T \in \mathcal{L}(X, Y)$ with $\|T\| = 1$ and $x \in S_X$ are such that

$$\|T(x)\| > 1 - \eta(\varepsilon),$$

there are $S \in \mathcal{L}(X, Y)$ with $\|S\| = 1$ and $x_0 \in S_X$ such that

$$\|S(x_0)\| = 1, \quad \|x_0 - x\| < \varepsilon, \quad \text{and} \quad \|S - T\| < \varepsilon.$$

When x can be chosen to coincide with x_0 in the previous definition, we say that (X, Y) has the Bishop–Phelps–Bollobás point property (**BPBpp**, for short); this property was defined and studied in [13, 15]. If instead of fixing the point x (as in the **BPBpp**), we fix the operator T, and we say that (X, Y) has the Bishop–Phelps–Bollobás operator property (see [12, 14]). That is, (X, Y) has the Bishop–Phelps–Bollobás operator property (**BPBop**, for short) if given $\varepsilon > 0$, there is $\eta(\varepsilon) > 0$ such that whenever $T \in \mathcal{L}(X, Y)$ with $\|T\| = 1$ and $x_0 \in S_X$ are such that $\|T(x_0)\| > 1 - \eta(\varepsilon)$, there is $x_1 \in S_X$ such that $\|T(x_1)\| = 1$ and $\|x_0 - x_1\| < \varepsilon$. Notice that the **BPBp**, **BPBpp**, and **BPBop** are uniform properties in the sense that η depends just on a given $\varepsilon > 0$. As we already mentioned before, we are interested to study the situations when η depends not only on ε, but also on the vector x or the operator T. Some of them were already studied by the authors of the present paper in [16] and here we are using a similar notation. We state now the definitions of the two local properties on which we will focus.

Definition 8.1.

(a) A pair (X, Y) has the $\mathbf{L}_{p,o}$ if given $\varepsilon > 0$ and $T \in \mathcal{L}(X, Y)$ with $\|T\| = 1$, then there is $\eta(\varepsilon, T) > 0$ such that whenever $x \in S_X$ satisfies

$$\|T(x)\| > 1 - \eta(\varepsilon, T),$$

there is $S \in \mathcal{L}(X, Y)$ with $\|S\| = 1$ such that

$$\|S(x)\| = 1 \quad \text{and} \quad \|S - T\| < \varepsilon.$$

(b) A pair (X, Y) has the $\mathbf{L}_{o,p}$ if given $\varepsilon > 0$ and $x \in S_X$, then there is $\eta(\varepsilon, x) > 0$ such that whenever $T \in \mathcal{L}(X, Y)$ with $\|T\| = 1$ satisfies

$$\|T(x)\| > 1 - \eta(\varepsilon, x),$$

there is $x_0 \in S_X$ such that

$$\|T(x_0)\| = 1 \quad \text{and} \quad \|x_0 - x\| < \varepsilon.$$

Let us clarify the notation: in the symbol $\mathbf{L}_{\square,\triangle}$, both $\square$ and $\triangle$ can be p or o, which are the initials of the words point and operator, respectively. If the pair of Banach spaces (X, Y) satisfies the $\mathbf{L}_{\square,\triangle}$, then it means that we fix $\square$ and η depends on $\triangle$.

In [16], properties $\mathbf{L}_{p,p}$ and $\mathbf{L}_{o,o}$ were addressed. Both of them are deeply related to geometric properties of the involved Banach spaces as, for instance, local uniform rotundity or some of the Kadec–Klee properties. In fact, it turns out that the $\mathbf{L}_{p,p}$ for linear functionals defined on a Banach space X is equivalent to the strong subdifferentiability of the norm of X (see [16, Theorem 2.3]). It is also a straightforward observation that if X is reflexive then $\mathbf{L}_{o,o}$ is dual to $\mathbf{L}_{p,p}$ in the sense that $(X, \mathbb{K})$ has the $\mathbf{L}_{o,o}$ if and only if $(X^*, \mathbb{K})$ has the $\mathbf{L}_{p,p}$ (see [16, Proposition 2.2]). Additionally, we would like to remark that it is not clear whether properties $\mathbf{L}_{p,o}$ and $\mathbf{L}_{o,p}$ imply reflexivity. Indeed, let us observe first that property $\mathbf{L}_{o,o}$ says that if ε and T are given, there exists $\eta = \eta(\varepsilon, T) > 0$ such that whenever x satisfies $\|T(x)\| > 1 - \eta$, there is a new norm one element x_0 such that it is close to x and T itself attains the norm at x_0. This means that if (X, Y) has the $\mathbf{L}_{o,o}$, then *every operator attains the norm* and, consequently, by the James's theorem, X must be reflexive (see comments just after [16, Definition 2.1]). Considering now property $\mathbf{L}_{p,o}$, although in this case η depends on ε and T, we get a new norm attaining operator S which is close to T and this does not give us any information whether T is also norm attaining or not. So, we cannot conclude that X is reflexive as in the $\mathbf{L}_{o,o}$ case. On the other hand, considering property $\mathbf{L}_{o,p}$, we have that η depends on given ε and x, and although T attains the norm in this case, not every operator satisfies condition $\|T(x)\| > 1 - \eta(\varepsilon, x)$, so again we cannot conclude that every operator attains the norm and apply James's theorem.

We describe now the contents of the paper. In first place, we obtain sufficient and necessary conditions for a pair $(X, \mathbb{K})$ to have the $\mathbf{L}_{o,p}$ in terms of some *rotundity* properties of X. Recall that a Banach space is strictly convex if $\|\frac{x+y}{2}\| < 1$ whenever $x, y \in S_X$, $x \neq y$, and that is locally uniformly rotund (LUR, for short) if for all $x, x_n \in S_X$, $\lim_n \|x_n + x\| = 2$ implies $\lim_n \|x_n - x\| = 0$. It is a well-known fact that if X is LUR, then is strictly convex. We prove that

$$\text{if } X \text{ is reflexive,} \quad X \text{ is LUR} \Rightarrow (X, \mathbb{K}) \text{ has the } \mathbf{L}_{o,p} \Rightarrow X \text{ is strictly convex.} \tag{8.1}$$

We also prove that there exists a dual relation between properties $\mathbf{L}_{p,o}$ and $\mathbf{L}_{o,p}$ in the functional case and, as a consequence, we get that if X is reflexive and X^* is locally uniformly rotund, then the pair $(X, \mathbb{K})$ satisfies the $\mathbf{L}_{p,o}$. As a consequence of (8.1) and the dual relation between $\mathbf{L}_{p,o}$ and $\mathbf{L}_{o,p}$ we see that, even for 2-dimensional spaces, there is a Banach space X such that the pair $(X, \mathbb{K})$ fails both properties. This establishes a difference between the local properties $\mathbf{L}_{p,o}$, $\mathbf{L}_{o,p}$ and $\mathbf{L}_{p,p}$, $\mathbf{L}_{o,o}$, since the latter properties hold in the finite-dimensional case. Concerning linear operators, we show that pairs of the form (X, ℓ_∞^2) and (Z, Z), where $\dim(X) \geqslant 2$ and Z is a 2-dimensional space, fail property $\mathbf{L}_{o,p}$. The situation with pairs like (X, ℓ_∞^2) changes for property $\mathbf{L}_{p,o}$: we prove that if Y has property β of Lindenstrauss with a finite index set I, then the pair (X, Y) satisfies the $\mathbf{L}_{p,o}$ whenever $(X, \mathbb{K})$ does. Nevertheless, this is no long true when I is infinite and we present a counterexample to prove this. Finally, we show that (ℓ_1, Y) and (c_0, Y) fail both properties for all Banach spaces Y. In the last part of the paper, we compare all of these properties with each other and also with the **BPBp**, **BPBpp**, and **BPBop**.

8.2 The results

In this section, we show the results we have for both properties $\mathbf{L}_{o,p}$ and $\mathbf{L}_{p,o}$. We start by proving some positive results. Notice that it is clear that the **BPBpp** implies the $\mathbf{L}_{p,o}$. Hence, there are some immediate examples of pairs of Banach spaces (X, Y) satisfying the $\mathbf{L}_{p,o}$ (see [13, 15] for positive results on the **BPBpp**). It is also clear that the **BPBop** implies the $\mathbf{L}_{o,p}$, although this does not provide many examples, since the **BPBop** holds only for the pairs $(\mathbb{K}, Y)$ for every Banach space Y and $(X, \mathbb{K})$ for uniformly convex Banach spaces X (see [13, 17]). Here, we get other examples of pairs (X, Y) satisfying the properties $\mathbf{L}_{p,o}$ and $\mathbf{L}_{o,p}$ (see Proposition 8.2, Corollary 8.5, and Theorems 8.9 and 8.11).

Proposition 8.2. *Let X be a Banach space.*
(i) *If X is reflexive and LUR, then the pair $(X, \mathbb{K})$ has the $\mathbf{L}_{o,p}$.*
(ii) *If X has the Radon–Nikodým property and $(X, \mathbb{K})$ has the $\mathbf{L}_{o,p}$, then X is strictly convex.*

Proof. (i) Otherwise, there are $\varepsilon_0 > 0$ and $x_0 \in S_X$ such that for every $n \in \mathbb{N}$, there is $x_n^* \in S_{X^*}$ with

$$1 \geqslant |x_n^*(x_0)| \geqslant 1 - \frac{1}{n}$$

such that whenever $x \in S_X$ satisfies $\|x - x_0\| < \varepsilon_0$, we have that $|x_n^*(x)| < 1$. Since X is reflexive, there is $x_n \in S_X$ such that $|x_n^*(x_n)| = 1$ for every $n \in \mathbb{N}$. For suitable modulus 1 constants c_n, we have that

$$1 \geqslant \left\| \frac{c_n x_n + x_0}{2} \right\| \geqslant \left| \frac{x_n^*(c_n x_n) + x_n^*(x_0)}{2} \right| \longrightarrow 1.$$

Since X is LUR, we see that $\|c_n x_n - x_0\| \longrightarrow 0$ as $n \to \infty$. Then we must have $|x_n^*(c_n x_n)| < 1$ for large enough n and this is a contradiction.

(ii) Let $\varepsilon > 0$ and $x, y \in S_X$ such that $\|x - y\| \geqslant \varepsilon$. We want to show that there is $\delta(\varepsilon, x, y) > 0$ such that $\frac{\|x+y\|}{2} \leqslant 1 - \delta(\varepsilon, x, y)$. Let Γ be the set of all bounded linear functionals in S_{X^*} that strongly expose B_{X^*}. Following the proof of [17, Theorem 2.1] (with slight modifications), we get that each $x^* \in \Gamma$ satisfies either

$$\operatorname{Re} x^*(x) \leqslant 1 - \min\left\{\eta\left(\frac{\varepsilon^2}{64}, x\right), \eta\left(\frac{\varepsilon^2}{64}, y\right), \frac{\varepsilon^2}{64}\right\}$$

or

$$\operatorname{Re} x^*(y) \leqslant 1 - \min\left\{\eta\left(\frac{\varepsilon^2}{64}, x\right), \eta\left(\frac{\varepsilon^2}{64}, y\right), \frac{\varepsilon^2}{64}\right\},$$

where $\eta(\cdot,\cdot)$ is the function in the definition of $\mathbf{L}_{o,p}$. Now, since X has the Radon–Nikodým property we have that Γ is dense in S_{X^*} (see [9, 21]) and, consequently,

$$\begin{aligned}\frac{\|x+y\|}{2} &= \sup\left\{\operatorname{Re}\frac{x^*(x)+x^*(y)}{2} : x^* \in \Gamma\right\} \\ &\leqslant \frac{2 - \min\{\eta(\frac{\varepsilon^2}{64}, x), \eta(\frac{\varepsilon^2}{64}, y), \frac{\varepsilon^2}{64}\}}{2} \\ &= 1 - \frac{1}{2}\min\left\{\eta\left(\frac{\varepsilon^2}{64}, x\right), \eta\left(\frac{\varepsilon^2}{64}, y\right), \frac{\varepsilon^2}{64}\right\}.\end{aligned}$$

Then, $\delta(\varepsilon, x, y) = \frac{1}{2}\min\{\eta(\frac{\varepsilon^2}{64}, x), \eta(\frac{\varepsilon^2}{64}, y), \frac{\varepsilon^2}{64}\}$. □

As we already commented in the Introduction, we do not know if reflexivity (or the Radon–Nikodým property) is a necessary condition for the $\mathbf{L}_{o,p}$ in the above proposition. However, if we assume that X is reflexive, we have some consequences of Proposition 8.2 (see Corollary 8.5). Before stating it, let us prove the following.

Proposition 8.3. *Let X be a Banach space. If $(X^*, \mathbb{K})$ has the $\mathbf{L}_{o,p}$, then $(X, \mathbb{K})$ has the $\mathbf{L}_{p,o}$.*

Proof. Assume $\varepsilon > 0$ and $x^* \in X^*$ with $\|x^*\| = 1$ are given. By hypothesis, we can take the constant $\eta(\varepsilon, x^*) > 0$ for the $\mathbf{L}_{o,p}$ of the pair $(X^*, \mathbb{K})$. Let $x \in S_X$ be such that $|x^*(x)| > 1 - \eta(\varepsilon, x^*)$. Using the canonical inclusion $\hat{} : X \longrightarrow X^{**}$, we have $|\hat{x}(x^*)| = |x^*(x)| > 1 - \eta(\varepsilon, x^*)$, and so there exists $x_1^* \in S_{X^*}$ such that $|\hat{x}(x_1^*)| = |x_1^*(x)| = 1$ and $\|x_1^* - x^*\| < \varepsilon$. This proves that $(X, \mathbb{K})$ has the $\mathbf{L}_{p,o}$. □

Proposition 8.4. *Let X be a reflexive Banach space. The pair $(X, \mathbb{K})$ has the $\mathbf{L}_{p,o}$ if and only if $(X^*, \mathbb{K})$ has the $\mathbf{L}_{o,p}$.*

Proof. From Proposition 8.3, we need to prove just the "only if" part. Assume $\varepsilon > 0$ and $x^* \in S_{X^*}$ are given. By hypothesis, there is the constant $\eta(\varepsilon, x^*) > 0$ for the $\mathbf{L}_{p,o}$ of the pair $(X, \mathbb{K})$. Let $x^{**} \in X^{**}$ with $\|x^{**}\| = 1$ be such that $|x^{**}(x^*)| > 1 - \eta(\varepsilon, x^*)$. Using the canonical inclusion $\hat{} : X \longrightarrow X^{**}$ and the reflexivity of X, there exists $x \in X$ such that $\hat{x} = x^{**}$. Hence, we have $|x^{**}(x^*)| = |x^*(x)| > 1 - \eta(\varepsilon, x^*)$, and so there exists $z \in S_X$ such that $|x^*(z)| = 1$ and $\|z - x\| < \varepsilon$. The bidual element $\hat{z}$ is the desired one for the $\mathbf{L}_{o,p}$ of the pair $(X^*, \mathbb{K})$. □

As a combination of Propositions 8.2, 8.3, and 8.4, we get the following result.

Corollary 8.5. *Let X be a reflexive Banach space.*
(i) *If X^* is LUR, then the pair $(X, \mathbb{K})$ has the $\mathbf{L}_{p,o}$.*
(ii) *If $(X, \mathbb{K})$ has the $\mathbf{L}_{p,o}$, then X^* is strictly convex.*

At this point, we would like to stress some open problems that we are not able to solve. The first one was mentioned above. The second one relies on the fact that those spaces X for which we can assure that $(X, \mathbb{K})$ has the $\mathbf{L}_{o,p}$ (resp., $\mathbf{L}_{p,o}$), satisfy also that $(X, \mathbb{K})$ has the $\mathbf{L}_{o,o}$ (resp., $\mathbf{L}_{p,p}$). Indeed, it was already observed (see the discussion above [16, Theorem 2.5]) that if X is reflexive and LUR, then $(X, \mathbb{K})$ has the $\mathbf{L}_{o,o}$.

Question 1. Does $\mathbf{L}_{o,p}$ (or $\mathbf{L}_{p,o}$) of the pair $(X, \mathbb{K})$ imply reflexivity of X?

Question 2. Does $\mathbf{L}_{o,p}$ (resp., $\mathbf{L}_{p,o}$) imply $\mathbf{L}_{o,o}$ (resp., $\mathbf{L}_{p,p}$) for the pair $(X, \mathbb{K})$?

It is known that, for finite-dimensional Banach spaces X and Y, the pair (X, Y) satisfies property $\mathbf{L}_{p,p}$ ([16, Proposition 2.9]). Besides this, it was proved in [12] that if X is finite dimensional, then the pair (X, Y) has the $\mathbf{L}_{o,o}$ for every Banach space Y. However, this is not the case for both properties $\mathbf{L}_{p,o}$ and $\mathbf{L}_{o,p}$ even for linear functionals defined on 2-dimensional spaces. This is an immediate consequence of Proposition 8.4 and item (ii) in Proposition 8.2. In what follows, we denote by ℓ_p^n the n-dimensional space endowed with the p-norm and $(e_i)_{i=1}^n$ their canonical basis.

Proposition 8.6. *The pairs $(\ell_1^n, \mathbb{K})$ and $(\ell_\infty^n, \mathbb{K})$ fail both $\mathbf{L}_{p,o}$ and $\mathbf{L}_{o,p}$ for $n \geqslant 2$.*

The next result shows that all the pairs of the form (X,X), for 2-dimensional Banach spaces X fails the $\mathbf{L}_{o,p}$ for linear operators.

Proposition 8.7. *Let X be a 2-dimensional Banach space. Then the pair (X,X) fails the $\mathbf{L}_{o,p}$.*

Proof. Consider $\{(v_1, v_1^*), (v_2, v_2^*)\}$ the Auerbach basis of the space X. Then, for every $x \in X$, we have that $x = v_1^*(x)v_1 + v_2^*(x)v_2$. Let us suppose by contradiction that the pair (X,X) satisfies the $\mathbf{L}_{o,p}$ with some function $\eta(\cdot,\cdot)$ and let $n \in \mathbb{N}$ be such that $\frac{1}{n} < \eta(\varepsilon_0, v_1)$ for a fixed positive number $\varepsilon_0 \in (0,1)$. Define $T_n : X \longrightarrow X$ by

$$T_n(x) := \left(1 - \frac{1}{n}\right)v_1^*(x)v_1 + v_2^*(x)v_2 \quad (x \in X).$$

We see that

$$\|T_n(x)\| = \left\|\left(1 - \frac{1}{n}\right)v_1^*(x)v_1 + v_2^*(x)v_2\right\| \leqslant \left(1 - \frac{1}{n}\right)\|(v_1^*(x)v_1 + v_2^*(x)v_2)\| + \frac{1}{n}\|v_2^*(x)v_2\|$$
$$\leqslant \left(1 - \frac{1}{n}\right)\|x\| + \frac{1}{n} \leqslant 1$$

for arbitrary $x \in B_X$. This implies that $\|T_n\| = 1 = \|T_n(v_2)\|$. Now, since

$$\|T_n(v_1)\| = 1 - \frac{1}{n} > 1 - \eta(\varepsilon_0, v_1),$$

there is $x_0 \in S_X$ such that $\|T_n(x_0)\| = 1$ and $\|x_0 - v_1\| < \varepsilon_0$. On the other hand, we have that

$$1 = \|T_n(x_0)\| \leqslant \left(1 - \frac{1}{n}\right)\|x_0\| + \frac{1}{n}|v_2^*(x_0)| \leqslant 1$$

which implies $|v_2^*(x_0)| = 1$. This gives us a contradiction since $1 > \varepsilon_0 > \|x_0 - v_1\| > |v_2^*(x_0)|$. □

We get another negative result for the property $\mathbf{L}_{o,p}$ when the range space is ℓ_∞^2.

Proposition 8.8. *Let X be a Banach space with* $\dim(X) \geqslant 2$. *Then (X, ℓ_∞^2) fails the $\mathbf{L}_{o,p}$.*

Proof. Let $x_1, x_2 \in S_X$ and $x_1^*, x_2^* \in S_{X^*}$ be such that $x_i^*(x_j) = \delta_{ij}$ for $i,j = 1,2$ (we may choose such elements by taking the Hahn–Banach extension of functionals of the Auerbach basis on a 2-dimensional subspace of X). We assume that the pair (X, ℓ_∞^2) has the $\mathbf{L}_{o,p}$ with some function $\eta(\cdot,\cdot)$ and consider $n \in \mathbb{N}$ such that $\frac{1}{n} < \eta(\varepsilon_0, x_1)$ for a fixed positive number $\varepsilon_0 \in (0,1)$. Define $T_n : X \longrightarrow \ell_\infty^2$ by

$$T_n(x) := \left(\left(1 - \frac{1}{n}\right)x_1^*(x), x_2^*(x)\right) \quad (x \in X).$$

Then $\|T_n\| \leqslant 1$ and $\|T_n(x_2)\|_\infty = 1$, which implies $\|T_n\| = 1$. Since

$$\|T_n(x_1)\|_\infty = 1 - \frac{1}{n} > 1 - \eta(\varepsilon_0, x_1),$$

there is $z \in S_X$ such that $\|T_n(z)\|_\infty = 1$ and $\|z - x_1\| < \varepsilon_0$. So, since

$$1 = \|T_n(z)\|_\infty = \max\left\{\left(1 - \frac{1}{n}\right)|x_1^*(z)|, |x_2^*(z)|\right\},$$

we have that $|x_2^*(z)| = 1$. Nevertheless, we have that $1 > \varepsilon_0 > \|z - x_1\| \geqslant |x_2^*(z) - x_2^*(x_1)| = |x_2^*(z)|$, which gives a contradiction. □

Taking into account Propositions 8.7 and 8.8 and Corollary 8.13 below, we leave the following open question.

Question 3. Are there spaces X, Y with $\dim(X)$, $\dim(Y) \geqslant 2$ such that (X, Y) satisfies property $\mathbf{L}_{o,p}$?

Although we have a negative result in Proposition 8.8 for the $\mathbf{L}_{o,p}$, the situation with property $\mathbf{L}_{p,o}$ is quite different. Indeed, we will prove that when we assume that the pair $(X, \mathbb{K})$ has the $\mathbf{L}_{p,o}$, so does the pair (X, ℓ_∞^2). In fact, we get a more general result for Banach spaces satisfying property β of Lindenstrauss (see [18]). We say that a Banach space Y has property β with a index set I and a constant $0 \leqslant \rho < 1$ if there is a set $\{(y_i, y_i^*) : i \in I\} \subset S_Y \times S_{Y^*}$ such that

- $y_i^*(y_i) = 1$ for all $i \in I$,
- $|y_i^*(y_j)| \leqslant \rho < 1$ for all $i, j \in I$ with $i \neq j$, and
- $\|y\| = \sup_{i \in I} |y_i^*(y)|$ for all $y \in Y$.

Examples of Banach spaces satisfying such a property are $c_0(I)$ and $\ell_\infty(I)$ by taking $\{(e_i, e_i^*) : i \in I\}$, the canonical biorthogonal system of these spaces. For the next result, we notice that in Definition 8.1(a) one can use $T \in B_{\mathcal{L}(X,Y)}$ instead of $T \in S_{\mathcal{L}(X,Y)}$ by a simple change of parameters.

Theorem 8.9. *Let X, Y be Banach spaces. Suppose that $(X, \mathbb{K})$ satisfies the $\mathbf{L}_{p,o}$ and assume that Y has property β with a finite index set I and constant ρ. Then the pair (X, Y) has the $\mathbf{L}_{p,o}$.*

Proof. The proof is similar to [15, Proposition 2.4], but we give the details for sake of completeness. Let I be a finite set and $\{(y_i, y_i^*) : i \in I\} \subset S_Y \times S_{Y^*}$ be the set of property β. Consider $\eta(\cdot,\cdot)$, the function for the pair $(X, \mathbb{K})$, which satisfies the $\mathbf{L}_{p,o}$. For each $\varepsilon > 0$ and $T \in S_{\mathcal{L}(X,Y)}$, we define

$$\psi(\varepsilon, T) = \min_{i \in I}\{\eta(\varepsilon, y_i^* \circ T)\} > 0.$$

Fixed $\varepsilon_0 > 0$ and $T_0 \in S_{\mathcal{L}(X,Y)}$, we choose $0 < \xi < \frac{\varepsilon_0}{4}$ such that

$$1 + \rho\left(\frac{\varepsilon_0}{4} + \xi\right) < \left(1 + \frac{\varepsilon_0}{4}\right)(1 - \xi). \tag{8.2}$$

Now, let $x_0 \in S_X$ be such that $\|T_0(x_0)\| > 1 - \psi(\xi, T_0)$. By the definition of property β and the construction of ψ, there exists $k \in I$ such that

$$|y_k^*(T_0(x_0))| > 1 - \psi(\xi, T_0) \geqslant 1 - \eta(\xi, y_k \circ T_0).$$

Hence, there exists a functional $x_1^* \in S_{X^*}$ such that $|x_1^*(x_0)| = 1$ and $\|x_1^* - y_k^* \circ T_0^*\| < \xi$. Now, we define $U : X \longrightarrow Y$ by

$$U(x) := T_0(x) + \left[\left(1 + \frac{\varepsilon_0}{4}\right)x_1^*(x) - y_k^* \circ T_0^*(x)\right]y_k \quad (x \in X).$$

We have that $\|U - T_0\| < \frac{\varepsilon_0}{4} + \xi < \frac{\varepsilon_0}{2}$. Moreover, for arbitrary $j \neq k$, we have that

$$\|y_j^* \circ U\| \leqslant 1 + \rho\left(\frac{\varepsilon_0}{4} + \xi\right) < \left(1 + \frac{\varepsilon_0}{4}\right)(1 - \xi) < 1 + \frac{\varepsilon_0}{4} \quad \text{and} \quad \|y_k^* \circ U\| = 1 + \frac{\varepsilon_0}{4}.$$

Then U attains its norm at x_0 and so the operator $V := U/\|U\|$ is the one we were looking for. □

The main difference between [15, Proposition 2.4] and Theorem 8.9 is the cardinality of the index set I. Indeed, in [15, Proposition 2.4], we see that the set I does not need to be finite, since if X is uniformly smooth, then the pair $(X, \mathbb{K})$ satisfies the **BPBpp** and so does the $\mathbf{L}_{p,o}$, which is, in this case, uniform, in the sense that η depends only on a given $\varepsilon > 0$. This gives that $\psi(\varepsilon, T) = \inf_{i \in I}\{\eta(\varepsilon, y_i^* \circ T)\}$, in the proof of Theorem 8.9, is strictly bigger than 0. Naturally, one may ask whether the same result holds for infinite index sets. It turns out that this is not the case. To see why this happens, we consider the Banach space $X = [\bigoplus_{i=2}^{\infty} \ell_i^2]_{\ell_2}$, the ℓ_2 direct sum of 2-dimensional ℓ_i-spaces. We have that X^* is a reflexive LUR Banach space (see, e. g., [20, Theorem 1.1]). Hence, the pair $([\bigoplus_{i=2}^{\infty} \ell_i^2]_{\ell_2}, \mathbb{K})$ satisfies property $\mathbf{L}_{p,o}$ by Corollary 8.5. Recall that ℓ_∞ satisfies property β with $I = \mathbb{N}$ and $\rho = 0$. Our counterexample is described in the next proposition.

Proposition 8.10. *The pair* $([\bigoplus_{i=2}^{\infty} \ell_i^2]_{\ell_2}, \ell_\infty)$ *does not satisfy the* $\mathbf{L}_{p,o}$.

Proof. We denote by E_i and $\tilde{E}_i$ the natural embeddings from ℓ_i^2 to X and $(\ell_i^2)^*$ to X^*. Also we denote by P_i the natural projections from ℓ_∞ to the ith coordinate. For $f_i^* = (1, 0) \in S_{(\ell_i^2)^*}$, we define $T \in S_{\mathcal{L}(X,\ell_\infty)}$ by $T(\cdot) = (\tilde{E}_i f_i^*(\cdot))_i$. Note that for each $z_i = (\frac{1}{2^{1/i}}, \frac{1}{2^{1/i}}) \in S_{\ell_i^2}$, the element $z_i^* = (\frac{1}{2^{1-\frac{1}{i}}}, \frac{1}{2^{1-\frac{1}{i}}})$ is the unique norm-one functional so that $z_i^*(z_i) = 1$. This shows that $\tilde{E}_i z_i^*$ is the unique element in S_{X^*} so that $\tilde{E}_i z_i^*(E_i z_i) = 1$, and then if an operator $S \in S_{\mathcal{L}(X,\ell_\infty)}$ attains its norm at $E_i z_i$, then there exists $j_0 \in \mathbb{N}$ and a modulus 1 scalar c so that $P_{j_0} S = c\tilde{E}_i z_i^*$. From the construction, we see that

$$\|T(E_i z_i)\| \longrightarrow 1 \quad \text{and} \quad \|P_j T - c\tilde{E}_i z_i^*\| > \frac{1}{2^{1-\frac{1}{i}}}$$

for any modulus 1 scalar c and $j \in \mathbb{N}$. This proves that $([\bigoplus_{i=2}^{\infty} \ell_i^2]_{\ell_2}, \ell_\infty)$ cannot satisfy the $\mathbf{L}_{p,o}$ □

Next, we give some results on stability concerning properties $\mathbf{L}_{p,o}$ and $\mathbf{L}_{o,p}$. Recall that a subspace Z of a Banach space X is one-complemented if Z is the range of a norm-one projection on X.

Theorem 8.11. *Let X, Y be Banach spaces, and let Z be a one-complemented subspace Z of X.*
(i) *If the pair (X, Y) has the $\mathbf{L}_{p,o}$, then so does (Z, Y).*
(ii) *If the pair (X, Y) has the $\mathbf{L}_{o,p}$, then so does (Z, Y).*

Proof. We denote by E and P the canonical embedding and projection between Z and X, respectively.

(i) Let $\varepsilon > 0$ and $T \in S_{\mathcal{L}(Z,Y)}$ be given. Assume that $z \in S_Z$ satisfy $\|T(z)\| > 1 - \eta(\varepsilon, T \circ P)$, where $\eta(\cdot,\cdot)$ is the function for the pair (X, Y) having the $\mathbf{L}_{p,o}$. Since $\|(T \circ P)(E(z))\| = \|T(z)\|$ and $\|T \circ P\| = \|T\|$, there exists $S \in S_{\mathcal{L}(X,Y)}$ such that $\|S(E(z))\| = 1$ and $\|S - T \circ P\| < \varepsilon$. Since $\|S \circ E - T\| \leqslant \|S - T \circ P\|$, we finish the proof.

(ii) Let $\varepsilon > 0$ and $z \in S_Z$ be given. Assume that $T \in S_{\mathcal{L}(Z,Y)}$ satisfy $\|T(z)\| > 1 - \eta(\varepsilon, E(z))$, where $\eta(\cdot,\cdot)$ is the function for the pair (X, Y) having the $\mathbf{L}_{o,p}$. Since $\|(T \circ P)(E(z))\| = \|T(z)\|$ and $\|T \circ P\| = \|T\|$, there exists $x \in S_X$ such that $\|x - E(z)\| < \varepsilon$ and $\|T \circ P(x)\| = 1$. Since $\|P(x) - z\| \leqslant \|x - E(z)\|$, we complete the proof. □

Proposition 8.12. *Let X and Y be Banach spaces.*
(i) *If the pair (X, Y) has the $\mathbf{L}_{o,p}$ for some Banach space Y, then so does $(X, \mathbb{K})$.*
(ii) *If the pair (X, Y) has the $\mathbf{L}_{p,o}$ for some Banach space Y, then so does $(X, \mathbb{K})$.*

Proof. (i) Let $\varepsilon > 0$ and $x \in S_X$ be given. By hypothesis, there is $\eta(\varepsilon, x) > 0$ for the pair (X, Y). Let $x^* \in X^*$ with $\|x^*\| = 1$ be such that $|x^*(x)| > 1 - \eta(\varepsilon, x)$. Define $T \in \mathcal{L}(X, Y)$ by $T(z) := x^*(z)y_0$ for $z \in X$ and for a fixed $y_0 \in S_Y$. Then $\|T\| = \|x^*\| = 1$ and $\|T(x)\| =$

$|x^*(x)| > 1 - \eta(\varepsilon, x)$. So, there is $x_0 \in S_X$ such that $\|T(x_0)\| = |x^*(x_0)| = 1$ and $\|x_0 - x\| < \varepsilon$. This proves that $(X, \mathbb{K})$ has the $\mathbf{L}_{o,p}$.

(ii) Let $\varepsilon > 0$ and $x^* \in X^*$ with $\|x^*\| = 1$ be given. Again, define $T(z) := x^*(z)y_0$ for $z \in X$ and for a fixed $y_0 \in S_Y$. Set $\eta(\varepsilon, x^*) := \eta(\varepsilon, T) > 0$. Let $x_0 \in S_X$ be such that $|x^*(x_0)| > 1 - \eta(\varepsilon, x^*)$. Then $\|T(x_0)\| > 1 - \eta(\varepsilon, T)$. So, there is $S \in \mathcal{L}(X, Y)$ with $\|S\| = 1$ such that $\|S(x_0)\| = 1$ and $\|S - T\| < \varepsilon$. Let $y_0^* \in S_{Y^*}$ be such that $y_0^*(S(x_0)) = \|S(x_0)\| = 1$. Set $x_1^* := S^*y_0^* \in S_{X^*}$. Then $|x_1^*(x_0)| = 1$ and $\|x_1^* - x^*\| < \varepsilon$. This means that the pair $(X, \mathbb{K})$ has the $\mathbf{L}_{p,o}$. □

By Proposition 8.6, we know that the pairs $(\ell_1^2, \mathbb{K})$ and $(\ell_\infty^2, \mathbb{K})$ fails both $\mathbf{L}_{o,p}$ and $\mathbf{L}_{p,o}$. So, as a consequence of Theorem 8.11 and Proposition 8.12, if X has ℓ_1^2 or ℓ_∞^2 as a one-complemented subspace, then the pair (X, Y) cannot have neither $\mathbf{L}_{o,p}$ nor $\mathbf{L}_{p,o}$ for all Banach spaces Y. Hence, we have the following consequence.

Corollary 8.13. *Let Y be a Banach space. The pairs (ℓ_1, Y) and (c_0, Y) fail both* $\mathbf{L}_{o,p}$ *and* $\mathbf{L}_{p,o}$.

We finish the paper by discussing some of the relations between the Bishop–Phelps–Bollobás properties we mentioned so far. There are two more of them we would like to consider that we did not discuss in the present article. They are the local versions of the **BPBp**, which we denote by $\mathbf{L}_\triangle$, where $\triangle$ means that the η depends on a fixed point x or on a fixed operator T. A pair of Banach spaces (X, Y) has the $\mathbf{L}_p$ if given $\varepsilon > 0$ and $x \in S_X$, then there is $\eta(\varepsilon, x) > 0$ such that whenever $T \in \mathcal{L}(X, Y)$ with $\|T\| = 1$ satisfies $\|T(x)\| > 1 - \eta(\varepsilon, x)$, there are $S \in \mathcal{L}(X, Y)$ with $\|S\| = 1$ and $x_0 \in S_X$ such that

$$\|S(x_0)\| = 1, \quad \|x_0 - x\| < \varepsilon, \quad \text{and} \quad \|S - T\| < \varepsilon. \tag{8.3}$$

On the other hand, (X, Y) has the $\mathbf{L}_o$ if given $\varepsilon > 0$ and $T \in S_{\mathcal{L}(X,Y)}$, there is $\eta(\varepsilon, T) > 0$ such that whenever $x \in S_X$ satisfies $\|T(x)\| > 1 - \eta(\varepsilon, T)$, there are $S \in \mathcal{L}(X, Y)$ with $\|S\| = 1$ and $x_0 \in S_X$ such that (8.3) holds. For more information about these properties, we refer the reader to [16, Section 3]. In the next remark, we compare the properties we have considered.

Remark 8.14. We have the following observations:

(i) All the implications below between the Bishop–Phelps–Bollobás properties hold.

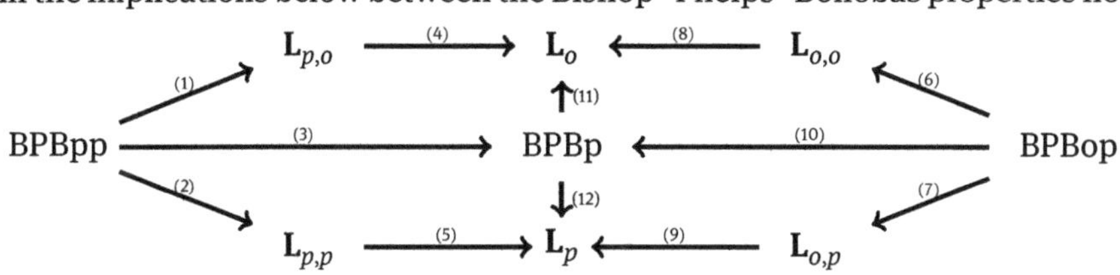

On the other hand, the reverse implications are not true.

(ii) The **BPBp** does not imply $\mathbf{L}_{\square,\triangle}$, where $\square$ and $\triangle$ can be p or o.

(iii) There is no relation between properties $\mathbf{L}_{o,p}$ and $\mathbf{L}_{p,o}$.
(iv) The $\mathbf{L}_{p,p}$ does not imply the $\mathbf{L}_{p,o}$, but we do not know whether the $\mathbf{L}_{p,o}$ implies (or not) the $\mathbf{L}_{p,p}$.
(v) The $\mathbf{L}_{o,o}$ does not imply the $\mathbf{L}_{o,p}$, but we do not know whether the $\mathbf{L}_{o,p}$ implies (or not) the $\mathbf{L}_{o,o}$.

We briefly discuss the statements in the above remark. It is clear that all the implications in (i) are satisfied, so let us show that the reverse implications do not hold. In [16, Section 5], it is proved that the reverse implications of (2), (3), (5), (6), (8), (10), (11), and (12) do not hold. The reverse implication of (4) (resp., (9)) fails since, for instance, the pairs $(\ell_1, \mathbb{K})$ or $(c_0, \mathbb{K})$ have the $\mathbf{L}_o$ (resp., $\mathbf{L}_p$) but fail the $\mathbf{L}_{p,o}$ (resp., $\mathbf{L}_{o,p}$). To show that the reverse implication of (7) fails, just take a pair $(X, \mathbb{K})$ with X reflexive and LUR but not uniformly convex. Analogously (reasoning with X^* instead of X) we see that the reverse implication of (1) does not hold. To see (ii), just note that $(X, \mathbb{K})$ has the **BPBp** for every Banach space X, which is clearly not true for any of the properties $\mathbf{L}_{\square,\triangle}$. For (iii), take X a uniformly smooth Banach space with $\dim(X) \geqslant 2$. Then we have that (X, ℓ_∞^2) has the **BPBpp** (see [15, Proposition 2.4]) and, consequently, the $\mathbf{L}_{p,o}$, but fails the $\mathbf{L}_{o,p}$ in virtue of Proposition 8.8. This shows that the $\mathbf{L}_{p,o}$ does not imply the $\mathbf{L}_{o,p}$. For the converse, take any finite-dimensional space X which is strictly convex but not smooth. Then X^* cannot be strictly convex and by Corollary 8.5(ii), the pair $(X, \mathbb{K})$ fails property $\mathbf{L}_{p,o}$. On the other hand, by using Proposition 8.2(i), it satisfies property $\mathbf{L}_{o,p}$. So, the $\mathbf{L}_{o,p}$ cannot imply the $\mathbf{L}_{p,o}$. Finally, for (iv) and (v), notice that the $\mathbf{L}_{p,p}$ cannot imply the $\mathbf{L}_{p,o}$ since $(\ell_1^2, \mathbb{K})$ has the $\mathbf{L}_{p,p}$ (see [16, Proposition 2.9]) but fails the $\mathbf{L}_{p,o}$ (see Proposition 8.6). The same example shows that $\mathbf{L}_{o,o}$ does not imply the $\mathbf{L}_{o,p}$.

Bibliography

[1] M. D. Acosta, *Denseness of norm attaining mappings*, RACSAM Rev. R. Acad. Cienc. Exactas Fís. Nat. Ser. A Mat. **100** (1–2) (2006), 9–30.
[2] M. D. Acosta, *The Bishop–Phelps–Bollobás property for operators on $C(K)$*, Banach J. Math. Anal. **10** (2) (2016), 307–319.
[3] M. D. Acosta, R. M. Aron, D. García and M. Maestre, *The Bishop–Phelps–Bollobás theorem for operators*, J. Funct. Anal. **294** (2008), 2780–2899.
[4] M. D. Acosta, M. Mastyło, and M. Soleimani-Mourchehkhorti, *The Bishop–Phelps–Bollobás and approximate hyperplane series properties*, J. Funct. Anal. **274** (9) (2018), 2673–2699.
[5] R. M. Aron, Y. S. Choi, S. K. Kim, H. J. Lee and M. Martín, *The Bishop–Phelps–Bollobás version of Lindenstrauss properties A and B*, Transl. Am. Math. Soc. **367** (2015), 6085–6101.
[6] E. Bishop and R. R. Phelps, *A proof that every Banach space is subreflexive*, Bull. Am. Math. Soc. **67** (1961), 97–98.
[7] E. Bishop and R. R. Phelps, *The support functionals of a convex set*, Proc. Symp. Pure Math. (Convexity), AMS, **7** (1963), 27–35.
[8] B. Bollobás, *An extension to the theorem of Bishop and Phelps*, Bull. Lond. Math. Soc. **2** (1970), 181–182.

[9] J. Bourgain, *On dentability and the Bishop–Phelps property*, Isr. J. Math. **28** (1977), 265–271.

[10] M. Chica, V. Kadets, M. Martín and S. Moreno-Pulido, *Bishop–Phelps–Bollobás moduli of a Banach space*, J. Math. Anal. Appl. **412** (2014), 697–719.

[11] D. H. Cho and Y. S. Choi, *The Bishop–Phelps–Bollobás theorem on bounded closed convex sets*, J. Lond. Math. Soc. **93** (2) (2016), 502–518.

[12] S. Dantas, *Some kind of Bishop–Phelps–Bollobás property*, Math. Nachr. **290** (5–6) (2017), 774–784.

[13] S. Dantas, V. Kadets, S. K. Kim, H. J. Lee and M. Martín, *On the pointwise Bishop–Phelps–Bollobás property for operators*, Can. J. Math., **71** (6) (2019), 1421–1443.

[14] S. Dantas, V. Kadets, S. K. Kim, H. J. Lee and M. Martín, There is no operatorwise version of the Bishop–Phelps–Bollobás property, *Linear and Multilinear Algebra*, https://doi.org/10.1080/03081087.2018.1560388.

[15] S. Dantas, S. K. Kim and H. J. Lee, *The Bishop–Phelps–Bollobás point property*, J. Math. Anal. Appl. **444** (2016), 1739–1751.

[16] S. Dantas, S. K. Kim, H. J. Lee and M. Mazzitelli, *Local Bishop–Phelps–Bollobás properties*, J. Math. Anal. Appl. **468** (1) (2018), 304–323.

[17] S. K. Kim and H. J. Lee, *Uniform convexity and the Bishop–Phelps–Bollobás property*, Can. J. Math. **66** (2014), 373–386.

[18] J. Lindenstrauss, *On operators which attain their norm*, Isr. J. Math. **1** (1963), 139–148.

[19] V. Lomonosov, *A counterexample to the Bishop–Phelps Theorem in complex spaces*, Isr. J. Math. **115** (1) (1998), 25–28.

[20] A. R. Lovaglia, *Locally uniformly convex Banach spaces*, Transl. Am. Math. Soc. **78** (1955), 225–238.

[21] C. Stegall, *Optimization and differentiation in Banach spaces*, Linear Algebra Appl. **84** (1986), 191–211.

Stephen Deterding

9 Bounded point derivations of fractional orders

Abstract: Let X be a compact subset of the complex plane and let $A_n(x_0)$ denote the annulus $\{x : 2^{-n-1} < |x - x_0| < 2^{-n}\}$. It is known that for a nonnegative integer t, the condition $\sum_{n=1}^{\infty} 2^{(1+\alpha)n}\gamma(A_n(x_0) \setminus X) < \infty$, where $\gamma(A_n(x_0) \setminus X)$ is the analytic capacity of $A_n(x_0) \setminus X$, implies the existence of a tth order bounded point derivation. By defining a bounded point derivation with nonintegral order using fractional derivatives, this result is extended to noninteger values of t.

Keywords: Point derivation, fractional derivative, analytic, boundary

MSC 2010: Primary 30H99, Secondary 26A33

9.1 Introduction

Let X be a compact subset of the complex plane and let $R(X)$ denote the uniform closure of rational functions with poles off X. An interesting topic of study in the theory of approximation by rational functions is how well differentiation is preserved under uniform convergence. It is not always the case that functions in $R(X)$ are differentiable; in fact, it is a result of Dolzhenko [2] that $R(X)$ contains a nowhere differentiable function whenever X is a compact nowhere dense set. One tool that has been used in the study of this problem is a bounded point derivation. For nonnegative integer values of t, we say that $R(X)$ admits a bounded point derivation of order t at x_0 if there exists a constant C such that for all rational functions f with poles off X, $|f^{(t)}(x_0)| \leq C\|f\|_{\infty}$. A bounded point derivation of order 0 is often called a bounded point evaluation. When they exist, bounded point derivations generalize the concept of the derivative to functions in $R(X)$ which may not be differentiable. The existence of bounded point derivations can be characterized using analytic capacity. We briefly review the definition; additional information about analytic capacity can be found in the book of Gamelin [3, Chapter VIII]. A function f is said to be admissible for X if:

1. f is analytic on $\hat{\mathbb{C}} \setminus X$.
2. $|f(z)| \leq 1$ on $\hat{\mathbb{C}} \setminus X$.
3. $f(\infty) = 0$.

The analytic capacity of the compact set X, denoted by $\gamma(X)$, is defined by

$$\gamma(X) = \sup|f'(\infty)|,$$

Stephen Deterding, West Liberty University, 208 University Dr, West Liberty, WV 26074, USA, e-mail: stephen.deterding@westliberty.edu, https://orcid.org/0000-0003-2498-0245

https://doi.org/10.1515/9783110656756-009

where the supremum is taken over all admissible functions f. $f'(\infty)$ is the derivative of f at the point at infinity in the extended complex plane, not the limit of $f'(x)$ as $x \to \infty$. One important property of analytic capacity, which follows directly from the definition, is that analytic capacity is monotone, that is, if $U \subseteq V$ then $\gamma(U) \leq \gamma(V)$. The connection between bounded point derivations and analytic capacity is given by the following theorem.

Theorem 9.1. *For a nonnegative integer t, $R(X)$ admits a t-th order bounded point derivation at x_0 if and only if*

$$\sum_{n=1}^{\infty} 2^{n(t+1)}\gamma(A_n(x_0) \setminus X) < \infty. \tag{9.1}$$

The case of $t = 0$ (bounded point evaluations) of this theorem was first proven by Mel'nikov [6] and the general case is due to Hallstrom [4]. The above theorem explains the significance of (9.1) when t is a nonnegative integer; however, less is known about the significance of

$$\sum_{n=1}^{\infty} 2^{n(\alpha+1)}\gamma(A_n(x_0) \setminus X) < \infty \tag{9.2}$$

for noninteger values of α. This question was first considered by O'Farrell [9] who showed that (9.2) is related to a Hölder continuity condition for the $(t-1)$-th derivatives of rational functions with poles off X, where t is the smallest integer greater than α. While this is one conclusion that can be drawn from (9.2), it is not the only possibility. Given the similarity between (9.1) and (9.2), it seems reasonable that (9.2) implies an extension of Theorem 9.1 to the case of derivatives of fractional orders. There has been an increased study of fractional derivatives as newer applications are discovered. Some examples of the many applications of fractional derivatives can be found in [1], [5], and [12]. Thus it is useful to consider the concept of bounded point derivations of a fractional order.

Unlike derivatives of integral orders, there are several different definitions for fractional derivatives, which are not all equivalent. We will consider two of the more commonly used definitions, the Riemann–Liouville and the Caputo fractional derivatives. Our main result relates Condition (9.2) to the boundedness of Riemann–Liouville fractional derivatives of functions in $R(X)$.

Theorem 9.2. *Let $\alpha \in \mathbb{C}$ and $a \in X$ and suppose that*

$$\sum_{n=1}^{\infty} 2^{n(\mathrm{Re}(\alpha)+1)}\gamma(A_n(x_0) \setminus X) < \infty. \tag{9.3}$$

Then for all rational functions f with poles off X, $|D_a^\alpha f(x_0)| \leq C\|f\|_\infty$, where $D_a^\alpha f(x_0)$ denotes the Riemann–Liouville derivative of order α evaluated at x_0 and the constant C does not depend on f.

We will also prove a similar result for Caputo fractional derivatives.

Theorem 9.3. *Let $\alpha \in \mathbb{C}$ with $0 < \mathrm{Re}(\alpha) < 1$ and let $a \in X$ and suppose that (9.3) holds. Then for all rational functions f with poles off X, $|{}^C D_a^\alpha f(x_0)| \le C\|f\|_\infty$, where ${}^C D_a^\alpha f(x_0)$ denotes the Caputo derivative of order α evaluated at x_0 and the constant C does not depend on f.*

In Theorem 9.2, α may be chosen with $\mathrm{Re}(\alpha) < 0$, so the above result shows the significance of a negative α in (9.3) As we will see, a derivative of a negative fractional order is a fractional integral.

9.2 Fractional derivatives

The study of fractional derivatives has a rich and extensive history. We will only discuss those topics which are relevant to the purposes of this paper and we refer the reader to resources such as [5] and [11] for a more detailed overview of fractional derivatives. The usual starting point in the study of fractional derivatives is the Riemann–Liouville fractional integral. Cauchy's formula for repeated integration states that the repeated integral

$$f^{(-n)}(x) = \int_a^x \int_a^{\sigma_1} \cdots \int_a^{\sigma_{n-1}} f(\sigma_n)\,d\sigma_n\,d\sigma_{n-1}\ldots d\sigma_1$$

can be expressed as a single integral

$$f^{(-n)}(x) = \frac{1}{(n-1)!}\int_a^x (x-t)^{n-1} f(t)\,dt. \tag{9.4}$$

The Riemann–Liouville integral of order α is defined by replacing the integer n in (9.4) with the complex number α. Let $\mathrm{Re}(\alpha) > 0$ and fix $a \in \mathbb{C}$. The Riemann–Liouville integral of order α of the function f at x_0, denoted by $D_a^{-\alpha} f(x_0)$, is defined as follows [5, p. 69]. (See also [11, p. 6].)

$$D_a^{-\alpha} f(w) = \frac{1}{\Gamma(\alpha)}\int_a^w \frac{f(z)}{(w-z)^{1-\alpha}}\,dz, \tag{9.5}$$

where the integral is taken over a path from a to w on which f is analytic.

The Riemann–Liouville fractional derivative of order α for $\mathrm{Re}(\alpha) \ge 0$ is defined using the Riemann–Liouville fractional integral [5, p. 70]. (See also [11, p. 18].) Let t be the smallest integer larger than $\mathrm{Re}(\alpha)$ and let $\beta = t - \alpha$. Then

$$D_a^\alpha f(w) = \frac{d^t}{dw^t} D_a^{-\beta} f(w) = \frac{d^t}{dw^t}\frac{1}{\Gamma(\beta)}\int_a^w \frac{f(z)}{(w-z)^{1-\beta}}\,dz. \tag{9.6}$$

Note that both the Riemann–Liouville integral and derivative depend on the choice of a; choosing a different value for a changes the value of the fractional derivatives. It also follows from the above definitions that for a non-negative integer n, and all $\alpha \in \mathbb{C}$, $D_a^\alpha (w-a)^n = \frac{\Gamma(n+1)}{\Gamma(n-\alpha+1)}(w-a)^{n-\alpha}$. (A generalization of this result can be found in [5, p. 71].) When $\operatorname{Re}(\alpha) < 0$, the derivative of order α is a fractional integral of order $-\alpha$. Notably the Riemann–Liouville derivative of a constant is not 0. Worse still, if $\operatorname{Re}(\alpha) > 0$ the derivative has a pole at $w = a$, and thus the Riemann–Liouville derivative of a constant is undefined at a. Thus if we wish to bound the Riemann–Liouville derivative at x_0 we must have $a \neq x_0$.

Alternatively, if $\operatorname{Re}(\alpha) > 0$, we can make use of the closely related Caputo derivative to solve this problem. The Caputo derivative of order α is obtained by first taking the tth derivative of the function and then applying the Riemann–Liouville integral of order $t - \alpha$. The precise definition of the Caputo derivative is given as follows. Let t be the smallest integer larger than $\operatorname{Re}(\alpha)$ and let $\beta = t - \alpha$. Then the Caputo derivative of f at a denoted by ${}^C D_a^\alpha f(w)$ is given by

$$ {}^C D_a^\alpha f(w) = D_a^{-\beta} f^{(t)}(w) = \frac{1}{\Gamma(\beta)} \int_a^w \frac{f^{(t)}(z)}{(w-z)^{1-\beta}} dz, \tag{9.7} $$

where the integral is taken over a path from a to w on which f is analytic. As we will soon see, the change in order means that the Caputo derivative is not equivalent to the Riemann–Liouville derivative.

While the Caputo derivative has the drawback that the function must possess a tth order derivative in order to have a Caputo derivative, it also means that the Caputo derivative of a constant is 0, thus eliminating the need to specify that $a \neq x_0$. For $\alpha \in \mathbb{C}$ and $a \in X$, we say that $R(X)$ admits a Riemann–Liouville bounded point derivation of order α at x_0 if there exists a constant C such that for all rational functions f with poles off X, we have $|D_a^\alpha f(x_0)| \leq C\|f\|_\infty$. If, in addition, $\operatorname{Re}(\alpha) > 0$ we say that $R(X)$ admits a Caputo bounded point derivation of order α at x_0 if there exists a constant C such that for all rational functions f with poles off X, we have $|{}^C D_a^\alpha f(x_0)| \leq C\|f\|_\infty$. In the definition of a Riemann–Liouville fractional bounded point derivation, $\operatorname{Re}(\alpha)$ may be negative. As we have seen, in this case the fractional derivative is a fractional integral; however, to be concise we will refer to it as a derivative of negative order.

The following theorem shows how the Riemann–Liouville and Caputo fractional derivatives are related to each other. (Compare with [5, equation (2.4.1) and Theorem 2.1].)

Theorem 9.4. *Let* $\operatorname{Re}(\alpha) > 0$ *and let t be the smallest integer larger than* $\operatorname{Re}(\alpha)$. *Then*

$$ {}^C D_a^\alpha f(w) = D_a^\alpha f(w) - D_a^\alpha \left[\sum_{k=0}^{t-1} \frac{f^{(k)}(a)}{k!}(w-a)^k \right]. $$

In particular, if $0 < \mathrm{Re}(\alpha) < 1$, *then*

$$^{C}D_a^{\alpha}f(w) = D_a^{\alpha}f(w) - D_a^{\alpha}[f(a)],$$

and thus if $f(a) = 0$, *then the Caputo and the Riemann–Liouville derivatives coincide.*

The Riemann–Liouville definition of fractional derivatives is closely related to the Cauchy integral formula in complex analysis. For derivatives of integral orders, the Cauchy integral formula states

$$f^{(n)}(w) = \frac{n!}{2\pi i}\int_C \frac{f(z)dz}{(z-w)^{n+1}}dz,$$

where C is a closed contour enclosing w. If we replace the integer n with an arbitrary α, then $(z-w)^{\alpha+1}$ no longer has a pole at w, but instead has a branch point. This means that the value of the contour integral now explicitly depends on the point where C crosses the branch cut for $(z-w)^{\alpha+1}$ and thus C cannot be deformed arbitrarily. The Cauchy integral formula for the Riemann–Liouville fractional derivative is obtained by taking this point to be the value of a in the definition. The following result is due to Nekrassov [8] when $a = 0$ and to Osler [10] in the general case.

Theorem 9.5. *Suppose that* f *is analytic in a finite sector of the complex plane with vertex at* a *and* $\oint f(z)dz = 0$ *along any closed path through* a. *Then*

$$D_a^{\alpha}f(w) = \frac{\Gamma(\alpha+1)}{2\pi i}\int_C \frac{f(z)}{(z-w)^{\alpha+1}}dz,$$

where the contour C *is a positively oriented closed contour which begins and ends at* a *and* f *is analytic on* C.

In [10], the proof of Theorem 9.5 is given for the case of $\mathrm{Re}(\alpha) < 1$. If $\mathrm{Re}(\alpha) \geq 1$, the proof follows by letting $\beta = t - \alpha$, where t is the smallest integer greater than α. Then applying Theorem 9.5 to (9.6) yields

$$D_a^{\alpha}f(w) = \frac{d^t}{dw^t}\left(\frac{\Gamma(1-\beta)}{2\pi i}\int_C \frac{f(z)}{(z-w)^{1-\beta}}dz\right),$$

and the result follows by differentiating under the integral.

A useful tool for analyzing integrals such as the one in Theorem 9.5 is the following result of Mel'nikov [7, Theorem 4] (See also [6] for the special case of an annulus) which shows how the Cauchy integral of a nice-enough function can be bounded by the analytic capacity of the set where the function is not analytic.

Theorem 9.6. *Let* C *be a closed curve that encloses a region* U. *Let* f *be any function continuous and bounded by* M_0 *on* U *and analytic on* $U \setminus K$, *where* K *is a compact subset* *of U. Then there is a constant A which only depends on the curve C such that*

$$\left|\int_C f(z)dz\right| \leq AM_0\gamma(K).$$

Note that if f is analytic on U, then Theorem 9.6 reduces to Cauchy's theorem. Thus Theorem 9.6 gives an upper bound for the failure of Cauchy's theorem for nonanalytic continuous functions.

9.3 The proofs of the main theorems

We begin with the proof of Theorem 9.2.

Proof. We will show that (9.3) implies the existence of a Riemann–Liouville fractional bounded point derivation. Since f is a rational function with poles off X, there exists an open neighborhood U of X such that f is analytic on U. For each n, let $B_n(x_0)$ be the ball centered at x_0 with radius 2^{-n}. Then there exists N such that $B_N(x_0)$ is entirely contained in U, and there exists $M < N$ such that $a \in B_M(x_0)$. It follows from the Cauchy integral formula for fractional derivatives (Theorem 9.5) that

$$D_a^\alpha f(x_0) = \frac{\Gamma(\alpha+1)}{2\pi i}\int_C \frac{f(z)}{(z-x_0)^{1+\alpha}}dz,$$

where the contour C consists of a path that starts at a, follows a contour that connects a to B_N, and travels around B_N in a counterclockwise direction but stops before completing an entire loop around B_N. From there, it follows another path that connects it back to a so that the closed contour does not contain or enclose any of the poles of f. (See Figure 9.1.) As f has a finite number of poles it is always possible to find such a path.

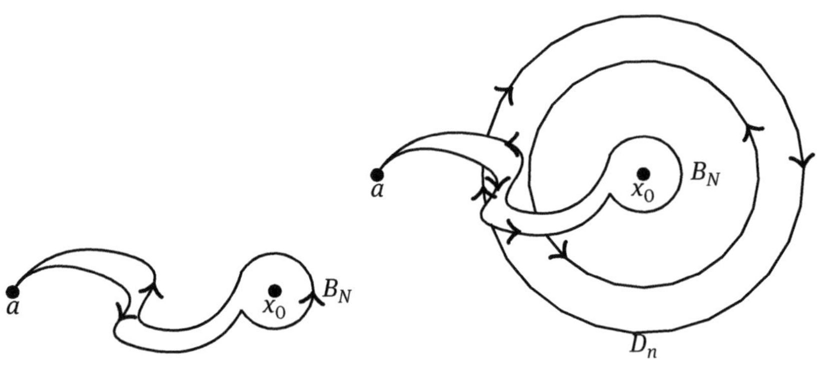

Figure 9.1: The contour of integration for Theorem 9.2.

We now modify the function f so that the modification $\tilde{f}$ is continuous on B_M, vanishes on the circle $|z - x_0| = 2^{-M}$, and $\tilde{f} = f$ on and inside C. Moreover, we can also make it so that $\|\tilde{f}\|_\infty \leq 2\|f\|_\infty$. Thus

$$D_a^\alpha f(x_0) = \frac{\Gamma(\alpha+1)}{2\pi i} \int_C \frac{\tilde{f}(z)}{(z-x_0)^{1+\alpha}} dz.$$

Let A_n denote the annulus $\{z : 2^{-(n+1)} < |z - x_0| < 2^{-n}\}$ and let $D_n = A_n \setminus C$ be oriented so that C is to the left of the boundary of D_n as shown in Figure 9.1. Then it follows that

$$D_a^\alpha f(x_0) = \frac{\Gamma(\alpha+1)}{2\pi i}\left(\sum_{n=M}^{N-1} \int_{D_n} \frac{\tilde{f}(z)}{(z-x_0)^{1+\alpha}} dz + \int_{|z-x_0|=2^{-M}} \frac{\tilde{f}(z)}{(z-x_0)^{1+\alpha}} dz\right).$$

Since $\tilde{f} = 0$ on $|z - x_0| = 2^{-M}$, it follows that

$$|D_a^\alpha f(x_0)| \leq \frac{\Gamma(\alpha+1)}{2\pi} \sum_{n=M}^{N-1} \left| \int_{D_n} \frac{\tilde{f}(z)}{(z-x_0)^{1+\alpha}} dz \right|.$$

On D_n, $\frac{1}{(z-x_0)^{1+\alpha}}$ is bounded by $2^{(n+1)(1+\mathrm{Re}(\alpha))}$. Hence by Mel'nikov's estimate (Theorem 9.6),

$$\left| \int_{D_n} \frac{\tilde{f}(z)}{(z-x_0)^{1+\alpha}} dz \right| \leq C 2^{n(1+\mathrm{Re}(\alpha))} \|\tilde{f}\|_\infty \gamma(D_n \setminus X),$$

where the constant C does not depend on f and since analytic capacity is monotone,

$$\left| \int_{D_n} \frac{\tilde{f}(z)}{(z-x_0)^{1+\alpha}} dz \right| \leq C 2^{n(1+\mathrm{Re}(\alpha))} \|\tilde{f}\|_\infty \gamma(A_n \setminus X).$$

Hence

$$|D_a^\alpha f(x_0)| \leq C \sum_{n=M}^{N-1} 2^{n(1+\mathrm{Re}(\alpha))} \|\tilde{f}\|_\infty \gamma(A_n \setminus X) < C\|\tilde{f}\|_\infty.$$

Since $\|\tilde{f}\|_\infty \leq 2\|f\|_\infty$, $|D_a^\alpha f(x_0)| \leq C\|f\|_\infty$, and hence there is a Riemann–Liouville fractional bounded point derivation of order α on $R(X)$ at x_0. □

We obtain Theorem 9.3 as a corollary of Theorem 9.2.

Proof. Since $0 < \mathrm{Re}(\alpha) < 1$, it follows from Theorem 9.4 that ${}^C D_a^\alpha f(x_0) = D_a^\alpha f(x_0) - D_a^\alpha [f(a)]$. Thus by Theorem 9.2, $|{}^C D_a^\alpha f(x_0)| \leq C\|f\|_\infty$. □

We have demonstrated that (9.3) is a sufficient condition for the existence of a fractional bounded point derivation of both the Riemann–Liouville and Caputo types. It would be interesting to know whether this condition is also necessary. That this is true when α is a positive integer is a result due to Hallstrom [4]. Hallstrom assumes that (9.1) does not hold and shows that this implies that there cannot be a bounded point derivation on $R(X)$ at x_0 by constructing a sequence of functions in $R(X)$ with unbounded derivatives at x_0. These sequences are partial sums consisting of the functions $h_n(z) = (z-x_0)^t f_n'(\infty) + (z-x_0)^{t-1}a_2 + \cdots + (z-x_0)a_t - (z-x_0)^{t+1}f_n(z)$, where f_n is a function that is analytic off $A_n \setminus X$ and a_n is the coefficient of the z^{-n} term of the Laurent series expansion of f at 0. If in these functions, t is replaced with a noninteger α, then the functions are no longer analytic and thus not in $R(X)$, so a different construction is required to prove necessity.

Bibliography

[1] M. Bashor and M. Dalir, *Applications of fractional calculus*, Appl. Math. Sci. **4** (21) (2010), 1021–1032.

[2] E. P. Dolzhenko, *Construction on a nowhere dense continuum of a nowhere differentiable function which can be expanded into a series of rational functions* (Russian), Dokl. Akad. Nauk SSSR **125** (1959), 970–973.

[3] T. W. Gamelin, *Uniform Algebras*, Prentice-Hall, Inc., Englewood Cliffs, N. J., 1969.

[4] A. P. Hallstrom, *On bounded point derivations and analytic capacity*, J. Funct. Anal. **3** (1969), 35–47.

[5] A. A. Kilbas, H. M. Srivastava and J. J. Trujillo, *Theory and Applications of Fractional Differential Equations*, North-Holland Mathematics Studies, **204**, Elsevier Science B. V., Amsterdam, 2006.

[6] M. S. Mel'nikov, *Estimates of the Cauchy integral along an analytic curve* (Russian), Mat. Sb. **71** (113) (1966), 503–515.

[7] M. S. Mel'nikov, *Analytic capacity and the Cauchy integral* (Russian), Dokl. Akad. Nauk SSSR **172** (1967), 26–29.

[8] P. A. Nekrassov, *General differentiation* (Russian), Mat. Sb. **14** (1888), 45–168.

[9] A. G. O'Farrell, *Analytic capacity, Hölder conditions, and τ-spikes*, Trans. Am. Math. Soc. **196** (1974), 415–424.

[10] T. J. Osler, *Leibniz rule for fractional derivatives generalized and an application to infinite series*, SIAM J. Appl. Math. **18** (1970), 658–674.

[11] B. Ross, A brief history and exposition of the fundamental theory of fractional calculus, in: B. Ross (ed.) *Fractional Calculus and Its Applications*, 1–36, Springer, Berlin, Heidelberg, 1975.

[12] J. A. Tenreiro Machado, M. F. Silva, R. S. Barbosa, et al., *Some applications of fractional calculus in engineering*. Math. Probl. Eng. **2010** (2010), 34 pp.

Gilles Godefroy

10 Invariant subspaces: some minimal proofs

This work is dedicated to the memory of Victor Lomonosov

Abstract: We show the existence of hyperinvariant subspaces for compact operators and of invariant subspaces for operators which admit a moment sequence on the Hilbert space, by using only minimizers of quadratic expressions.

Keywords: Lomonosov theorem, moment sequences, quadratic functionals

MSC 2010: Primary 47A15, Secondary 46B20

10.1 Introduction

Among the important results shown by Victor Lomonosov, the most celebrated one is probably the existence of closed nontrivial hyperinvariant subspaces for any nonzero compact operator on any complex Banach space ([9], see Theorem 10.35 in [13]). Lomonosov's original proof used Schauder's fixed-point theorem, and later on H. M. Hilden showed how to dispense with this nonlinear argument through a proper use of the spectral radius formula (see [12]). Minimal vectors were used by S. Ansari and P. Enflo [1] to prove Lomonosov's theorem in Hilbert spaces. We refer to Chapter 7 in [6] for a full display of this technique and its applications. Also, A. Atzmon showed in [2] that any operator on a reflexive real Banach space (in particular on a real Hilbert space) which admits a moment sequence has a non-trivial closed invariant subspace – the converse being trivially true.

This short note contains no new result, but focuses on delivering proofs which are as elementary as possible, in the frame of real or complex Hilbert spaces. Hence we will provide proofs of Lomonosov's and Atzmon's theorems which use nothing else than the most basic Hilbertian tool: nearest points in closed convex sets. In particular, weak topologies are not needed. Moreover, our proofs do not proceed by contradiction. Therefore, this note presents self-contained (and somewhat constructive) proofs of these theorems which can be taught at the undergraduate level.

We denote the scalar product on the Hilbert space by $\langle\ ,\ \rangle$. The Hilbert space is equipped with its usual Euclidean norm. The closed ball of center x and radius $r \geq 0$ is denoted $B(x,r)$. A linear continuous operator A on a Banach space X admits an invariant subspace if there exists a closed linear subspace $M \subset X$ with $\{0\} \neq M \neq X$ and $A(M) \subset M$. We say that M is hyperinvariant if moreover $T(M) \subset M$ for all T such that $TA = AT$.

Gilles Godefroy, Institut de Mathématiques de Jussieu, Case 247, 4 place Jussieu, 75005 Paris, France, e-mail: gilles.godefroy@imj-prg.fr

https://doi.org/10.1515/9783110656756-010

10.2 Lomonosov's theorem

In this section, $\mathcal{H}$ denotes the *complex* infinite-dimensional separable Hilbert space, and $L(\mathcal{H})$ denotes the algebra of continuous linear operators on $\mathcal{H}$. Let $T \in L(\mathcal{H})$ be an operator with dense range. Following [1], we first define minimal vectors and establish their elementary properties. Pick $x \in \mathcal{H}$ a nonzero vector, and $0 < \epsilon < \|x\|$. With this notation, we have the following.

Lemma 10.1. *There exists a unique $y \in \mathcal{H}$ which satisfies the following two conditions:*
(i) $\|T(y) - x\| \le \epsilon$.
(ii) $\|y\| = \inf\{\|v\|;\ \|T(v) - x\| \le \epsilon\}$.

Moreover, $y \neq 0$ and $\|T(y) - x\| = \epsilon$.

Indeed it suffices to consider the non-empty closed convex set $C = T^{-1}(B(x, \epsilon))$, which does not contain 0 since $\epsilon < \|x\|$, and to pick the unique $y \in C$ of minimal norm.

We call this vector y the *minimal vector* for (T, x, ϵ). We keep the same notation in our next lemma, which gives two useful properties of y.

Lemma 10.2. *The vector y satisfies:*

If $v \in \mathcal{H}$ and $\langle v, y\rangle = 0$, then

$$\big\langle T(v), T(y) - x \big\rangle = 0. \tag{10.1}$$

Moreover,

$$\big\langle x - T(y), x \big\rangle \ge \epsilon^2. \tag{10.2}$$

Proof. The proof relies on Euclidean geometry. We pick any nonzero $u \in \mathcal{H}$ and we define the real function g by

$$g(t) = \big\|T(y + tu) - x\big\|^2 = \big\|T(y) - x + tT(u)\big\|^2.$$

The function g is differentiable and $g'(0) = 2\,\mathrm{Re}(\langle T(y)-x, T(u)\rangle)$. If we have $g'(0) < 0$, then g is decreasing on some interval $[0, t_0]$. It follows that $T(y + tu) \in B(x, \epsilon)$ for $0 \le t \le t_0$. Therefore, $\|y + tu\| \ge \|y\|$ for $0 \le t \le t_0$ by minimality of y, and thus $\mathrm{Re}(\langle y, u\rangle) \ge 0$. Hence, if

$$\mathrm{Re}\big(\big\langle T(y) - x, T(u)\big\rangle\big) = \mathrm{Re}\big(\big\langle T^*\big(T(y) - x\big), u\big\rangle\big) \le 0$$

then $\mathrm{Re}(\langle y, u\rangle) \ge 0$. It follows that there exists $\delta > 0$ such that

$$y = -\delta T^*\big(T(y) - x\big).$$

Condition (1) follows immediately since $\langle T(v), T(y) - x\rangle = \langle v, T^*(T(y) - x)\rangle$. For showing (2), we compute

$$\epsilon^2 = \|x - T(y)\|^2 = \big\langle x - T(y), x\big\rangle + \big\langle T(y) - x, T(y)\big\rangle = \big\langle x - T(y), x\big\rangle - (\|y\|^2)/\delta. \qquad \square$$

We now prove Lomonosov's theorem on complex Hilbert spaces, through a simplified version of Ansari–Enflo's proof ([1]).

Theorem 10.3. *Let $A \in L(\mathcal{H})$ be a compact operator, $A \neq 0$. Then A has a non-trivial closed hyperinvariant subspace.*

Proof. We recall the spectral radius formula:

$$\rho(A) = \mathrm{Sup}\{|\lambda| : \lambda \in \sigma(A)\} = \lim \|A^n\|^{1/n}.$$

If $\rho(A) > 0$, then compactness of A shows that $\rho(A)$ is an eigenvalue of A, and the corresponding eigenspace is hyperinvariant. If A does not have dense range, then the closure of its range is hyperinvariant. Hence we may and do assume that A has dense range and $\rho(A) = 0$.

Pick $x \in \mathcal{H}$ such that $A(x) \neq 0$, and set

$$\epsilon = \frac{\|A(x)\|}{2\|A\|}.$$

We note that $0 < \epsilon \leq \|x\|/2$ and that $\|z\| \geq \|A(x)\|/2$ for all $z \in A(B(x, \epsilon))$.

We denote by y_n the minimal vector for (A^n, x, ϵ), and we set

$$t = \inf\left\{\frac{\|y_{n-1}\|}{\|y_n\|}; n \geq 1\right\}.$$

We now show that $t = 0$. Indeed, by definition of t we have $\|y_{n-1}\| \geq t\|y_n\|$ for all n and thus $\|y_1\| \geq t^{n-1}\|y_n\|$. Moreover, by Lemma 10.1 we have $\|A(A^{n-1}(y_n)) - x\| = \epsilon$ and thus $\|A^{n-1}(y_n)\| \geq \|y_1\|$ since y_1 is minimal. Thus

$$\|A^{n-1}(y_n)\| \geq t^{n-1}\|y_n\|$$

and since $\rho(A) = 0$, the spectral radius formula shows that $t = 0$. Since $t = 0$, there is a subsequence $n(k)$ such that

$$\lim_k \frac{\|y_{n(k)-1}\|}{\|y_{n(k)}\|} = 0. \tag{10.3}$$

Since $A^{n(k)}(y_{n(k)-1}) = A[A^{n(k)-1}(y_{n(k)-1})] \in A(B(x, \epsilon))$ and A is a compact operator, the sequence $A^{n(k)}(y_{n(k)-1})$ has a norm-convergent subsequence (which we still denote by the same notation) to some $s \neq 0$.

We denote $\mathrm{Com}(A) = \{T \in L(\mathcal{H}); AT = TA\}$. We claim that the vector space $M = \{T(s);\ T \in \mathrm{Com}(A)\}$ is not dense in $\mathcal{H}$. This will conclude the proof since then $\overline{M}$ is clearly a hyperinvariant subspace for A.

For showing our claim, we use again Euclidean geometry. Pick any $T \in \mathrm{Com}(A)$. We consider the orthogonal decomposition

$$T(y_{n(k)-1}) = \alpha_k y_{n(k)} + v_k$$

with $\langle y_{n(k)}, v_k \rangle = 0$. We have

$$|\alpha_k| \|y_{n(k)}\|^2 = |\langle T(y_{n(k)-1}), y_{n(k)} \rangle| \leq \|T\| \|y_{n(k)-1}\| \|y_{n(k)}\|$$

and it follows from (3) that lim $\alpha_k = 0$. Since $TA = AT$, we have

$$TA^{n(k)}(y_{n(k)-1}) = \alpha_k A^{n(k)}(y_{n(k)}) + A^{n(k)}(v_k).$$

We now have by (1) that

$$\langle A^{n(k)}(v_k), A^{n(k)}(y_{n(k)}) - x \rangle = 0$$

and thus finally

$$\langle TA^{n(k)}(y_{n(k)-1}), A^{n(k)}(y_{n(k)}) - x \rangle = \alpha_k \langle A^{n(k)}(y_{n(k)}), A^{n(k)}(y_{n(k)}) - x \rangle.$$

Since $\lim_k \alpha_k = 0$ and $A^{n(k)}(y_{n(k)}) \in B(x, \epsilon)$ for all k, this shows that

$$\lim_k \langle TA^{n(k)}(y_{n(k)-1}), A^{n(k)}(y_{n(k)}) - x \rangle = 0$$

and thus

$$\lim_k \langle T(s), A^{n(k)}(y_{n(k)}) - x \rangle = 0.$$

On the other hand, we have by (2) that for all k,

$$\langle x, A^{n(k)}(y_{n(k)}) - x \rangle \leq -\epsilon^2$$

and since $\|A^{n(k)}(y_{n(k)}) - x\| = \epsilon$ for all k, it follows that $\|T(s) - x\| \geq \epsilon$. Since $T \in \mathrm{Com}(A)$ was arbitrary, this concludes the proof. □

Remarks.

(1) The above proof provides information on the set $\mathcal{L}$ of nontrivial hyperinvariant subspaces. It shows indeed that if A is a one-to-one compact operator on $\mathcal{H}$ with dense range and $\rho(A) = 0$, then $\cup\{M;\ M \in \mathcal{L}\}$ is dense in $\mathcal{H}$, while $\cap\{M;\ M \in \mathcal{L}\} = \{0\}$.

(2) Lomonosov's theorem is valid as well on Banach spaces, and it actually states that if $K \neq 0$ is a compact operator, V is a nonscalar operator such that $VK = KV$ and T is such that $TV = VT$, then T has a nontrivial invariant subspace. On Banach spaces, it is optimal in this form: indeed it does not extend to a chain of 4 operators ([15]). Moreover, there exist finitely strictly singular operators without nontrivial invariant subspace ([5]), hence compactness is needed in full generality.

(3) It is still open whether operators without invariant subspaces exist on the Hilbert space, or even on reflexive spaces. We refer in particular to [7] and [8] for recent progress in this direction.

10.3 Atzmon's theorem

In this section, we denote by H the *real* separable infinite dimensional Hilbert space. We recall that $T \in L(H)$ has a moment sequence if there exist $z_0 \in H$ and $z_1 \in H$, both nonzero, and $\mu \geq 0$ a positive measure on $\mathbb{R}$ such that for all integers $n \geq 0$,

$$\langle z_1, T^n(z_0)\rangle = \int u^n d\mu(u).$$

Such a couple (z_0, z_1) is called a moment pair. We refer to Chapter 9 in [6] for results on moment sequences and their applications. The following result is due to A. Atzmon ([2]).

Theorem 10.4. *Any operator $T \in L(H)$ which has a moment sequence has a non-trivial invariant subspace.*

Proof. We denote

$$\mathcal{P}_+ = \{P \in \mathbb{R}[X];\ P(t) \geq 0 \text{ for all } t \in \mathbb{R}\}.$$

The set $D = \{P(T)(z_0) : P \in \mathcal{P}_+\}$ is a convex cone which is contained in the half-space $z_1^{-1}([0, +\infty))$ defined by z_1, hence its closure $C = \overline{D}$ is a proper closed convex cone. We denote its boundary by ∂C. Note that if $P \in \mathcal{P}_+$, then $P(T)(D) \subset D$, and thus $P(T)(C) \subset C$.

If $u \in \partial C \setminus \{0\}$, we can pick $v \notin C$ such that $\|u - v\| \leq \|u\|/3$. Let $z \in C$ be such that $\|v - z\| = \inf\{\|v - x\| : x \in C\}$. Then we have $z \neq 0$, and if $w = v - z$, one has $w \neq 0$, $\langle w, z\rangle = 0$ and $\langle w, x\rangle \leq 0$ for all $x \in C$.

Pick any $Q \in \mathbb{R}[X]$. We consider the quadratic polynomial $f_Q : \mathbb{R} \to \mathbb{R}$ defined by

$$f_Q(s) = \langle w, [\mathrm{Id} + sQ(T)]^2(z)\rangle.$$

Since $P(X) = [1 + sQ(X)]^2 \in \mathcal{P}_+$, we have $f_Q(s) \leq 0$ for all $s \in \mathbb{R}$, and also $f_Q(0) = \langle w, z\rangle = 0$. Therefore $f_Q'(0) = 0$. But expanding the square shows that

$$f_Q'(0) = 2\langle w, Q(T)(z)\rangle.$$

The subspace $M = \overline{\{Q(T)(z);\ Q \in \mathbb{R}[X]\}}$ is nontrivial since it contains z and it is contained in $\mathrm{Ker}(w)$, and clearly $T(M) \subset M$. □

Remarks.

(1) Atzmon's Theorem 10.4 above extends to all real Banach spaces ([3], see also [4]). The proof is similar, but the nearest point argument has to be replaced by a proper use of the Bishop–Phelps theorem. It is interesting to notice that such proofs provide little information on the set of invariant subspaces, beyond the fact that this set is nonempty. This is somehow natural since conversely, any operator with a nontrivial invariant subspace clearly has moment pairs.

(2) All operators $T \in L(X)$ on a real Banach space X which satisfy for some $C \geq 0$ and some compact subset K of $\mathbb{R}$ that

$$\|P(T)\| \leq C \sup\{|P(x)|\,;\ x \in K\} \tag{10.4}$$

for all polynomials $P \in \mathbb{R}[\xi]$ have a moment sequence, and thus an invariant subspace. More generally, it can be shown with the help of Lomonosov's work on Burnside's theorem [11] that if $C = 1$ in (4), we may replace the operator norm of $P(T)$ by the essential norm, that is, the norm in the Calkin algebra $L(X)/K(X)$ and reach the same conclusion [4]. It follows for instance that a compact perturbation of a self-adjoint operator on a *real* Hilbert space has a moment sequence, and thus an invariant subspace [14]. It is not known if this result due to A. Simonic extends to *complex* Hilbert spaces.

(3) Our arguments use crucially the real structure, and the order structure on $\mathbb{R}$. We recall that Bishop–Phelps theorem does not extend to complex Banach spaces [10], and also that Von Neumann's inequality states that (4) is satisfied with $C = 1$ for every contraction on the complex Hilbert space $\mathcal{H}$, provided that we take K to be the unit disc in the complex plane. However, our techniques unfortunately fall short to apply to arbitrary contractions on $\mathcal{H}$.

Bibliography

[1] S. I. Ansari and P. Enflo, *Extremal vectors and invariant subspaces*, Trans. Am. Math. Soc. **350** (2) (1998), 539–558.

[2] A. Atzmon, *The existence of translation- invariant subspaces of symmetric self-adjoint sequence spaces on $\mathbb{Z}$*, J. Funct. Anal.. **178** (2000), 372–380.

[3] A. Atzmon and G. Godefroy, *An application of the smooth variational principle to the existence of nontrivial invariant subspaces*, Note aux C. R. A. S. Paris **332** (1) (2001), 151–156.

[4] A. Atzmon, G. Godefroy and N. J. Kalton, *Invariant subspaces and the exponential map*, Positivity **8** (2004), 101–107.

[5] I. Chalendar, E. Fricain, A. I. Popov, D. Timotin and V. G. Troitsky, *Finitely strictly singular operators between James spaces*, J. Funct. Anal. **256** (4) (2009), 1258–1268.

[6] I. Chalendar and J. Partington, *Modern Approaches to the Invariant-subspace Problem*, Cambridge Tracts in Mathematics, 188, 2011.

[7] S. Grivaux and M. Roginskaya, *A general approach to Read's type constructions of operators without non-trivial invariant closed subspaces*, Proc. Lond. Math. Soc. (3) **109** (3) (2014), 596–652.

[8] S. Grivaux, E. Matheron and Q. Menet, *Linear Dynamical Systems on Hilbert Spaces: Typical Properties and Explicit Examples*, Memoir of the Amer. Math. Soc., 2018.

[9] V. Lomonosov, *Invariant subspaces for operators commuting with compact operators*, Funct. Anal. Appl. **7** (1973), 213–214.

[10] V. Lomonosov, *A counterexample to the Bishop–Phelps theorem in complex spaces*, Isr. J. Math. **115** (2000), 25–48.

[11] V. Lomonosov, *Positive functionals on general operator algebras*, J. Math. Anal. Appl. **245** (2000), 221–224.

[12] A. J. Michaels, *Hilden's simple proof of Lomonosov's invariant subspace theorem*, Adv. Math. **25** (1977), 56–58.
[13] W. Rudin, *Functional Analysis*, 2nd edition, McGraw-Hill International Editions, 1991.
[14] A. Simonic, *An extension of Lomonosov's techniques to non-compact operators*, Trans. Am. Math. Soc. **348** (1996), 975–995.
[15] V. Troitsky, *Lomonosov's theorem cannot be extended to chain of four operators*, Proc. Am. Math. Soc. **128**, (2) (2000), 521–525.

Sophie Grivaux

11 On the Hypercyclicity Criterion for operators of Read's type

Abstract: Let T be a so-called *operator of Read's type* on a (real or complex) separable Banach space, having no nontrivial invariant subset. We prove in this note that $T \oplus T$ is then hypercyclic, that is, that T satisfies the Hypercyclicity Criterion.

Keywords: Invariant subspace/subset problem, operators of Read's type, hypercyclic operators, Hypercyclicity Criterion

MSC 2010: 47A15, 47A16

11.1 The invariant subspace problem

Given a (real or complex) infinite-dimensional separable Banach space X, the invariant subspace problem for X asks whether every bounded operator T on X admits a nontrivial invariant subspace, that is, a closed subspace M of X with $M \neq \{0\}$ and $M \neq X$ such that $T(M) \subseteq M$. It was answered in the negative in the 1980s, first by Enflo [11] and then by Read [24], who constructed examples of separable Banach spaces supporting operators without nontrivial closed invariant subspace. One of the most famous open questions in modern operator theory is the Hilbertian version of the invariant subspace problem, but it is also widely open in the reflexive setting: to this day, all the known examples of operators without nontrivial invariant subspace live on nonreflexive Banach spaces.

Read provided several classes of operators on $\ell_1(\mathbb{N})$ having no nontrivial invariant subspace [25], [26], [29]. In the work [28], he gave examples of such operators on $c_0(\mathbb{N})$ and $X = \bigoplus_{\ell_2} J$, the ℓ_2-sum of countably many copies of the James space J; since J is quasi reflexive (i. e., has codimension 1 in its bidual J^{**}), the space X has the property that X^{**}/X is separable. This approach was further developed in [17], where it was shown that whenever Z is a nonreflexive separable Banach space admitting a Schauder basis, the ℓ_p-sums $X = \bigoplus_{\ell_p} Z$ of countably many copies of Z $(1 \leq p < +\infty)$ as well as the c_0-sum $X = \bigoplus_{c_0} Z$ support an operator without nontrivial invariant closed subspace. Actually, these spaces support an operator without

Acknowledgement: This work was supported in part by the project FRONT of the French National Research Agency (grant ANR-17-CE40-0021) and by the Labex CEMPI (ANR-11-LABX-0007-01).
I am grateful to Gilles Godefroy and Quentin Menet for interesting comments on a first version of this note.

Sophie Grivaux, CNRS, Univ. Lille, UMR 8524 – Laboratoire Paul Painlevé, France, e-mail: sophie.grivaux@univ-lille.fr; URL: http://math.univ-lille1.fr/ grivaux/

https://doi.org/10.1515/9783110656756-011

nontrivial invariant closed *subset*. This generalizes a result of Read, who exhibited in [27] the first known example of an operator (on the space $\ell_1(\mathbb{N})$) without a nontrivial invariant closed subset. The most recent counterexample to the invariant subspace problem is given in the joint work by Gallardo-Gutiérrez and Read [13], which happens to be Read's last article: the authors give an example of a quasi-nilpotent operator T on $\ell_1(\mathbb{N})$ with the property that whenever f is the germ of a holomorphic function at 0, the operator $f(T)$ has no nontrivial invariant closed subspace.

On the other hand, many powerful techniques have been developed in the past decade to show that operators enjoying certain additional properties have nontrivial invariant subspaces. Among these, some of the most interesting have been developed by Lomonosov: his best-known result in this direction, striking for its simplicity and effectiveness, states that every operator on a Banach space commuting with a nonzero compact operator admits a nontrivial invariant subspace [22]. Another important work of Lomonosov concerns the generalizations of the Burnside inequality obtained in [20] and [21] (see [19] for a simpler proof, relying on nonlinear arguments from [22]). The Lomonosov inequality from [20] runs as follows.

The Lomonosov inequality
Let X be a complex separable Banach space, and let $\mathcal{A}$ be a weakly closed subalgebra of $\mathcal{B}(X)$ with $\mathcal{A} \neq \mathcal{B}(X)$. There exist two nonzero elements x^* and x^{**} of X^* and X^{**}, respectively, such that $|\langle x^{**}, A^*x^*\rangle| \leq \|A\|_e$ for every $A \in \mathcal{A}$.

Here, $\|A\|_e$ denotes the essential norm of A, which is the distance of A to the space of compact operators on X.

This inequality is a powerful tool and has been used in many contexts to prove the existence of nontrivial invariant subsets or subspaces for certain classes of operators (see, for instance, [3], [12], [9], [17]). It is one of the main results which support the conjecture that adjoint operators on infinite-dimensional dual Banach spaces have nontrivial invariant subspaces.

It would be impossible to mention here all the beautiful existence results for invariant subspaces proved in the past decade. We refer to the books [23] and [8] for a description of many of these. We conclude this introduction by mentioning the important work [1] of Argyros and Haydon, who constructed an example of a space X on which any operator is the sum of a multiple of the identity and a compact operator. As a consequence of the Lomonosov theorem [22], every operator on X has a nontrivial invariant subspace. Subsequent work of Argyros and Motakis [2] shows the existence of reflexive separable Banach spaces on which any operator has a nontrivial invariant subspace. Again, the Lomonosov theorem is brought to use in the proof, although the spaces of [2] do support operators which are not the sum of a multiple of the identity and a compact operator.

11.2 Hypercyclic operators and the Hypercyclicity Criterion

Let us now shift our point of view, and consider the invariant subspace and subset problems from the point of view of orbit behavior. It is not difficult to see that $T \in \mathcal{B}(X)$ has no nontrivial invariant subspace if and only if every nonzero vector $x \in X$ is *cyclic* for T: the linear span in X of the orbit $\{T^n x\,;\, n \geq 0\}$ of the vector x under the action of T is dense in X. In a similar way, T has no nontrivial invariant closed subset if and only if every vector $x \neq 0$ is *hypercyclic*, that is, the orbit $\{T^n x\,;\, n \geq 0\}$ itself is dense in X. An operator is called *hypercyclic* if it admits a hypercyclic vector (in which case it admits a dense G_δ set of such vectors).

The study of hypercyclicity and related notions fits into the framework of *linear dynamics*, which is the study of the dynamical systems given by the action of a bounded operator on a separable Banach space. It has been the object of many investigations in the past years, as testified by the two books [18] and [6] which retrace important recent developments in this direction. One of the main open problems in hypercyclicity theory was solved in 2006 by De la Rosa and Read [10]. They constructed an example of a hypercyclic operator T on a Banach space X such that the direct sum $T \oplus T$ of T with itself on $X \oplus X$ is not hypercyclic. In other words, although there exists $x \in X$ with the property that for every $u \in X$ and every $\varepsilon > 0$, there exists $n \geq 0$ such that $\|T^n x - u\| < \varepsilon$, there is no pair (x, y) of vectors of X such that for every $(u, v) \in X \times X$ and every $\varepsilon > 0$, there exists $n \geq 0$ which simultaneously satisfies $\|T^n x - u\| < \varepsilon$ and $\|T^n y - v\| < \varepsilon$. Further examples of such operators (hypercyclic but not *topologically weakly mixing*) were constructed by Bayart and Matheron in [5] on many classical spaces such as the spaces $\ell_p(\mathbb{N})$, $1 \leq p < +\infty$ and $c_0(\mathbb{N})$.

The question of the existence of hypercyclic operators T such that $T \oplus T$ is not hypercyclic arose in connection with the so-called *Hypercyclicity Criterion*, which is certainly the most effective tool for proving that a given operator is hypercyclic. Despite its somewhat intricate form, which we recall below, it is very easy to use.

The Hypercyclicity Criterion

Let $T \in \mathcal{B}(X)$. Suppose that there exist two dense subsets D and D' of X, a strictly increasing sequence $(n_k)_{k\geq 0}$ of integers, and a sequence $(S_{n_k})_{k\geq 0}$ of maps from D' into X satisfying the following three assumptions:

(i) $T^{n_k} x \to 0$ as $k \to +\infty$ for every $x \in D$;

(ii) $S_{n_k} y \to 0$ as $k \to +\infty$ for every $y \in D'$;

(iii) $T^{n_k} S_{n_k} y \to y$ as $k \to +\infty$ for every $y \in D'$.

Then T is hypercyclic, as well as $T \oplus T$.

The Hypercyclicity Criterion admits many equivalent formulations, which we will not detail here. An important result, due to Bès and Peris [7], shows that T satisfies the Hypercyclicity Criterion if and only if $T \oplus T$ is hypercyclic as an operator on the Banach space $X \oplus X$. This criterion is thus deeper than one may think at first glance. Many sufficient conditions implying the Hypercyclicity Criterion have been proved over the years, always in the spirit that "hypercyclicity plus some regularity assumption implies the Hypercyclicity Criterion": see [18, Chapter 3]. For instance, hypercyclicity plus the existence of a dense set of vectors with bounded orbit implies that the Hypercyclicity Criterion is satisfied ([14], see also [15, Section 5] for generalizations). This phenomenon is well known in dynamics: an irregular behavior of some orbits (density) combined with the regular behavior of some other orbits (typically, periodicity) implies chaos; see, for instance, [4].

11.3 Operators without nontrivial invariant subsets and the Hypercyclicity Criterion

In the light of this observation (and also of the fact that Read had a hand in the construction of operators without nontrivial invariant subsets, as well as in the construction of hypercyclic operators which are not weakly topologically mixing!), the following question comes naturally to mind.

Question 11.1. Does there exist a bounded operator T on a Banach space X which simultaneously satisfies

(a) T has no nontrivial invariant subset, that is, all nonzero vectors $x \in X$ are hypercyclic for T;

(b) $T \oplus T$ is not hypercyclic as an operator on $X \oplus X$?

One may be tempted to guess that operators whose set of hypercyclic vectors is too large are somehow less likely to satisfy the Hypercyclicity Criterion than others (since the usual regularity assumptions may be missing), or one may be inclined to believe that such operators should indeed satisfy the criterion (as the set of hypercyclic vectors is so large, there is every chance that there exists a pair (x, y) of vectors of X whose orbits are independent enough for $x \oplus y$ to have a dense orbit under the action of $T \oplus T$). Both arguments are plausible, and it is difficult to get a deeper intuition in Question 11.1, besides saying that it is probably hard!

Our aim in this note is to prove the following modest result, which shows that all the known examples of operators without nontrivial invariant closed subset do satisfy the Hypercyclicity Criterion.

Theorem 11.2. *Let T be an operator of Read's type, acting on a (real or complex) separable Banach space, and having no nontrivial invariant subset. Then $T \oplus T$ is hypercyclic, that is, T satisfies the Hypercyclicity Criterion.*

What are *operators of Read's type*? We group under this rather vague denomination all the operators which satisfy certain structure properties, appearing in the constructions carried out by Read, and common to almost all the operators which have no (or few) nontrivial invariant subspaces or subsets. All the operators constructed by Read in [24–29], as well as the operators from [16] and [17], fall within this category (Enflo's examples are of a different type). See [17, Section 2] for an informal description of the properties of operators of Read's type. As will be seen in Section 11.4 below, only two of the properties of operators of Read's type are involved in the proof of Theorem 11.2, so that it could potentially be applied to much wider classes of operators.

11.4 Proof of Theorem 11.2

We will carry out this proof in the context of [17], and will in particular use the notation introduced in [17, Section 2.2]. Read's type constructions involve two sequences $(f_j)_{j\geq 0}$ and $(e_j)_{j\geq 0}$ of vectors, defined inductively. The sequence $(f_j)_{j\geq 0}$ is a Schauder basis of the space X. When X is a classical space like $\ell_1(\mathbb{N})$ or $c_0(\mathbb{N})$, $(f_j)_{j\geq 0}$ is simply the canonical basis of X. The vectors e_j, $j \geq 0$, are defined in such a way that $e_0 = f_0$ and $\operatorname{span}[e_0, \ldots, e_j] = \operatorname{span}[f_0, \ldots, f_j]$ for every $j \geq 1$. They are thus linearly independent and span a dense subspace of X. The operator T is then defined by setting $Te_j = e_{j+1}$ for every $j \geq 0$; this definition makes sense since the vectors e_j are linearly independent. The whole difficulty of the construction is to define the vectors e_j in such a way that T extends to a bounded operator on X, and that T has no nontrivial invariant subspace (or subset). Observe that $T^j e_0 = e_j$ for every $j \geq 0$, that is, that $(e_j)_{j\geq 0}$ is the orbit of e_0 under the action of T. In particular, e_0 is by construction a cyclic vector for T.

The vectors e_j are defined differently, depending on whether j belongs to what is called in [16] or [17] a *working interval* or a *lay-off interval*. Lay-off intervals lie between the working intervals, and if $I = [\nu + 1, \nu + l]$ is such a lay-off interval of length l, e_j is defined for $j \in I$ as

$$e_j = 2^{-\frac{1}{\sqrt{l}}(\frac{l}{2}+\nu+1-j)} f_j$$

and $Tf_j = 2^{-\frac{1}{\sqrt{l}}} f_{j+1}$ for every $\nu + 1 \leq j < \nu + l$.

The working intervals are of three types: (a), (b), and (c). The (c)-working intervals appear only in the case where one is interested in constructing operators without nontrivial invariant subset. These are the only working intervals which will be relevant here. One of their roles is to ensure that e_0 is not only cyclic, but hypercyclic for T. There is at each step n of the construction a whole family of (c)-working intervals, which is called in [16] and [17] the (c)-fan. The first of these intervals has the form $[c_{1,n}, c_{1,n} + v_n]$, where v_n is the index corresponding to the end of the last (b)-working interval constructed at step n, and $c_{1,n}$ is extremely large with respect to v_n. In order to simplify the notation, we set $c_n = c_{1,n}$ for every $n \geq 0$. Thus $[v_n + 1, c_n - 1]$ is the lay-off interval which precedes the first (c)-working interval. For $j \in [c_n, c_n + v_n]$, the vector e_j is defined as

$$e_j = \gamma_n f_j + p_n(T)e_0$$

where $\gamma_n > 0$ is extremely small and p_n is a polynomial with suitably controlled degree, and such that $|p_n| \leq 2$ (the polynomial p_n is denoted by $p_{1,n}$ in [16] and [17]; again we simplify the notation). Here, the modulus $|p|$ of a polynomial p is defined as the sum of the moduli of its coefficients.

Thus, in particular, $e_{c_n} = T^{c_n}e_0 = \gamma_n f_{c_n} + p_n(T)e_0$ and $\|e_{c_n} - p_n(T)e_0\| = \gamma_n$. The family $(p_n)_{n\geq 1}$ is chosen in such a way that for every polynomial p with $|p| \leq 2$ and every $\varepsilon > 0$, there exists $n \geq 1$ such that $\|p_n(T)e_0 - p(T)e_0\| < \varepsilon$. Hence there exists for every polynomial p with $|p| \leq 2$ and every $\varepsilon > 0$ an integer n such that $\|T^{c_n}e_0 - p(T)e_0\| < \varepsilon$.

An important observation is that this property actually extends to *all* polynomials p, regardless of the size of their moduli $|p|$. The simple argument is given already in the proof of [16, Theorem 1.1] and in [17, Section 3.1], but we recall it briefly for the sake of completeness: let p be any polynomial, and fix $\varepsilon > 0$. Let j be an integer such that $|p| \leq 2^j$. Then we know that there exists an integer n_1 such that $\|T^{c_{n_1}}e_0 - 2^{-j}p(T)e_0\| < \varepsilon 2^{-2j}$. There also exists an integer n_2 such that $\|T^{c_{n_2}}e_0 - 2T^{c_{n_1}}e_0\| < \varepsilon 2^{-(2j-1)}$. Then it follows that $\|T^{c_{n_2}}e_0 - 2^{-(j-1)}p(T)e_0\| < \varepsilon 2^{-2(j-1)}$. Continuing in this fashion, we obtain that there exists an integer n_j such that $\|T^{c_{n_j}}e_0 - p(T)e_0\| < \varepsilon$, which proves our claim: there exists for every polynomial p and every $\varepsilon > 0$ an integer n such that $\|T^{c_n}e_0 - p(T)e_0\| < \varepsilon$.

Before moving over to the proof of Theorem 11.2, we recall the following result from [14], which provides a useful sufficient condition for the Hypercyclicity Criterion to be satisfied.

Theorem 11.3 ([14]). *Let T be a bounded operator on a separable Banach space X. Suppose that for every pair (U, V) of nonempty open subsets of X, and for every neighborhood W of 0, there exists a polynomial p such that $p(T)(U) \cap W$ and $p(T)(W) \cap V$ are simultaneously nonempty. If T is hypercyclic, then T satisfies in fact the Hypercyclicity Criterion.*

Theorem 11.3 can be rewritten in somewhat more concrete terms as the following.

Proposition 11.4. *Let T be a hypercyclic operator on a separable Banach space X, and let x_0 be a cyclic vector for T. If there exist a sequence $(q_k)_{k\geq 0}$ of polynomials and a sequence $(w_k)_{k\geq 0}$ of vectors of X such that*

$$q_k(T)x_0 \to 0, \quad w_k \to 0, \quad \textit{and} \quad q_k(T)w_k \to x_0$$

as $k \to +\infty$, then T satisfies the Hypercyclicity Criterion.

We are now ready for the proof of Theorem 11.2.

Proof of Theorem 11.2. Let $(n_k)_{k\geq 0}$ be a strictly increasing sequence of integers such that

$$\|T^{c_{n_k}} e_0 - 4^k e_0\| < 1 \quad \text{for every } k \geq 1.$$

Write c_{n_k} as $c_{n_k} = i_{n_k} + j_{n_k}$ where $i_{n_k} = \lfloor c_{n_k}/2 \rfloor$ and $j_{n_k} = c_{n_k} - \lfloor c_{n_k}/2 \rfloor$.

Since c_n is extremely large with respect to v_n at each step n of the construction of T, i_{n_k} belongs to the lay-off interval $[v_{n_k} + 1, c_{n_k} - 1]$ for every k. Thus

$$\|T^{i_{n_k}} e_0\| = \|e_{i_{n_k}}\| = 2^{-\frac{1}{\sqrt{c_{n_k} - v_{n_k} - 1}}\left(\frac{1}{2}(c_{n_k} - v_{n_k} - 1) + v_{n_k} + 1 - i_{n_k}\right)}$$

and $\|T^{i_{n_k}} e_0\| \to 1$ as $k \to +\infty$. Exactly the same argument shows that $\|T^{j_{n_k}} e_0\| \to 1$ as $k \to +\infty$.

Set $w_k = 2^{-k} T^{i_{n_k}} e_0$ and $q_k(T) = 2^{-k} T^{j_{n_k}}$ for every $k \geq 1$. Then $w_k \to 0$ and $q_k(T)e_0 \to 0$. Moreover, $q_k(T)w_k = 4^{-k} T^{c_{n_k}} e_0 \to e_0$. Since the vector e_0 is hypercyclic for T, the assumptions of Proposition 11.4 are in force, and T satisfies the Hypercyclicity Criterion. □

Remark 11.5. The same argument shows that the hypercyclic operators from [16], which have few nontrivial invariant subsets but still do have some nontrivial invariant subspaces, also satisfy the Hypercyclicity Criterion. The fact that the operators of Read's type on a separable infinite-dimensional complex Hilbert space H from [16] have a non-trivial invariant subspace relies on the Lomonosov inequality from [20]: there exists a pair (x, y) of nonzero vectors of H such that $|\langle T^n x, y\rangle| \leq \|T^n\|_e$ for every integer n. Since the operators T are by construction compact perturbations of power-bounded (forward) weighted shifts with respect to a fixed Hilbertian basis $(f_j)_{j\geq 0}$ of H, $\sup_{n\geq 0} |\langle T^n x, y\rangle| < +\infty$, and the closure of the orbit of x under the action of T is a nontrivial closed invariant subset for T. Moreover, the operator T has the following property (called (P1) in [17]): all closed invariant subsets of T are actually closed invariant subspaces. Therefore, T has a nontrivial closed invariant subspace. See [17, Section 7.2] for details and more general results.

We conclude this note with the following question, which may help to shed a light on Question 11.1.

Question 11.6. Let T be one of the operators from [10] or [5] which are hypercyclic but do not satisfy the Hypercyclicity Criterion. What can be said about the size of the set $\mathrm{HC}(T)$ of hypercyclic vectors for T? Is it "large," or rather "small"? Is its complement Haar-null, for instance?

Bibliography

[1] S. A. Argyros and R. G. Haydon, *A hereditarily indecomposable $\mathscr{L}_\infty$-space that solves the scalar-plus-compact problem*, Acta Math. **206** (1) (2011), 1–54, MR2784662.

[2] S. A. Argyros and P. Motakis, *A reflexive hereditarily indecomposable space with the hereditary invariant subspace property*, Proc. Lond. Math. Soc. (3) **108** (6) (2014), 1381–1416, MR3218313.

[3] A. Atzmon, G. Godefroy and N. J. Kalton, *Invariant subspaces and the exponential map*, Positivity **8** (2) (2004), 101–107, MR2097081.

[4] J. Banks, J. Brooks, G. Cairns, G. Davis and P. Stacey, *On Devaney's definition of chaos*, Am. Math. Mon. **99** (4) (1992), 332–334, MR1157223.

[5] F. Bayart and É Matheron, *Hypercyclic operators failing the Hypercyclicity Criterion on classical Banach spaces*, J. Funct. Anal. **250** (2) (2007), 426–441, MR2352487.

[6] F. Bayart and É Matheron, *Dynamics of Linear Operators*, Cambridge Tracts in Mathematics, **179**, Cambridge University Press, Cambridge, 2009, MR2533318.

[7] J. Bès and A. Peris, *Hereditarily hypercyclic operators*, J. Funct. Anal. **167** (1) (1999), 94–112, MR1710637.

[8] I. Chalendar and J. R. Partington, *Modern Approaches to the Invariant-subspace Problem*, Cambridge Tracts in Mathematics, **188**, Cambridge University Press, Cambridge, 2011, MR2841051.

[9] B. Chevreau, I. B. Jung, E. Ko and C. Pearcy, *Operators that admit a moment sequence. II*, Proc. Am. Math. Soc. **135** (6) (2007), 1763–1767, MR2286086.

[10] M. de la Rosa and C. Read, *A hypercyclic operator whose direct sum $T \oplus T$ is not hypercyclic*, J. Oper. Theory **61** (2) (2009), 369–380, MR2501011.

[11] P. Enflo, *On the invariant subspace problem for Banach spaces*, Acta Math. **158** (3–4) (1987), 213–313, MR892591.

[12] C. Foias, C. Pearcy and L. Smith, *Weak orbit-transitivity on Hilbert space*, Acta Sci. Math. (Szeged) **76** (1–2) (2010), 155–164, MR2668412.

[13] E. A. Gallardo-Gutiérrez and C. J. Read, *Operators having no non-trivial closed invariant subspaces on ℓ_1: a step further*, Proc. Lond. Math. Soc. **118** (3) (2019), 649–674.

[14] S. Grivaux, *Hypercyclic operators, mixing operators, and the bounded steps problem*, J. Oper. Theory **54** (1) (2005), 147–168, MR2168865.

[15] S. Grivaux, É Matheron and Q. Menet, *Linear dynamical systems on Hilbert spaces: typical properties and explicit examples*, to appear in Mem. Am. Math. Soc. 2018, available at http://front.math.ucdavis.edu/1703.01854.

[16] S. Grivaux and M. Roginskaya, *On Read's type operators on Hilbert spaces*, Int. Math. Res. Not. IMRN 2008, Art. ID rnn 083, 42, MR2439560.

[17] S. Grivaux and M. Roginskaya, *A general approach to Read's type constructions of operators without non-trivial invariant closed subspaces*, Proc. Lond. Math. Soc. (3) **109** (3) (2014), 596–652, MR3260288.

[18] K-G. Grosse-Erdmann and A. Peris Manguillot, *Linear Chaos*, Universitext, Springer, 2011, MR2919812.

[19] M. Lindström and G. Schlüchtermann, *Lomonosov's techniques and Burnside's theorem*, Can. Math. Bull. **43** (1) (2000), 87–89, MR1749953.

[20] V. Lomonosov, *An extension of Burnside's theorem to infinite-dimensional spaces*, Isr. J. Math. **75** (2–3) (1991), 329–339, MR1164597.

[21] V. Lomonosov, *Positive functionals on general operator algebras*, J. Math. Anal. Appl. **245** (1) (2000), 221–224, MR1756586.

[22] V. I. Lomonosov, *Invariant subspaces of the family of operators that commute with a completely continuous operator*, Funk. Anal. Priložen. **7** (3) (1973), 55–56, MR0420305.

[23] H. Radjavi and P. Rosenthal, *Invariant Subspaces*, 2nd edition, Dover Publications, Inc., Mineola, NY, 2003, MR2003221.
[24] C. J. Read, *A solution to the invariant subspace problem*, Bull. Lond. Math. Soc. **16** (4) (1984), 337–401, MR749447.
[25] C. J. Read, *A solution to the invariant subspace problem on the space l_1*, Bull. Lond. Math. Soc. **17** (4) (1985), 305–317, MR806634.
[26] C. J. Read, *A short proof concerning the invariant subspace problem*, J. Lond. Math. Soc. (2) **34** (2) (1986), 335–348, MR856516.
[27] C. J. Read, *The invariant subspace problem for a class of Banach spaces. II. Hypercyclic operators*, Isr. J. Math. **63** (1) (1988), 1–40, MR959046.
[28] C. J. Read, *The invariant subspace problem on some Banach spaces with separable dual*, Proc. Lond. Math. Soc. (3) **58** (3) (1989), 583–607, MR988104.
[29] C. J. Read, *Quasinilpotent operators and the invariant subspace problem*, J. Lond. Math. Soc. (2) **56** (3) (1997), 595–606, MR1610408.

A. J. Guirao, S. Lajara, and S. Troyanski

12 Three-space problem for strictly convex renormings

Abstract: We establish a weak version of the three-space property for strictly convex renormings, by using a topological characterization of Orihuela, Smith, and Troyanski of the class of strictly convexifiable Banach spaces.

Keywords: Strictly convex norm, LUR norm, three-space properties

MSC 2010: Primary 46B20, 46B03

A norm $\|\cdot\|$ on a Banach space X is said to be *strictly convex* (or *rotund*) if, given $x, y \in X$ such that $\|x\| = \|y\| = \|(x+y)/2\|$ we have $x = y$, or equivalently, if the unit sphere of X in that norm does not contain any nondegenerate segment. If for every $x \in X$ and every sequence $(x_n)_n$ in X such that $\lim_n \|x_n\| = \lim_n \|(x_n + x)/2\| = \|x\|$, we have $\lim_n \|x_n - x\| = 0$, then we say that the norm $\|\cdot\|$ is *locally uniformly rotund* (LUR, for short).

In the paper [3] (see, e. g., [2, Theorem VII.3.1]), it was proved that the existence of an equivalent LUR norm is a three-space property, that is, a Banach space X admits an equivalent LUR norm whenever there exists a closed subspace Y of X such that both Y and X/Y are LUR renormable. There, it was also proved that the space X is strictly convexifiable if there exists a closed strictly convexifiable subspace Y of X such that X/Y is LUR renormable. Later on, it was shown in [5] that a Banach space X admits an equivalent norm whose dual norm is LUR provided that there exists a closed subspace Y of X such that both Y and X/Y have an equivalent norm with dual LUR norm.

The existence of an equivalent strictly convex norm is not a three-space property: in the paper [4], Haydon gave an example of an Asplund space X (namely, a space of continuous functions on a tree) and a closed subspace Y of X such that Y has an

Acknowledgement: A. J. Guirao was supported by MTM2017-83262-C2-1-P (AEI/FEDER, UE) and 19368/PI/14 (Fundación Séneca). S. Lajara was supported by MICINN projects MTM2014-54182-P and MTM2015-65825-P (Spain), and by MTM2017-86182-P (AEI/FEDER, UE). S. Troyanski was supported by MICINN project MTM2014-54182-P (Spain), by MTM2017-86182-P (AEI/FEDER, UE), and by the Bulgarian National Scientific Fund, Grant KP-06-H22/4, 04.12.2018.

A. J. Guirao, Instituto Universitario de Matemática Pura y Aplicada, Universitat Politècnica de València, Camino de Vera, s/n, 46022 Valencia, Spain, e-mail: anguisa2@mat.upv.es
S. Lajara, Departamento de Matemáticas, Universidad de Castilla-La Mancha, Escuela Técnica Superior de Ingenieros Industriales, 02071 Albacete, Spain, e-mail: sebastian.lajara@uclm.es
S. Troyanski, Institute of Mathematics and Informatics, Bulgarian Academy of Science, bl. 8, acad. G. Bonchev str., 1113 Sofia, Bulgaria; and Departamento de Matemáticas, Universidad de Murcia, Campus de Espinardo, 30100 Murcia, Spain, e-mail: stroya@um.es

https://doi.org/10.1515/9783110656756-012

equivalent LUR norm and the quotient X/Y is strictly convexifiable, while X admits no equivalent strictly convex norm. The three-space problem for strictly convex dual renormings has also a negative answer. Talagrand [9] (cf. [2, Theorem VII.3.5]) proved the existence of a Banach space X that contains $\mathcal{C}([0,1])$ as a closed subspace, and such that X admits no equivalent Gâteaux smooth norm and $X/\mathcal{C}([0,1])$ is isomorphic to $c_0([0,1])$. Thanks to [2, Proposition II.1.6] and [2, Theorem II.7.4], the spaces $\mathcal{C}([0,1])^*$ and $(X/\mathcal{C}([0,1]))^*$ have respectively an equivalent dual strictly convex norm and an equivalent dual LUR norm. However, as X admits no equivalent Gâteaux smooth renorming, according to a well-known result of Šmulyan (cf. [2, Proposition II.1.6]) it follows that the space X^* does not have any equivalent dual strictly convex norm. For more information on three-space problems in Banach space theory, we refer to the monograph [1].

The class of strictly convexifiable Banach spaces has been characterized in linear topological terms in [8]. The main ingredient in this characterization is the so-called property $(*)$, defined below. Recall that a subspace M of the dual X^* of a Banach space X is said to be 1-*norming* if $\sup\{f(x) : f \in M, \|f\| \le 1\} = \|x\|$ for all $x \in X$.

Definition 12.1. Let X be a Banach space and M be a 1-norming subspace of X^*. We say that $(X, \sigma(X,M))$ has property $(*)$ if there exists a countable collection of families of $\sigma(X,M)$-open half-spaces of X, $\mathcal{H} = \{\mathcal{H}_n\}_n$, such that for every $x, y \in X$ there exists a number $n \in \mathbb{N}$ satisfying:
(i) $\{x,y\} \cap (\cup\mathcal{H}_n) \neq \emptyset$ and,
(ii) for every $H \in \mathcal{H}_n$, the set $\{x,y\} \cap H$ contains no more than one element.

In this situation, we also say that the couple $\{x,y\}$ has property $(*)$ with respect to $\mathcal{H}$.

The aforementioned characterization of strictly convexifiable spaces (Theorem 2.7 in [8]), states that if M is a 1-norming subspace of X^*, then X admits an equivalent $\sigma(X,M)$-lower semicontinuous (in short, $\sigma(X,M)$-l. s. c.) strictly convex norm if, and only if, $(X, \sigma(X,M))$ has property $(*)$.

The aim of this note is to use this characterization of strictly convexifiable Banach spaces to obtain the following result.

Theorem 12.2. *Let X be a Banach space, let $M \subset X^*$ be a 1-norming subspace, let Y be a $\sigma(X,M)$-closed subspace of X and let $Q : X \to X/Y$ denote the canonical quotient map. Assume that Y has an equivalent $\sigma(X,M)$-l. s. c. strictly convex norm and there exists a 1-norming subspace $N \subset (X/Y)^*$ such that:*
(1) *$Q^*(N) \subset M$ and*
(2) *X/Y has an equivalent $\sigma(X/Y,N)$-l. s. c. LUR norm.*

Then X admits an equivalent $\sigma(X,M)$-l. s. c. strictly convex norm.

As an immediate application of the previous theorem, we obtain the aforementioned result in [3] that a Banach space X is strictly convexifiable provided that there exists a strictly convexifiable closed subspace Y of X such that X/Y is LUR renormable. Another consequence of Theorem 12.2 is the following result, which is essentially optimal in view of Talagrand's counterexample.

Corollary 12.3. *Let Y be a closed subspace of a Banach space X such that Y admits an equivalent norm with dual LUR norm. Then X has an equivalent norm with dual strictly convex norm if, and only if, so does X/Y.*

Proof. The space $(X/Y)^*$ is isometrically isomorphic to the annhilator $Y^\perp$ of Y, which is a w^*-closed subspace of X^*, and Y^* can be identified with the quotient $X^*/Y^\perp$. Thus, Theorem 12.2 applies with X^* instead of X, $Y^\perp$ instead of Y, $M = X$, and $N = Y$. □

An analogue of the former corollary replacing dual strictly convex renormings by dual strictly convex renormings with the Kadets property was proved in [7].

Now, we shall proceed with the proof of Theorem 12.2. Apart from the characterization of strictly convexifiable spaces in [8], we shall use the following slight generalization of a result from [6] concerning extensions of norms (cf. [2, Lemma II.8.1]).

Lemma 12.4. *Let X be a Banach space, M a 1-norming subspace of X^* and Y a $\sigma(X,M)$-closed subspace of X. If $|\cdot|$ is an equivalent $\sigma(X,M)$-l. s. c. norm on Y, then $|\cdot|$ can be extended to an equivalent $\sigma(X,M)$-l. s. c. norm on X.*

Proof. Let $\|\cdot\|$ denote the original norm on X, and assume without loss of generality that

$$|y| \leq \|y\| \quad \text{for all } y \in Y. \tag{12.1}$$

Set $A_X = B_{(X,\|\cdot\|)}$ and $A_Y = B_{(Y,|\cdot|)}$, and define

$$B = \overline{\operatorname{co}(A_X \cup A_Y)}^{\sigma(X,M)}.$$

Then the Minkowski functional of B defines an equivalent norm $|||\cdot|||$ on X, which is $\sigma(X,M)$-l. s. c. (as B is $\sigma(X,M)$-closed).

We shall prove that $B \cap Y = A_Y$. It is clear that $A_Y \subseteq B \cap Y$, thus we only need to see that

$$B \cap Y \subseteq A_Y. \tag{12.2}$$

Pick a vector $z \in B \cap Y$, and let $(z_\alpha)_{\alpha\in\Lambda}$ be a net in the set $\operatorname{co}(A_X \cup A_Y)$ such that $z_\alpha \xrightarrow{\sigma(X,M)} z$. Then there exist scalars $\lambda_\alpha \in [0,1]$ and vectors $x_\alpha \in A_X$ and $y_\alpha \in A_Y$ such that

$$z_\alpha = \lambda_\alpha x_\alpha + (1-\lambda_\alpha)y_\alpha \quad \text{for all } \alpha \in \Lambda.$$

We can assume without loss of generality that $x_\alpha \xrightarrow{w^*} u$, $y_\alpha \xrightarrow{w^*} v$ and $\lambda_\alpha \to \lambda$, for some $u, v \in X^{**}$ and $\lambda \in [0,1]$. Hence,

$$z = \lambda u + (1-\lambda)v.$$

Since the subspace Y is $\sigma(X,M)$-closed, we actually have $v \in Y$, and taking into account that the norm $|\cdot|$ is $\sigma(X,M)$-l. s. c. and $|y_\alpha| \leq 1$ for all α, we obtain $|v| \leq 1$. Thus,

$$v \in A_Y.$$

Moreover, as M is 1-norming, the norm $\|\cdot\|$ is $\sigma(X,M)$-lower semicontinuous as well, and bearing in mind that $\|x_\alpha\| \leq 1$ for all α, we get $\|u\| \leq 1$.

Now, we distinguish three cases, according to the values of λ. If $\lambda = 0$, then $z = v \in A_Y$, and we are done. If $\lambda = 1$, then $z = u$. In particular, u lies in Y, and using (12.1) we obtain $|u| \leq \|u\| \leq 1$, that is, $z \in A_Y$. Finally, suppose that $0 < \lambda < 1$. Since $v \in Y$ we also have $u \in Y$, and a new appeal to (12.1) yields $|u| \leq \|u\| \leq 1$, that is, $u \in A_Y$. As v also belongs to A_Y, thanks to the convexity of A_Y we obtain $z \in A_Y$, and (12.2) is proved. □

Before proving Theorem 12.2, it will be convenient to stress the following fact (see the proof of *(i)* ⇒ *(ii)* of [8, Theorem 2.7]).

Remark 12.5. Let $(X, \|\cdot\|)$ be a Banach space and let $M \subset X^*$ be a 1-norming subspace for $(X, \|\cdot\|)$, and set, for each $s \in \mathbb{Q}^+$,

$$\mathcal{H}_s = \{H_{f,s} : f \in M\},$$

where $H_{f,s}$ denotes the $\sigma(X,M)$-open half-space of X defined as

$$H_{f,s} = \{x \in X : f(x) > s\}.$$

If the norm $\|\cdot\|$ is strictly convex, then any couple $\{x,y\}$ satisfies $(*)$ with respect to the collection $\mathcal{H} = \{\mathcal{H}_s\}_{s\in\mathbb{Q}^+}$.

Proof of Theorem 12.2. Let $\|\cdot\|$ be the original norm of X, let $|\cdot|_1$ be an equivalent $\sigma(X,M)$-l. s. c. strictly convex norm on Y and $\|\cdot\|_2$ an equivalent $\sigma(X/Y,N)$-l. s. c. LUR norm on X/Y. According to the previous lemma, there exists an equivalent $\sigma(X,M)$-l. s. c. norm $\|\cdot\|_1$ on X such that $\|y\|_1 = |y|_1$ for every $y \in Y$.

Since the norm $\|\cdot\|_1$ is $\sigma(X,M)$-l. s. c., we have that M is a 1-norming subspace for $(X, \|\cdot\|_1)$. Therefore, for any $v \in Y$ we can choose a sequence of functionals $\{h_{v,k}\}_k \subset M$ satisfying

$$\|h_{v,k}\|_1 \leq 1 \quad \text{and} \quad 0 \leq \|v\|_1 - h_{v,k}(v) < \frac{1}{k} \quad \text{for all } k \in \mathbb{N}. \tag{12.3}$$

Analogously, bearing in mind that $\|\cdot\|_2$ is an equivalent $\sigma(X/Y,N)$-l. s. c. norm on X/Y, we have that N is a 1-norming subspace for $(X/Y, \|\cdot\|_2)$. Thus, for every $w \in X$ there is a sequence $\{g_{w,k}\}_k \subset N$ such that

$$\|g_{w,k}\|_2 \leq 1 \quad \text{and} \quad 0 \leq \|Qw\|_2 - g_{w,k}(Qw) < \frac{1}{k} \quad \text{for all } k \in \mathbb{N}. \tag{12.4}$$

On the other hand, according to Bartle–Graves theorem (see, e. g., [2, Lemma 3.2]), there exists a continuous map $B : X/Y \to X$ such that

$$w - BQw \in Y \quad \text{whenever } w \in X. \tag{12.5}$$

In particular, thanks to (12.3), for every $w \in X$ we can select a sequence $\{h_{w,k}\}_k \subset M$ satisfying

$$\|h_{w,k}\|_1 \le 1 \quad \text{and} \quad 0 \le \|w - BQw\|_1 - h_{w,k}(w - BQw) < \frac{1}{k} \quad \text{for all } k \in \mathbb{N}. \tag{12.6}$$

We notice that, since $\|\cdot\|_1$ is an equivalent norm on X,

$$\sup\{\|h_{w,k}\| : w \in X, k \in \mathbb{N}\} < \infty. \tag{12.7}$$

Let us write

$$f_{w,n,k} = Q^* g_{w,k} + \frac{1}{n} h_{w,k}, \quad w \in X, n, k \in \mathbb{N}.$$

Now, for $n, k \in \mathbb{N}$, $q \in \mathbb{Q}^+$ and $r \in \mathbb{Q}$ we put

$$X_{n,k}^{q,r} = \left\{w \in X : q - \frac{1}{n} < \|Qw\|_2 < q \text{ and } r - \frac{1}{n} < -h_{w,k}(BQw) < r\right\},$$

and given $n, k \in \mathbb{N}$, $s \in \mathbb{Q}^+$ and $w \in X$, we define the following half-spaces of X:

$$F_{w,n,k,s} = \{f_{w,n,k} > s\}, \quad G_{w,k,s} = \{Q^* g_{w,k} > s\} \quad \text{and} \quad H_{w,k,s} = \{h_{w,k} > s\}.$$

Finally, for $n, k \in \mathbb{N}$, $q, s \in \mathbb{Q}^+$ and $r \in \mathbb{Q}$, we write

$$\mathcal{F}_{n,k,s}^{q,r} = \{F_{w,n,k,s} : w \in X_{n,k}^{q,r}\},$$
$$\mathcal{G}_{k,s} = \{G_{w,k,s} : w \in X\},$$
$$\mathcal{H}_{k,s} = \{H_{w,k,s} : w \in X\},$$

and we denote by $\mathcal{F}$ be the union of the above three families.

Observe that, according to our hypothesis we have that the functionals $Q^* g_{w,k}$ lie in M. Thus, the functionals $f_{w,n,k}$ belong to M as well. In particular, all the half-spaces in $\mathcal{F}$ are $\sigma(X, M)$-open.

Now, we shall show that any pair $\{x, y\} \subset X$ satisfies $(*)$ with respect to $\mathcal{F}$. Assume that this is not so for some pair $\{x, y\}$. Then the pair $\{Qx, Qy\}$ fails $(*)$ with respect to the collection $\{\mathcal{G}'_{k,s}\}_{k \in \mathbb{N}, s \in \mathbb{Q}^+}$, where $\mathcal{G}'_{k,s}$ denotes the family made up of all $(X/Y, N)$-open half-spaces of X/Y of the form $\{g_{w,k} > s\}$, $w \in X$. Since $\|\cdot\|_2$ is a LUR (and in particular, a strictly convex) norm on X/Y and the subspace N is norming for X/Y, according to Remark 12.5 it follows that

$$Qx = Qy.$$

Next, we consider two cases:

Case I: $Qx = 0$. Then $x, y \in Y$. By our assumption, $\{x, y\}$ fails condition $(*)$ with respect to the family $\{\mathcal{H}'_{k,s}\}_{k\in\mathbb{N},s\in\mathbb{Q}^+}$, where $\mathcal{H}'_{k,s}$ stands for the family of $\sigma(X, M)$-open half-spaces of Y of the form $\{h_{w,k} > s\}$, $w \in X$. Taking into account that the norm $|\cdot|_1$ is strictly convex, we get $x = y$.

Case II: $Qx \neq 0$. Assume, without loss of generality, that

$$\|x - BQx\|_1 \geq \|y - BQy\|_1.$$

According to the definition of the set $X^{q,r}_{n,k}$, and inequality (12.7), we can find a sequence $\{k_n\}_n \subset \mathbb{N}$ with $k_n > n^2$ for all n, and sequences $\{q_n\}_n \subset \mathbb{Q}^+$ and $\{r_n\}_n \subset \mathbb{Q}$ such that

$$x \in X^{q_n,r_n}_{n,k_n} \quad \text{for all } n \in \mathbb{N}.$$

From (12.4), we get $\lim_n g_{x,k_n}(Qx) = \|Qx\|_2 > 0$, and using again (12.7) we deduce that $f_{x,n,k_n}(x) > 0$ for n big enough. We may assume that $f_{x,n,k_n}(x) > 0$ for all $n \in \mathbb{N}$. Choose a sequence $\{s_n\}_n \subset \mathbb{Q}^+$ such that

$$s_n < f_{x,n,k_n}(x) < s_n + \frac{1}{n^2} \quad \text{for all } n \in \mathbb{N}. \tag{12.8}$$

Pick $n \in \mathbb{N}$, and set

$$\mathcal{F}_n = \mathcal{F}^{q_n,r_n}_{n,k_n,s_n}.$$

It is clear that $F_{x,n,k_n,s_n} \in \mathcal{F}_n$ and that $x \in F_{x,n,k_n,s_n}$. Since the pair $\{x, y\}$ fails condition $(*)$ with respect to the collection $\mathcal{F}_n$, we deduce the existence of a vector $w_n \in X^{q_n,r_n}_{n,k_n}$ such that x and y both lie in the half-space $F_{w_n,n,k_n,s_n} = \{u \in X : f_{w_n,n,k_n}(u) > s_n\}$. In particular, due to the convexity of this half-space, the midpoint $z = (x + y)/2$ also belongs to it, that is,

$$f_{w_n,n,k_n}(z) > s_n. \tag{12.9}$$

We claim that

$$\|x - BQx\|_1 \leq \|z - BQw_n\| + \frac{4}{n}. \tag{12.10}$$

Indeed, from (12.4) and the facts that $Qz = Qx$ and $k_n > n^2$ we get

$$g_{w_n,k_n}(Qz) \leq \|Qz\|_2 = \|Qx\|_2 < g_{x,k_n}(Qx) + \frac{1}{n^2}.$$

Using subsequently this inequality, (12.9) and (12.8) we obtain

$$\begin{aligned}
g_{x,k_n}(Qx) + \frac{1}{n}h_{w_n,k_n}(z) &\geq f_{w_n,n,k_n}(z) - \frac{1}{n^2} \\
&> s_n - \frac{1}{n^2} \\
&> f_{x,n,k_n}(x) - \frac{2}{n^2} \\
&= g_{x,k_n}(Qx) + \frac{1}{n}h_{x,k_n}(x) - \frac{2}{n^2}.
\end{aligned}$$

Therefore,

$$h_{x,k_n}(x) < h_{w_n,k_n}(z) + \frac{2}{n}. \tag{12.11}$$

On the other hand, as $x, w_n \in X_{n,k_n}^{q_n,r_n}$ and $k_n > n$ it follows that

$$-h_{x,k_n}(BQx) < r_n + \frac{1}{k_n} < -h_{w_n,k_n}(BQw_n) + \frac{1}{n}.$$

Adding up this inequality and (12.11), it follows that

$$h_{x,k_n}(x - BQx) < h_{w_n,k_n}(z - BQw_n) + \frac{3}{n}.$$

But, because of (12.6) we have

$$h_{x,k_n}(x - BQx) \geq \|x - BQx\|_1 - \frac{1}{k_n} \geq \|x - BQx\|_1 - \frac{1}{n}.$$

Hence,

$$\|x - BQx\|_1 \leq \frac{1}{n} + h_{x,k_n}(x - BQx) < h_{w_n,k_n}(z - BQw_n) + \frac{4}{n} \leq \|z - BQw_n\|_1 + \frac{4}{n},$$

and inequality (12.10) is proved.

Now, using (12.4) and inequality $k_n > n$ we get

$$\|Qx + Qw_n\|_2 \geq g_{w_n,k_n}(Qx) + g_{w_n,k_n}(Qw_n) \geq g_{w_n,k_n}(Qx) + \|Qw_n\|_2 - \frac{1}{n}.$$

Taking into account that $w_n, x \in X_{n,k_n}^{q_n,r_n}$, we have $\|Qw_n\|_2 > q_n > \|Qx\|_2 - \frac{1}{n}$, thus

$$\|Qw_n + Qx\|_2 > g_{w_n,k_n}(Qx) + \|Qx\|_2 - \frac{2}{n}.$$

Moreover, thanks to (12.9) we have $g_{w_n,k_n}(Qz) > s_n - \frac{1}{n}h_{w_n,k_n}(z)$. Therefore, by (12.8) we get

$$\liminf_n g_{w_n,k_n}(Qz) \geq \lim_n s_n = \|Qx\|_2,$$

and so,

$$\liminf_n \|Qw_n + Qx\|_2 \geq 2\|Qx\|_2.$$

On the other hand, bearing in mind again that $x, w_n \in X_{n,k_n}^{q_n,r_n}$ we get $\lim_n \|Qw_n\|_2 = \|Qx\|_2$. Since $\|\cdot\|_2$ is a LUR norm on X/Y, it follows that $\lim_n \|Qw_n - Qz\|_2 = 0$, and due to the continuity of the map B, we deduce that

$$\lim_n \|BQw_n - BQz\|_2 = 0.$$

Thus, letting n go to infinity in (12.10) we obtain

$$\|z - BQz\|_1 \geq \|x - BQx\|_1 \geq \|y - BQy\|_1.$$

Now, taking into account that

$$z - BQz = \frac{1}{2}[(x - BQx) + (y - BQy)]$$

and the strict convexity of the norm $\|\cdot\|_1$ we deduce that $x - BQx = y - BQy$, which implies that $x = y$. Therefore, the space X admits an equivalent $\sigma(X, M)$-l. s. c. strictly convex norm. □

Bibliography

[1] J. M. F. Castillo and M. González, *Three-space Problems in Banach Space Theory*, Lecture Notes in Mathematics, **1667**, Springer-Verlag, Berlin, 1997.

[2] R. Deville, G. Godefroy and V. Zizler, *Smoothness and Renormings in Banach spaces*, Monographs and Surveys in Pure and Appl. Math., **64**, Pitman, 1993.

[3] G. Godefroy, S. Troyanski, J. Whitfield and V. Zizler, *Three-space problem for locally uniformly rotund renormings of Banach spaces*, Proc. Am. Math. Soc. **94** (4) (1985), 647–652.

[4] R. Haydon, *Trees in renorming theory*, Proc. Lond. Math. Soc. **78** (3) (1999), 541–584.

[5] M. Jiménez-Sevilla and J. P. Moreno, *Renorming Banach spaces with the Mazur Intersection Property*, J. Funct. Anal. **144** (1997), 486–504.

[6] K. John and V. Zizler, *On extension of rotund norms*, Bull. Acad. Polon. Sci. Ser. Sci. Math. Astron. Phys. **24** (1976), 705–707.

[7] S. Lajara, *Average locally uniform rotundity and a class of nonlinear maps*, Nonlinear Anal. **74** (2011), 1937–1944.

[8] J. Orihuela, R. Smith, and S. Troyanski, *Strictly convex norms and topology*, Proc. Lond. Math. Soc. **104** (3) (2012), 197–222.

[9] M. Talagrand, *Renormages de quelques $C(K)$*, Isr. J. Math. **54** (1986), 327–334.

Vladimir Kadets, Ginés López, Miguel Martín, and Dirk Werner

13 Norm attaining operators of finite rank

Dedicated to the memory of Victor Lomonosov.

Abstract: We provide sufficient conditions on a Banach space X in order that there exist norm attaining operators of rank at least two from X into any Banach space of dimension at least two. For example, a rather weak such condition is the existence of a nontrivial cone consisting of norm attaining functionals on X. We go on to discuss density of norm attaining operators of finite rank among all operators of finite rank, which holds for instance when there is a dense linear subspace consisting of norm attaining functionals on X. In particular, we consider the case of Hilbert space valued operators where we obtain a complete characterization of these properties. In the final section, we offer a candidate for a counterexample to the complex Bishop–Phelps theorem on c_0, the first such counterexample on a certain complex Banach space being due to V. Lomonosov.

Keywords: Norm attaining operators, cones of norm attaining functionals

MSC 2010: Primary 46B04, Secondary 46B20, 46B87

13.1 Introduction

Shortly after Bishop and Phelps's papers ([6], [7]) on the density of norm attaining functionals on a Banach space had appeared, Lindenstrauss, in his seminal work [24],

Acknowledgement: The research of the first author was done in the framework of the Ukrainian Ministry of Science and Education Research Program 0118U002036. The research of the second and third authors was partially supported by projects MTM2015-65020-P (MINECO/FEDER, UE) and PGC2018-093794-B-I00 (MCIU/AEI/FEDER, UE) and by the grant FQM-185 (Junta de Andalucía/FEDER, UE). Part of the research was done during several stays of the first and fourth authors to the University of Granada which have been supported by the project MTM2015-65020-P (MINECO/FEDER, UE).
The authors would like to thank Gilles Godefroy, Rafael Payá, and Beata Randrianantoanina for kindly answering several inquiries related to the topics of this manuscript and for providing valuable references.

Vladimir Kadets, School of Mathematics and Informatics, V. N. Karazin Kharkiv National University, pl. Svobody 4, 61022 Kharkiv, Ukraine, e-mail: vova1kadets@yahoo.com, https://orcid.org/0000-0002-5606-2679
Ginés López, Miguel Martín, Departamento de Análisis Matemático, Facultad de Ciencias, Universidad de Granada, 18071 Granada, Spain, e-mails: glopezp@ugr.es, mmartins@ugr.es, https://orcid.org/0000-0002-3689-1365, https://orcid.org/0000-0003-4502-798X
Dirk Werner, Department of Mathematics, Freie Universität Berlin, Arnimallee 6, D-14195 Berlin, Germany, e-mail: werner@math.fu-berlin.de, https://orcid.org/0000-0003-0386-9652

https://doi.org/10.1515/9783110656756-013

launched the study of norm attaining operators. Let us recall that a bounded linear operator $T\colon X \longrightarrow Y$ between Banach spaces, $T \in \mathcal{L}(X,Y)$, is called *norm attaining* if there is some $x_0 \in X$ with $\|x_0\| = 1$ and $\|Tx_0\| = \|T\|$; in this case, we write $T \in \mathrm{NA}(X,Y)$. Lindenstrauss introduced the following properties (A) and (B) of a Banach space: X has (A) if $\mathrm{NA}(X,Z)$ is dense in $\mathcal{L}(X,Z)$ for all Z; and Y has (B) if $\mathrm{NA}(W,Y)$ is dense in $\mathcal{L}(W,Y)$ for all W. Among many other results, he showed that reflexive spaces have (A) as does ℓ_1, and that c_0, ℓ_∞ and finite dimensional polyhedral spaces are examples of Banach spaces with property (B) and, finally, that there are Banach spaces (X,Y) such that $\mathrm{NA}(X,Y)$ is not dense in $\mathcal{L}(X,Y)$. Major progress was made by Bourgain [9] who proved that spaces with the RNP have property (A) and provided a certain converse result. The problem of whether Hilbert spaces have property (B) was left open in [24], and it was solved only 25 years later by Gowers [17, Appendix], who showed that none of the spaces ℓ_p for $1 < p < \infty$ have (B). This result was pushed out by M. Acosta [1, 2] showing that neither infinite-dimensional $L_1(\mu)$ spaces nor any strictly convex infinite dimensional Banach space have property (B). Finally, let us comment that even though there are many Banach spaces X for which all compact linear operators from them can be approximated by norm attaining (finite rank) operators [21] (including $X = C_0(L)$ and $X = L_1(\mu)$), it was proved in [29] that there exists a compact operator between certain Banach spaces which cannot be approximated by norm attaining operators. For more information and background on the topic of norm attaining operators, we refer to the survey papers [3] and [30].

One should observe that none of the negative results summed up above says anything about operators of finite rank. Actually, it is one of the major open questions in the theory whether all finite dimensional Banach spaces have Lindenstrauss's property (B); equivalently, whether every finite rank operator between Banach spaces can be approximated by (finite rank) norm attaining operators. The aim of our paper is to contribute to this problem, in particular for rank-two operators.

In the case of linear functionals, it is clear from the Hahn–Banach theorem that $\mathrm{NA}(X)$, the set of norm attaining functionals, is always nonempty. For operators of rank two, it is not clear at all whether there are norm attaining ones. We are going to investigate this problem in detail, both in general and in the particular case when the range space is the two-dimensional Hilbert space ℓ_2^2.

Let us outline the contents of the paper. We devote Section 13.2 to the study of the existence of norm attaining operators of finite rank. For certain Banach spaces, we prove that there are norm attaining operators of finite rank to all range spaces using known sufficient conditions. A first new result says that whenever $\mathrm{NA}(X)$ contains n-dimensional subspaces for a Banach space X, there are rank n norm attaining operators from X into any Banach space Y with dimension at least n. There are Banach spaces X for which $\mathrm{NA}(X)$ contains no two-dimensional subspaces (this is proved in [34], [22], or [23] building on the ingenious ideas of Charles Read [33]), so this does not solve the existence problem for all domain spaces. However, we also show that it is sufficient for the existence of norm attaining rank two operators that $\mathrm{NA}(X)$ contains a nontrivial cone. We do not know whether this condition is also necessary; neither do we know whether every Banach space shares this property.

Section 13.3 contains a new result on the density of norm attaining finite rank operators from a Banach space X: this is the case if $\mathrm{NA}(X)$ contains "sufficiently many" linear subspaces. In particular, this holds if $\mathrm{NA}(X)$ contains a dense linear subspace (for instance, if $\mathrm{NA}(X)$ is actually a linear space itself). We use this result to recover the known results from [21] and [30] about the density of norm attaining compact operators on a Banach space whose dual satisfies an appropriate version of the approximation property like, for example, $C_0(L)$ spaces, $L_1(\mu)$ spaces, preduals of ℓ_1, among many others. But it can also be used to get some new results. Among other examples, we show that all finite rank operators from X can be approximated by norm attaining operators in the following cases: X is a finite-codimensional proximinal subspace of c_0 or of $\mathcal{K}(\ell_2)$, X is a c_0-sum of reflexive spaces.

The special case when the range space is a two-dimensional Hilbert space is studied in Section 13.4. Here, we characterize the norm attaining rank-two operators in $\mathcal{L}(X, \ell_2^2)$ in terms of the geometry of the dual norm on the set $\mathrm{NA}(X)$. As a consequence, we show that the set of norm attaining rank-two operators in $\mathcal{L}(X, \ell_2^2)$ is not empty if and only if there are $f \in \mathrm{NA}(X)$ and $g \in X^*$ with $\|f\| = 1$ and $0 < \|g\| \leqslant 1$ such that $\|f + tg\| \leqslant \sqrt{1+t^2}$ for all $t \in \mathbb{R}$ and if and only if there is $f \in \mathrm{NA}(X)$ with $\|f\| = 1$ such that the operator $f \otimes (1,0) \in \mathcal{L}(X, \ell_2^2)$ is not an extreme point in the unit ball of $\mathcal{L}(X, \ell_2^2)$. We do not know if such a norm attaining functional f can be found on every Banach space X.

The last part of the paper, Section 13.5, is devoted to commenting on V. Lomonosov's solution of the complex Bishop–Phelps problem, which is explained in the first few paragraphs of that section. We present a subset of the complex space c_0 which might be a candidate for a bounded, closed, convex subset without modulus attaining (complex) functionals, that is, a possible Lomonosov type example in c_0.

We finish this Introduction with the needed notation. We have already explained the notation $\mathcal{L}(X, Y)$, $\mathrm{NA}(X, Y)$, and $\mathrm{NA}(X)$. In addition, we define $\mathrm{NA}_1(X) := \{f \in \mathrm{NA}(X)\colon \|f\| = 1\}$. For $k \in \mathbb{N}$ with $k \geqslant 2$, we also use the notation $\mathcal{L}^{(k)}(X, Y)$ (resp., $\mathrm{NA}^{(k)}(X, Y)$) for the subset of $\mathcal{L}(X, Y)$ (resp., $\mathrm{NA}(X, Y)$) consisting of operators of rank k. As usual, $B_X = \{x \in X\colon \|x\| \leqslant 1\}$ stands for the closed unit ball of X, $S_X = \{x \in X\colon \|x\| = 1\}$ for its unit sphere and, less canonically, $U_X = \{x \in X\colon \|x\| < 1\}$ for its open unit ball. Further needed notation is the following: $M^\perp$ is the annihilator in X^* of a closed subspace M of X, $J_X\colon X \longrightarrow X^{**}$ denotes the canonical isometric inclusion of a Banach space into its bidual, $\mathcal{K}(X, Y)$ is the space of compact linear operators between X and Y, $\operatorname{cone}\{f, g\}$ stands for the cone generated by f and g, that is, $\operatorname{cone}\{f, g\} = \{af + bg\colon a, b \geqslant 0\}$.

If $x_0 \in X$ is a nonzero vector, any functional $f \in S_{X^*}$ with $f(x_0) = \|x_0\|$ is called a *supporting functional* at x_0. (Obviously, the supporting functionals are precisely the norm attaining ones.) If $x_0 \in S_X$ admits a unique supporting functional, it is called a *smooth point*. A theorem due to Mazur guarantees that for separable (in particular finite-dimensional) X the set of smooth points is dense in S_X [19, Th. 20F].

Finally, let us remark that the spaces considered in this paper are Banach spaces over the reals, with the exception of Section 13.5 where complex Banach spaces are the issue.

13.2 Existence of norm attaining finite rank operators

Before asking for the density of finite rank norm attaining operators, we should ask for the existence of such operators. It is not clear, to the best of our knowledge, that for all Banach spaces X and Y of dimension at least two, there exists a norm attaining operator from X to Y with finite rank greater than one. Our goal in this section is to discuss known and new sufficient conditions for the existence of norm attaining finite rank operators. In particular, we will focus on the rank-two case. So the leading question here is the following.

Problem 13.1. *Is* $\mathrm{NA}^{(2)}(X,Y)$ *nonempty for all Banach spaces* X *and* Y *of dimension at least two?*

An obvious comment here is that for the above problem, it is enough to deal with range spaces Y of dimension two. The next comment, though easy as well, is more surprising.

Remark 13.2. (a) *If* X *is a Banach space and* $\mathrm{NA}^{(2)}(X,\ell_2^2)$ *is nonempty, then* $\mathrm{NA}^{(2)}(X,Y)$ *is also nonempty for every Banach space* Y *of dimension at least two.*

Indeed, we can assume that Y is a two-dimensional Banach space. Now, take $T \in \mathrm{NA}^{(2)}(X,\ell_2^2)$ with $\|T\| = 1$ and pick $x_0 \in S_X$ such that $\|T(x_0)\| = 1$. Fix an isomorphism S from ℓ_2^2 onto Y with $\|S\| = 1$. Then S attains its norm, so there is $z \in S_{\ell_2^2}$ such that $\|S(z)\| = 1$. Consider a rotation operator π on ℓ_2^2 such that $\pi(T(x_0)) = z$. Then $S\pi T \in \mathrm{NA}^{(2)}(X,Y)$ and so $\mathrm{NA}^{(2)}(X,Y)$ is nonempty.

(b) In fact, *if there exist* $n \in \mathbb{N}$ *and Banach spaces* X *and* Y *of dimension at least* n *such that* $\mathrm{NA}^{(n)}(X,Y) = \emptyset$, *then*

$$\mathrm{NA}(X,\ell_2) \subset \bigcup_{k\leqslant n-1} \mathcal{L}^{(k)}(X,\ell_2),$$

that is, every norm attaining operator from X *to* ℓ_2 *has rank at most* $n-1$.

Indeed, if there exists $T \in \mathrm{NA}(X,\ell_2)$ whose rank is greater than or equal to n (or even has infinite rank), then composing with a suitable orthogonal projection P from ℓ_2 to an n-dimensional subspace H_n of ℓ_2, we get that $PT \in \mathrm{NA}^{(n)}(X,H_n) \cong \mathrm{NA}^{(n)}(X,\ell_2^n)$. By an argument completely identical to the one given in item (a), this provides that $\mathrm{NA}^{(n)}(X,Y) \neq \emptyset$.

The above arguments explain the key role of ℓ_2^2 to solve the problem of deciding whether $\mathrm{NA}^{(2)}(X, Y)$ is nonempty for all Banach spaces X and Y with $\dim(Y) \geqslant 2$. However, in the following we will study the problem for arbitrary range spaces Y. In Section 13.4, we will study the particular case of $\mathrm{NA}^{(2)}(X, \ell_2^2)$ and we will even give characterizations of the statement that this set is nonempty.

But let us return to the general case of Problem 13.1. First, we try to focus on the range space. If Y is a polyhedral two-dimensional Banach space, then as a result of Lindenstrauss [24, Proposition 3], $\mathrm{NA}(X, Y)$ is dense in $\mathcal{L}(X, Y)$ for all Banach spaces X, so the result is clear. Next, if Y is not polyhedral and not strictly convex either, then it is easy to construct a norm attaining rank-two operator from any Banach space X of dimension greater than one into Y (indeed, go first onto ℓ_∞^2 and then use that S_Y contains a segment to produce an injective operator from ℓ_∞^2 into Y that carries a whole maximal face of $S_{\ell_\infty^2}$ to the unit sphere of Y; see the proof of Proposition 13.17). For strictly convex range spaces Y, we do not know the answer, even for Y being a two-dimensional Hilbert space, and actually this case will be studied in depth in Section 13.4, as announced before.

In this section, we will mainly focus on the domain space. Our first comment is that, since compact operators are completely continuous, every compact operator whose domain is a reflexive space attains its norm (see [30, p. 270] for an argument). Next, there is an easy argument to get rank-two norm-attaining operators from a given Banach space X having a one-complemented reflexive subspace of dimension greater than one to arbitrary range spaces Y. Indeed, let $P\colon X \longrightarrow X$ be a norm-one projection such that $P(X) = Z$ is reflexive and $\dim(Z) \geqslant 2$. Now, every finite rank (actually compact) operator from a reflexive space is norm-attaining, so we just have to compose an arbitrary rank-two operator $S\colon Z \longrightarrow Y$ with the operator P viewed as $P\colon X \longrightarrow P(X) = Z$ to get that $T = SP \in \mathcal{L}(X, Y)$ has rank-two and attains its norm (indeed, $\|T\| = \|S\|$ and there exists $z \in S_Z$ such that $\|Sz\| = \|T\|$, but $z = P(z) \in S_X$ and so $\|Tz\| = \|SPz\| = \|Sz\| = \|T\|$). Actually, the same proof shows that every compact operator which factors through P is norm attaining, but this is the same as requiring that the kernel of the operator contains the kernel of P.

Result 13.3 (Folklore). *Let X be a Banach space, let $P \in \mathcal{L}(X, X)$ be a norm-one projection such that $P(X)$ is reflexive, and let Y be an arbitrary Banach space. Then every compact operator $T\colon X \longrightarrow Y$ for which $\ker P \subset \ker T$ attains its norm.*

But to have one-complemented closed subspaces of dimension greater than one is a quite strong requirement, and there are even Banach spaces without norm-one projections apart from the trivial ones (the identity and rank-one projections); see [8] for a finite-dimensional example.

Anyway, a quick glance at the proof of the above result makes one realize that the only properties of the norm-one projection P that we have used are that $P(X)$ is reflexive and that $P(B_X) = B_{P(X)}$, but not that $P^2 = P$. As $P(X) = X/\ker P$, we may try to consider general quotient maps instead of projections. Let X be a Banach space, let Z

be a closed subspace of X, and let Y be an arbitrary Banach space. Suppose that X/Z is reflexive (this is usually referred to by saying that Z is a *factor reflexive* subspace of X) and suppose also that the quotient map $q\colon X \longrightarrow X/Z$ satisfies that $q(B_X) = B_{X/Z}$, then every compact operator $T\colon X \longrightarrow Y$ such that $Z \subset \ker T$ attains its norm. Indeed, we may write $T = \widetilde{T} \circ q$ where $\|\widetilde{T}\| = \|T\|$ and $\widetilde{T}$ is compact, so there is $\xi \in B_{X/Z}$ such that $\|\widetilde{T}(\xi)\| = \|T\|$, but $\xi = q(x)$ for some $x \in B_X$ by hypothesis, so $\|T(x)\| = \|T\|$.

How to get the condition that $q(B_X) = B_{X/Z}$? This is just the proximinality of Z in X. Recall that a (closed) subspace M of X is called *proximinal* if for each $x \in X$ there is some $m \in M$ such that $\|x - m\| = \operatorname{dist}(x, M)$. We refer to the book [35] for background. Clearly, M is proximinal in X if and only if $q(B_X) = B_{X/M}$, see [35, Theorem 2.2] for instance. Therefore, we have shown the following.

Result 13.4 (Folklore). *Let X be a Banach space, let Z be a factor reflexive proximinal subspace of X, and let Y be an arbitrary Banach space. Then every compact operator T: $X \longrightarrow Y$ for which $Z \subset \ker T$ attains its norm.*

The problem of whether every infinite-dimensional Banach space contains a two-codimensional proximinal subspace [35, Problem 2.1] was open until a celebrated example was recently given by Read [33]: there is a Banach space $\mathcal{R}$ containing no finite-codimensional proximinal subspaces of codimension greater than one (and then it contains no proximinal factor reflexive subspaces of infinite codimension either, use [31, Proposition 2.3]). We refer the reader to [22, 23, 34] for more information on Read type spaces. Therefore, Result 13.4 does not provide a complete positive solution of Problem 13.1.

Our next step is to get a slightly weaker sufficient condition for norm attainment than the one given in Result 13.4, which is new as far as we know. Namely, it is easy to see that if Z is a factor reflexive proximinal subspace of a Banach space X, then $Z^\perp \subset \mathrm{NA}(X)$ (see [5, Lemma 2.2] for instance), but the converse result is not true (see [20, Section 2] or [34, Section 2] for a discussion of this). Our result is that the condition $Z^\perp \subset \mathrm{NA}(X)$ is enough to get the conclusion of Result 13.4.

Proposition 13.5. *Let X be a Banach space, let Z be a closed subspace of X such that $Z^\perp \subset \mathrm{NA}(X)$, and let Y be an arbitrary Banach space. Then every compact operator $T \in \mathcal{L}(X, Y)$ for which $Z \subset \ker T$ attains its norm.*

Proof. As $Z^\perp \subset \mathrm{NA}(X)$, it is immediate from James's theorem that X/Z is reflexive (see the proof of [5, Lemma 2.2]). As $Z \subset \ker T$, the operator T factors through X/Z, that is, there is an operator $\widetilde{T}\colon X/Z \longrightarrow Y$ such that $T = \widetilde{T} \circ q$, and it is clear that $\|\widetilde{T}\| = \|T\|$ and that $\widetilde{T}$ is compact whenever T is. Then $\widetilde{T}$ attains its norm (it is compact defined on a reflexive space), so also the adjoint $\widetilde{T}^*$ attains its norm. That is, there is $y^* \in S_{Y^*}$ such that $\|\widetilde{T}^* y^*\| = \|T\|$. Now, the functional $x^* = T^* y^* = [q^* \widetilde{T}^*](y^*) \in X^*$ vanishes on Z, so it belongs to $Z^\perp \subset \mathrm{NA}(X)$. This implies that there is $x \in S_X$ such that

$$|x^*(x)| = \|x^*\| = \|[q^* \widetilde{T}^*](y^*)\| = \|q^*(\widetilde{T}^* y^*)\| = \|\widetilde{T}^*(y^*)\| = \|T\|,$$

where we have used the immediate fact that q^* is an isometric embedding as q is a quotient map. Therefore, $\|T\| = |[T^*y^*](x)| = |y^*(Tx)|$ and so $\|Tx\| = \|T\|$, as desired. □

Observe that the proposition above can also be written in the following more suggestive form.

Corollary 13.6. *Let X, Y be Banach spaces and let $T \in \mathcal{L}(X,Y)$ be a compact operator. If $[\ker T]^{\perp} \subset \mathrm{NA}(X)$, then T attains its norm.*

The following obvious consequence gives a solution to Problem 13.1 in most Banach spaces.

Corollary 13.7. *Let X be a Banach space. If $\mathrm{NA}(X)$ contains two-dimensional subspaces, then $\mathrm{NA}^{(2)}(X,Y)$ is nonempty for any Banach space Y of dimension at least two.*

What happens with Read's space $\mathcal{R}$? (Un)fortunately, Corollary 13.7 does not apply since $\mathrm{NA}(\mathcal{R})$ does not contain two-dimensional subspaces, as was shown by Rmoutil [34, Theorem 4.2]. Actually, Rmoutil used the fact that if Z is a finite-codimensional closed subspace of a Banach space X such that X/Z is strictly convex, then $Z^{\perp} \subset \mathrm{NA}(X)$ if and only if Z is proximinal (see [34, Lemma 3.1]). Then he showed, for $X = \mathcal{R}$, that if $Z^{\perp}$ is contained in $\mathrm{NA}(\mathcal{R})$, then $\mathcal{R}/Z$ is strictly convex and so Z is proximinal, and hence it has codimension one. Actually, $\mathcal{R}^{**}$ is strictly convex [22, Theorem 4], so all quotients of $\mathcal{R}$ are strictly convex.

What to do then with $\mathcal{R}$? Well, $\mathcal{R}$ is not smooth (this follows from the formula for the directional derivative of its norm given in [33, Lemma 2.5]), so the following easy observation applies to it.

Observation 13.8. *If X is a nonsmooth Banach space, then $\mathrm{NA}^{(2)}(X,Y)$ is non-empty for any Banach space Y with $\dim(Y) \geqslant 2$.*

Indeed, there are $x_0 \in S_X$ and linearly independent $f,g \in S_{X^*}$ such that $f(x_0) = g(x_0) = 1$. Consider two linearly independent vectors y_1 and y_2 of S_Y and, replacing y_2 by $-y_2$ if necessary, let

$$\alpha_0 := \|y_1 + y_2\| = \max\{\|ay_1 + by_2\| : |a|, |b| \leqslant 1\}.$$

Now, define the operator $T \in \mathcal{L}(X,Y)$ by $Tx = \frac{1}{\alpha_0}(f(x)y_1 + g(x)y_2)$ for all $x \in X$ and observe that T has rank two, that $\|T\| \leqslant 1$, and that $\|Tx_0\| = 1$. Thus $T \in \mathrm{NA}^{(2)}(X,Y)$, giving the result.

Observation 13.8 solves Problem 13.1 for $\mathcal{R}$. But, are the already presented results applicable to solve the problem for all Banach spaces? The answer is no since we may construct a smooth renorming $\widetilde{\mathcal{R}}$ of $\mathcal{R}$ such that $\mathrm{NA}(\widetilde{\mathcal{R}}) = \mathrm{NA}(\mathcal{R})$ [22, Example 12], and neither Corollary 13.7 nor Observation 13.8 apply. Nevertheless, there is something these two results have in common: in both cases, the set of norm attaining functionals contains nontrivial cones. This also happens in $\mathrm{NA}(\widetilde{\mathcal{R}})$ (as it coincides with $\mathrm{NA}(\mathcal{R})$), and this will be the key to obtain the main new existence result about norm attaining rank-two operators.

Theorem 13.9. *Let X be a real Banach space, let $f_1, f_2 \in S_{X^*}$ be linearly independent, $Z = \ker f_1 \cap \ker f_2$ and $\operatorname{cone}\{f_1, f_2\} \subset \mathrm{NA}(X)$. Then, for every real two-dimensional normed space E there is a norm attaining surjective operator $T\colon X \longrightarrow E$ with $\ker T = Z$.*

Before providing the proof of this result, let us give some consequences and comments.

Observe that Theorem 13.9 implies the following sufficient condition for the existence of norm attaining operators of rank two.

Corollary 13.10. *Let X be a Banach space. If there exist two linearly independent $f, g \in X^*$ such that $\operatorname{cone}\{f, g\} \subset \mathrm{NA}(X)$, then $\mathrm{NA}^{(2)}(X, Y) \neq \emptyset$ for every Banach space Y of dimension at least two.*

We do not know whether the condition is necessary as well, and we do not know any Banach space that fails it.

Problem 13.11. *Does $\mathrm{NA}(X)$ contain nontrivial cones for every infinite-dimensional Banach space X?*

Problem 13.12. *Let X be a Banach space and suppose that $\mathrm{NA}^{(2)}(X, Y) \neq \emptyset$ for every Banach space Y of dimension at least two. Does this imply that $\mathrm{NA}(X)$ contains a nontrivial cone?*

As promised before, we stated Theorem 13.9; this result solves the problem of the existence of rank-two norm attaining operators for $\widetilde{\mathcal{R}}$.

Example 13.13. *The Read space $\mathcal{R}$ given in [33] and its smooth renorming $\widetilde{\mathcal{R}}$ given in [22, Example 12], satisfy that their set of norm attaining functionals contains non-trivial cones (but no nontrivial subspaces). Therefore, for every Banach space Y of dimension at least two, both $\mathrm{NA}^{(2)}(\mathcal{R}, Y)$ and $\mathrm{NA}^{(2)}(\widetilde{\mathcal{R}}, Y)$ are nonempty.*

Indeed, as $\mathcal{R}$ is not smooth, taking linearly independent $f_1, f_2 \in \mathcal{R}^*$ and $x_0 \in S_{\mathcal{R}}$ such that $f_1(x_0) = 1 = f_2(x_0)$, it is immediate that $\operatorname{cone}\{f_1, f_2\} \subset \mathrm{NA}(\mathcal{R})$. For the space $\widetilde{\mathcal{R}}$, we just have to observe that $\mathrm{NA}(\widetilde{\mathcal{R}}) = \mathrm{NA}(\mathcal{R})$, as shown in [22, Example 12].

It is now time to present the proof of Theorem 13.9. We first need some preliminary results. We recall that we denote the open unit ball of a Banach space X by U_X.

Lemma 13.14. *Let E be a two-dimensional normed space, let $e_1 \in S_E$, $e_1^* \in S_{E^*}$ such that $e_1^*(e_1) = 1$, and let $e_2 \in S_E \cap \ker e_1^*$. For $0 < \tau < 1$, denote by $T_\tau\colon E \longrightarrow E$ the norm-one linear operator such that $T_\tau(e_1) = e_1$ and $T_\tau(e_2) = \tau e_2$. Then, for every compact subset $K \subset E$ such that $\sup |e_1^*(K)| < 1$ there is $0 < \tau < 1$ such that $T_\tau(K) \subset U_E$.*

Proof. Let $\tau_n = \frac{1}{n}$. Then the operators T_{τ_n} converge pointwise to the operator $T = e_1^* \otimes e_1$. Now, pointwise convergence on E implies uniform convergence on K. Since, by hypothesis, U_E is an open neighborhood of $T(K) = \{e_1^*(x)e_1 : x \in K\}$, there is $n \in \mathbb{N}$ such that $T_{\tau_n}(K) \subset U_E$. □

Lemma 13.15. *Under the conditions of the previous lemma, let additionally e_1 be a smooth point of S_E, and let $h_1, h_2 \in E^*$ be two linearly independent functionals such that $e_1^* = \frac{1}{2}(h_1 + h_2)$ and $h_1(e_1) = h_2(e_1) = 1$. Denote*

$$A = \{x \in E: \max\{|h_1(x)|, |h_2(x)|\} \leqslant 1\}.$$

Then there is $0 < \tau < 1$ such that $T_\tau(A) \subset B_E$.

Proof. Since e_1 is a smooth point of B_E, it follows by geometrical reasoning in the plane that there is a neighborhood V of e_1 such that $A \cap V \subset B_E$ (i. e., the parallelogram A touches S_E at e_1 from the inside of B_E). Note that e_1 is a strongly exposed point of A; it is strongly exposed by $e_1^* = \frac{1}{2}(h_1 + h_2)$. Hence, there is some $\delta > 0$ such that

$$A_\delta := \{x \in A: |e_1^*(x)| > 1 - \delta\} \subset A \cap V \subset B_E.$$

Let us apply Lemma 13.14 to $K = A \setminus A_\delta = \{x \in A: |e_1^*(x)| \leqslant 1 - \delta\}$. We obtain some $0 < \tau < 1$ so that $T_\tau(K) \subset U_E$. Since $T_\tau(A_\delta) \subset A_\delta \subset B_E$, this gives us the desired inclusion $T_\tau(A) \subset B_E$. □

Lemma 13.16. *Let Y be a two-dimensional normed space, $x_1, x_2 \in S_Y$ be linearly independent smooth points, and let the corresponding supporting functionals $x_1^*, x_2^* \in S_{Y^*}$, $x_1^*(x_1) = x_2^*(x_2) = 1$, be also linearly independent. Let $y \in S_Y$ be of the form $y = a_1x_1 + a_2x_2$ with $a_i > 0$, $i = 1, 2$. Then every supporting functional $f \in S_{Y^*}$ at y belongs to* cone$\{x_1^*, x_2^*\}$.

Proof. We again argue geometrically. Denote $b_1 = x_1^*(x_2)$, $b_2 = x_2^*(x_1)$. Evidently, $|b_i| < 1$, $i = 1, 2$. Denote by x_3 the point at which $x_1^*(x_3) = x_2^*(x_3) = 1$ and consider the triangle Δ whose vertices are x_1, x_2, and x_3. Then $y \in \Delta$, and the supporting line $\ell = \{x \in Y: f(x) = 1\}$ contains y. Now, x_1, x_2 lie on one side of ℓ (actually, all points of B_Y do), whereas x_3 lies on the opposite side of ℓ, that is,

$$f(x_1) \leqslant 1, \quad f(x_2) \leqslant 1, \quad \text{and} \quad f(x_3) \geqslant 1.$$

Since $x_1^*, x_2^* \in Y^*$ are linearly independent, there is a (unique) representation of f as $f = c_1x_1^* + c_2x_2^*$. Let us substitute this representation into the previous inequalities:

$$c_1 + c_2b_2 \leqslant 1, \tag{13.1}$$

$$c_1b_1 + c_2 \leqslant 1, \tag{13.2}$$

$$c_1 + c_2 \geqslant 1. \tag{13.3}$$

From (13.1) and (13.3) together with $b_2 \neq 1$, we deduce that $c_2 \geqslant 0$, and likewise from (13.2) and (13.3) that $c_1 \geqslant 0$. □

We are now able to present the pending proof.

Proof of Theorem 13.9. Denote by q the corresponding quotient map $q: X \longrightarrow X/Z$; then q^* maps $(X/Z)^*$ isometrically onto $Z^\perp = \operatorname{span}\{f_1, f_2\} \subset X^*$. Let $x_1, x_2 \in S_X$ be points at which $f_1(x_1) = f_2(x_2) = 1$, $\tilde{x}_1 = q(x_1)$, $\tilde{x}_2 = q(x_2)$ and let $\tilde{f}_1, \tilde{f}_2 \in S_{(X/Z)^*}$ be those functionals for which $q^*(\tilde{f}_i) = f_i$, $i = 1, 2$. Then $\tilde{f}_1(\tilde{x}_1) = \tilde{f}_2(\tilde{x}_2) = 1$, so, in particular, $\tilde{x}_1, \tilde{x}_2 \in S_{X/Z}$.

We will consider two cases.

Case 1: $\tilde{x}_1$ and $\tilde{x}_2$ are smooth points of $S_{X/Z}$. In this case, since $\tilde{f}_1$, $\tilde{f}_2$ are linearly independent, $\tilde{x}_1$, $\tilde{x}_2$ are linearly independent as well. Let $\ell \subset X/Z$ be the straight line connecting $\tilde{x}_1$ with $\tilde{x}_2$, let $\tilde{f}_3 \in S_{(X/Z)^*}$ be the norm-one functional taking a positive constant value $\alpha < 1$ on ℓ and let $\tilde{x}_3 \in S_{X/Z}$ be a point at which $\tilde{f}_3(\tilde{x}_3) = 1$. Select a point $e_1 \in S_E$, a supporting functional $e_1^* \in S_{E^*}$ at e_1, $e_2 \in S_E \cap \ker e_1^*$ and $T_\tau: E \longrightarrow E$ as in Lemma 13.14. Choose $t \in (\alpha, 1)$, and denote by $R: X/Z \longrightarrow E$ the linear operator such that $R(\tilde{x}_1 - \tilde{x}_2) = e_2$ and $R(\tilde{x}_1) = te_1$. Applying Lemma 13.14 to $K = \{e \in R(B_{X/Z}): |e_1^*(e)| \leqslant t\}$, we obtain some $0 < \tau < 1$ such that $T_\tau(K) \subset U_E$. We also note for future use that $e_1^*(R(\tilde{x}_1)) = e_1^*(R(\tilde{x}_2)) = t$, consequently $R^* e_1^* = \frac{t}{\alpha}\tilde{f}_3$ and $e_1^*(R(\tilde{x}_3)) = \frac{t}{\alpha} > 1$.

Our goal is to demonstrate that $T = T_\tau \circ R \circ q: X \longrightarrow E$ is the operator we are looking for. Consider the composition $T_\tau \circ R: X/Z \longrightarrow E$. It is a bijection, which ensures that $\ker T = \ker q = Z$. The property $e_1^* \circ T_\tau = e_1^*$ implies that

$$\|T_\tau \circ R\| \geqslant |e_1^*((T_\tau \circ R)\tilde{x}_3)| = e_1^*(R(\tilde{x}_3)) > 1.$$

Let $y \in S_{X/Z}$ be a point at which $\|(T_\tau \circ R)(y)\| = \|T_\tau \circ R\|$. Then y must belong to $\operatorname{cone}\{\tilde{x}_1, \tilde{x}_2\} \cup (-\operatorname{cone}\{\tilde{x}_1, \tilde{x}_2\})$ because otherwise we would have $|\tilde{f}_3(y)| \leqslant \alpha$ and thus $|e_1^*(Ry)| = \frac{t}{\alpha}|\tilde{f}_3(y)| \leqslant t$ so that $(T_\tau \circ R)y \in T_\tau(K) \subset U_E$, which contradicts the estimate $\|(T_\tau \circ R)(y)\| > 1$.

Replacing y by $-y$, if necessary, we may assume that $y \in S_{X/Z} \cap \operatorname{cone}\{\tilde{x}_1, \tilde{x}_2\}$. Let $g \in S_{E^*}$ be a supporting functional for $[T_\tau \circ R](y)$, that is, $g([T_\tau \circ R](y)) = \|T_\tau \circ R\|$. Then

$$\frac{(T_\tau \circ R)^* g}{\|T_\tau \circ R\|}$$

is a supporting functional at y, so, by Lemma 13.16,

$$\frac{(T_\tau \circ R)^* g}{\|T_\tau \circ R\|} \in \operatorname{cone}\{\tilde{f}_1, \tilde{f}_2\}$$

and consequently

$$q^*\left(\frac{(T_\tau \circ R)^* g}{\|T_\tau \circ R\|}\right) \in \operatorname{cone}\{f_1, f_2\}.$$

Since $\operatorname{cone}\{f_1, f_2\}$ consists only of norm attaining functionals, there is $x \in S_X$ such that

$$\left[q^*\left(\frac{(T_\tau \circ R)^* g}{\|T_\tau \circ R\|}\right)\right](x) = 1.$$

From this, we get that

$$\begin{aligned}\|T(x)\| &= \|(T_\tau \circ R \circ q)(x)\| \geqslant g((T_\tau \circ R \circ q)(x)) \\ &= \left(q^* \left((T_\tau \circ R)^* g\right)\right)(x) = \|T_\tau \circ R\| .\end{aligned}$$

On the other hand, $\|T\| \leqslant \|T_\tau \circ R\|$, which means that T attains its norm at x.

Case 2: At least one of $\widetilde{x}_1, \widetilde{x}_2$ is not a smooth point of $S_{X/Z}$. Without loss of generality, we may assume that $\widetilde{x}_1$ is not a smooth point, so there are two linearly independent functionals $g_1, g_2 \in S_{(X/Z)^*}$ with $g_1(\widetilde{x}_1) = g_2(\widetilde{x}_1) = 1$.

Select a smooth point $e_1 \in S_E$, and let $e_1^* \in S_{E^*}$ be the supporting functional at e_1. Select $e_2 \in S_E \cap \ker e_1^*$ and define $T_\tau: E \longrightarrow E$ as in Lemma 13.14. Denote $g = \frac{1}{2}(g_1 + g_2)$ and choose $y \in S_{X/Z} \cap \ker g$. Denote by $R: X/Z \longrightarrow E$ the linear operator satisfying $R(\widetilde{x}_1) = e_1$ and $R(y) = e_2$. Then $[R^* e_1^*](\widetilde{x}_1) = 1$, $[R^* e_1^*](y) = 0$, so $R^* e_1^* = g$. Let us consider those $h_1, h_2 \in E^*$ for which $R^* h_1 = g_1$ and $R^* h_2 = g_2$. These $h_1, h_2 \in E^*$ satisfy all the conditions of Lemma 13.15, so for the corresponding set

$$A = \{x \in E\colon \max\{|h_1(x)|, |h_2(x)|\} \leqslant 1\}$$

there is $0 < \tau < 1$ such that $T_\tau(A) \subset B_E$.

Let us demonstrate that $T = T_\tau \circ R \circ q\colon X \longrightarrow E$ attains its norm at x_1. Indeed,

$$\begin{aligned}T(B_X) &= T_\tau(R(q(B_X))) \subset T_\tau(R(B_{X/Z})) \\ &\subset T_\tau(R(\{\widetilde{x} \in X/Z\colon \max\{|g_1(\widetilde{x})|, |g_2(\widetilde{x})|\} \leqslant 1\})) \\ &= T_\tau(A) \subset B_E,\end{aligned}$$

which gives us that $\|T\| \leqslant 1$. But, on the other hand,

$$T(x_1) = T_\tau(R(\widetilde{x}_1)) = T_\tau(e_1) = e_1,$$

so $\|T(x_1)\| = 1$. □

Our last goal in this section is to present all the implications proved so far in the particular case of rank-two operators, and to discuss the possibility of reversing them.

Let X and Y be Banach spaces of dimension at least two, and let Z be a closed subspace of X of codimension two. Consider the following properties:

(a) Z is the kernel of a norm-one projection.
(b) Z is proximinal in X.
(c) $Z^\perp \subset \mathrm{NA}(X)$.
(d) Every $T \in \mathcal{L}(X, Y)$ with $\ker T \supset Z$ is norm attaining.
(e) Every $T \in \mathcal{L}(X, Y)$ with $\ker T = Z$ is norm attaining.
(◇) There exists $T \in \mathcal{L}(X, Y)$ with $\ker T = Z$ which is norm attaining.
(f) There are linearly independent $f, g \in Z^\perp$ such that $\operatorname{cone}\{f, g\} \subset Z^\perp \cap \mathrm{NA}(X)$.
(g) There are linearly independent $f, g \in Z^\perp \cap S_{X^*}$ and $x \in S_X$ such that $f(x) = 1 = g(x)$.

Then, the following implications hold:

$$\begin{array}{ccccccccccccc}
(a) & \Longrightarrow & (b) & \Longrightarrow & (c) & \overset{(\star)}{\Longrightarrow} & (e) & \overset{(\star\star)}{\Longrightarrow} & \boxed{(\diamond)} & \overset{(\star\star\star)}{\Longleftarrow} & (f) & \Longleftarrow & (g) \\
 & & & & \Updownarrow & & & & & & & & \\
 & & & & (d) & & & & & & & &
\end{array} \tag{13.4}$$

We now discuss these implications and the possibility of reversing them.

It is immediate that (a) implies (b), but the converse result is obviously false, even for finite-dimensional (non-Hilbertian) spaces. It is well known that (b) implies (c), but not conversely; see [20, Section 2] or [34, Section 2] for a discussion of this. The implication (c) $\Rightarrow$ (d) is Proposition 13.5, and the reverse implication is obvious using rank-one operators. Next, the implications (d) $\Rightarrow$ (e) $\Rightarrow$ ($\diamond$) are obvious.

On the other side of condition ($\diamond$), we have that (g) $\Rightarrow$ (f) since, obviously, $\operatorname{cone}\{f, g\} \subset \mathrm{NA}(X)$ if (g) holds, but the converse result is not true as follows by taking X to be a smooth reflexive Banach space. That (f) implies ($\diamond$) is exactly our Theorem 13.9.

So it remains to discuss the possible converses of the implications ($\star$), ($\star\star$), and ($\star\star\star$).

Let us start by discussing the possibility of the reciprocal result to implication ($\star$) above to be true. We have two different behaviors, depending on whether the range space is strictly convex or not (i. e., whether the unit sphere of the range space does not or does contain nontrivial segments).

For nonstrictly convex range spaces, we have the following positive result.

Proposition 13.17. *Let X be a Banach space, let E be a two-dimensional space which is not strictly convex and let Z be a two-codimensional closed subspace of X. If every $T \in \mathcal{L}(X, E)$ with $\ker T = Z$ attains its norm, then $Z^\perp \subset \mathrm{NA}(X)$.*

Proof. Let us start with the simpler case of $E = \ell_\infty^2$. Fix $\varphi \in Z^\perp$ with $\|\varphi\| = 1$; our aim is to show that $\varphi \in \mathrm{NA}(X)$. To get this, consider $\psi \in Z^\perp$ with $\|\psi\| = 1$ such that $Z^\perp = \operatorname{span}\{\varphi, \psi\}$ and define $T\colon X \longrightarrow \ell_\infty^2$ by $Tx = (\varphi(x), \frac{1}{2}\psi(x))$ for all $x \in X$. Then $\|T\| = 1$ and $\ker T = Z$, so $T \in \mathrm{NA}(X, \ell_\infty^2)$ by hypothesis. But then, clearly, $\varphi \in \mathrm{NA}(X)$, as desired.

Now, suppose that E is a two-dimensional nonstrictly convex space. Then we may find a bijective norm-one operator $U\colon \ell_\infty^2 \longrightarrow E$ such that $U(1, t) \in S_E$ for every $t \in [-1, 1]$ (we just have to use the segment contained in S_E). Now, fix $\varphi \in Z^\perp$ with $\|\varphi\| = 1$, our aim is to show that $\varphi \in \mathrm{NA}(X)$. Again, we consider $\psi \in Z^\perp$ with $\|\psi\| = 1$ such that $Z^\perp = \operatorname{span}\{\varphi, \psi\}$ and this time we define the operator $T\colon X \longrightarrow E$ by $Tx = U(\varphi(x), \frac{1}{2}\psi(x))$ for all $x \in X$. On the one hand, $\|T\| = 1$: consider a sequence $\{x_n\}_{n\in\mathbb{N}}$ in S_X such that $\varphi(x_n) \longrightarrow 1$ and, passing to a subsequence, we also have that $\psi(x_n) \longrightarrow t_0 \in [-1, 1]$ and so

$$\|Tx_n\| = \left\| U\left(\varphi(x_n), \frac{1}{2}\psi(x_n)\right) \right\| \longrightarrow \|U(1, t_0)\| = 1.$$

On the other hand, $\ker T = Z$, so T attains its norm by hypothesis. That is, there is $x \in S_X$ such that $1 = \|Tx\| = \|U(\varphi(x), \frac{1}{2}\psi(x))\|$. But as $\|U\| = 1$, this implies that $\|(\varphi(x), \frac{1}{2}\psi(x))\|_\infty = 1$ and this immediately gives that $|\varphi(x)| = 1$, that is, $\varphi \in \mathrm{NA}(X)$, as desired. □

When the range space is strictly convex, the above proof is not valid and, actually, the result is false as the following counterexample shows.

Example 13.18. Let E be a two-dimensional strictly convex Banach space. Then there are a Banach space X and a two-codimensional closed subspace Z of X satisfying that every operator $T \in \mathcal{L}(X, E)$ with $\ker T = Z$ attains its norm, but $Z^\perp$ is not contained in $\mathrm{NA}(X)$.

Proof. Take a two-dimensional Banach space W whose unit sphere S_W contains a segment $[a, b]$, where the endpoints a, b are extreme points of the sphere, the number of extreme points is countable, and the endpoints a, b are smooth point of the sphere. Let $X = \ell_1$ and define an operator $U: X = \ell_1 \longrightarrow W$ that maps the vectors of the unit basis $\{e_n\}$ onto all the extreme points of S_W with the exception of $\pm a$ and $\pm b$. Then $Z = \ker U$ is a two-codimensional closed subspace of X, whose annihilator $Z^\perp$ is not contained in $\mathrm{NA}(X)$, because if one takes $f \in W^*$ which attains its norm on $[a, b]$, then $U^* f \in Z^\perp$ does not attain its norm. On the other hand, as $\overline{U(B_X)} = B_W$, X/Z is isometrically isomorphic to W by virtue of the injectivization $\widetilde{U} \in \mathcal{L}(X/Z, W)$ of U which satisfies $U = \widetilde{U} q$. So, if one takes an arbitrary norm-one operator $T: X \longrightarrow E$ with $\ker T = Z$, then T factors through U (or, what is the same, through the composition of the quotient map q and $\widetilde{U}$). That is, $T = \widetilde{T} U$ for some norm-one operator $\widetilde{T}: W \longrightarrow E$, so the image $T(B_X)$ of the closed unit ball is a linear copy (under $\widetilde{T}$) of $U(B_X) = B_W \setminus ([a, b] \cup [-a, -b])$.

We shall argue that $\|T(a)\| \neq 1$. Otherwise, one could pick some $e^* \in S_{E^*}$ with $e^*(\widetilde{T}(a)) = 1$. It follows that $w^* := \widetilde{T}^*(e^*)$ is a supporting functional at a of norm one. Any supporting functional at $\frac{1}{2}(a + b)$ also supports a, and by smoothness of a it has to coincide with w^*. Consequently, $w^*(a) = w^*(\frac{1}{2}(a + b)) = w^*(b) = 1$ and so $\widetilde{T}(a)$, $\widetilde{T}(\frac{1}{2}(a + b))$ and $\widetilde{T}(b)$ lie on a nontrivial segment of S_E, which is impossible when E is strictly convex; likewise $\|\widetilde{T}(b)\| \neq 1$. Hence there exists an extreme point w of B_W different from $\pm a, \pm b$ for which $\|\widetilde{T}(w)\| = 1$. This w is of the form $w = U(e_n)$ for some n, and we see that $T = \widetilde{T} U$ attains its norm at this e_n. □

We would like to emphasize a question related to the example above, which asks about the possibility of the implication (e) ⇒ (f) being true when the range space is strictly convex (it is true for nonstrictly convex range spaces by Proposition 13.17). The failure of (f) ⇒ (e) will be shown shortly in Example 13.20.

Problem 13.19. *Let X be a Banach space, let Z be a closed subspace of X of codimension two and let E be a two-dimensional strictly convex space. Suppose that every $T \in \mathcal{L}(X,E)$ with $\ker T = Z$ attains its norm, does then $Z^\perp \cap \mathrm{NA}(X)$ contain a nontrivial cone?*

To show the failure of the converse implication to $(\star\star)$ in diagram (13.4), the next example works. Note that it also shows that (f) does not imply (e).

Example 13.20. *There exists a rank-two operator $T \in \mathcal{L}(\ell_1, \ell_2^2)$ such that $[\ker T]^\perp \cap \mathrm{NA}(\ell_1)$ contains a nontrivial cone, but T does not attain its norm.*

Indeed, let $T \in \mathcal{L}(\ell_1, \ell_2^2)$ be an operator such that

$$T(B_{\ell_1}) = \operatorname{conv}\left\{\pm u_1, \pm\frac{1}{2}u_2\right\} \setminus \{\pm u_1\}$$

where $\{u_1, u_2\}$ is the canonical basis of ℓ_2^2. This operator can easily be constructed by mapping the unit vector basis of ℓ_1 onto a countable dense subset of the union of the half-open segments $(-u_1, \frac{1}{2}u_2] \cup (u_1, \frac{1}{2}u_2] \subset \ell_2^2$. Then $\|T\| = 1$, but the norm is not attained. Nevertheless, the functionals from the cone in $(\ell_2^2)^*$ generated by $2u_2^* \pm u_1^*$ attain their maxima on $T(B_{\ell_1})$ at the point $\frac{1}{2}u_2$, so the image of this cone under the isomorphic embedding T^* is contained in $[\ker T]^\perp$ and consists of norm attaining functionals.

Finally, it follows from the next example that the converse implication to $(\star\star\star)$ in diagram (13.4) fails in general.

Example 13.21. *Let $X = \ell_1$ and let E be an arbitrary two-dimensional space. Then there is $T \in \mathrm{NA}^{(2)}(X,E)$ such that $[\ker T]^\perp \cap \mathrm{NA}(X)$ does not contain nontrivial cones.*

Indeed, let $\{z_n\colon n \geqslant 2\}$ be a dense subset of the open unit ball of E, let u_0 be a smooth point of S_E whose unique support functional is called $u_0^* \in S_{E^*}$, and define $T\colon \ell_1 \longrightarrow E$ by means of the unit vector basis $\{e_n\}$ of ℓ_1 by

$$T(e_1) = u_0, \quad T(e_n) = z_n \quad \text{for } n \geqslant 2.$$

Clearly, T is onto and attains its norm (at e_1), so $T \in \mathrm{NA}^{(2)}(X,E)$. On the other hand, since $T(B_{\ell_1})$ is dense in B_E, the adjoint operator $T^*\colon E^* \longrightarrow (\ell_1)^*$ is an isometric embedding. Also, $T^*(E^*) \subset [\ker T]^\perp$ and the dimensions of both subspaces are equal to 2, so we have $[\ker T]^\perp = T^*(E^*)$. Now, consider an arbitrary non-zero $h \in [\ker T]^\perp \cap \mathrm{NA}(X)$. Let us write $h = T^*y^*$, where $y^* \in E^*$, and let $x = (x_1, x_2, \dots) \in B_{\ell_1}$ be such that $\|h\| = h(x)$. Then

$$\|y^*\| = \|h\| = h(x) = y^*(Tx) \leqslant \|y^*\| \, \|Tx\| \leqslant \|y^*\|,$$

so $\|Tx\| = 1$ and y^* attains its norm at Tx. Taking into account that $\|x\| = \sum_{n=1}^\infty |x_n| = 1$, that $\|z_n\| < 1$ and that

$$\|Tx\| \leqslant |x_1| + \sum_{n \geqslant 2} |x_n| \, \|z_n\|,$$

we see that the equality $\|Tx\| = 1$ may happen only if $x = x_1 e_1$ with $|x_1| = 1$. Therefore, $Tx = \pm u_0$ and y^* attains its norm at Tx, so y^* is proportional to u_0^*. We have demonstrated that $[\ker T]^\perp \cap \mathrm{NA}(X) \subset \operatorname{span} T^*(u_0^*)$, so $\mathrm{NA}(X)$ does not contain two linearly independent elements of $[\ker T]^\perp$.

13.3 Density of norm attaining finite rank operators

After discussing the existence of norm attaining finite rank operators, it is now time to study positive results for the density of such operators. An easy observation is pertinent, namely, we may restrict ourselves to consider finite-dimensional codomain spaces if we are interested in results valid for all codomain spaces: if X has the property that for all Y, all finite rank operators $T: X \longrightarrow Y$ can be approximated by norm attaining operators, then all such T can be approximated by norm attaining finite rank operators. Indeed, if $T: X \longrightarrow Y$ has finite rank, then we may view $T: X \longrightarrow T(X)$, approximate T here, and then compose the approximating sequence with the isometric inclusion operator from $T(X)$ into Y.

The leading question here is the following open problem.

Problem 13.22. *Is it true that every finite rank operator can be approximated by norm attaining (finite rank) operators?*

As in the previous section, we will focus on the domain spaces. So, the general aim in this section is to provide partial answers to the following question.

Problem 13.23. *Find sufficient conditions on a Banach space X so that every finite rank operator whose domain is X can be approximated by (finite rank) norm attaining operators.*

First, it is immediate that Lindenstrauss's property (A) on a Banach space X implies that a finite rank operator whose domain is X can be approximated by norm attaining finite rank operators (just restrict the codomain to the range space, use property (A) there and inject the range space again into the codomain). Therefore, some positive solutions to the problem above are the known sufficient conditions for property (A) like the Radon–Nikodým property, the property alpha, or the fact that the unit ball contains a set of uniformly strongly exposed points which generates the ball by taking the closed convex hull. We refer to the already cited survey paper [3] for more information.

If one looks for less restrictive conditions valid for finite rank operators but not necessarily for all operators, there are such conditions for compact operators. A detailed account of these properties is given in the survey paper [30]. But all of the known results of this kind need some sort of approximation property of the dual space, since they actually provide that every compact operator can be approximated by norm attaining finite rank operators.

Our main aim here is to try to provide a sufficient condition for the density of norm attaining finite rank operators which does not require the approximation property of the dual of the domain space. Here is the result which follows directly from Proposition 13.5.

Theorem 13.24. *Let X be a Banach space satisfying that for every $n \in \mathbb{N}$, every $\varepsilon > 0$, and all $x_1^*, \dots, x_n^* \in B_{X^*}$, there are $y_1^*, \dots, y_n^* \in B_{X^*}$ such that $\|x_i^* - y_i^*\| < \varepsilon$ for every $i = 1, \dots, n$ and*

$$\operatorname{span}\{y_1^*, \dots, y_n^*\} \subset \operatorname{NA}(X).$$

Then every finite rank operator whose domain is X can be approximated by finite rank norm attaining operators.

If, moreover, X^* has the approximation property, then every compact operator whose domain is X can be approximated by finite rank norm attaining operators.

Before providing the proof of the theorem, let us state the main consequence which follows immediately from it.

Corollary 13.25. *Let X be a Banach space such that there is a norm dense linear subspace of X^* contained in $\operatorname{NA}(X)$. Then, for every Banach space Y, every operator $T \in \mathcal{L}(X, Y)$ of finite rank can be approximated by finite rank norm attaining operators.*

If, moreover, X^* has the approximation property, then every compact operator whose domain is X can be approximated by finite rank norm attaining operators.

We do not know whether there are Banach spaces satisfying the conditions of Theorem 13.24 but not the ones of Corollary 13.25.

We may now give the pending proof.

Proof of Theorem 13.24. It is enough to show that $\operatorname{NA}(X, F)$ is dense in $\mathcal{L}(X, F)$ for every finite-dimensional space F. We fix an arbitrary finite-dimensional Banach space F and consider an Auerbach basis $\{e_1, \dots, e_n\}$ of F [13, Theorem 4.5] with biorthogonal functionals $\{e_1^*, \dots, e_n^*\}$ in F^*. Given a norm-one operator $T \in \mathcal{L}(X, F)$ and $\varepsilon > 0$, let $x_i^* = T^* e_i^* \in B_{X^*}$ for $i = 1, \dots, n$, and observe that $T = \sum_{i=1}^n x_i^* \otimes e_i$. By hypothesis, we may find $y_1^*, \dots, y_n^* \in B_{X^*}$ such that $\|x_i^* - y_i^*\| < \varepsilon/n$ and $\operatorname{span}\{y_1^*, \dots, y_n^*\} \subset \operatorname{NA}(X)$. We write $S = \sum_{i=1}^n y_i^* \otimes e_i \in \mathcal{L}(X, F)$ and first observe that $\|T - S\| < \varepsilon$. On the other hand, as S vanishes on $\bigcap_{i=1}^n \ker y_i^*$ we have that $[\ker S]^\perp \subset \operatorname{span}\{y_1^*, \dots, y_n^*\} \subset \operatorname{NA}(X)$. This gives that $S \in \operatorname{NA}(X, F)$ by Proposition 13.5.

Let us show the moreover part: if X^* has the approximation property, then every compact operator whose domain is X can be approximated by finite rank operators (see [25, Theorem 1.e.5] for instance) and the result now follows from the first part of the proof. □

As a consequence of this result, we may recover some results stated in [21] and [30, Section 3] on norm attaining compact operators. The main tool provided in [21] to get solutions to Problem 13.23 is the following easy observation.

Corollary 13.26 ([21, Lemma 3.1]). *Let X be a Banach space such that for all $x_1^*, \dots, x_n^* \in B_{X^*}$ and every $\varepsilon > 0$, there is a norm-one projection $P \in \mathcal{L}(X, X)$ of finite rank such that $\max_i \|x_i^* - P^*(x_i^*)\| < \varepsilon$. Then every compact operator whose domain is X can be approximated by norm attaining operators of finite rank.*

This result can also easily be deduced from Theorem 13.24 as the hypotheses imply that X^* has the approximation property and that the subspace $P^*(X^*)$ is contained in $\mathrm{NA}(X)$ (indeed, $[P^*(x^*)](B_X) = x^*(P(B_X))$ is compact as P is a finite rank projection, and $P(B_X) = B_{P(X)}$ since $\|P\| = 1$).

This result applies to $X = C_0(L)$ for every locally compact Hausdorff space L [21, Proposition 3.2] and also to $L_1(\mu)$ for every finite positive measure μ (see [10, Lemma 3.12] for a detailed proof). For $C_0(L)$, we do not know whether it is actually true that $\mathrm{NA}(C_0(L))$ contains a dense linear subspace. In the case of $L_1(\mu)$ for a localizable measure μ (see, e. g., [14, Definition 211G] for the definition), the subspace of $L_1(\mu)^* = L_\infty(\mu)$ of those functions in $L_\infty(\mu)$ taking finitely many values, that is, the subspace of step functions, is clearly contained in $\mathrm{NA}(L_1(\mu))$ and it is dense in $L_\infty(\mu)$. Of course, the hypothesis of being localizable may be dropped, as every $L_1(\mu)$ space is isometrically isomorphic to an $L_1(\nu)$-space where ν is localizable (this follows, for instance, by Maharam's theorem). Let us comment that the fact that norm attaining compact operators from an $L_1(\mu)$ space are dense in the space of compact operators was proved in [12, p. 6].

Let us state these two results.

Corollary 13.27 ([21, Proposition 3.2]). *Let L be a locally compact Hausdorff topological space. Then $X = C_0(L)$ satisfies the hypotheses of Theorem 13.24. Therefore, every compact linear operator whose domain is $C_0(L)$ can be approximated by finite rank norm attaining operators.*

Corollary 13.28 (extension of [12, p. 6]). *Let μ be a positive measure. Then there is a dense linear subspace of $L_1(\mu)^*$ which is contained in $\mathrm{NA}(L_1(\mu))$. As a consequence, every compact linear operator whose domain is $L_1(\mu)$ can be approximated by finite rank norm attaining operators.*

Another known case in which Corollary 13.26 applies is the case of isometric preduals of ℓ_1 [30, Corollary 3.8]. Here, we are also able to get dense lineability of the set of norm attaining functionals.

Corollary 13.29 (extension of [30, Corollary 3.8]). *Let X be an isometric predual of ℓ_1. Then, there is a norm dense linear subspace of X^* contained in $\mathrm{NA}(X)$. Therefore, for every Banach space Y, every compact operator $T \in \mathcal{L}(X, Y)$ can be approximated by finite rank norm attaining operators.*

Proof. We just have to justify the existence of a dense linear subspace of X^* contained in $\mathrm{NA}(X)$, the rest of the results follows from Corollary 13.25. Indeed, it is shown in the proof of [30, Corollary 3.8] (based on results by Gasparis from 2002) that there is a sequence of finite rank norm-one projections $Q_n \in \mathcal{L}(X, X)$ such that the sequence $\{Q_n^*\}_{n\in\mathbb{N}}$ has increasing ranges and converges pointwise to the identity of X^*. Then $\bigcup_{n\in\mathbb{N}} Q_n^*(X^*)$ is a subspace contained in $\mathrm{NA}(X)$ since each Q_n is a norm-one projection of finite rank; this subspace is dense by the pointwise convergence of $\{Q_n^*\}$ to the identity. □

Another easy case in which Corollary 13.26 applies is when a Banach space X has a shrinking monotone Schauder basis [30, Corollary 3.10] but actually the result follows from Corollary 13.25 as $\mathrm{NA}(X)$ contains a dense linear subspace in this case [4, Theorem 3.1].

Corollary 13.30 ([30, Corollary 3.10] and [4, Theorem 3.1]). *Let X be a Banach space. If X has a shrinking monotone Schauder basis, then* $\mathrm{NA}(X)$ *contains a dense linear subspace. Therefore, every compact operator whose domain is X can be approximated by norm attaining finite rank operators.*

This applies, in particular, to closed subspaces of c_0 with a monotone Schauder basis, as shown in [29, Corollary 12] using a result of G. Godefroy and P. Saphar from 1988.

Example 13.31 ([29, Corollary 12]). Let X be a closed subspace of c_0 with a monotone Schauder basis. Then $\mathrm{NA}(X)$ contains a dense linear subspace. Therefore, every compact operator whose domain is X can be approximated by norm attaining finite rank operators.

Next, we get the following result as an obvious consequence of Corollary 13.25 (and the Bishop–Phelps theorem).

Corollary 13.32. *Let X be a Banach space. If* $\mathrm{NA}(X)$ *is a linear subspace of X^*, then finite rank operators with domain X can be approximated by finite rank norm attaining operators.*

If, moreover, X^* has the approximation property, then actually compact operators with domain X can be approximated by finite rank norm attaining operators.

Of course, the result above applies to c_0, but also when X is a finite-codimensional proximinal subspace of c_0, as shown in [15, Remark b on p. 180]. Besides, the noncommutative case also holds: $\mathrm{NA}(\mathcal{K}(\ell_2))$ is also a linear space (see [16, Lemma]), so Corollary 13.32 applies to it. Moreover, this linearity property of the set of norm attaining operators passes down to every finite-codimensional proximinal subspace of $\mathcal{K}(\ell_2)$; see [15, Section 3]. Finally, if X is a c_0-sum of reflexive spaces, then clearly $\mathrm{NA}(X)$ is a linear subspace of X^*. Let us state all the examples we have presented so far.

Examples 13.33. The following spaces satisfy that their sets of norm attaining functionals are vector spaces:
(a) c_0 and its finite-codimensional proximinal subspaces;
(b) $\mathcal{K}(\ell_2)$ and its finite-codimensional proximinal subspaces;
(c) c_0-sums of reflexive spaces.

Therefore, a finite rank operator whose domain is any of the spaces above can be approximated by norm attaining finite rank operators.

In the simplest case of closed subspaces of c_0, we do not know whether the hypothesis of finite codimension or the hypothesis of proximinality can be dropped in (a) above. What is easy to show is that there is a closed hyperplane of c_0 whose set of norm attaining functionals is not a vector space (see [15, Remark b on p. 180] again).

Problem 13.34. *Let X be a closed subspace of c_0. Is it true that every finite rank operator whose domain is X can be approximated by norm attaining (finite rank) operators?*

Let us note that there are compact operators whose domains are closed subspaces of c_0 which cannot be approximated by norm attaining operators [29, Proposition 3].

Corollary 13.26 depends heavily on the fact that the norm of the projections is 1 and fails if one considers renormings. By contrast, Theorem 13.24 and its consequence Corollary 13.25 only depend on the set of norm attaining functionals itself, so both remain valid for renormings which conserve this set. In [11, Theorem 9.(4)], it is shown that every separable Banach space X admits a smooth renorming $\widetilde{X}$ such that $\mathrm{NA}(X) = \mathrm{NA}(\widetilde{X})$, and this result has recently been extended to weakly compactly generated spaces (WCG spaces) [18, Proposition 2.3]. Therefore, if Theorem 13.24 applies for a WCG space X, then so it does for the corresponding $\widetilde{X}$. In particular, we get the following examples.

Examples 13.35. Let X be a WCG Banach space which is equal to $C_0(L)$, is equal to $L_1(\mu)$, satisfies that $X^* = \ell_1$, or is a finite-codimensional proximinal subspace of c_0 or of $\mathcal{K}(\ell_2)$. (In the latter cases, X is of course separable.) Let $\widetilde{X}$ be the equivalent smooth renorming of X given in [18, Proposition 2.3] such that $\mathrm{NA}(\widetilde{X}) = \mathrm{NA}(X)$. Then every compact operator whose domain is $\widetilde{X}$ can be approximated by norm attaining (for the norm of $\widetilde{X}$) finite rank operators.

We do not even know whether the particular case of $\widetilde{c_0}$ can be deduced from previously known results.

Although it is not directly related to finite rank operators, we would like to finish the section by providing a condition which extends the known result by Lindenstrauss [24, Theorem 1] that reflexive spaces have property (A), that is, an operator whose domain is a reflexive space can be approximated by norm attaining operators (this fact is actually used in the proof below).

Proposition 13.36. *Let X, Y be Banach spaces. Then every operator $T \in \mathcal{L}(X,Y)$ for which $[\ker T]^\perp \subset \mathrm{NA}(X)$ can be approximated by norm attaining operators (whose kernels contain $\ker T$).*

Proof. We follow the lines of the proof of Proposition 13.5. As $[\ker T]^\perp \subset \mathrm{NA}(X)$, it is immediate from James's theorem that $X/\ker T$ is reflexive (see the proof of [5, Lemma 2.2]). Now, T factors through $X/\ker T$, that is, there is an operator $\widetilde{T}$: $X/\ker T \longrightarrow Y$ such that $T = \widetilde{T} \circ q$, and it is clear that $\|\widetilde{T}\| = \|T\|$. By the result of J. Lindenstrauss just mentioned, [24, Theorem 1], there is a sequence $\{\widetilde{S}_n\}_{n\in\mathbb{N}}$ of norm attaining operators from $X/\ker T$ into Y which converges in norm to $\widetilde{T}$. On the one hand, the same argument as the one given in Proposition 13.5 allows us to see that for every $n \in \mathbb{N}$, the operator $S_n := \widetilde{S}_n \circ q\colon X \longrightarrow Y$ attains its norm. On the other hand, it is clear that $\|S_n - T\| \leqslant \|\widetilde{S}_n - \widetilde{T}\| \longrightarrow 0$, so $\{S_n\}_{n\in\mathbb{N}}$ converges to T. □

13.4 Norm attaining operators onto a two-dimensional Hilbert space

Our aim in this section is to study the special case when the range space is a (two-dimensional) Hilbert space, where some specific tools can be used, for instance, we may rotate every point of the unit sphere to any other one. As shown in Remark 13.2, this study is actually equivalent to the study of the existence of norm attaining operators of rank two into all Banach spaces of dimension greater than or equal to two.

We will eventually provide some characterizations of the fact that an operator from a Banach space onto a two-dimensional Hilbert space attains its norm and also a characterization of when norm attaining operators onto a two-dimensional Hilbert space are dense.

Let us observe that the existence of a norm attaining operator T of rank at least 2 from a Banach space X to a Hilbert space H gives the existence of a surjective norm attaining operator from X onto ℓ_2^2 (just composing T with a convenient orthogonal projection).

Let $T\colon X \longrightarrow \ell_2^2$ be an operator of rank two. One can identify T with a pair of linearly independent functionals $(f,g) \in X^* \times X^*$. Throughout this section, we will make this identification without further reference. Note that, obviously, $\|(\sigma f, \sigma' g)\| \leqslant \|(f,g)\|$ if $|\sigma|, |\sigma'| \leqslant 1$ so, in particular,

$$\|(\pm f, \pm g)\| = \|(f,g)\| \quad \text{and} \quad \max\{\|f\|, \|g\|\} \leqslant \|(f,g)\|.$$

We will also use these facts frequently in this section without recalling them.

Our first goal in this section is to characterize when $\|(f,g)\| \leqslant 1$ in terms of the functionals f and g, especially in the case when $\|f\| = 1$ and f, g are linearly independent. We next will use this idea to produce pairs of functionals of this form.

The desired characterization of when $\|(f,g)\| \leqslant 1$ is the following.

Proposition 13.37. *Let X be a Banach space, and let $f \in S_{X^*}$ and $g \in B_{X^*}$ be linearly independent. Then $\|(f,g)\| \leqslant 1$ if and only if*

$$\|f + tg\| \leqslant \sqrt{1+t^2} \qquad \textit{for all } t \in \mathbb{R}. \tag{13.5}$$

We need the following easy lemma.

Lemma 13.38. *Let X be a Banach space. Fix $z_0 \in X$ and linearly independent $f,g \in X^*$, and consider $M = \ker f \cap \ker g$. Then*

$$\operatorname{dist}(z_0, M) = \sup_{(t,s)\in\mathbb{R}^2\setminus\{(0,0)\}} \frac{|tf(z_0)+sg(z_0)|}{\|tf+sg\|}.$$

Proof. If we consider $z_0 = J_X(z_0)$ as an element of X^{**}, we have that

$$\operatorname{dist}(z_0, M) = \operatorname{dist}(J_X(z_0), J_X(M)) = \|J_X(z_0)|_{M^\perp}\|,$$

where $J_X(z_0)|_{M^\perp}$ denotes the restriction of $J_X(z_0)$ to the subspace $M^\perp$ of X^*. But $M^\perp$ is the subspace of X^* generated by f and g, hence

$$\operatorname{dist}(z_0, M) = \sup_{x^*\in M^\perp\setminus\{0\}} \frac{|x^*(z_0)|}{\|x^*\|} = \sup_{(t,s)\in\mathbb{R}^2\setminus\{(0,0)\}} \frac{|tf(z_0)+sg(z_0)|}{\|tf+sg\|},$$

and we are done. □

We can now give the pending proof.

Proof of Proposition 13.37. As f and g are linearly independent, there are $x_0, x_1 \in X$ such that $f(x_0) = 1$, $g(x_0) = 0$ and $g(x_1) = 1$, $f(x_1) = 0$. We then have that $X = M \oplus \operatorname{span}\{x_0, x_1\}$ with $M = \ker f \cap \ker g$. Now, for $T := (f,g)$, $\|T\| \leqslant 1$ if, and only if,

$$\|(\lambda,\mu)\|_2 = \|T(m+\lambda x_0+\mu x_1)\|_2 \leqslant \|m+\lambda x_0+\mu x_1\| \qquad \text{for all } \lambda,\mu \in \mathbb{R},\ m \in M.$$

The above is equivalent to

$$\sqrt{\lambda^2+\mu^2} \leqslant \operatorname{dist}(\lambda x_0+\mu x_1, M) \qquad \text{for all } \lambda,\mu \in \mathbb{R}. \tag{13.6}$$

If we define $|(\lambda,\mu)| = \operatorname{dist}(\lambda x_0+\mu x_1, M)$ for all $\lambda,\mu \in \mathbb{R}$, we get a norm on $\mathbb{R}^2$. Now, using Lemma 13.38, we see that

$$|(\lambda,\mu)| = \sup_{(t,s)\in\mathbb{R}^2\setminus\{(0,0)\}} \frac{|\lambda t+\mu s|}{\|tf+sg\|}.$$

We deduce that the dual norm of the above norm is given by $|(\lambda,\mu)|^* = \|\lambda f+\mu g\|$ for all $\lambda,\mu \in \mathbb{R}$, since by definition $|\,.\,|$ is the dual norm of $|\,.\,|^*$. Taking dual norms in the inequality (13.6), we get that this inequality is equivalent to

$$\|\lambda f+\mu g\| \leqslant \sqrt{\lambda^2+\mu^2} \qquad \text{for all } \lambda,\mu \in \mathbb{R}.$$

The last inequality can be rephrased by saying that $\|g\| \leqslant 1$ and

$$\|f + tg\| \leqslant \sqrt{1+t^2} \qquad \text{for all } t \in \mathbb{R}. \qquad \square$$

Remark 13.39. We observe that (13.5) implies that $g(x_0) = 0$ whenever $f(x_0) = 1 = \|x_0\| = \|f\|$.

To facilitate the notation, we introduce the following vocabulary.

Definition 13.40. Given $f \in S_{X^*}$, we call an element $g \in B_{X^*} \setminus \{0\}$ such that (f, g) has rank two and $\|(f, g)\| = 1$, a *mate* of f. This is equivalent to requiring that $\|f + tg\| \leqslant \sqrt{1+t^2}$ for all $t \in \mathbb{R}$, by Proposition 13.37.

Observe that if $g \in B_{X^*} \setminus \{0\}$ is a mate of $f \in S_{X^*}$, one has that

$$\lim_{t\to 0} \frac{\|f + tg\| - 1}{t} = 0 \quad \text{and} \quad \limsup_{t\to 0} \frac{\|f + tg\| - 1}{t^2} \leqslant \frac{1}{2} < \infty.$$

The last condition suggests another formulation of the existence of mates, which will be shown next.

Proposition 13.41. *Let X be a Banach space and $f \in S_{X^*}$. Then f has a mate if and only if there exist $h \in B_{X^*} \setminus \{0\}$ and $K, \varepsilon > 0$ such that*

$$\|f + th\| \leqslant 1 + Kt^2 \qquad \text{for all } t \in (-\varepsilon, \varepsilon),$$

equivalently,

$$\limsup_{t\to 0} \frac{\|f + th\| - 1}{t^2} < \infty.$$

In fact, given $f \in S_{X^*}$ and $h \in B_{X^*} \setminus \{0\}$ such that $\limsup_{t\to 0} \frac{\|f+th\|-1}{t^2} < \infty$, there exists $0 < s \leqslant 1$ such that sh is a mate of f.

Proof. The proof of the necessity of the limsup condition is given in the previous comment. For the sufficiency, assume that $h \in B_{X^*} \setminus \{0\}$, $K, \varepsilon > 0$ are such that $\|f + th\| \leqslant 1 + Kt^2$ for all $t \in (-\varepsilon, \varepsilon)$. (Note that this implies that f and h are linearly independent.) It is enough to show that there exists $0 < s \leqslant 1$ such that $\|f + tsh\| \leqslant \sqrt{1+t^2}$ for all $t \in \mathbb{R}$. If not, there is a sequence $\{t_n\}$ in $\mathbb{R}$ such that

$$\left\|f + \frac{t_n}{n} h\right\| > \sqrt{1+t_n^2} \qquad \text{for all } n \in \mathbb{N}. \tag{13.7}$$

Now,

$$1 + \frac{|t_n|}{n}\|h\| \geqslant \left\|f + \frac{t_n}{n} h\right\| > \sqrt{1+t_n^2}$$

for all $n \in \mathbb{N}$, and we deduce that

$$\frac{|t_n|}{n} < \frac{2\|h\|}{n^2 - \|h\|^2} \leqslant \frac{2}{n^2 - 1}$$

for $n > 1$. Then, $t_n/n \longrightarrow 0$. From (13.7), we get that

$$\frac{\left\|f + \frac{t_n}{n}h\right\|^2 - 1}{t_n^2/n^2} \geqslant n^2$$

and so

$$\limsup_{t\to 0} \frac{\|f + th\|^2 - 1}{t^2} = +\infty.$$

But $\|f + th\|^2 - 1 = (\|f + th\| + 1)(\|f + th\| - 1)$ and $\lim_{t\to 0} \|f + th\| + 1 = 2$, and so

$$\limsup_{t\to 0} \frac{\|f + th\| - 1}{t^2} = +\infty.$$

This completes the proof. □

Now, we can formulate a first positive result about the existence of mates.

Lemma 13.42. *Let X be a Banach space. If $f \in S_{X^*}$ is not an extreme point of B_{X^*}, then f has a mate.*

Proof. Suppose $f = \frac{1}{2}(f_1 + f_2)$, with $f_j \in B_{X^*}$ and $g := \frac{1}{2}(f_1 - f_2) \neq 0$. Clearly, $\|g\| \leqslant 1$, and f and g are linearly independent. We shall show that g is a mate of f using Proposition 13.37. For $t \in \mathbb{R}$, we have

$$\|f + tg\| = \left\|\frac{1}{2}f_1 + \frac{1}{2}f_2 + \frac{t}{2}f_2 - \frac{t}{2}f_1\right\| = \left\|\frac{1-t}{2}f_1 + \frac{1+t}{2}f_2\right\|.$$

The latter norm is $\leqslant 1$ for $|t| \leqslant 1$ and is $\leqslant \frac{|t|-1}{2} + \frac{1+|t|}{2} = |t|$ if $|t| \geqslant 1$. In either case, we have $\|f + tg\| \leqslant \sqrt{1 + t^2}$. □

As an immediate consequence, if X^* is not strictly convex, then there is $f \in S_{X^*}$ with a mate (this is a not very surprising result, see Proposition 13.45). If actually X is not smooth, then we get a more interesting result.

Corollary 13.43. *Let X be nonsmooth Banach space. Then there is $f \in \mathrm{NA}_1(X)$ with a mate.*

Proof. Suppose $x \in S_X$ is such that there are distinct $f_1, f_2 \in S_{X^*}$ with $f_1(x) = f_2(x) = 1$. Then $f := \frac{1}{2}(f_1 + f_2)$ has norm 1 and attains its norm at x, but is not an extreme point of the dual unit ball and Lemma 13.42 applies. □

We may also provide a characterization of mates in terms of extreme points of the space of operators.

Proposition 13.44. *Let X be a Banach space and $f \in S_{X^*}$. There is a mate $g \in B_{X^*} \setminus \{0\}$ for f if, and only if, the operator $(f,0)$ is not an extreme point of $B_{\mathcal{L}(X,\ell_2^2)}$.*

Proof. Suppose $g \neq 0$ and $\|(f,g)\| \leqslant 1$. Then also $\|(f,-g)\| \leqslant 1$, and so $(f,0) = \frac{1}{2}((f,g) + (f,-g))$ is not an extreme point of the unit ball of $\mathcal{L}(X,\ell_2^2)$.

Conversely, if $(f,0)$ is not an extreme point of the unit ball of $\mathcal{L}(X,\ell_2^2)$, then there is a nontrivial convex combination in the unit ball of $\mathcal{L}(X,\ell_2^2)$ representing $(f,0)$, say $(f,0) = \frac{1}{2}((f_1,g_1) + (f_2,g_2))$ where $(f,0) \neq (f_1,g_1)$. If $f_1 = f_2 = f$, then necessarily $g_1 \neq 0$, and hence $\|(f,g_1)\| \leqslant 1$ and f and g_1 are linearly independent; if not, then f is not an extreme point of the unit ball, and hence has a mate by Lemma 13.42. □

We can also ask if there exists some Banach space X of dimension at least two such that there is no mate for any element in S_{X^*}, equivalently $(f,0)$ is an extreme point of $B_{\mathcal{L}(X,\ell_2^2)}$ for every $f \in S_{X^*}$. From Lemma 13.42, we know that the dual of such an example cannot be strictly convex. Indeed, there is no such space whatsoever.

Proposition 13.45. *Let X be a Banach space with $\dim(X) \geqslant 2$. Then there exists $f \in S_{X^*}$ with a mate. Actually, given linearly independent $f', g' \in X^*$ such that $\|(f',g')\| = 1$, there is a rotation π on ℓ_2^2 such that $(f',g') = \pi \circ (f,g)$, $f \in S_{X^*}$ and g is a mate for f.*

Proof. Consider the rank two operator $T = (f',g') \in \mathcal{L}^{(2)}(X,\ell_2^2)$ with $\|T\| = 1$. Then $T^* \in \mathrm{NA}(\ell_2^2, X^*)$ and so $T^{**} \in \mathrm{NA}(X^{**}, \ell_2^2)$, so there is $x_0^{**} \in S_{X^{**}}$ such that $\|T^{**}(x_0^{**})\| = \|x_0^{**}\| = 1$. Now, we compose T with a rotation π' on ℓ_2^2 to get a new operator $S = \pi' T$ with $\|S\| = \|S^{**}(x_0^{**})\| = 1$ and $S^{**}(x_0^{**}) = (1,0)$. Of course, S still has rank two and is represented by

$$(f,g) = (\cos(\varphi) \cdot f' + \sin(\varphi) \cdot g',\ -\sin(\varphi) \cdot f' + \cos(\varphi) \cdot g')$$

for suitable $\varphi \in (-\pi,\pi]$. Then we have that $f \in B_{X^*}$, $g \in B_{X^*} \setminus \{0\}$ satisfy that $x_0^{**}(f) = 1$ and $x_0^{**}(g) = 0$. Therefore, $\|f\| = 1$, (f,g) has rank two, and $\|(f,g)\| = 1$. That is, g is a mate for f. □

We now use all the previous ideas to study norm attaining operators. First, the next result explains the link between norm attaining operators and the existence of mates. It says that, up to rotation and rescaling, norm attaining operators onto ℓ_2^2 are pairs of the form (f,g) where $f \in \mathrm{NA}_1(X)$ and g is a mate of f.

Theorem 13.46. *Let X be a Banach space and let $T \in \mathcal{L}^{(2)}(X,\ell_2^2)$ with $\|T\| = 1$. Then the following assertions are equivalent:*

(i) $T \in \mathrm{NA}^{(2)}(X,\ell_2^2)$.

(ii) *There are $f,g \in X^* \setminus \{0\}$, $x_0 \in S_X$ and a rotation π on ℓ_2^2 such that $f \in \mathrm{NA}_1(X)$ with $f(x_0) = 1$, $g(x_0) = 0$, $\|(f,g)\| \leqslant 1$, and $T = \pi \circ (f,g)$.*

(iii) *There are $f \in \mathrm{NA}_1(X)$ with a mate $g \in B_{X^*} \setminus \{0\}$ and a rotation on ℓ_2^2 such that $T = \pi \circ (f,g)$.*

Proof. (i) ⇒ (ii). Suppose $T = (f', g') \in \mathrm{NA}^{(2)}(X, \ell_2^2)$, say $\|T\| = \|T(x_0)\| = 1$ for some $x_0 \in S_X$. Using a rotation as in the proof of Proposition 13.45, we get a new operator $S = \pi T$ with $\|S\| = \|S(x_0)\| = 1$ and $S(x_0) = (1, 0)$. Now, $S = (f, g)$ satisfies $f(x_0) = 1$, $g(x_0) = 0$. Hence $\|f\| = 1$ and $f \in \mathrm{NA}_1(X)$, but $g \neq 0$ since S has rank two as well, that is, g is a mate for f. The converse implication (ii) ⇒ (i) is clear as $S = (f, g)$ attains its norm at x_0 and so does $T = \pi S$.

Finally, (ii) ⇒ (iii) is immediate and (iii) ⇒ (ii) follows from Remark 13.39. □

The following corollary summarizes the results of this section so far.

Corollary 13.47. *Let X be a Banach space with $\dim(X) \geqslant 2$. Then the following assertions are equivalent:*
(i) $\mathrm{NA}^{(2)}(X, \ell_2^2) \neq \emptyset$.
(ii) *There is $f \in \mathrm{NA}_1(X)$ with a mate.*
(iii) *There are $f \in \mathrm{NA}_1(X)$ and $g \in B_{X^*} \setminus \{0\}$ such that $\|f + tg\| \leqslant \sqrt{1 + t^2}$ for all $t \in \mathbb{R}$.*
(iv) *There are $f \in \mathrm{NA}_1(X)$, $g \in B_{X^*} \setminus \{0\}$, and $\varepsilon > 0$ such that $\|f + tg\| \leqslant 1 + \frac{t^2}{2}$ for all $t \in (-\varepsilon, \varepsilon)$.*
(v) *There are $f \in \mathrm{NA}_1(X)$, $h \in B_{X^*} \setminus \{0\}$, and $\varepsilon, K > 0$ such that $\|f + th\| \leqslant 1 + Kt^2$ for all $t \in (-\varepsilon, \varepsilon)$.*
(vi) *There are $f \in \mathrm{NA}_1(X)$ and $h \in B_{X^*} \setminus \{0\}$ such that $\limsup_{t \to 0} \frac{\|f + th\| - 1}{t^2} < \infty$.*
(vii) *There is $f \in \mathrm{NA}_1(X)$ such that $(f, 0)$ is not an extreme point in the unit ball of $\mathcal{L}(X, \ell_2^2)$.*

The above conditions hold automatically if X is nonsmooth.

Proof. The equivalence between (i) and (ii) follows from Theorem 13.46. The implication (ii) ⇒ (iii) is Proposition 13.37 and the implications (iii) ⇒ (iv) ⇒ (v) ⇒ (vi) are trivial. The implication (vi) ⇒ (ii) is Proposition 13.41. Finally, the equivalence between (ii) and (vii) is Proposition 13.44.

The validity of the conditions in the nonsmooth case is remarked in Proposition 13.43. □

We note from Corollary 13.47.vii that $\mathrm{NA}^{(2)}(X, \ell_2^2) = \emptyset$ if, and only if, $(f, 0)$ is an extreme point of $B_{\mathcal{L}(X, \ell_2^2)}$ for every $f \in \mathrm{NA}_1(X)$, which implies that every $f \in \mathrm{NA}_1(X)$ is an extreme point of B_{X^*}. Again, we see that if X is not smooth there are norm attaining operators from X onto ℓ_2^2.

The proof of Proposition 13.45 implies the following positive result. We already know the result from Proposition 13.5 (or even from Theorem 13.9 which shows that it is valid even with a weaker hypothesis), but we include this alternative proof here for completeness.

Corollary 13.48. *Suppose X is a Banach space for which $\mathrm{NA}(X)$ contains a two-dimensional subspace. Then $\mathrm{NA}^{(2)}(X, \ell_2^2) \neq \emptyset$.*

Proof. Suppose f' and g' are linearly independent so that $\operatorname{span}\{f',g'\} \subset \mathrm{NA}(X)$ and $\|(f',g')\| = 1$. It was shown in the proof of Proposition 13.45 how to obtain some $f \in S_{X^*}$ with a mate by performing a rotation; note that this f is a linear combination of f' and g', and thus, it is norm attaining by the assumption. Hence, $\mathrm{NA}^{(2)}(X, \ell_2^2) \neq \emptyset$ by Theorem 13.46. □

Our final goal in the section is to discuss the density of norm attaining operators whose range is a two-dimensional Hilbert space in terms of mates.

Proposition 13.49. *Let X be a Banach space. Then the following are equivalent:*
(i) *$\mathrm{NA}(X, \ell_2^2)$ is dense in $\mathcal{L}(X, \ell_2^2)$.*
(ii) *For every $f \in S_{X^*}$ and $g \in B_{X^*} \setminus \{0\}$ such that $\|(f,g)\| = 1$ there are sequences $\{f_n\}$ in $\mathrm{NA}_1(X)$ and $\{g_n\}$ in $B_{X^*} \setminus \{0\}$ and a rotation π on ℓ_2^2 such that $\|f_n + tg_n\| \leqslant \sqrt{1+t^2}$ for all $t \in \mathbb{R}$ and all $n \in \mathbb{N}$, and $\lim_n (f_n, g_n) = \pi \circ (f,g)$.*

Proof. (i) ⇒ (ii): Let f and g be as in (ii), and consider the rank-two operator $T = (f,g)$ with $\|T\| = 1$. By (i), there is a sequence of norm attaining operators $T_n' = (f_n', g_n')$ converging to T. The T_n' are also of rank two, at least eventually; and we may assume that $\|T_n'\| = 1$ for all $n \in \mathbb{N}$ as well. Pick $x_n \in S_X$ such that $\|T_n'(x_n)\| = 1$; that is,

$$f_n'(x_n)^2 + g_n'(x_n)^2 = 1.$$

Let us consider a rotation π_{φ_n} by some angle $\varphi_n \in [-\pi, \pi]$ mapping $T_n'(x_n) = (f_n'(x_n), g_n'(x_n))$ to $(1,0)$. By passing to a subsequence, we may suppose that $\{\varphi_n\}$ converges to some φ, and then writing $T_n = (f_n, g_n) := \pi_\varphi \circ (f_n', g_n')$ for every $n \in \mathbb{N}$, we have that the sequence $\{T_n\}$ converges to $\pi_\varphi \circ (f,g)$. Note that T_n belongs to $\mathrm{NA}(X, \ell_2^2)$ for every $n \in \mathbb{N}$ since every T_n' does and, therefore, we have the desired inequality by Proposition 13.37 and Theorem 13.46.

(ii) ⇒ (i): By the Bishop–Phelps theorem, one can approximate rank 1 operators by norm attaining ones; and by the rotation argument in the proof of Proposition 13.45, it is enough to show that operators $T = (f,g)$ in $S_{\mathcal{L}(X,\ell_2^2)}$ with $f \in S_{X^*}$ and $g \in B_{X^*} \setminus \{0\}$ can be approximated. From (ii), there are sequences $\{f_n\}$ in $\mathrm{NA}_1(X)$ and $\{g_n\}$ in $B_{X^*} \setminus \{0\}$ such that $\|f_n + tg_n\| \leqslant \sqrt{1+t^2}$ for all $t \in \mathbb{R}$ and $\lim_n (f_n, g_n) = \pi \circ T$, for some rotation π. By Theorem 13.46, $T_n = (f_n, g_n)$ is norm attaining, hence also $\pi^{-1} \circ T_n \in \mathrm{NA}(X, \ell_2^2)$ and $\pi^{-1} \circ T_n \longrightarrow T$. □

We remark that there are sufficient conditions on a Banach space X expounded in Section 13.3 to assure that each finite rank operator from X can be approximated by norm attaining finite rank operators; in particular, this is true for X a $C_0(L)$ space, an $L_1(\mu)$ space, a predual of ℓ_1, or a proximinal subspace of c_0 or of $\mathcal{K}(\ell_2)$ of finite codimension. Therefore, for these domain spaces X, item (ii) of Proposition 13.49 holds.

13.5 A question about Lomonosov's example

When one speaks about norm attaining functionals, there is no big difference between the real and the complex case, because a complex functional on a complex space attains its norm if, and only if, the real part of the functional does, and besides, a complex functional on a complex Banach space is completely determined by its real part. Therefore, if X is a complex space, then the set of complex-linear functionals on X which attain their maximum modulus coincides with the set of those complex-linear functionals on X whose real parts attain their maximum, so this set is dense by the Bishop–Phelps theorem (compare with the situation which occurs when we consider real-linear operators from X to $\mathbb{C} \equiv \ell_2^2$, see Section 13.4).

But in the same papers [6, 7] that deal with norm attaining functionals, Bishop and Phelps considered an analogous question about functionals that attain their maximum on a given set C. It is proved that, for a closed bounded convex subset C of a real Banach space, the set of maximum attaining functionals is dense in X^* (in [6] this was just a remark at the end of the paper, saying that the proof may be done in the same way as for norm attaining functionals, and in [7] the result is given with all details).

Passing to complex functionals, one cannot speak about the maximal value on a subset C, but it is natural to ask if $\sup_{x\in C} |f(x)|$ is actually a maximum. Let us fix some terminology. For a given subset $C \neq \emptyset$ of a complex Banach space X, a nonzero complex functional $f \in X^*$ is said to be a *modulus support functional* for C if there is a point $y \in C$ (called the corresponding *modulus support point* of C) such that $|f(y)| = \sup_{x\in C} |f(x)|$. The natural question [32] whether for every closed bounded convex subset of a complex Banach space the corresponding set of support functionals is dense in X^* remained open until 2000, when Victor Lomonosov [26, 27] constructed his striking example of a closed bounded convex subset of the predual space of H^∞ which does not admit any modulus support functionals. A similar construction can be made [28] in the predual A_* of every dual algebra A of operators on a Hilbert space which is not self-adjoint (i. e., there is an operator $T \in A$ such that $T^* \notin A$), contains the identity operator and such that the spectral radius of every operator in A coincides with its norm.

By a weak compactness argument, examples of such kind cannot live in a reflexive space. Moreover, they do not exist in spaces with the Radon–Nikodým property by Bourgain's result [9]; see the argument at the end of page 340 of [32]. Therefore, in most classical spaces like $L_p[0,1]$ with $1 < p < \infty$ or ℓ_p with $1 \leqslant p < \infty$ the complex version of the Bishop–Phelps theorem for subsets is valid. The spaces ℓ_∞, $L_\infty[0,1]$ and $C[0,1]$ have subspaces isometric to any given separable space, which makes it possible to transfer Lomonosov's example to these spaces. For the remaining two classical spaces, c_0 and $L_1[0,1]$, the validity of the complex version of the Bishop–Phelps theorem for subsets is an open question.

In the case of the complex space c_0, we have an easy way to define a concrete closed, bounded, and convex subset S for which we do not know whether its set of modulus support functionals is dense, and not even whether it is nonempty. The first author discussed this example with several colleagues, in particular with Victor Lomonosov, but to no avail. So we decided to use this occasion to appeal to a wider circle of people interested in the subject by publishing the example here.

Let $\mathbb{D} = \{z \in \mathbb{C}: |z| < 1\}$ be the open unit disk $e_n \in c_0$ be the elements of the canonical basis, and $e_n^* \in \ell_1$ be the corresponding coordinate functionals. For every $z \in \mathbb{D}$, consider $\varphi_z = \sum_{n=1}^{\infty} z^n e_n \in c_0$. The set $S \subset c_0$ in question is

$$S = \overline{\operatorname{conv}}\{\varphi_z : z \in \mathbb{D}\}. \tag{13.8}$$

Remark that, identifying each element $a = (a_1, a_2, \ldots) \in \ell_1$ with the function f_a on the unit disk by the rule $f_a(\zeta) = \sum_{n=1}^{\infty} a_n \zeta^n$ for all $\zeta \in \mathbb{D}$, we identify $c_0^* = \ell_1$ with the corresponding algebra $\tilde{\ell}_1$ of analytic functions vanishing at zero and having an absolutely convergent series of Taylor coefficients, equipped with the norm $\|f_a\| = \|a\|_1 = \sum_{n=1}^{\infty} |a_n|$. Taking into account that, in the duality of c_0 and ℓ_1,

$$a(\varphi_z) = \sum_{n=1}^{\infty} a_n z^n = f_a(z),$$

we may identify each element φ_z with the evaluation functional δ_z at the point z on $\tilde{\ell}_1$. Having a look at the papers [26, 27], one can see that our S is basically the same as in Lomonosov's example, with the difference that the algebra H^∞ is substituted by $\tilde{\ell}_1$. For every $a \in \ell_1$, one has that

$$\sup_{x \in S} |a(x)| = \sup_{z \in \mathbb{D}} |f_a(z)|,$$

which is the spectral radius of the element $f_a \in \tilde{\ell}_1$. Lomonosov uses in his example that in H^∞ the spectral radius of every element is equal to its norm. In $\tilde{\ell}_1$, this is not the case, which does not permit us to use Lomonosov's argument in our case. Nevertheless, many features survive, which makes the existence of modulus support functionals very questionable.

At first, we remark that by the maximum modulus principle for analytic functions, $\sup_{z \in \mathbb{D}} |f_a(z)|$ cannot be attained, so none of the points φ_z is a modulus support point. Digging deeper, assume that $y = (y_1, y_2, \ldots) \in S$ is a modulus support point that corresponds to the modulus support functional $b = (b_1, b_2, \ldots) \in \ell_1 \setminus \{0\}$, that is,

$$|b(y)| = \sup_{x \in S} |b(x)| = \sup_{z \in \mathbb{D}} |f_b(z)|.$$

Pick elements $w_n \in \operatorname{conv}\{\varphi_z : z \in \mathbb{D}\}$ that converge to y, $w_n = \sum_{k \in \mathbb{N}} w_{n,k} \varphi_{z_k}$, $z_k \in \mathbb{D}$, where $w_{n,k} \geqslant 0$, $\sum_{k \in \mathbb{N}} w_{n,k} = 1$, and for every $n \in \mathbb{N}$ there is an $m(n)$ such that $w_{n,k} = 0$ for all $k > m(n)$.

Consider the corresponding probability measures $\mu_n = \sum_{k \in \mathbb{N}} w_{n,k} \delta_{z_k} \in C(\bar{\mathbb{D}})^*$. By the separability of $C(\bar{\mathbb{D}})$, passing to a subsequence, we may assume without loss of generality that the sequence $\{\mu_n\}$ converges in the weak-$*$ topology of $C(\bar{\mathbb{D}})^*$ to a Borel probability measure μ on $\bar{\mathbb{D}}$. This μ is related to y as follows: for every $j \in \mathbb{N}$, one has that

$$\int_{\bar{\mathbb{D}}} z^j \, d\mu(z) = \lim_{n\to\infty} \int_{\bar{\mathbb{D}}} z^j \, d\mu_n(z) = \lim_{n\to\infty} e_j^*(w_n) = e_j^*(y) = y_j,$$

so

$$\int_{\bar{\mathbb{D}}} z^j \, d\mu(z) \xrightarrow[j\to\infty]{} 0.$$

By a similar argument,

$$\int_{\bar{\mathbb{D}}} f_b(z)\, d\mu(z) = \lim_{n\to\infty} \int_{\bar{\mathbb{D}}} f_b(z)\, d\mu_n(z) = \lim_{n\to\infty} b(w_n) = b(y),$$

and consequently

$$\left| \int_{\bar{\mathbb{D}}} f_b(z)\, d\mu(z) \right| = |b(y)| = \sup_{z\in\bar{\mathbb{D}}} |f_b(z)| .$$

Denoting $r = \sup_{z\in\bar{\mathbb{D}}} |f_b(z)|$, we deduce from the above property that

$$\operatorname{supp} \mu \subset \{v \in \bar{\mathbb{D}} : |f_b(v)| = r\} \subset \bar{\mathbb{D}} \setminus \mathbb{D}$$

and, moreover, the function f_b must take a constant value α on $\operatorname{supp}\mu$ with $|\alpha| = r$. These conditions on μ and b are very restrictive, and we do not know whether such a wild pair of animals exists.

We finish the section by emphasizing the question we have been discussing here.

Problem 13.50. *Let S be the subset of the complex space c_0 given in (13.8). Are the modulus support functionals for S dense in c_0^*?*

Bibliography

[1] M. D. Acosta, *Denseness of norm-attaining operators into strictly convex spaces*, Proc. R. Soc. Edinb., Sect. A **129** (1999), 1107–1114.

[2] M. D. Acosta, Norm-attaining operators into $L_1(\mu)$, in: K. Jarosz (ed.), *Function Spaces. Proceedings of the 3rd Conference, Edwardsville, IL, USA, May 19–23, 1998*, 1–11 Contemporary Math., **232**, 1999.

[3] M. D. Acosta, *Denseness of norm attaining mappings*, RACSAM, Rev. R. Acad. Cienc. Exactas Fís. Nat., Ser. A Mat. **100** (2006), 9–30.

[4] M. D. Acosta, A. Aizpuru, R. Aron and F. García-Pacheco, *Functionals that do not attain their norm*, Bull. Belg. Math. Soc. Simon Stevin **14** (2007), 407–418.

[5] P. Bandyopadhyay and G. Godefroy, *Linear structures in the set of norm-attaining functionals on a Banach space*, J. Convex Anal. **13** (2006), 489–497.

[6] E. Bishop and R. R. Phelps, *A proof that every Banach space is subreflexive*, Bull. Am. Math. Soc. **67** (1961), 97–98.

[7] E. Bishop and R. R. Phelps, *The support functionals of a convex set*, Proc. Symp. Pure Math. **7** (1963), 27–35.

[8] A. Bosznay and B. Garay, *On norms of projections*, Acta Sci. Math. (Szeged) **50** (1986), 87–92.

[9] J. Bourgain, *On dentability and the Bishop–Phelps property*, Isr. J. Math. **28** (1977), 265–271.

[10] S. Dantas, D. García, M. Maestre and M. Martín, *The Bishop–Phelps–Bollobás property for compact operators*, Can. J. Math. **70** (2018), 53–73.

[11] G. Debs, G. Godefroy and J. Saint Raymond, *Topological properties of the set of norm-attaining linear functionals*, Can. J. Math. **47** (1995), 318–329.

[12] J. Diestel and J. J. Uhl, *The Radon–Nikodým theorem for Banach space valued measures*, Rocky Mt. J. Math. **6** (1976), 1–46.

[13] M. Fabian, P. Habala, P. Hájek, V. Montesinos and V. Zizler, *Banach Space Theory*, CMS Books in Mathematics, Springer, 2011.

[14] D. H. Fremlin, *Measure Theory, vol. 2*, Torres Fremlin, 2001.

[15] G. Godefroy and V. Indumathi, *Proximinality in subspaces of c_0*, J. Approx. Theory **101** (1999), 175–181.

[16] G. Godefroy, V. Indumathi and F. Lust-Piquard, *Strong subdifferentiability of convex functionals and proximinality*, J. Approx. Theory **116** (2002), 397–415.

[17] W. T. Gowers, *Symmetric block bases of sequences with large average growth*, Isr. J. Math. **69** (1990), 129–151.

[18] A. J. Guirao, V. Montesinos and V. Zizler, *Remarks on the set of norm-attaining functionals and differentiability*, Stud. Math. **241** (2018), 71–86.

[19] R. B. Holmes, *Geometric Functional Analysis and Its Applications*, Springer, 1975.

[20] V. Indumathi, *Proximinal subspaces of finite codimension in general normed spaces*, Proc. Lond. Math. Soc. **45** (1982), 435–455.

[21] J. Johnson and J. Wolfe, *Norm attaining operators*, Stud. Math. **65** (1979), 7–19.

[22] V. M. Kadets, G. López and M. Martín, *Some geometric properties of Read's space*, J. Funct. Anal. **274** (2018), 889–899.

[23] V. M. Kadets, G. López, M. Martín and D. Werner, *Equivalent norms with an extremely nonlineable set of norm attaining functionals*, J. Inst. Math. Jussieu **19** (2020), 259–279.

[24] J. Lindenstrauss, *On operators which attain their norm*, Isr. J. Math. **1** (1963), 139–148.

[25] J. Lindenstrauss and L. Tzafriri, *Classical Banach Spaces I: Sequence Spaces*, Springer, 1977.

[26] V. Lomonosov, *A counterexample to the Bishop–Phelps theorem in complex spaces*, Isr. J. Math. **115** (2000), 25–28.

[27] V. Lomonosov, *On the Bishop–Phelps theorem in complex spaces*, Quaest. Math. **23** (2000), 187–191.

[28] V. Lomonosov, *The Bishop–Phelps theorem fails for uniform non-selfadjoint dual operator algebras*, J. Funct. Anal. **185** (2001), 214–219.

[29] M. Martín, *Norm-attaining compact operators*, J. Funct. Anal. **267** (2014), 1585–1592.

[30] M. Martín, *The version for compact operators of Lindenstrauss properties A and B*, RACSAM, Rev. R. Acad. Cienc. Exactas Fís. Nat., Ser. A Mat. **110** (2016), 269–284.

[31] D. Narayana and T. S. S. R. K. Rao, *Transitivity of proximinality and norm attaining functionals*, Colloq. Math. **104** (2006), 1–19.

[32] R. R. Phelps, *The Bishop–Phelps theorem in complex spaces: an open problem*, Lect. Notes Pure Appl. Math. **136** (1991), 337–340.

[33] C. J. Read, *Banach spaces with no proximinal subspaces of codimension* 2, Isr. J. Math. **223** (2018), 493–504.
[34] M. Rmoutil, *Norm-attaining functionals need not contain 2-dimensional subspaces*, J. Funct. Anal. **272** (2017), 918–928.
[35] I. Singer, *The Theory of Best Approximation and Functional Analysis*, Conference Board of the Mathematical Sciences Regional Conference Series in Applied Mathematics, **13**, Society for Industrial and Applied Mathematics, Philadelphia, Pa., 1974.

Seychelle S. Khan, Mutasim Mim, and Mikhail I. Ostrovskii

14 Isometric copies of ℓ_∞^n and ℓ_1^n in transportation cost spaces on finite metric spaces

To Victor Lomonosov

Abstract: Main results: (a) If a metric space contains $2n$ elements, the transportation cost space on it contains a 1-complemented isometric copy of ℓ_1^n. (b) An example of a finite metric space whose transportation cost space contains an isometric copy of ℓ_∞^4. Transportation cost spaces are also known as Arens–Eells, Lipschitz-free, or Wasserstein 1 spaces.

Keywords: Arens–Eells space, Banach space, duality in linear programming, earth mover distance, Edmonds matching algorithm, Kantorovich–Rubinstein distance, Lipschitz-free space, perfect matching, transportation cost, Wasserstein distance

MSC 2010: Primary 52A21, Secondary 05C70, 30L05, 46B07, 46B85, 91B32

14.1 Introduction

The introduced below notions go back at least to Kantorovich and Gavurin [10]. We use the terminology and notation of [13]. History of the notions introduced below as well as related terminology (Arens–Eells space, earth mover distance, Kantorovich–Rubinstein distance, Lipschitz-free space, Wasserstein distance) is discussed in [13, Section 1.6] and references therein.

Definition 14.1. Let (M, d) be a metric space. Consider a real-valued finitely supported function f on M with a zero sum, that is,

$$\sum_{v \in M} f(v) = 0. \tag{14.1}$$

Acknowledgement: The contribution of the first two authors is the outcome of their research done under the supervision of the third author. The authors gratefully acknowledge the support by the National Science Foundation grant NSF DMS-1700176. The first author was supported by GAAP. The authors thank the referee for corrections and suggestions which helped to improve the presentation. The third-named author also would like to thank Florin Catrina and Beata Randrianantoanina for useful discussions. During the final stage of the work on this paper, the third author was on research leave supported by St. John's University.

Seychelle S. Khan, Mutasim Mim, Mikhail I. Ostrovskii, Department of Mathematics and Computer Science, St. John's University, 8000 Utopia Parkway, Queens, NY 11439, USA, e-mails: seychelle.khan16@my.stjohns.edu, mutasim.mim16@my.stjohns.edu, ostrovsm@stjohns.edu

https://doi.org/10.1515/9783110656756-014

A natural and important interpretation of such a function is the following: $f(v) > 0$ means that $f(v)$ units of a certain product are produced or stored at point v; $f(v) < 0$ means that $(-f(v))$ units of the same product are needed at v. The number of units can be any real number. With this in mind, f may be regarded as a *transportation problem*. For this reason, we denote the vector space of all real-valued functions finitely supported on M with a zero sum by TP(M), where TP stands for *transportation problems*.

One of the standard norms on the vector space TP(M) is related to the *transportation cost* and is defined in the following:

A *transportation plan* is a plan of the following type: we intend to deliver

- a_1 units of the product from x_1 to y_1,
- a_2 units of the product from x_2 to y_2,
- ...
- a_n units of the product from x_n to y_n,

where $a_1, \dots, a_n$ are nonnegative real numbers, and $x_1, \dots, x_n, y_1, \dots, y_n$ are elements of M, which do not have to be distinct.

This transportation plan is said to *solve the transportation problem* f if

$$f = a_1(\mathbf{1}_{x_1} - \mathbf{1}_{y_1}) + a_2(\mathbf{1}_{x_2} - \mathbf{1}_{y_2}) + \dots + a_n(\mathbf{1}_{x_n} - \mathbf{1}_{y_n}), \tag{14.2}$$

where $\mathbf{1}_u(x)$ for $u \in M$ is the *indicator function* defined as

$$\mathbf{1}_u(x) = \begin{cases} 1 & \text{if } x = u, \\ 0 & \text{if } x \neq u. \end{cases}$$

The *cost* of transportation plan (14.2) is defined as $\sum_{i=1}^{n} a_i d(x_i, y_i)$. We introduce the *transportation cost norm* (or just *transportation cost*) $\|f\|_{\mathrm{TC}}$ of a transportation problem f as the minimal cost of transportation plans solving f. It is easy to see that the transportation plan of the minimum cost exists. We introduce the *transportation cost space* TC(M) on M as the completion of TP(M) with respect to the norm $\|\cdot\|_{\mathrm{TC}}$.

It is worth mentioning that the norm of an element in TC(M) can be computed using linear programming; see [12] and [15], and also related historical comments in [15, pp. 221–223].

Arens and Eells [1] observed that if we pick a *base point* O in the space M, then the *canonical embedding* of M into $(\mathrm{TP}(M), \|\cdot\|_{\mathrm{TC}})$ given by the formula:

$$v \mapsto \mathbf{1}_v - \mathbf{1}_O \tag{14.3}$$

is an isometric embedding. This observation can be easily derived from the following characterization of optimal transportation plans.

Let $0 \leq C < \infty$. A real-valued function l on a metric space (M, d) is called *C-Lipschitz* if

$$\forall x, y \in M \quad |l(x) - l(y)| \leq Cd(x, y).$$

The *Lipschitz constant* of a function l on a metric space containing at least two points is defined as

$$\mathrm{Lip}(l) = \max_{x,y\in M,\ x\neq y} \frac{|l(x) - l(y)|}{d(x,y)}.$$

Theorem 14.2 ([10]). *A plan*

$$f = a_1(\mathbf{1}_{x_1} - \mathbf{1}_{y_1}) + a_2(\mathbf{1}_{x_2} - \mathbf{1}_{y_2}) + \cdots + a_n(\mathbf{1}_{x_n} - \mathbf{1}_{y_n}) \tag{14.4}$$

is optimal if and only if there exist a 1*-Lipschitz real-valued function* l *on* M *such that*

$$l(x_i) - l(y_i) = d(x_i, y_i) \tag{14.5}$$

for all pairs x_i, y_i *for which* $a_i > 0$.

The mentioned above observation of Arens and Eells makes transportation cost spaces an important object in the theory of metric embeddings; see [14, Chapter 10] and [13, Section 1.4]. This theory makes it very important to study the conditions of isometric embeddability of spaces ℓ_∞^n into $\mathrm{TC}(M)$.

Problems on isometric embeddability of spaces ℓ_1^n and ℓ_∞^n into $\mathrm{TC}(M)$ are also motivated by the following definitions, the first of which goes back to Kantorovich and Gavurin [10].

Definition 14.3. Let $f_1,\dots,f_n$ be nonzero transportation problems in $\mathrm{TP}(M)$ and $x_1,\dots,x_n$ be their normalizations, that is, $x_i = f_i/\|f_i\|_{\mathrm{TC}}$.

We say that transportation problems $f_1,\dots,f_n$ are *completely unrelated*, if

$$\left\|\sum_{i=1}^n a_i x_i\right\|_{\mathrm{TC}} = \sum_{i=1}^n |a_i|$$

for every collection $\{a_i\}_{i=1}^n$ of real numbers.

We say that transportation problems $f_1,\dots,f_n$ are *completely intertwined*, if

$$\left\|\sum_{i=1}^n a_i x_i\right\|_{\mathrm{TC}} = \max_{1\le i\le n} |a_i|$$

for every collection $\{a_i\}_{i=1}^n$ of real numbers.

Remark 14.4. The notion of completely unrelated problems has a natural meaning in applications: we cannot decrease the total cost by combining the transportation plans for a set of completely unrelated transportation problems.

The notion of completely intertwined problems describes the very unusual situation: we have several transportation problems $\{x_i\}_{i=1}^n$ such that each of them has cost 1 and the sum $\sum_{i=1}^n \theta_i x_i$ (of n summands with cost 1 each) has cost 1 for every collection $\theta_i = \pm 1$.

It is clear that problems are completely unrelated if and only if their normalizations are isometrically equivalent to the unit vector basis of ℓ_1^n and problems are completely intertwined if and only if their normalizations are isometrically equivalent to the unit vector basis of ℓ_∞^n.

The main goal of this paper is to study embeddability of ℓ_1^n and ℓ_∞^n into $\mathrm{TC}(M)$ for finite metric spaces M. The following theorem is our main result.

Theorem 14.5. *If a metric space M contains $2n$ elements, then $\mathrm{TC}(M)$ contains a 1-complemented subspace isometric to ℓ_1^n. If the space M is such that triangle inequalities for all distinct triples in M are strict, then $\mathrm{TC}(M)$ does not contain a subspace isometric to ℓ_1^{n+1}.*

Remark 14.6. It can be easily seen from the proof that in the case where a finite metric space M contains more than $2n$ elements, the space $\mathrm{TC}(M)$ also contains a 1-complemented subspace isometric to ℓ_1^n. This is not completely obvious only if $|M|$ is odd. In this case, we add to M one point in an arbitrary way, apply Theorem 14.5, and then observe that all elements of standard basis of the constructed space, except one, are contained in $\mathrm{TC}(M)$.

Theorem 14.5 solves [6, Problem 3.3] by strengthening [6, Theorem 3.1] which states that for M with $2n$ elements the space $\mathrm{TC}(M)$ contains a 2-complemented subspace 2-isomorphic to ℓ_1^n.

Problems of isometric embeddability of ℓ_1 into $\mathrm{TC}(M)$ on infinite metric spaces M were considered in [4, 13].

The existing knowledge on embeddability of ℓ_∞^n is very limited. The most important sources in this direction are [2] and [9]. In Section 14.3, we present a special case of one of the results of [9] in the form which, in our opinion, helps to understand the phenomenon. Bourgain [2] proved (see also a presentation in [14, Section 10.4]) that $\mathrm{TC}(\ell_1)$ contains almost isometric copies of ℓ_∞^n for all n.

Our contribution to the case of ℓ_∞^n (Section 14.3) consists in examples of relatively small finite metric spaces M_3 and M_4 such that $\mathrm{TC}(M_3)$ and $\mathrm{TC}(M_4)$, respectively, contain ℓ_∞^3 and ℓ_∞^4 isometrically. The reason for our interest to M_3 is that it is smaller than M_4. We do not know whether such finite metric spaces can be constructed for ℓ_∞^n with $n \geq 5$.

In this connection, it is natural to recall the well-known fact that the spaces ℓ_1^2 and ℓ_∞^2 are isometric. It is easy to see that the standard proof of this can be stated as the following.

Observation 14.7. *The transportation problems f_1 and f_2 are completely unrelated if and only if the transportation problems $g_1 = \frac{1}{2}(f_1 + f_2)$ and $g_2 = \frac{1}{2}(f_1 - f_2)$ are completely intertwined.*

14.2 Proof of Theorem 14.5

We use terminology of [5]. Consider the metric space M as a weighted complete graph with $2n$ elements; we denote it also $G = (V(G), E(G))$, the weight of an edge is the distance between its ends. We consider matchings containing n edges in this graph, such matchings are called *perfect matchings* or 1-*factors*. We pick among all perfect matchings a matching of minimum weight (the *weight* of a matching is defined as the sum of weights of its edges). Let $e_1 = u_1v_1, \dots, e_n = u_nv_n$ be a perfect matching of minimum weight. We claim that the transportation problems $f_1 = \mathbf{1}_{u_1} - \mathbf{1}_{v_1}, \dots, f_n = \mathbf{1}_{u_n} - \mathbf{1}_{v_n}$ are completely unrelated.

We need to show that for any set $\{a_i\}_{i=1}^n$ of real numbers we have

$$\left\|\sum_{i=1}^n a_i(\mathbf{1}_{u_i} - \mathbf{1}_{v_i})\right\|_{\mathrm{TC}} = \sum_{i=1}^n |a_i| d(u_i, v_i).$$

Assume for simplicity that all a_i are positive (all other cases can be done similarly, we can just interchange u_i and v_i for those i for which $a_i < 0$).

The inequality

$$\left\|\sum_{i=1}^n a_i(\mathbf{1}_{u_i} - \mathbf{1}_{v_i})\right\|_{\mathrm{TC}} \le \sum_{i=1}^n |a_i| d(u_i, v_i)$$

is obvious. To prove the inverse inequality, assume the contrary, that is,

$$\left\|\sum_{i=1}^n a_i(\mathbf{1}_{u_i} - \mathbf{1}_{v_i})\right\|_{\mathrm{TC}} < \sum_{i=1}^n |a_i| d(u_i, v_i).$$

In such a case, there exist transportation plans for $f = \sum_{i=1}^n a_i(\mathbf{1}_{u_i} - \mathbf{1}_{v_i})$ with lower costs than the straightforward plan (by the *straightforward plan* we mean the plan in which a_i units are moved from u_i to v_i for each $i = 1, \dots, n$). Let

$$\sum_{j=1}^m b_j(\mathbf{1}_{x_j} - \mathbf{1}_{y_j}), \tag{14.6}$$

where $b_j > 0$ for $j = 1, \dots, m$, be an optimal plan for f, that is, a plan satisfying

$$\|f\|_{\mathrm{TC}} = \sum_{j=1}^m b_j d(x_j, y_j).$$

It is known (see [16, Proposition 3.16]) that such plans exist and that there exists an optimal plan satisfying the following condition:

(A) Each x_j is one of $\{u_i\}_{i=1}^n$ and each y_j is one of $\{v_i\}_{i=1}^n$.

Since plan (14.6) is different from the straightforward plan and satisfies condition **(A)**, there are $n(0), n(1) \in \{1, \dots, n\}$, $n(0) \neq n(1)$ such that some amount $c_0 > 0$ of the product is moved according to (14.6) from $u_{n(0)}$ to $v_{n(1)}$. Then there exist $n(2) \in \{1, \dots, n\}$, $n(2) \neq n(1)$ such that some amount $c_1 > 0$ of the product is moved according to (14.6) from $u_{n(1)}$ to $v_{n(2)}$. We continue in an obvious way. Since we consider finite sets, there is $k < n$ such that $n(k+1) \in \{n(0), n(1), \dots, n(k)\}$. Without loss of generality, we may assume (changing the notation if necessary) that $n(k+1) = n(0)$.

Let $c = \min_{0 \le i \le k} c_i$. Then $c > 0$ and part of the transportation done according to the plan (14.6) is: c units of the product are moved

- from $u_{n(0)}$ to $v_{n(1)}$,
- from $u_{n(1)}$ to $v_{n(2)}$,
- $\dots$,
- from $u_{n(k)}$ to $v_{n(0)}$.

It is clear that if we modify this part of the plan to: c units of the product are moved

- from $u_{n(0)}$ to $v_{n(0)}$,
- from $u_{n(1)}$ to $v_{n(1)}$,
- $\dots$,
- from $u_{n(k)}$ to $v_{n(k)}$;

we get another transportation plan for f.

To clarify the main idea of the proof, first we consider the case where $\{u_i v_i\}_{i=1}^n$ is a unique perfect matching of the minimum weight, that is, all other perfect matchings have strictly larger weights.

In this case, we show that the cost of the modified (two paragraphs above) transportation plan is strictly smaller than the cost of (14.6), and get a contradiction with the assumption that (14.6) is an optimal plan.

To show this, it suffices to prove that

$$\sum_{i=0}^{k} c d(u_{n(i)}, v_{n(i)}) < \sum_{i=0}^{k} c d(u_{n(i)}, v_{n(i+1)}), \tag{14.7}$$

recall that $n(k+1) = n(0)$. Inequality (14.7) is an immediate consequence of the assumption that the perfect matching $\{u_i v_i\}_{i=1}^n$ has a strictly smaller weight than the perfect matching

$$\{u_{n(i)} v_{n(i+1)}\}_{i=1}^{k} \cup \{u_i v_i\}_{i \in R}, \quad \text{where } R = \{1, \dots, n\} \backslash \{n(0), n(1), \dots, n(k)\},$$

so the proof is completed under the additional assumption of the uniqueness of the minimum weight perfect matching.

Let us turn to the general case. In this case, we can claim only a nonstrict inequality in (14.7). This does not lead to an immediate contradiction, but we can finish the argument in the following way. Since $c > 0$, the nonstrict version of (14.7) proves the following lemma.

Lemma 14.8. *If an optimal transportation plan for f satisfies* **(A)** *and does not coincide with the straightforward plan, then we can construct another optimal transportation plan satisfying* **(A)** *in which the total amount of the product which is moved as in the straightforward plan, that is, from u_i to v_i is strictly larger.*

With this lemma, we can complete the proof in the general case as follows. Consider optimal transportation plans for f satisfying the condition **(A)**. Such plans can be regarded as $n \times n$ matrices with nonnegative entries in which the entry $s_{i,j}$ is the amount of the product which is to be moved from u_i to v_j. It is clear that the set of such optimal plans is closed in any usual topology on the set of matrices. If it contains the straightforward plan, we are done. If it does not, we get a contradiction as follows. It is easy to check that among all optimal plans satisfying condition **(A)** there is a plan for which the sum $\sum_{i=1}^n s_{i,i}$ is the maximal possible. If this plan does not coincide with the straightforward, then by Lemma 14.8, there is an optimal plan satisfying **(A)** for which the sum $\sum_{i=1}^n s_{i,i}$ is larger, contrary to the maximality assumption. This contradiction proves the existence in $\mathrm{TC}(M)$ of the subspace isometric to ℓ_1^n.

Now, assume that M is such that all triangle inequalities in M are strict. Let $f_1, \ldots, f_k$ be completely unrelated transportation problems on M.

Lemma 14.9. *The functions f_i have disjoint supports.*

This lemma is essentially known [13, Lemma 3.3], for convenience of the reader we provide a proof.

Proof. Assume the contrary, let $v \in M$ be in the supports of both f_i and f_j, $i \neq j$. Without loss of generality, we assume that $f_i(v) > 0$ and $f_j(v) < 0$, changing signs of f_i and f_j if needed (the change of signs does not affect complete unrelatedness).

To get a contradiction, it suffices to show that $\|f_i + f_j\|_{\mathrm{TC}} < \|f_i\|_{\mathrm{TC}} + \|f_j\|_{\mathrm{TC}}$. This can be done in the following way. In an optimal plan for f_i some amount of units, denote it $\alpha > 0$, is moved from v to some $u \in M$. In an optimal plan for f_j some amount of units, denote it $\beta > 0$, is moved to v from some $w \in M$ (w can be the same as u).

Let $\gamma = \min\{\alpha, \beta\}$. Now we combine the optimal plans for f_i and f_j with the following exception: we move γ units of the product directly from w to u. Since, by our assumption, $d(w,u) < d(w,v)+d(v,u)$, the cost of the obtained plan is $< \|f_i\|_{\mathrm{TC}}+\|f_j\|_{\mathrm{TC}}$. □

Finally, since support of each function f_i contains at least two points, we get that $k \leq n$. This proves the last statement of Theorem 14.5.

It remains to show that there is a projection of norm 1 onto the subspace spanned by $\{f_i\}_{i=1}^n$. We show that a linear operator P is a norm-1 projection onto the subspace spanned by $\{\mathbf{1}_{u_i} - \mathbf{1}_{v_i}\}_{i=1}^n$ if and only if it can be represented in the form

$$P(f) = \sum_{i=1}^{n} l_i(f) \frac{f_i}{\|f_i\|_{\mathrm{TC}}}, \tag{14.8}$$

where:
- $f_i = \mathbf{1}_{u_i} - \mathbf{1}_{v_i}$.
- l_i are Lipschitz functions, and $l_i(f_j) = \delta_{i,j}\|f_j\|_{\mathrm{TC}} = \delta_{i,j} d(u_j, v_j)$ ($\delta_{i,j}$ is the Kronecker delta).
- $\|Pf\|_{\mathrm{TC}} \le \|f\|_{\mathrm{TC}}$ for every $f \in \mathrm{TC}(M)$ of the form $f = \mathbf{1}_w - \mathbf{1}_z$ for $w, z \in M$.

Since $\{f_i\}_{i=1}^n$ are linearly independent and the dual of $\mathrm{TC}(M)$ is the space of the Lipschitz functions on M, which take value 0 at the base point (see [14, Theorem 10.2]), any projection onto the subspace spanned by $\{f_i\}_{i=1}^n$ is of the form (14.8) for some Lipschitz functions $\{l_i\}$ satisfying $l_i(f_j) = \delta_{i,j}\|f_j\|_{\mathrm{TC}} = \delta_{i,j} d(u_j, v_j)$.

It remains to show the condition $\|Pf\|_{\mathrm{TC}} \le \|f\|_{\mathrm{TC}}$ for $f \in \mathrm{TC}(M)$ of the form $f = \mathbf{1}_w - \mathbf{1}_z$ implies that $\|P\| \le 1$. This follows from our definitions and observations made above: In fact, since for any $g \in \mathrm{TC}(M)$ there exists a transportation plan of minimal cost, we can represent g as a sum $g = \sum_{i=1}^m g_i$, where g_i are of the form $g_i = b_i(\mathbf{1}_{w_i} - \mathbf{1}_{z_i})$, $b_i \in \mathbb{R}$, and $\|g\|_{\mathrm{TC}} = \sum_{i=1}^m \|g_i\|_{\mathrm{TC}}$. Therefore, we get

$$\|Pg\|_{\mathrm{TC}} = \left\| P\left(\sum_{i=1}^{m} g_i\right)\right\|_{\mathrm{TC}} \le \sum_{i=1}^{m} \|Pg_i\|_{\mathrm{TC}} \le \sum_{i=1}^{m} \|g_i\|_{\mathrm{TC}} = \|g\|_{\mathrm{TC}},$$

and thus $\|P\| \le 1$.

Our approach to the construction of suitable functions l_i is based on the duality theorem of linear programming and the Edmonds [7] algorithm for the minimum weight perfect matching problem. We use the description of the algorithm in the form given in [11, Theorem 9.2.1], where it is shown that the minimum weight perfect matching problem on a complete graph G with even number of vertices and weight $w : E(G) \to \mathbb{R}$, $w \ge 0$, can be reduced to the following linear program. (An *odd cut* in G is the set of edges joining a subset of $V(G)$ of odd cardinality with its complement, a *trivial odd cut* is a set of edges joining one vertex with its complement. If x is a real-valued function on $E(G)$ and A is a set of edges, we write $x(A) = \sum_{e\in A} x(e)$.)
- **(LP1)** minimize $w^\top \cdot x$ (where $x : E(G) \to \mathbb{R}$)
- subject to
  (1) $x(e) \ge 0$ for each $e \in E(G)$
  (2) $x(C) = 1$ for each trivial odd cut C
  (3) $x(C) \ge 1$ for each nontrivial odd cut C.

We introduce a variable y_C for each odd cut C.

The dual program of the program **(LP1)** is:

- **(LP2)** maximize $\sum_C y_C$
- subject to

(D1) $y_C \geq 0$ for each nontrivial odd cut C

(D2) $\sum_{C \text{ containing } e} y_C \leq w(e)$ for every $e \in E(G)$.

The duality in linear programming [15, Section 7.4] (see also a summary in [11, Chapter 7]) states that the optima **(LP1)** and **(LP2)** are equal. (In the general case, we need to require the existence of vectors satisfying the constraints and finiteness of one of the optima.)

This means that the total length of the minimum weight perfect matching coincides with the sum of entries of the optimal solution of the dual problem.

We complete our proof of the existence of norm-1 projection P of the desired form by proving the following two lemmas.

Lemma 14.10. *Suppose that there is an optimal dual solution satisfying $y_C \geq 0$ for all odd cuts C **including trivial ones**. Then there exist functions l_i for which P defined by* (14.8) *is a norm-1 projection.*

Lemma 14.11. *If the weight function $w : E(G) \to \mathbb{R}$ corresponds to a metric on $V(G)$ (this means that $w(uv) = d(u, v)$ for some metric d on $V(G)$), then there is an optimal dual solution satisfying $y_C \geq 0$ for all odd cuts, including trivial ones.*

Proof of Lemma 14.10. Let $\mathcal{M}$ be the minimum weight perfect matching, then $e \in \mathcal{M}$ is of the form u_iv_i. We introduce the function $l_i : V(G) \to \mathbb{R}$ by

$$l_i(w) = \begin{cases} 0 & \text{if } w = u_i \\ \sum_{C \text{ contains } u_iv_i \text{ and separates } u_i \text{ and } w} y_C & \text{if } w \neq u_i. \end{cases} \tag{14.9}$$

We claim that the function l_i has the following desired properties:

1. l_i is 1-Lipschitz.
2. $l_i(v_i) - l_i(u_i) = d(v_i, u_i)$.
3. $l_i(v_j) - l_i(u_j) = 0$ if $j \neq i$.
4. $\sum_{i=1}^n |l_i(w) - l_i(z)| \leq d(w, z)$ for every $w, z \in M = V(G)$.

The discussion following (14.8) implies that these conditions imply that the obtained P is a norm-1 projection.

Proofs of 1–4:

1. $|l_i(w) - l_i(z)| \leq \sum_{C \text{ separates } w \text{ and } z} y_C \leq w(wz) = d(w, z)$, where in the first inequality we used the definition of l_i, in the second we used (D2). Observe also that item 1 follows from the stronger inequality in item 4, which we prove below.

2. $l_i(v_i) - l_i(u_i) = d(v_i, u_i)$.
   The corresponding argument is shown in [11, p. 371]. We reproduce it. We have

$$w(\mathcal{M}) = \sum_{e\in\mathcal{M}} w(e) \geq \sum_{e\in\mathcal{M}} \sum_{C \text{ containing } e} y_C = \sum_C |\mathcal{M} \cap C| y_C \geq \sum_C y_C, \tag{14.10}$$

   where in the first inequality we used (D2) and in the second inequality we used $|\mathcal{M} \cap C| \geq 1$ for each odd cut.
   If y_C is an optimal dual solution, we get that the leftmost and the rightmost sides in (14.10) coincide and, therefore,

$$w(e) = \sum_{C \text{ containing } e} y_C \tag{14.11}$$

   for each $e \in \mathcal{M}$ and

$$|\mathcal{M} \cap C| = 1 \text{ for each nontrivial odd cut } C \text{ satisfying } y_C > 0. \tag{14.12}$$

   Equality (14.11) implies $l_i(v_i) - l_i(u_i) = \sum_{C \text{ containing } u_i v_i} y_C - 0 = w(u_i v_i) = d(u_i, v_i)$.
3. $l_i(v_j) - l_i(u_j) = 0$ if $j \neq i$.
   This equality follows from (14.12). In fact, equality (14.12) implies that none of the cuts with $y_C > 0$ containing $u_i v_i$ can contain $u_j v_j$ for $j \neq i$, and thus $l_i(v_j) = l_i(u_j)$ for all $j \neq i$.
4. $\sum_{i=1}^n |l_i(w) - l_i(z)| \leq d(w, z)$ for every $w, z \in M$.
   To prove this inequality, we observe that $|l_i(w) - l_i(z)| \leq \sum_{C\in S_i(w,z)} y_C$, where $S_i(w, z)$ is the set of cuts C with $y_C > 0$ which simultaneously separate u_i from v_i and w from z. It is important to observe that (14.12) implies that the sets $\{S_i(w, z)\}_{i=1}^n$ are disjoint. Therefore, by (D2), $\sum_{i=1}^n |l_i(w) - l_i(z)| \leq d(w, z)$. □

Proof of Lemma 14.11. We follow the presentation in [11, Section 9.2] of the Edmonds algorithm for construction of an optimal dual solution. To prove the lemma, it suffices to show that the assumption that w corresponds to a metric implies that when we run the algorithm we maintain $y_C \geq 0$ in each step, even for trivial odd cuts.

We decided not to copy the whole Section 9.2 at a price that we expect readers (who do not remember the algorithm) to have [11, Section 9.2] handy.

The beginning of the algorithm can be described as follows: we assign the number $y_C = \frac{1}{2} \min_{u,v} d(u, v)$ to all trivial cuts C and set $y_C = 0$ for all nontrivial cuts C. This function on the set of all odd cuts satisfies the conditions (D1) and (D2). Such functions are called *dual solutions*. For a dual solution y, we form a graph G_y whose vertex set is $V(G)$ and edge set is defined by

$$E_y = \left\{ e \in E(G) : \sum_{C \text{ containing } e} y_C = w(e) \right\}.$$

It is clear that with y_C defined as above we get a graph G_y which can contain any number of edges between 1 and $\frac{n(n-1)}{2}$.

In each step of the Edmonds algorithm, we construct not only the function y_C, but also a set $\mathcal{H}$ of odd cardinality subsets of $V(G)$ satisfying four conditions listed in [11, (P-1)–(P-4), p. 372]. We list only the first two conditions, because the contents of the last two conditions does not affect our modification of the argument in [11, Section 9.2].

(P-1) $\mathcal{H}$ is nested, that is, if $S, T \in \mathcal{H}$, then either $S \subset T$ or $T \subset S$ or $S \cap T = \emptyset$.

(P-2) $\mathcal{H}$ contains all singletons of $V(G)$.

At the end of the first step described above, the set $\mathcal{H}$ is let to be the set of singletons (and all of the desired conditions are satisfied).

After that the following step is repeated and the function y_C is modified until the graph G'_y (described below) becomes a graph having perfect matching.

Let $S_1, \dots, S_k$ be the (inclusionwise) maximal members of $\mathcal{H}$. It follows from (P-1) that $S_1, \dots, S_k$ are mutually disjoint and from (P-2) that they form a partition of $V(G)$. Let G'_y denote the graph obtained from G_y by contracting each S_i to a single vertex s_i. Since $|V(G)|$ is even, but S_j is odd, it follows that $k := |V(G'_y)|$ is even.

Suppose that G'_y does not have a perfect matching. Let $A(G'_y)$, $C(G'_y)$, and $D(G'_y)$ be the sets of the Gallai–Edmonds decomposition for G'_y (see [11, Section 3.2]).

We use the notation $A(G'_y) = \{s_1, \dots, s_m\}$ and denote the components of the subgraph of G'_y induced by $D(G'_y)$ by $H_1, \dots, H_{m+d}$, where d is the number of vertices which are not matched in a maximum matching in G'_y. Let

$$T_i = \bigcup_{s_j \in V(H_i)} S_j.$$

Now we modify the dual solution y as follows (by $\nabla(S)$ we denote the set of edges connecting a vertex set S with its complement):

$$\begin{aligned} y^t_{\nabla(S_j)} &= y_{\nabla(S_j)} - t \quad (1 \le j \le m), \\ y^t_{\nabla(T_i)} &= y_{\nabla(S_i)} + t \quad (1 \le i \le m + d), \\ y^t_C &= y_C, \quad \text{otherwise.} \end{aligned}$$

In this formula, t is chosen as the minimum of three numbers, t_1, t_2, t_3, defined as

$$t_1 = \min\{y_{\nabla(S_j)} : 1 \le j \le m, |S_j| > 1\},$$

$$t_2 = \min\left\{w(e) - \sum_{e \in C} y_C : e \in \nabla(T_1) \cup \dots \cup \nabla(T_{m+d}) \backslash (\nabla(S_1) \cup \dots \cup \nabla(S_m))\right\},$$

$$t_3 = \frac{1}{2} \min\left\{w(e) - \sum_{e \in C} y_C : e \in (\nabla(T_i) \cap \nabla(T_j)),\ 1 \le i < j \le m + d\right\}.$$

It is clear from the definition of t_1 that negative coefficients can appear only for those S_j which are singletons. So suppose that S_j is a singleton, $S_j = \{v\}$. To complete the proof of Lemma 14.11, it remains to show that $t_3 \le y_{\nabla(v)}$, and so $y^t_{\nabla(v)}$ is still non-negative.

Because of the positive surplus condition in [11, Theorem 3.2.1 (c)], the vertex v is connected in G_y with at least two of the sets $\{T_i\}_{i=1}^{m+d}$, suppose that these are sets T_{i_1} and T_{i_2}. Let $u \in T_{i_1}$ and $w \in T_{i_2}$ be adjacent to v in G_y. Let $\{U_p\}_{p=1}^{\tau}$ be the elements of $\mathcal{H}$ containing u and let $\{W_q\}_{q=1}^{\sigma}$ be the elements of $\mathcal{H}$ containing w. Since the edges uv and wv are in G_y, we have

$$w(uv) = y_{\nabla(v)} + \sum_{p=1}^{\tau} y_{\nabla U_p} \tag{14.13}$$

$$w(wv) = y_{\nabla(v)} + \sum_{q=1}^{\sigma} y_{\nabla W_q} \tag{14.14}$$

On the other hand, the definition of t_3 and our choice of $S_1, \dots, S_k$ imply that

$$\begin{aligned} t_3 &\le \frac{1}{2}\left(w(uw) - \sum_{p=1}^{\tau} y_{\nabla U_p} - \sum_{q=1}^{\sigma} y_{\nabla W_q} \right) \\ &\le \frac{1}{2}\left(\left(w(uv) - \sum_{p=1}^{\tau} y_{\nabla U_p} \right) + \left(w(vw) - \sum_{q=1}^{\sigma} y_{\nabla W_q} \right) \right) \\ &= y_{\nabla(v)}, \end{aligned}$$

where in the second inequality we use the triangle inequality for the distance corresponding to weight w, and in the last equality we use (14.13) and (14.14). □

14.3 Isometric copies of ℓ_∞^n in TC(M)

As is well known the spaces $\{\ell_\infty^n\}$ admit low-distortion and even isometric embeddings into some transportation cost spaces. This follows from the basic property of TC(M): it contains an isometric copy of M (see (14.3)).

Another related fact is the following immediate consequence of the Bourgain discretization theorem (see [3], [8], [14, Section 9.2]): for sufficiently large m, the transportation cost space on the set of integer points in ℓ_∞^n with absolute values of coordinates $\le m$ contains an almost-isometric copy of ℓ_∞^n.

In the next example, we need the following well-known fact (see [16, Section 3.3], [13, Section 1.6]): If (M, d) is a complete metric space, then TC(M) contains the vector space of differences between finite positive compactly supported measures μ and ν on M with the same total masses and $\|\mu - \nu\|_{TC}$ is equal to the quantity $\mathcal{T}_1(\mu, \nu)$ defined in the following way.

A *coupling* of a pair of finite positive Borel measures (μ, ν) with the same total mass on M is a Borel measure π on $M \times M$ such that $\mu(A) = \pi(A \times M)$ and $\nu(A) = \pi(M \times A)$

for every Borel measurable $A \subset M$. The set of couplings of (μ, ν) is denoted $\Pi(\mu, \nu)$. We define

$$\mathcal{T}_1(\mu, \nu) := \inf_{\pi \in \Pi(\mu,\nu)} \left(\iint_{M\times M} d(x,y)\, d\pi(x,y) \right).$$

The result of Godefroy and Kalton [9, Theorem 3.1] has the following special case.

Example 14.12. Let us consider the following (nondiscrete) transportation problems on the unit cube $[0,1]^n$ with its ℓ_∞-distance:

P_i: "available" is the Lebesgue measure on the face $x_i = 0$, "needed" is the Lebesgue measure on the face $x_i = 1$.

Then the problems $\{P_i\}_{i=1}^n$ span a subspace isometric to ℓ_∞^n in $\mathrm{TC}([0,1]^n)$.

It is clear that P_i has cost 1, and actually any measure-preserving transportation from bottom to top does the job. The easiest transportation plan is to move each point from the face $x_i = 0$ to the point with the same coordinates, changing only x_i from 0 to 1.

It is not that easy to see that $\sum_{i=1}^n \theta_i P_i$ has cost 1. This can be done as follows. By symmetry, it suffices to consider the case where all $\theta_i = 1$. In this case, we move each point from the surface with "availability" to the surface with "need" in the direction of the diagonal $(1, \dots, 1)$. It is easy to see that it will be a bijection between points of "availability" and "need." The cost can be computed as the following integral:

$$\begin{aligned} n\int_0^1 t(-d(1-t)^{n-1}) &= n(n-1)\int_0^1 t(1-t)^{n-2}dt \\ &= n(n-1)\int_0^1 ((1-t)^{n-2} - (1-t)^{n-1})dt \\ &= n(n-1)\left(\frac{(1-t)^n}{n} - \frac{(1-t)^{n-1}}{n-1}\right)\Big|_0^1 = 1. \end{aligned}$$

We are interested in constructing finite metric spaces M for which $\mathrm{TC}(M)$ contains ℓ_∞^n isometrically. So far, we succeeded to do this only for $n = 3$ and $n = 4$ (the case $n = 2$ is easy, see Observation 14.7).

Example 14.13 (Finite M with $\mathrm{TC}(M)$ containing ℓ_∞^3 isometrically). The set M which we consider is a subset of the surface of the cube $[0,1]^3$ endowed with its ℓ_∞ distance. Transportation problem P_i is described in the following way: "available" is $\frac{1}{6}$ at each midpoint of the edge in the face $x_i = 0$ and $\frac{1}{3}$ at the center of the face; "needed" is at the similar points with $x_i = 1$.

The transportation cost for P_i is 1—just shift from $x_i = 0$ to $x_i = 1$. Again by symmetry, it suffices to show that the cost of $P_1 + P_2 + P_3$ is 1.

Consider faces with $x_i = 0$ as colored "red" and faces with $x_i = 1$ as colored "blue." It is clear that availability and need on two-dimensional faces which are on the boundary between blue and red cancel each other. There will be 6 points of availability left. Three of them are on edges, and three are centers of faces. The value is $\frac{1}{3}$ at each. So to achieve cost 1 it suffices to match red and blue vertices in such a way that the distance between any two matched vertices is $\frac{1}{2}$.

This is possible. To achieve this, we match red points which are centers of edges with blue vertices which are centers of faces and red points which are centers of faces with blue vertices which are centers of edges.

Our example in dimension 4 is even more symmetric.

Example 14.14 (Finite M with $\mathrm{TC}(M)$ containing ℓ_∞^4 isometrically). The set M which we consider is a subset of the surface of the cube $[0,1]^4$ endowed with its ℓ_∞ distance. Transportation problem P_i is described in the following way: "available" is $\frac{1}{6}$ at the center of each of each 2-dimensional face of the face $x_i = 0$; "needed" is at the similar points with $x_i = 1$.

The transportation cost for P_i is 1—just shift from $x_i = 0$ to $x_i = 1$. Again by symmetry, it suffices to show that the cost of $P_1 + P_2 + P_3 + P_4$ is 1.

As in the above discussion with blue and red, we see that half of the availability and need will cancel each other.

The remaining availability of value $\frac{1}{3}$ will be concentrated at 6 centers of 2-dimensional faces of 3-dimensional faces. Each of these centers will have coordinates $\frac{1}{2}, \frac{1}{2}, 1, 1$ in some order. The need of value $\frac{1}{3}$ will be concentrated at 6 points with coordinates $\frac{1}{2}, \frac{1}{2}, 0, 0$. Cancellation will occur at points with coordinates $\frac{1}{2}, \frac{1}{2}, 0, 1$.

To get the transportation plan of cost 1, we need to find a matching between points with coordinates $\frac{1}{2}, \frac{1}{2}, 1, 1$, and points with coordinates $\frac{1}{2}, \frac{1}{2}, 0, 0$, such that the distance between each pair of matched vertices is $\frac{1}{2}$. Such matching is obvious.

Bibliography

[1] R. F. Arens and J. Eells, Jr., *On embedding uniform and topological spaces*, Pac. J. Math. **6** (1956), 397–403.

[2] J. Bourgain, *The metrical interpretation of superreflexivity in Banach spaces*, Isr. J. Math. **56** (2) (1986), 222–230.

[3] J. Bourgain, Remarks on the extension of Lipschitz maps defined on discrete sets and uniform homeomorphisms, in: *Geometrical Aspects of Functional Analysis (1985/86)*, 157–167, Lecture Notes in Math., **1267**, Springer, Berlin, 1987.

[4] M. Cúth and M. Johanis, *Isometric embedding of ℓ_1 into Lipschitz-free spaces and ℓ_∞ into their duals*, Proc. Am. Math. Soc. **145** (8) (2017), 3409–3421.

[5] R. Diestel, *Graph theory*, 5th edition, Graduate Texts in Mathematics, **173**, Springer-Verlag, Berlin, 2017.

[6] S. J. Dilworth, D. Kutzarova and M. I. Ostrovskii, *Lipschitz free spaces on finite metric spaces*, Can. J. Math. **72** (3) (2020), 774–804. DOI:10.4153/S0008414X19000087.
[7] J. Edmonds, *Maximum matching and a polyhedron with* 0, 1*-vertices*, *J. Res. Nat. Bur. Standards, Sect. B* **69B** (1965), 125–130.
[8] O. Giladi, A. Naor and G. Schechtman, *Bourgain's discretization theorem*, Ann. Math. Fac. Sci. Toulouse **XXI** (4) (2012), 817–837 (see also a later correction in arXiv:1110.5368v2).
[9] G. Godefroy and N. J. Kalton, *Lipschitz-free Banach spaces*, Stud. Math. **159** (1) (2003), 121–141.
[10] L. V. Kantorovich and M. K. Gavurin, Application of mathematical methods in the analysis of cargo flows (Russian), in: *Problems of Improving of Transport Efficiency*, 110–138, USSR Academy of Sciences Publishers, Moscow, 1949.
[11] L. Lovász and M. D. Plummer, *Matching Theory*, corrected reprint of the 1986 original, AMS Chelsea Publishing, Providence, RI, 2009.
[12] J. Matoušek and B. Gärtner, *Understanding and Using Linear Programming*, Springer, Berlin, New York, 2007.
[13] S. Ostrovska and M. I. Ostrovskii, *Generalized transportation cost spaces*, Mediterr. J. Math. **16** (6) (2019), Paper No. 157.
[14] M. I. Ostrovskii, *Metric Embeddings: Bilipschitz and Coarse Embeddings into Banach Spaces*, de Gruyter Studies in Mathematics, **49**, Walter de Gruyter & Co., Berlin, 2013.
[15] A. Schrijver, *Theory of Linear and Integer Programming*, Wiley–Interscience Series in Discrete Mathematics. A Wiley–Interscience Publication, John Wiley & Sons, Ltd., Chichester, 1986.
[16] N. Weaver, *Lipschitz Algebras*, 2nd edition, World Scientific Publishing Co. Pte. Ltd., Hackensack, NJ, 2018.

Edward Kissin, Victor S. Shulman, and Yurii V. Turovskii

15 From Lomonosov lemma to radical approach in joint spectral radius theory

To the memory of Victor Lomonosov, a man who moved mountains in mathematics

Abstract: In this paper, we discuss the infinite-dimensional generalizations of the famous theorem of Berger–Wang (generalized Berger–Wang formulas) and give an operator-theoretic proof of I. Morris's theorem about coincidence of three essential joint spectral radius, related to these formulas. Further, we develop Banach-algebraic approach based on the theory of topological radicals, and obtain some new results about these radicals.

Keywords: Invariant subspace, joint spectral radius, topological radical

MSC 2010: Primary 47A15, Secondary 47L10

15.1 Introduction

15.1.1 Banach-algebraic consequences of Lomonosov lemma

The famous Lomonosov lemma [12] states:

> *If an algebra A of operators on a Banach space X contains a nonzero compact operator T, then either A has a nontrivial closed invariant subspace* (IS, for brevity) *or it contains a compact operator with a nonzero point in spectrum.*

An immediate consequence of this result is that *any algebra of compact quasinilpotent operators has an IS*; the standard technique gives then that such an algebra is triangularizable.

M. G. Krein proposed to call compact quasinilpotent operators *Volterra operators*; respectively, a set of operators is called *Volterra* if all its elements are Volterra operators. Thus *any Volterra algebra has an IS*. This result was extended by the second

Acknowledgement: This paper is in final form and no version of it will be submitted for publication elsewhere.

Edward Kissin, STORM, London Metropolitan University, 166-220 Holloway Road, London N7 8DB, Great Britain, e-mail: e.kissin@londonmet.ac.uk
Victor S. Shulman, Yurii V. Turovskii, Department of Mathematics, Vologda State University, Vologda, Russia, e-mails: shulman.victor80@gmail.com, yuri.turovskii@gmail.com

https://doi.org/10.1515/9783110656756-015

author [17] as follows:

> *Any Volterra algebra A has an IS which is also invariant for all operators commuting with A* (such subspaces are called *hyperinvariant*).

Besides of the Lomonosov's technique, the proof used estimations of the norms of products for elements of a Volterra algebra A; in fact, it was proved in [17] that *the joint spectral radius* $\rho(M)$ *of any finite set* $M \subseteq A$ *equals* 0, that is, A is *finitely quasinilpotent.*

This result can be considered as an application of the invariant subspace theory to the theory of joint spectral radius. Conversely, the second part of the proof in [17] is an application of the joint spectral radius technique to the invariant subspace theory (again via Lomonosov's theorem about Volterra algebras): if $M = \{T_1, \dots, T_n\}$ and $\rho(M) = 0$, then

$$\rho\left(\sum_{i=1}^{n} T_i S_i\right) = 0$$

for all operators S_i commuting with every operator from M. So the algebra generated by a Volterra algebra A and its commutant has a nonzero Volterra ideal. The interaction of these theories remained to be fruitful in subsequent studies.

The notion of the joint spectral radius of a bounded subset M in a normed algebra A was introduced by Rota and Strang [16]. To give precise definition, let us set $\|M\| = \sup\{\|a\| : a \in M\}$, the *norm* of M, and $M^n = \{a_1 \cdots a_n : a_1, \dots, a_n \in M\}$, the *n-power* of M. The number

$$\rho(M) := \lim \|M^n\|^{1/n} = \inf \|M^n\|^{1/n}$$

is called a *(joint) spectral radius* of M. If $\rho(M) = 0$ then we say that M is *quasinilpotent.*

In [29], the third author, using the joint spectral radius approach, obtained the solution of Volterra semigroup problem posed by Heydar Radjavi: it was proved in [29] that *any Volterra semigroup generates a Vollterra algebra* and, therefore, has an IS by Lomonosov's theorem. Further, in [18] it was proved that *any Volterra Lie algebra has an IS*; this result can be regarded as an infinite-dimensional extension of Engel theorem, playing the fundamental role in the theory of finite-dimensional Lie algebras.

One of the main technical tools obtained and applied in [18] was an infinite-dimensional extension of the Berger–Wang theorem [4], a fundamental result of the finite dimensional linear dynamics [8]. This theorem establishes the equality

$$\rho(M) = r(M), \tag{15.1}$$

for any bounded set M of matrices, where

$$r(M) := \lim \sup\{\rho(a) : a \in M^n\}^{1/n};$$

the number $r(M)$ called a *BW-radius* of M. In [18] the equality (15.1) was proved for any precompact set M of compact operators on an *infinite-dimensional* Banach space.

To see the importance of validity of (15.1) for precompact sets of compact operators, note that it easily implies that if G is a Volterra semigroup then $\rho(M) = 0$, for each precompact $M \subset G$ (because clearly $r(M) = 0$). This result proved in [29] played a crucial role in the solution of Volterra semigroup problem. But it should be said that the proof of (15.1) in [18] used the results of [29].

Other results on invariant subspaces of operator semigroups, Lie algebras and Jordan algebras were obtained in this way in [18, 21, 9].

15.1.2 The generalized *BW*-formula

To move further, we have to introduce some "essential radii" $\rho_e(M)$, $\rho_f(N)$ and $\rho_\chi(M)$ of a set M of operators on a Banach space X. They are defined in the same way as $\rho(M)$ but by using seminorms $\|\cdot\|_e$, $\|\cdot\|_f$ and $\|\cdot\|_\chi$, instead of the operator norm $\|\cdot\|$.

Let $B(X)$ be the algebra of all bounded linear operators on X, and $K(X)$ the ideal of all compact operators. The essential norm $\|T\|_e$ of an operator $T \in B(X)$ is just the norm of the image $T + K(X)$ of T in the quotient $B(X)/K(X)$; in other words,

$$\|T\|_e = \inf\{\|T - S\| : S \text{ is a compact operator}\}.$$

Similarly,

$$\|T\|_f = \inf\{\|T - S\| : S \text{ is a finite rank operator}\}.$$

The Hausdorff norm $\|T\|_\chi$ is defined as $\chi(TX_\odot)$, the Hausdorff measure of noncompactness of the image of the unit ball $X_\odot$ of X under T. Recall that, for any bounded subset E of X, the value $\chi(E)$ is equal to the infimum of such $t > 0$ that E has a finite t-net.

It is easy to check that $\|T\|_\chi \le \|T\|_e \le \|T\|_f$ and, therefore,

$$\rho_\chi(M) \le \rho_e(M) \le \rho_f(M), \tag{15.2}$$

for each bounded set $M \subset B(X)$. The number $\rho_\chi(M)$ is called the *Hausdorff radius*, $\rho_e(M)$ the *essential radius*, and $\rho_f(M)$ the *f-essential radius* of M.

In what follows, for a set M in a normed algebra A and a closed ideal J of A, we write M/J for the image of M in A/J under the canonic quotient map $q_J : A \longrightarrow A/J$:

$$M/J := q_J(M).$$

So we write $\rho_e(M) = \rho(M/K(X))$. This reflects the fact that essential radius $\rho_e(M)$ is the usual joint spectral radius of the image of M in the Calkin algebra $B(X)/K(X)$.

In [20], the following extension of (15.1) to precompact sets of general (not necessarily compact) operators was obtained:

$$\rho(M) = \max\{\rho_\chi(M), r(M)\}. \tag{15.3}$$

It was proved under assumption that X is reflexive (or, more generally, that M consists of weakly compact operators). We call this equality the *generalized* BW-*formula*.

Furthermore, in the short communication [19] a Banach algebraic version of the generalized BW-formula was announced (see (15.15) below) which, being applied to the algebra $B(X)$, shows that

$$\rho(M) = \max\{\rho_e(M), r(M)\}, \tag{15.4}$$

for all Banach spaces. The proof of (15.3) in full generality was firstly presented in the arXive publication [26]; the journal version appeared in [24].

Several months after presentation of [26], I. Morris in arXive publication [14] gave another proof of (15.3) based on the multiplicative ergodic theorem of Tieullen [27] and deep technique of the theory of cohomology of dynamical systems. The main result of [14] establishes an equality similar to (15.3) for operator valued cocycles of dynamical systems. It was also proved in [14] that

$$\rho_\chi(M) = \rho_e(M) = \rho_f(M) \tag{15.5}$$

for any precompact set $M \subset B(X)$. The journal publication of these results appeared in [13].

Here, we give another operator-theoretic proof of (15.5) and then discuss related Banach-algebraic results and constructions connected with the different joint spectral radius formulas. It will be shown that topological radicals present a convenient tool in the search of an optimal joint spectral radius formula.

15.2 Coincidence of Hausdorff and essential radii

In this section, we are going to prove the equality $\rho_\chi(M) = \rho_e(M)$, for any precompact set in $B(X)$; the proof of the equality $\rho_e(M) = \rho_f(M)$ will be presented in the next section.

15.2.1 An estimation of the Hausdorff radius for multiplication operators

At the beginning, we transfer some results of [18] from operators to elements of the Calkin algebra $B(X)/K(X)$. We use the following link of Hausdorff norm with the Hausdorff measure of noncompactness.

Lemma 15.1. *Let M be a precompact subset of $B(X)$. Then $\chi(MW) \le \|M\|_\chi \|W\|$ for any bounded subset W of X, and $\|M\|_\chi = \chi(MX_\odot)$.*

The inequality in Lemma 15.1 was obtained in [18, Lemma 5.2]. The equality $\|M\|_\chi = \chi(MX_\odot)$ is obvious for a finite $M \subseteq B(X)$ by definition, due to $\chi(G \cup K) = \max\{\chi(G), \chi(K)\}$ for bounded subsets of X If M is precompact then $\|M\|_\chi = \sup\{\|N\|\chi\colon N \subseteq M \text{ is finite}\}$, and the result follows.

For $T \in B(X)$, let L_T and R_T denote the left and right multiplication operators on $B(X)$: $\mathrm{L}_T P = TP$ and $\mathrm{R}_T P = PT$ for each $P \in B(X)$.

For $M \subseteq B(X)$, put $\mathrm{L}_M := \{\mathrm{L}_T : T \in M\}$ and $\mathrm{R}_M := \{\mathrm{R}_T : T \in M\}$. If M is a set in a Banach algebra A, we define L_M and R_M similarly. By [24, Lemma 2.1],

$$r(\mathrm{L}_M \mathrm{R}_M) = r(M)^2 \quad \text{and} \quad \rho(\mathrm{L}_M \mathrm{R}_M) = \rho(M)^2 \tag{15.6}$$

for every bounded set M in A.

Lemma 15.2. *Let M be a bounded subset of $B(X)$. Then*

$$\|\mathrm{L}_{M/K(X)} \mathrm{R}_{M/K(X)}\|_\chi \le 16\|M\|_\chi \|M/K(X)\|.$$

Proof. Let $T, S \in B(X)$. It is clear that

$$\begin{aligned}\|\mathrm{L}_{T/K(X)} \mathrm{R}_{S/K(X)}\|_\chi &= \chi((T/K(X))(B(X)/K(X))_\odot(S/K(X))) \\ &\le \chi(T(B(X))_\odot S) = \|\mathrm{L}_T \mathrm{R}_S\|_\chi.\end{aligned}$$

By [18, Lemma 6.4], $\|\mathrm{L}_T \mathrm{R}_S\|_\chi \le 4(\|T^*\|_\chi \|S\| + \|S\|_\chi \|T\|)$ for any $T, S \in B(X)$. As $\|T^*\|_\chi \le 2\|T\|_\chi$ by [7], and $\|T - P\|_\chi = \|T\|_\chi$, $\|S - F\|_\chi = \|S\|_\chi$, for any $P, F \in K(X)$, we obtain that

$$\begin{aligned}\|\mathrm{L}_{T/K(X)} \mathrm{R}_{S/K(X)}\|_\chi &\le \inf_{P,F \in K(X)} \|\mathrm{L}_{T-P} \mathrm{R}_{S-F}\|_\chi \\ &\le 8 \inf_{P,F \in K(X)} (\|T\|_\chi \|S - F\| + \|S\|_\chi \|T - P\|) \\ &= 8(\|T\|_\chi \|S/K(X)\| + \|S\|_\chi \|T/K(X)\|).\end{aligned}$$

Therefore,

$$\begin{aligned}\|\mathrm{L}_{M/K(X)} \mathrm{R}_{M/K(X)}\|_\chi &\le 8 \sup_{T,S \in M} (\|T\|_\chi \|S/K(X)\| + \|S\|_\chi \|T/K(X)\|) \\ &\le 16 \sup_{T \in M} \|T\|_\chi \sup_{S \in M} \|S/K(X)\| = 16\|M\|_\chi \|M/K(X)\|.\end{aligned}$$ □

15.2.2 Semigroups in the Calkin algebra

Let $M \subseteq B(X)$, and let $\mathrm{SG}(M)$ be the semigroup generated by M. The same notation is used if M is a subset of an arbitrary Banach algebra.

Proposition 15.3. *Let M be a precompact subset of $B(X)$. If $\mathrm{SG}(M/K(X))$ is bounded and $\rho_\chi(M) < 1$ then $\mathrm{SG}(M/K(X))$ is precompact.*

Proof. Let $G_n = \cup\{M^k/K(X) : k > n\}$ for each $n \geq 0$. As $L_{M^k/K(X)}R_{M^k/K(X)}$ is a precompact set in $B(B(X)/K(X))$, then by Lemmas 15.1 and 15.2,

$$\begin{aligned}\chi(G_{2k}) &= \chi((M^k/K(X))G_0(M^k/K(X))) = \chi(\mathrm{L}_{M^k/K(X)}\mathrm{R}_{M^k/K(X)}G_0)\\ &\leq \|\mathrm{L}_{M^k/K(X)}\mathrm{R}_{M^k/K(X)}\|_\chi\|G_0\| \leq 16\|M^k\|_\chi\|M^k/K(X)\|\|G_0\|\\ &\leq (16\|G_0\|^2)\|M^k\|_\chi.\end{aligned}$$

As $\rho_\chi(M) < 1$, there is $m > 0$ such that $\|M^m\|_\chi < 1/2$. Then for $n > 2km$, we have that

$$\chi(G_n) \leq \chi(G_{2km}) \leq (16\|G_0\|^2)(1/2)^k \to 0 \quad \text{under } k \to \infty.$$

This shows that $\chi(G_n) \to 0$ under $n \to 0$. As $\mathrm{SG}(M/K(X))\backslash G_n$ is precompact,

$$\chi(\mathrm{SG}(M/K(X))) = \chi(G_n)$$

for every n. Therefore, $\chi(\mathrm{SG}(M/K(X))) = 0$, that is, $\mathrm{SG}(M/K(X))$ is precompact. □

Let A be a Banach algebra and $M \subseteq A$. Let $\mathrm{LIM}(M)$ be the set of limit points of all sequences (a_k) with $a_k \in M^{n_k}$, $n_k \to \infty$ when $k \to \infty$. It follows from [18, Corollary 6.12] that *if* $\rho(M) = 1$ *and* $\mathrm{SG}(M)$ *is precompact then* $\mathrm{LIM}(M) = \mathrm{LIM}(M)^2$ *and it has a nonzero idempotent.* We will use this fact in the proof of Theorem 15.5 (the part Case 1).

An element $a \in A$ is called *n-leading* for M if $a \in M^n$ and $\|a\| \geq \|\bigcup_{k<n} M^k\|$; a sequence $(a_k) \subseteq A$ is called *leading* for M, if a_k is n_k-leading for M, where $n_k \to \infty$, and $\|a_k\| \to \infty$ under $k \to \infty$.

Let $\mathrm{ld}^n(M)$ be the set of all n-leading elements for M, $\mathrm{ld}(M) = \bigcup_{n\geq 2} \mathrm{ld}^n(M)$ and $\mathrm{ld}_{[1]}(M) = \{a/\|a\| : a \in \mathrm{ld}(M)\}$.

Lemma 15.4. *Let M be a precompact set of $B(X)$. If $\|M^m\|_\chi\|M^m/K(X)\| < 1$, for some $m > 0$ then $\mathrm{ld}_{[1]}(M/K(X))$ is precompact.*

Proof. Let $G_n = \{a/\|a\| : a \in \mathrm{ld}^i(M/K(X)), i \geq n\}$ for any $n > 0$. Let $n = 2km + j$, where $0 \leq j < 2m$. Then, for $N = M^m$ and $B_{(1)} = (B(X)/K(X))_\odot$, we obtain that

$$G_n \subseteq (N^k/K(X))B_{(1)}(N^k/K(X)). \tag{15.7}$$

Indeed, if $T/K(X) \in \mathrm{ld}^i(M/K(X))$ where $i \geq n$, then

$$T = T_1T_2T_3/K(X)$$

for some $T_1/K(X), T_3/K(X) \in N^k/K(X)$ and $T_2/K(X) \in M^{i-2km}/K(X)$. As $T/K(X)$ is an i-leading element for $M/K(X)$, then

$$\|T_2/K(X)\| \leq \|T/K(X)\|.$$

This proves (15.7).

Let $t = \|N\|_\chi \|N/K(X)\|$. As $N^k/K(X)$ is a precompact set, we get from Lemmas 15.1 and 15.2 that

$$\begin{aligned}\chi(G_n) &\le \chi(\mathrm{L}_{N^k/K(X)}\mathrm{R}_{N^k/K(X)}B_{(1)}) \le \|\mathrm{L}_{N^k/K(X)}\mathrm{R}_{N^k/K(X)}\|_\chi \\ &\le 16\|N^k\|_\chi \|N^k/K(X)\| \le 16(\|N\|_\chi \|N/K(X)\|)^k = 16t^k\end{aligned}$$

whence $\chi(G_n) \to 0$ under $n \to \infty$. As $\mathrm{ld}_{[1]}(M/K(X))\backslash G_n$ is precompact,

$$\chi(\mathrm{ld}_{[1]}(M/K(X))) = \chi(G_n)$$

for every n. Therefore, $\chi(\mathrm{ld}_{[1]}(M/K(X))) = 0$, that is, $\mathrm{ld}_{[1]}(M/K(X))$ is a precompact set. □

15.2.3 Hausdorff radius equals essential radius

Theorem 15.5. *Let M be a precompact subset of $B(X)$. Then*

$$\rho_e(M) = \rho_\chi(M). \tag{15.8}$$

Proof. Let $\rho_e(M) = 1$. Assume, aiming at the contrary, that $\rho_\chi(M) < 1$. We consider two cases.

Case 1. $\mathrm{SG}(M/K(X))$ *is bounded.*

By Proposition 15.3, $\mathrm{SG}(M/K(X))$ is precompact. As $\rho(M/K(X)) = 1$, then $\mathrm{LIM}(M/K(X))$ has a nonzero idempotent by [18, Corollary 6.12]. On the other hand, let $T \in \mathrm{SG}(M)$ be an arbitrary operator such that $T/K(X) \in \mathrm{LIM}(M/K(X))$. Then there is a sequence (T_k) with $T_k/K(X) \in (M/K(X))^{n_k}$ for $n_k \to \infty$ and $T_k/K(X) \to T/K(X)$ under $k \to \infty$. Hence $\|T_k - T\|_\chi \to 0$ under $k \to \infty$. As $q := \rho_\chi(M) < 1$ then $\|T_k\|_\chi \le q^{n_k} \to 0$ under $n_k \to \infty$. So T is a compact operator. Hence $\mathrm{LIM}(M/K(X)) = (0)$, a contradiction. This shows that $\rho_\chi(M) = \rho_e(M)$ holds in Case 1.

Case 2. $\mathrm{SG}(M/K(X))$ *is not bounded.*

It follows easily from definition that in this case there exists a leading sequence for $M/K(X)$. Let $(T_k/K(X))_{k=1}^\infty$ be such a sequence. For brevity, set $a_k = T_k/K(X)$ for each k. Then

$$G := \{a_k/\|a_k\| : k \in \mathbb{N}\} \subseteq \mathrm{ld}_{[1]}(M/K(X)).$$

Let $\rho_\chi(M) = t_1 < 1$ and $t_1 < t_2 < 1$. It follows from the condition $\rho_e(M) = 1$ that, for any $\varepsilon > 0$ with $t_2(1+\varepsilon) < 1$, there is $n_1 > 0$ such that $\|M^n/K(X)\|^{1/n} < 1+\varepsilon$ for all $n > n_1$, and also there is $n_2 > 0$ such that $\|M^n\|_\chi^{1/n} < t_2$. Then

$$\|M^n\|_\chi \|M^n/K(X)\| < (t_2(1+\varepsilon))^n < 1$$

for any $n > \max\{n_1, n_2\}$. By Lemma 15.4, G is precompact. Let $b := S/K(X)$ be a limit point of G. One may assume that

$$\|b - a_k/\|a_k\|\| \to 0 \quad \text{under } k \to \infty.$$

It is clear that $\|b\| = 1$. We have

$$\begin{aligned} \|S\|_\chi &\le \|S - T_k/\|a_k\|\|_\chi + \|T_k/\|a_k\|\|_\chi \\ &\le \|b - a_k/\|a_k\|\| + \|T_k\|_\chi/\|a_k\|. \end{aligned} \tag{15.9}$$

As $\rho_\chi(M) < 1$, $\{\|T_k\|_\chi : k \in \mathbb{N}\}$ is a bounded set. As $\|a_k\| \longrightarrow_k \infty$, we get $\|T_k\|_\chi/\|a_k\| \to_k 0$. We obtain from (15.9) that $\|S\|_\chi = 0$, that is, S is a compact operator. Hence immediately $b = 0$, a contradiction. Thus $\rho_e(M) = \rho_\chi(M)$ in any case. □

15.3 Banach-algebraic approach to the joint spectral radius formulas

15.3.1 *BW*-ideals

Now we present the Banach-algebraic approach to the formulas for the joint spectral radius. Let us consider a Banach algebra A instead of $B(X)$. Let $\mathrm{BW}(A)$ denote the set of all closed ideals J of A such that

$$\rho(M) = \max\{\rho(M/J), r(M)\} \quad \text{for all precompact } M \subseteq A. \tag{15.10}$$

The ideals J for which (15.10) holds are called BW-*ideals*. Clearly, if $I \subset J$, $J \in \mathrm{BW}(A)$ then $I \in \mathrm{BW}(A)$. It is known that $\mathrm{BW}(A)$ has maximal elements; moreover, it was proved in [24, Lemma 5.2] that if $J = \overline{\cup J_\lambda}$ where (J_λ) is a linearly ordered set of BW-ideals of A then $J \in \mathrm{BW}(A)$.

Let us call an increasing transfinite sequence $(J_\alpha)_{\alpha\le\gamma}$ of closed ideals in a Banach algebra A an *increasing transfinite chain of closed ideals* if $J_\beta = \overline{\bigcup_{\alpha<\beta} J_\alpha}$ for any limit ordinal $\beta \le \gamma$, and a decreasing transfinite sequence $(I_\alpha)_{\alpha\le\gamma}$ – a *decreasing transfinite chain of closed ideals* if $I_\beta = \cap_{\alpha<\beta} I_\alpha$ for any limit ordinal $\beta \le \gamma$.

By [24], if $I \subset J$ are closed ideals of A, $I \in \mathrm{BW}(A)$ and $J/I \in \mathrm{BW}(A/I)$ then $J \in \mathrm{BW}(A)$. This implies the transfinite stability for BW-ideals.

Proposition 15.6. *If in increasing transfinite chain $(J_\alpha)_{\alpha\le\gamma}$ of closed ideals in a Banach algebra A, the ideal J_0 belongs to BW(A) and $J_{\alpha+1}/J_\alpha \in BW(A/J_\alpha)$, for all α, then $J_\gamma \in BW(A)$.*

Every BW-ideal J of a Banach algebra turns out to be a *Berger–Wang algebra* in the sense that the equality

$$\rho(M) = r(M) \tag{15.11}$$

holds for any precompact set M of J. It follows from (15.11) and [28, Proposition 3.5] that *every semigroup consisting of quasinilpotent elements of a Berger–Wang algebra generates a finitely quasinilpotent subalgebra.*

Since the Jacobson radical $\mathrm{Rad}(A)$ of every Banach algebra A consists of quasinilpotents, then for a Berger–Wang Banach algebra A, $\mathrm{Rad}(A)$ *is compactly quasinilpotent, that is,* $\rho(M) = 0$ for any precompact set M of $\mathrm{Rad}(A)$.

15.3.2 First Banach-algebraic formulas for the joint spectral radius

A natural analogue of compact operators in the Banach algebra context was proposed by K. Vala [30] who proved that the map $T \mapsto S_1TS_2$ on the algebra $B(X)$ is compact if and only the operators S_1 and S_2 are compact. So an element a of a normed algebra A is called *compact (finite rank)* if the operator $\mathrm{L}_a\mathrm{R}_a : x \mapsto axa$ on A is compact (finite rank). A normed algebra A is called *bicompact* if all operators $\mathrm{L}_a\mathrm{R}_b : x \mapsto axb$ $(a, b \in A)$ are compact. An ideal of A is called *bicompact* if it is bicompact as a normed algebra.

It follows from [24, Corollary 4.8] that for every bicompact ideal J of A the equality (15.10) holds. Since, by [30], $K(X)$ is a bicompact ideal of $B(X)$, this result widely extends the generalized BW-formula (15.4) (which is the same as (15.3) by virtue of Theorem 15.5).

A normed algebra A is called *hypocompact (hypofinite)* if every nonzero quotient A/J has a nonzero compact (finite rank) element. An ideal is *hypocompact (hypofinite)* if it is hypocompact (hypofinite) as a normed algebra.

Each bicompact algebra is hypocompact, and any hypocompact ideal can be composed from bicompact blocks:

Proposition 15.7 ([24, Proposition 3.8]). *For any hypocompact closed ideal I of a Banach algebra A, there is a transfinite increasing chain $(J_\alpha)_{\alpha\le\gamma}$ of closed ideals of A such that $J_1 = (0)$ and $J_\gamma = I$, and every quotient space $J_{\alpha+1}/J_\alpha$ is a bicompact ideal of A/J_α.*

Theorem 15.8 ([24, Theorem 4.11]). *The formula* (15.10) *holds for every hypocompact closed ideal J of A.*

Indeed, as every closed bicompact ideal of a Banach algebra A is a BW-ideal, the result follows from Propositions 15.6 and 15.7.

A Banach algebra is called *scattered* if its elements have countable spectra. It follows from [25, Theorem 8.15] that every hypocompact algebra is scattered.

15.3.3 Compact quasinilpotence, and coincidence of essential and f-essential joint spectral radii

Recall that a Banach algebra A is *compactly quasinilpotent* if $\rho(M) = 0$ for any precompact subset M of A.

The following result shows that any compactly quasinilpotent ideal can be considered as inessential when one calculates the joint spectral radius.

Theorem 15.9 ([22, Theorem 4.18]). $\rho(M) = \rho(M/J)$ *for each compactly quasinilpotent ideal and precompact set* $M \subset A$.

In particular, all compactly quasinilpotent ideals are BW-ideals.

Theorem 15.10 ([24, Theorem 3.14]). *If a Banach algebra A is hypocompact and consists of quasinilpotents, then it is compactly quasinilpotent.*

The following result shows that the reverse inclusion fails.

Proposition 15.11. *There are compactly quasinilpotent Banach algebras without nonzero hypocompact ideals.*

Proof. Let V be the algebra $\ell^1(w)$, where the weight $w = (w_k)_{k=1}^{\infty}$ satisfies the condition

$$\lim(w_{k+1}/w_k) = 0 \tag{15.12}$$

(for instance, one can take $w_k = 1/k^k$). It follows easily from (15.12) that such a weight is radical, that is, $\lim_{k\to\infty} w_k^{1/k} = 0$. Therefore, all elements of V are quasinilpotent.

Let A be the projective tensor product $V\hat{\otimes}B$ of V and any commutative Banach algebra B without nonzero compact elements (for instance, one may take for B the algebra $C[0,1]$ of continuous functions on $[0,1] \subseteq \mathbb{R}$).

Let us write elements $v \in V$ as

$$v = \sum_{k=1}^{\infty} \lambda_k e_k, \tag{15.13}$$

where e_k is the sequence $(\alpha_1, \alpha_2, \ldots)$ with $\alpha_i = 1$ if $i = k$, and 0 otherwise. It follows that $e_k \neq 0$ for all k, so that V has no nonzero nilpotents. Indeed, if $v^m = 0$ and λ_n is the first nonzero coefficient in the expansion (15.13) then clearly $e_{mn} = 0$, a contradiction.

To see that the algebra V is compact, note that the set of all compact elements in any Banach algebra is closed and with each element contains the algebra generated by it. So it suffices to show that the element e_1 is compact, because V is topologically generated by e_1.

Let $V_{\odot}$ be the unit ball of V:

$$V_{\odot} = \left\{ \sum_{k=1}^{\infty} \lambda_k e_k : \sum_{k=1}^{\infty} |\lambda_k| w_k \leq 1 \right\}.$$

We are going to show that L_{e_1} is a compact operator. For this, it suffices to show that, for each $\varepsilon > 0$, the set $e_1 V_\odot$ contains a finite ε-net.

For each n, let P_n be the natural projection on the linear span of $\{e_1, \ldots, e_n\}$, and let $K_n = P_n V_\odot$ and $K_n^\perp = (1 - P_n)V_\odot$. Then

$$e_1 V_\odot \subset e_1 K_n + e_1 K_n^\perp.$$

The set $e_1 K_n$ is compact for each n, so in any case it contains a finite $(\varepsilon/2)$-net. Now it suffices to show that $\|e_1 a\| \leq \varepsilon/2$ for each $a \in K_n^\perp$ if n is sufficiently large.

By (15.12), there is n such that $w_{k+1} < \varepsilon w_k/2$ for all $k \geq n$. Then for each $a = \sum_{k=n+1}^{\infty} \lambda_k e_k \in K_n^\perp$, we have

$$\begin{aligned}\|e_1 a\| &= \left\| \sum_{k=n+1}^{\infty} \lambda_k e_{k+1} \right\| = \sum_{k=n+1}^{\infty} |\lambda_k| w_{k+1} < \varepsilon \left(\sum_{k=n+1}^{\infty} |\lambda_k| w_k \right)/2 = \varepsilon\|a\|/2 \\ &\leq \varepsilon/2.\end{aligned}$$

Thus $e_1 V_\odot$ contains a finite ε-net for every $\varepsilon > 0$, that is, $e_1 V_\odot$ is a compact set, whence V is a compact algebra consisting of quasinilpotents.

Applying Theorem 15.10, we see that the algebra V is compactly quasinilpotent. By [22, Theorem 4.29], the same is true for the tensor product of V and any Banach algebra. Thus the algebra A is compactly quasinilpotent. It remains to show that A is not hypocompact. Each element of A has the form

$$a = \sum_{k=1}^{\infty} e_k \otimes b_k, \quad \text{where } \sum_{k=1}^{\infty} \|b_k\| w_k < \infty.$$

Suppose that an element $c \in A$ is compact. Since A is commutative, the operator $\mathrm{L}_{c^2} = \mathrm{L}_c \mathrm{R}_c$ is compact. Thus setting $a = c^2$ we have that the set $aA_\odot$ is precompact. Let $B_\odot$ be the unit ball of B. If $aA_\odot$ is a precompact subset of A then, in particular, the set

$$\{a(e_1 \otimes b) : b \in B_\odot\} = \left\{ \sum_{k=1}^{\infty} e_{k+1} \otimes b_k b : b \in B_\odot \right\}$$

is precompact. In particular, all sets

$$E_k := \{e_{k+1} \otimes b_k b : b \in B_\odot\}$$

are precompact because the natural projection of $V \hat{\otimes} B$ onto the subspace $e_j \otimes B$ is bounded. Each set E_k is homeomorphic to $b_k B_\odot$ whence b_k is a compact element of B. Since B has no nonzero compact elements, $b_k = 0$ for any $k > 0$, whence $a = 0$, that is, $c^2 = 0$.

Let us show that $c = 0$. Indeed, if $c \neq 0$ let

$$c = \sum_{k=1}^{\infty} e_k \otimes d_k,$$

and let d_m be the first nonzero element among all d_k. Then

$$0 = c^2 = \sum_{k=2m}^{\infty} e_k \otimes \sum_{i+j=k} d_i d_j \quad \text{whence } d_m^2 = 0.$$

Therefore, d_m is a compact element of B (since B is commutative, $d_m b d_m = d_m^2 b = 0$, for all $b \in B$). Since B has no nonzero compact elements, $d_m = 0$, a contradiction.

We proved that A has no nonzero compact elements. It follows that A has no bicompact and hypocompact ideals. □

Theorem 15.12. *Let A be a Banach algebra. Then there are the largest hypocompact ideal $\mathcal{R}_{\mathrm{hc}}(A)$, the largest hypofinite ideal $\mathcal{R}_{\mathrm{hf}}(A)$, the largest compactly quasinilpotent ideal $\mathcal{R}_{\mathrm{cq}}(A)$ and the largest scattered ideal $\mathcal{R}_{\mathrm{sc}}(A)$.*

For the proofs, see [24, Corollary 3.10], [22, Theorem 4.18], and [25, Theorem 8.10]. Now we return to our initial problem.

Theorem 15.13. *Let M be a precompact subset of $B(X)$. Then*

$$\rho_f(M) = \rho_e(M). \tag{15.14}$$

Proof. As $K(X)$ is a bicompact algebra by [30], the algebra $K(X)/\overline{F(X)}$ is also bicompact. As spectral projections of compact operators are in $F(X)$, it is easy to see that $K(X)/\overline{F(X)}$ consists of quasinilpotents. Then it is compactly quasinilpotent by Theorem 15.10. Therefore, $K(X)/\overline{F(X)}$ is a compactly quasinilpotent ideal of $B(X)/\overline{F(X)}$. Using Theorem 15.9 applied to $J = K(X)/\overline{F(X)}$, we have that

$$\begin{aligned}\rho_f(M) &= \rho(M/\overline{F(X)}) = \rho((M/\overline{F(X)})/(K(X)/\overline{F(X)})) \\ &= \rho(M/K(X)) = \rho_e(M)\end{aligned}$$

for a precompact subset M of $B(X)$. □

15.3.4 The largest BW-ideal problem and topological radicals

Let A be a Banach algebra. As it was already noted, the set of all BW-ideals has maximal elements. However, it is not known whether $\overline{I+J} \in \mathrm{BW}(A)$ if $I, J \in \mathrm{BW}(A)$. So the problem of existence of the largest BW-ideal is open.

On the other hand, the largest BW-ideal problem disappears if one only considers ideals defined by some natural properties — as, for example, the ideals $\mathcal{R}_{\mathrm{hc}}(A)$, $\mathcal{R}_{\mathrm{hf}}(A)$, $\mathcal{R}_{\mathrm{cq}}(A)$ and $\mathcal{R}_{\mathrm{sc}}(A)$ defined in Theorem 15.12. To formulate this precisely, we turn to the theory of topological radicals. We recall some definitions and results of this theory; a reader can refer to the works [6, 22, 10, 23, 11, 24, 25, 5] for additional information.

In what follows, the term *ideal* will mean *a two-sided ideal*. In general, radicals can be defined on classes of rings and algebras; the topological radicals are defined on classes of normed algebras. A radical is an *ideal map*, that is, a map that assigns to each algebra its ideal, while a topological radical is a *closed ideal map*, it assigns to a normed algebra its closed ideal. In correspondence with our subject here, we restrict our attention to the class of all Banach algebras.

We begin with the most important and convenient class of topological radicals. A *hereditary topological radical* on the class of all Banach algebras is a closed ideal map $\mathcal{P}$ which assigns to each Banach algebra A a closed two-sided ideal $\mathcal{P}(A)$ of A and satisfies the following conditions:

(H1) $f(\mathcal{P}(A)) \subset \mathcal{P}(B)$ for a continuous surjective homomorphism $f: A \longrightarrow B$;
(H2) $\mathcal{P}(A/\mathcal{P}(A)) = (0)$;
(H3) $\mathcal{P}(J) = J \cap \mathcal{P}(A)$ for any ideal J of A.

It can be seen from (H2) that every radical $\mathcal{P}$ accumulates some special property in the ideal $\mathcal{P}(A)$ of an algebra A which is called *the $\mathcal{P}$-radical* of A.

For the proof of the following theorem, see [22, Theorem 4.25], [23, Theorems 3.58, and 3.59], [25, Section 8].

Theorem 15.14. *The maps $\mathcal{R}_{\mathrm{cq}}: A \longmapsto \mathcal{R}_{\mathrm{cq}}(A)$, $\mathcal{R}_{\mathrm{hc}}: A \longmapsto \mathcal{R}_{\mathrm{hc}}(A)$, $\mathcal{R}_{\mathrm{hf}}: A \longmapsto \mathcal{R}_{\mathrm{hf}}(A)$ and $\mathcal{R}_{\mathrm{sc}}: A \longmapsto \mathcal{R}_{\mathrm{sc}}(A)$ are hereditary topological radicals.*

The maps $\mathcal{R}_{\mathrm{hc}}$, $\mathcal{R}_{\mathrm{hf}}$, $\mathcal{R}_{\mathrm{cq}}$, and $\mathcal{R}_{\mathrm{sc}}$ are called the *hypocompact, hypofinite, compactly quasinilpotent*, and *scattered radical,* respectively.

It follows immediately from Axiom (H3) that hereditary radicals satisfy the conditions:

(I1) $\mathcal{P}(\mathcal{P}(A)) = \mathcal{P}(A)$;
(I2) $\mathcal{P}(J)$ of an ideal J of A is an ideal of A which is contained in the radical $\mathcal{P}(A)$.

If a closed ideal map $\mathcal{P}$ on the class of all Banach algebras satisfies (H1), (H2) and, instead of (H3), also (I1) and (I2) then $\mathcal{P}$ is called a *topological radical* (see [6]).

If an ideal map [a closed ideal map] $\mathcal{P}$ satisfies (H1), it is called a *preradical* [a *topological preradical*].

A closed ideal map $\mathcal{P}$ is called an *under topological radical* (UTR) if it satisfies all axioms of topological radicals, besides possibly of (H2), and an *over topological radical* (OTR) if it satisfies all axioms, apart from possibly of (I1) (see [6, Definition 6.2]).

Given a preradical $\mathcal{P}$, an algebra A is called *$\mathcal{P}$-radical* if $A = \mathcal{P}(A)$, and *$\mathcal{P}$-semisimple* if $\mathcal{P}(A) = 0$. It follows from (H1) for a topological preradical $\mathcal{P}$ that $\mathcal{P}$-radical and $\mathcal{P}$-semisimple algebras are invariant with respect to topological isomorphisms.

Let $\mathcal{P}$ be a topological radical. It follows easily from the definition that quotients of $\mathcal{P}$-radical algebras are $\mathcal{P}$-radical, and ideals of $\mathcal{P}$-semisimple algebras are $\mathcal{P}$-semisimple. Moreover, the class of all $\mathcal{P}$-radical ($\mathcal{P}$-semisimple) algebras is stable with respect to extensions: *If J is a $\mathcal{P}$-radical* ($\mathcal{P}$-semisimple) *ideal* of A *and the quotient A/J is $\mathcal{P}$-radical* ($\mathcal{P}$-semisimple) *then A itself is also $\mathcal{P}$-radical* ($\mathcal{P}$-semisimple).

The proof of the following properties *of transfinite stability* can be found in [25, Theorem 4.18].

Proposition 15.15. *Let $\mathcal{P}$ be a topological radical, A a Banach algebra, and let $(I_\alpha)_{\alpha\le\gamma}$ and $(J_\alpha)_{\alpha\le\gamma}$ be decreasing and increasing transfinite chains of closed ideals of A. Then*
(1) *If A/I_1 and all quotients $I_\alpha/I_{\alpha+1}$ are $\mathcal{P}$-semisimple then A/I_γ is $\mathcal{P}$-semisimple;*
(2) *If J_1 and all quotients $J_{\alpha+1}/J_\alpha$ are $\mathcal{P}$-radical then J_γ is $\mathcal{P}$-radical.*

15.4 Around joint spectral radius formulas and radicals

15.4.1 Comparison of joint spectral radius formulas

It follows from Theorem 15.8 that for any Banach algebra A and precompact set $M \subseteq A$, the equality

$$\rho(M) = \max\{\rho(M/\mathcal{R}_{\mathrm{hc}}(A)), r(M)\} \tag{15.15}$$

holds. Since $\mathcal{R}_{\mathrm{hf}}(A) \subseteq \mathcal{R}_{\mathrm{hc}}(A)$, we certainly have

$$\rho(M) = \max\{\rho(M/\mathcal{R}_{\mathrm{hf}}(A)), r(M)\}, \tag{15.16}$$

for any precompact set in A. Obviously $\overline{F(X)} \subseteq \mathcal{R}_{\mathrm{hf}}(B(X)) \subseteq \mathcal{R}_{\mathrm{hc}}(B(X))$, so the inequalities

$$\rho(M/\mathcal{R}_{\mathrm{hc}}(B(X))) \le \rho(M/\mathcal{R}_{\mathrm{hf}}(B(X))) \le \rho_f(M) = \rho_e(M) \tag{15.17}$$

are always true for all precompact $M \subset B(X)$; recall that $\rho_f(M) = \rho_e(M)$ by Theorem 15.13.

The inequality $\rho(M/\mathcal{R}_{\mathrm{hc}}(B(X))) \le \rho_f(M)$ in (15.17) can be strict. For example, if X is an Argyros–Haydon space then $\rho(M/\mathcal{R}_{\mathrm{hc}}(B(X))) = 0$ for each precompact set $M \subseteq B(X)$ while $\rho_f(M)$ can be nonzero by virtue of semisimplicity $K(X)$. This shows that *even in the operator case the joint spectral radius formula* (15.15) *is stronger than the generalized BW-formula* (15.3).

In general, the inequality $\rho(M/\mathcal{R}_{\mathrm{hc}}(A)) \le \rho(M/\mathcal{R}_{\mathrm{hf}}(A))$ can be also strict. To see this, let V be the radical compact Banach algebra $\ell_1(w)$ considered in Proposition 15.11. As we saw, V has no non-zero nilpotent elements. Therefore, the only finite-rank element v in V is 0. Indeed, the multiplication operator $\mathrm{L}_v\mathrm{R}_v = \mathrm{L}_{v^2}$ is quasinilpotent. So if it has finite rank then it is nilpotent:

$$\mathrm{L}_{v^2}^m = 0.$$

Applying the operator $\mathrm{L}^m_{v^2}$ to v, we have that $v^{2m+1} = 0$, that is, v is nilpotent, whence $v = 0$.

Let now A be the unitization of V. Since all finite-rank elements of A must lie in V, it follows from the above that

$$\mathcal{R}_{\mathrm{hf}}(V) = \mathcal{R}_{\mathrm{hf}}(A) = (0).$$

On the other hand, A is hypocompact, whence $A = \mathcal{R}_{\mathrm{hc}}(A)$. For $M = \{1\}$ we have that

$$\rho(M/\mathcal{R}_{\mathrm{hc}}(A)) = 0 \neq 1 = \rho(M/\mathcal{R}_{\mathrm{hf}}(A)).$$

As usual, in the class of all C*-algebras the situation is simpler.

Theorem 15.16. *If A is a C*-algebra, then $\mathcal{R}_{\mathrm{hf}}(A) = \mathcal{R}_{\mathrm{hc}}(A)$.*

Proof. Indeed, if an element a of A is compact then a^*a is compact and therefore its spectral projections are finite-rank elements and, therefore, belong to $\mathcal{R}_{\mathrm{hf}}(A)$. Since a^*a is a limit of linear combinations of its spectral projections we have that

$$a^*a \in \mathcal{R}_{\mathrm{hf}}(A).$$

But it is known (see, e. g., [15, Proposition 1.4.5]) that the closed ideal generated by a^*a contains a. Thus $\mathcal{R}_{\mathrm{hf}}(A)$ contains all compact elements of A. Now let $(J_\alpha)_{\alpha\le\gamma}$ be an increasing transfinite chain of closed ideals with bicompact quotients, and $J_\gamma = \mathcal{R}_{\mathrm{hc}}(A)$. Assume by induction that J_α are contained in $\mathcal{R}_{\mathrm{hf}}(A)$ for all $\alpha < \lambda$. If the ordinal λ is limit, then clearly J_λ is also contained in $\mathcal{R}_{\mathrm{hf}}(A)$. Otherwise, we have that $\lambda = \beta + 1$ for some β. If $a \in J_\lambda$, then a/J_β is a compact element of A/J_β whence $a/J_\beta \in \mathcal{R}_{\mathrm{hf}}(A/J_\beta)$ and $a \in \mathcal{R}_{\mathrm{hf}}(A)$. Therefore, by induction, $\mathcal{R}_{\mathrm{hc}}(A) = J_\gamma = \mathcal{R}_{\mathrm{hf}}(A)$. □

The class of hypocompact C^*-algebras is contained in the class of all GCR algebras (algebras of type I) and this inclusion is strict: it suffices to note that even the algebra $C([0,1])$ is not hypocompact. Moreover, there is an analogue of (15.3) that holds for all C^*-algebras A satisfying some natural restrictions on the space $\mathrm{Prim}(A)$ of all primitive ideals of A:

$$\rho(M) = \max\{\rho(M/\mathcal{R}_{\mathrm{gcr}}(A)), r(M)\}, \tag{15.18}$$

where $\mathcal{R}_{\mathrm{gcr}}(A)$ is the largest GCR ideal of A. The map $A \longmapsto \mathcal{R}_{\mathrm{gcr}}(A)$ is a hereditary topological radical on the class of all C*-algebras. It follows from (15.18) that any GCR-algebra is a Berger–Wang algebra. The proof and more information can be found in [25, Section 10].

Apart from (15.15), another version of the joint spectral radius formula was established in [26]:

$$\rho(M) = \max\{\rho^\chi(M), r(M)\} \tag{15.19}$$

holds for every precompact set M in A, where $\rho^{\chi}(M)$ is defined as $\rho_{\chi}(\mathrm{L}_M\mathrm{R}_M)^{1/2}$. Unlike $\rho(M/\mathcal{R}_{\mathrm{hc}}(A))$ and $\rho(M/\mathcal{R}_{\mathrm{hf}}(A))$, the value $\rho_{\chi}(M)$ is not of the form $\rho(M/J)$, but it deserves some interest because it is natural to regard $\rho^{\chi}(M)$ as a Banach algebraic analogue of $\rho_{\chi}(M)$. By Theorem 15.5 and [24, Lemma 4.7],

$$\rho^{\chi}(M) = \rho_{\chi}(\mathrm{L}_M\mathrm{R}_M)^{1/2} = \rho_e(\mathrm{L}_M\mathrm{R}_M)^{1/2} \le \rho(M/J)^{1/2}\rho(M)^{1/2} \tag{15.20}$$

for every precompact set M in A and every bicompact ideal J of A.

In general, $\rho^{\chi}(M) \neq \rho(M/\mathcal{R}_{\mathrm{hc}}(A))$. Indeed, if X is an Argyros–Haydon space [1] then $B(X)$ is a one-dimensional extension of $K(X)$. So the algebra $B(X)$ is hypocompact and $\rho(M/\mathcal{R}_{\mathrm{hc}}(B(X))) = 0$ for each precompact set $M \subseteq B(X)$. On the other hand, for $M = \{1\}$, we see that $\mathrm{L}_M\mathrm{R}_M$ is the identity operator on the infinite-dimensional space $B(X)$, whence $\rho_{\chi}(\mathrm{L}_M\mathrm{R}_M) = 1$ and $\rho^{\chi}(M) = 1$.

15.4.2 *BW*-radicals

In line with the above discussion, we are looking for such radicals $\mathcal{P}$ that $\mathcal{P}(A)$ is a BW-ideal for each A; it is natural to call them BW-*radicals*. Clearly, we are interested in "large" BW-radicals, so that we have to compare them.

The order for ideal maps, in particular for topological radicals, is introduced in the usual way: $\mathcal{P} \le \mathcal{R}$ means that $\mathcal{P}(A) \subseteq \mathcal{R}(A)$ for every algebra A. For instance, it is obvious that

$$\mathcal{R}_{\mathrm{hf}} \le \mathcal{R}_{\mathrm{hc}} \quad \text{and} \quad \mathcal{R}_{\mathrm{cq}} \le \mathrm{Rad},$$

where Rad is the *Jacobson radical* $A \longmapsto \mathrm{Rad}(A)$ (recall that for a Banach algebra A, $\mathrm{Rad}(A)$ can be defined as the largest ideal of A consisting of quasinilpotents). It is well known that Rad is hereditary.

As usual, we write $\mathcal{P} < \mathcal{R}$ if $\mathcal{P} \le \mathcal{R}$ and there is an algebra A such that $\mathcal{P}(A) \neq \mathcal{R}(A)$. For example,

$$\mathcal{R}_{\mathrm{hf}} < \mathcal{R}_{\mathrm{hc}} < \mathcal{R}_{\mathrm{sc}} \quad \text{and} \quad \mathrm{Rad} < \mathcal{R}_{\mathrm{sc}}.$$

It is known that, for any family $\mathcal{F}$ of topological radicals, there exists the smallest upper bound $\vee\mathcal{F}$ and the largest lower bound $\wedge\mathcal{F}$ of $\mathcal{F}$ in the class of all topological radicals; clearly, $\vee\mathcal{F}$ and $\wedge\mathcal{F}$ need not belong to $\mathcal{F}$ itself. If $\mathcal{F} = \{\mathcal{P}, \mathcal{R}\}$, we write $\mathcal{P} \vee \mathcal{R}$ for $\vee\mathcal{F}$ and $\mathcal{P} \wedge \mathcal{R}$ for $\wedge\mathcal{F}$. We will describe later a constructive way for obtaining the radicals $\vee\mathcal{F}$ and $\wedge\mathcal{F}$.

The following theorem establishes that there is the largest BW-radical.

Theorem 15.17 ([24, Theorem 5.9]). *Let $\mathcal{F}$ be the family of all BW-radicals and $\mathcal{R}_{\mathrm{bw}} = \vee\mathcal{F}$. Then $\mathcal{R}_{\mathrm{bw}}$ is a BW-radical; any topological radical $\mathcal{P} \le \mathcal{R}_{\mathrm{bw}}$ is a BW-radical.*

The proof uses the structure of radical ideals in $\vee\mathcal{F}$, and transfinite stability of the class of BW-ideals (see Proposition 15.6).

To show the utility of $\mathcal{R}_{\mathrm{bw}}$, consider the following example. It follows from Theorems 15.8 and 15.9 that $\mathcal{R}_{\mathrm{hc}}$ and $\mathcal{R}_{\mathrm{cq}}$ are BW-radicals. So, for any Banach algebra A, $\mathcal{R}_{\mathrm{hc}}(A)$ and $\mathcal{R}_{\mathrm{cq}}(A)$ are BW-ideals. They can differ; moreover, it can be deduced from Proposition 15.11 that there is a Banach algebra A such that $\mathcal{R}_{\mathrm{hc}}(A)$ and $\mathcal{R}_{\mathrm{cq}}(A)$ are both nonzero, but have zero intersection. The existence of $\mathcal{R}_{\mathrm{bw}}$ implies that $\overline{\mathcal{R}_{\mathrm{hc}}(A) + \mathcal{R}_{\mathrm{cq}}(A)}$ is a BW-ideal, because both summands are contained in $\mathcal{R}_{\mathrm{bw}}(A)$. Now one can further extend this BW-ideal by building an increasing transfinite chain (J_α) of closed ideals such that

$$J_0 = (0) \quad \text{and} \quad J_{\alpha+1}/J_\alpha = \overline{\mathcal{R}_{\mathrm{hc}}(A/J_\alpha) + \mathcal{R}_{\mathrm{cq}}(A/_\alpha)} \quad \text{for all } \alpha.$$

In the correspondence with Proposition 15.6 we conclude that all J_α are BW-ideals. It is obvious that there is an ordinal γ such that $J_{\gamma+1} = J_\gamma$. It turns out that $J_\gamma = (\mathcal{R}_{\mathrm{hc}} \vee \mathcal{R}_{\mathrm{cq}})(A)$. To see it and much more, we consider the details of a construction of radicals $\vee\mathcal{F}$ and $\wedge\mathcal{F}$ in the following subsection. Of course, we have that $\mathcal{R}_{\mathrm{hc}} \vee \mathcal{R}_{\mathrm{cq}} \le \mathcal{R}_{\mathrm{bw}}$, so that the formula

$$\rho(M) = \max\{\rho(M/(\mathcal{R}_{\mathrm{hc}} \vee \mathcal{R}_{\mathrm{cq}})(A)), r(M)\} \tag{15.21}$$

is valid, for any precompact set M in A.

It seems that in the Banach algebra context the best candidate for the joint spectral radius formula is

$$\rho(M) = \max\{\rho(M/\mathcal{R}_{\mathrm{bw}}(A)), r(M)\}. \tag{15.22}$$

But a priori there can exist a Banach algebra A with nontrivial BW-ideals and with $\mathcal{R}_{\mathrm{bw}}(A) = 0$ — the disadvantage of formula (15.22) is that the largest BW-radical is defined not directly, since the family of BW-radicals is not completely described. However, in radical context the formula (15.22) is certainly optimal. In particular, it is stronger than formula (15.15) because the largest BW-radical contains the hypocompact radical for any Banach algebra, and the inclusion can be strict as the above example shows.

In what follows, we gather some facts for the better understanding of the nature of the radical $\mathcal{R}_{\mathrm{bw}}$.

15.4.3 Procedures and operations

Here, we describe some ways to construct radicals from preradicals that only partially satisfy the axioms.

Procedures are mappings from one class of ideal maps to another class of ideal maps. The important examples are the following. If $\mathcal{P}$ and $\mathcal{R}$ are topological preradicals satisfying (I1) and (I2), for any algebra A, let $(I_\alpha)_{\alpha\le\gamma}$ and $(J_\alpha)_{\alpha\le\delta}$ be transfinite chains such that

$$J_\alpha = A,\quad J_{\alpha+1} = \mathcal{P}(J_\alpha);\quad I_0 = (0),\quad I_{\alpha+1} = q_{I_\alpha}^{-1}(\mathcal{R}(A/I_\alpha)), \tag{15.23}$$

where $q_{I_\alpha} : A \longrightarrow A/I_\alpha$ is the standard quotient map. Then the maps $\mathcal{P}_{(\alpha)^\circ}: A \longmapsto J_\alpha$ and $\mathcal{R}_{(\alpha)^*}: A \longmapsto I_\alpha$ are topological preradicals satisfying (I1) and (I2). So $\mathcal{P} \longmapsto \mathcal{P}_{(\alpha)^\circ}$ and $\mathcal{R} \longmapsto \mathcal{R}_{(\alpha)^*}$ are procedures (α*-superposition* and α*-convoluton procedures*). The transfinite chains of ideals in (15.23) stabilize at some steps $\gamma = \gamma(A)$ and $\delta = \delta(A)$, that is,

$$I_\gamma = I_{\gamma+1} \quad\text{and}\quad J_{\delta+1} = J_\delta.$$

Set $\mathcal{P}^\circ: A \longmapsto J_\delta$ and $\mathcal{R}^*: A \longmapsto I_\gamma$. Then $\mathcal{P} \longmapsto \mathcal{P}^\circ$ and $\mathcal{R} \longmapsto \mathcal{R}^*$ are called *superposition* and *convolution procedures*, respectively; $\mathcal{P}^\circ$ *satisfies* (I1) and $\mathcal{R}^*$ *satisfies* (H2) (see [6, Theorems 6.6 and 6.10]).

The following two ways of getting new ideal maps are very useful in the theory. If $\mathcal{F}$ is a family of UTRs, then

$$\mathsf{H}_\mathcal{F}: A \longmapsto \mathsf{H}_\mathcal{F}(A) := \overline{\sum_{\mathcal{R}\in\mathcal{F}} \mathcal{R}(A)}$$

is a UTR; if $\mathcal{F}$ consists of OTRs, then

$$\mathsf{B}_\mathcal{F}: A \longmapsto \mathsf{B}_\mathcal{F}(A) := \bigcap_{\mathcal{R}\in\mathcal{F}} \mathcal{R}(A)$$

is an OTR (see [25, Theorem 4.1]).

Now we extend the action of operations $\vee$ and $\wedge$ introduced in the preceding subsection. Let $\mathcal{F}$ be a family of topological preradicals satisfying (I1) and (I2). Set

$$\vee\mathcal{F} = (\mathsf{H}_\mathcal{F})^* \quad\text{and}\quad \wedge\mathcal{F} = (\mathsf{B}_\mathcal{F})^\circ. \tag{15.24}$$

Then $\vee\mathcal{F}$ is the smallest OTR larger than or equal to each $\mathcal{P} \in \mathcal{F}$; and $\wedge\mathcal{F}$ is the largest UTR smaller than or equal to each $\mathcal{P} \in \mathcal{F}$. In particular, if $\mathcal{F}$ consists of UTRs then $\vee\mathcal{F}$ is the smallest topological radical that is no less than each $\mathcal{P} \in \mathcal{F}$; if $\mathcal{F}$ consists of OTRs then $\wedge\mathcal{F}$ is the largest topological radical that does not exceed each $\mathcal{P} \in \mathcal{F}$ (see [25, Remark 4.2 and Corollary 4.3]).

Theorem 15.18 ([25, Theorem 8.15]). Rad $\vee\,\mathcal{R}_{\mathrm{hc}} = \mathcal{R}_{\mathrm{sc}}$.

If a family $\mathcal{F}$ consists of hereditary topological radicals, then

$$\wedge\mathcal{F} = \mathsf{B}_\mathcal{F}$$

is the largest hereditary topological radical that does not exceed each $P \in \mathcal{F}$ (see [24, Lemma 3.2]).

As $\mathcal{R}_{\mathrm{hc}}$ and Rad are hereditary topological radicals, then it follows from Theorem 15.10 that $\mathsf{B}_{\{\mathcal{R}_{\mathrm{hc}},\mathrm{Rad}\}}$ coincides with the hereditary topological radical $\mathcal{R}_{\mathrm{hc}} \wedge \mathrm{Rad}$ and that $\mathcal{R}_{\mathrm{hc}} \wedge \mathrm{Rad} \le \mathcal{R}_{\mathrm{cq}}$. Moreover, it follows from Proposition 15.11 that

$$\mathsf{B}_{\{\mathcal{R}_{\mathrm{hc}},\mathrm{Rad}\}} = \mathcal{R}_{\mathrm{hc}} \wedge \mathrm{Rad} < \mathcal{R}_{\mathrm{cq}}. \tag{15.25}$$

15.4.4 Convolution and superposition operations

In this subsection, we prove two useful lemmas.

For an ideal map $\mathcal{P}$ and a closed ideal I of a Banach algebra A, it is convenient to define an ideal $\mathcal{P} * I$ of A by setting

$$\mathcal{P} * I = q_I^{-1}(\mathcal{P}(A/I))$$

where $q_I\colon A \longrightarrow A/I$ is the standard quotient map. Clearly, $I \subseteq \mathcal{P} * I$. If $\mathcal{P}$ and $\mathcal{R}$ are topological preradicals satisfying (I1) and (I2), define the *convolution* $\mathcal{P} * \mathcal{R}$ and *superposition* $\mathcal{P} \circ \mathcal{R}$ by

$$\mathcal{P} * \mathcal{R}(A) = q_{\mathcal{R}}^{-1}(\mathcal{P}(A/\mathcal{R}(A))) \quad \text{and} \quad \mathcal{P} \circ \mathcal{R}(A) = \mathcal{P}(\mathcal{R}(A)) \tag{15.26}$$

for every algebra A, where $q_{\mathcal{R}}\colon A \longrightarrow A/\mathcal{R}(A)$ is the standard quotient map. Then $\mathcal{P} * \mathcal{R}$ and $\mathcal{P} \circ \mathcal{R}$ are topological preradicals satisfying (I1) and (I2) (see [25, Subsection 4.2]); the convolution operation for preradicals is associative (see [25, Lemma 4.10]). If $\mathcal{P}$ and $\mathcal{R}$ are UTRs, then so is $\mathcal{P} * \mathcal{R}$; if $\mathcal{P}$ and $\mathcal{R}$ are OTRs then so is $\mathcal{P} \circ \mathcal{R}$ (see [25, Corollary 4.11]).

We underline that one may define the convolution $\mathcal{P} * \mathcal{R}$ as above if $\mathcal{P}$ is an ideal map and $\mathcal{R}$ is a closed ideal map.

Lemma 15.19. *If $\mathcal{P}$ is a preradical, $\mathcal{R}$ and $\mathcal{S}$ are closed ideal maps and $\mathcal{R} \le \mathcal{S}$, then $\mathcal{P} * \mathcal{R} \le \mathcal{P} * \mathcal{S}$ and $\mathsf{H}_{\{\mathcal{P},\mathcal{R}\}} \le \mathcal{P} * \mathcal{S}$.*

Proof. Let A be a Banach algebra, $J = \mathcal{R}(A)$ and $I = \mathcal{S}(A)$. Let $q_J : A \longrightarrow A/J$, $q_I : A \longrightarrow A/I$ and $q : A/J \longrightarrow A/I$ be the standard quotient maps. Then $q \circ q_J = q_I$ and $q(\mathcal{P}(A/J)) \subseteq \mathcal{P}(A/I)$. Therefore,

$$\mathcal{P} * \mathcal{R}(A) = q_J^{-1}(\mathcal{P}(A/J)) \subseteq q_J^{-1}q^{-1}q(\mathcal{P}(A/J)) \subseteq q_I^{-1}(\mathcal{P}(A/I)) = \mathcal{P} * \mathcal{S}(A).$$

Hence $\mathcal{P} * \mathcal{R} \le \mathcal{P} * \mathcal{S}$.

Further, $\mathcal{R}(A) = J \subseteq I$ and $q_I(\mathcal{P}(A)) \subseteq \mathcal{P}(A/I)$ whence $\mathcal{P}(A) \subseteq q_I^{-1}(\mathcal{P}(A/I))$. Hence

$$\mathsf{H}_{\{\mathcal{P},\mathcal{R}\}}(A) = \overline{\mathcal{P}(A) + \mathcal{R}(A)} \subseteq q_I^{-1}(\mathcal{P}(A/I)) = \mathcal{P} * \mathcal{S}(A),$$

that is, $\mathsf{H}_{\{\mathcal{P},\mathcal{R}\}} \le \mathcal{P} * \mathcal{S}$. □

The implication $\mathcal{P} \le \mathcal{S} \Longrightarrow \mathcal{P} * \mathcal{R} \le \mathcal{S} * \mathcal{R}$ is obvious.

Lemma 15.20. *If $\mathcal{P}$ and $\mathcal{R}$ are UTRs, then the radical $\mathcal{P} \vee \mathcal{R}$ is equal to $(\mathcal{P} * \mathcal{R})^*$; if $\mathcal{P}$ and $\mathcal{R}$ are OTRs then the radical $\mathcal{P} \wedge \mathcal{R}$ is equal to $(\mathcal{P} \circ \mathcal{R})^\circ$.*

Proof. Let $\mathcal{P}$ and $\mathcal{R}$ be UTRs. By Lemma 15.19, $\mathrm{H}_{\{\mathcal{P},\mathcal{R}\}} \le \mathcal{P} * \mathcal{R} \le (\mathcal{P} * \mathcal{R})^*$ whence

$$\mathcal{P} \vee \mathcal{R} = (\mathsf{H}_{\{\mathcal{P},\mathcal{R}\}})^* \le (\mathcal{P} * \mathcal{R})^{**} = (\mathcal{P} * \mathcal{R})^*.$$

On the other hand, $\mathcal{P} * \mathcal{R} \le \mathsf{H}_{\{\mathcal{P},\mathcal{R}\}} * \mathcal{R} \le \mathsf{H}_{\{\mathcal{P},\mathcal{R}\}} * \mathsf{H}_{\{\mathcal{P},\mathcal{R}\}}$ by Lemma 15.19. Therefore,

$$(\mathcal{P} * \mathcal{R})^* \le (\mathsf{H}_{\{\mathcal{P},\mathcal{R}\}} * \mathsf{H}_{\{\mathcal{P},\mathcal{R}\}})^* = ((\mathsf{H}_{\{\mathcal{P},\mathcal{R}\}})_{(2)^*})^* = (\mathsf{H}_{\{\mathcal{P},\mathcal{R}\}})^* = \mathcal{P} \vee \mathcal{R}.$$

Let $\mathcal{P}$ and $\mathcal{R}$ be OTRs. Then $(\mathcal{P} \circ \mathcal{R})^\circ \le \mathcal{P} \circ \mathcal{R} \le \mathsf{B}_{\{\mathcal{P},\mathcal{R}\}}$ whence

$$(\mathcal{P} \circ \mathcal{R})^\circ \le (\mathsf{B}_{\{\mathcal{P},\mathcal{R}\}})^\circ = \mathcal{P} \wedge \mathcal{R}.$$

On the other hand, $\mathsf{B}_{\{\mathcal{P},\mathcal{R}\}} \circ \mathsf{B}_{\{\mathcal{P},\mathcal{R}\}} \le \mathcal{P} \circ \mathsf{B}_{\{\mathcal{P},\mathcal{R}\}} \le \mathcal{P} \circ \mathcal{R}$. Therefore,

$$\mathcal{P} \wedge \mathcal{R} = (\mathsf{B}_{\{\mathcal{P},\mathcal{R}\}})^\circ = (\mathsf{B}_{\{\mathcal{P},\mathcal{R}\}} \circ \mathsf{B}_{\{\mathcal{P},\mathcal{R}\}})^\circ \le (\mathcal{P} \circ \mathcal{R})^\circ.$$

□

15.4.5 Scattered *BW*-radical

Here, we will show that the restriction of $\mathcal{R}_{\mathrm{bw}}$ to the class of scattered algebras is closely related to radicals of somewhat less mysterious nature. Namely it coincides with the topological radical $\mathcal{R}_{\mathrm{hc}} \vee \mathcal{R}_{\mathrm{cq}}$ constructed earlier.

Theorem 15.21. *Let A be a scattered Banach algebra. Then $\mathcal{R}_{\mathrm{bw}}(A) = (\mathcal{R}_{\mathrm{hc}} \vee \mathcal{R}_{\mathrm{cq}})(A)$.*

Proof. Clearly, $\mathcal{R}_{\mathrm{hc}} \vee \mathcal{R}_{\mathrm{cq}} \le \mathcal{R}_{\mathrm{bw}}$. Let $I = (\mathcal{R}_{\mathrm{hc}} \vee \mathcal{R}_{\mathrm{cq}})(A)$, $J = \mathcal{R}_{\mathrm{bw}}(A)$, $B = A/I$ and $K = J/I$. Then B is a scattered, $\mathcal{R}_{\mathrm{hc}} \vee \mathcal{R}_{\mathrm{cq}}$-semisimple algebra and $K \subseteq \mathcal{R}_{\mathrm{bw}}(B)$ is a closed ideal of B. Assume to the contrary that $K \neq (0)$.

As K is a BW-ideal, it is a Berger–Wang algebra. So

$$\mathrm{Rad}(K) = \mathcal{R}_{\mathrm{cq}}(K) = K \cap \mathcal{R}_{\mathrm{cq}}(B) \quad \text{(we used heredity of } \mathcal{R}_{\mathrm{cq}}\text{)}.$$

But $\mathcal{R}_{\mathrm{cq}}(B) \subseteq (\mathcal{R}_{\mathrm{hc}} \vee \mathcal{R}_{\mathrm{cq}})(B) = (0)$. Therefore, $\mathrm{Rad}(K) = (0)$, whence K is a semisimple algebra.

Since B is scattered, K is also scattered. By Barnes' theorem [3] (see also [2, Theorem 5.7.8] with another proof) K has a nonzero socle. Since the socle is generated by finite-rank projections, it is a hypocompact (even hypofinite) ideal and, therefore, is contained in $\mathcal{R}_{\mathrm{hc}}(K)$. Since $\mathcal{R}_{\mathrm{hc}}$ is a hereditary radical, then

$$(0) \neq \mathcal{R}_{\mathrm{hc}}(K) = K \cap \mathcal{R}_{\mathrm{hc}}(B).$$

But $\mathcal{R}_{\mathrm{hc}}(B) = (0)$, a contradiction. Hence $K = (0)$, that is, $J = I$. □

Theorem 15.22. *The radical* $\mathcal{R}_{hc} \vee \mathcal{R}_{cq}$ *is hereditary.*

Proof. Set $\mathcal{P} = \mathcal{R}_{hc} \vee \mathcal{R}_{cq}$. Let A be a Banach algebra and I its closed ideal. If $\mathcal{P}(I) \neq I \cap \mathcal{P}(A)$, let $B = A/\mathcal{P}(I)$, $J = I/\mathcal{P}(I)$ and $K = \mathcal{P}(A)/\mathcal{P}(I)$. Then J is a $\mathcal{P}$-semisimple ideal of B. Therefore,

$$\mathcal{R}_{hc}(J) = \mathcal{R}_{cq}(J) = (0) \tag{15.27}$$

and K is an ideal of B contained in $\mathcal{P}(B)$. Hence $K \in \mathrm{BW}(B)$. As $\mathcal{P}(B) \subseteq \mathcal{R}_{sc}(B)$ and $\mathcal{R}_{sc}$ is hereditary, K is a scattered algebra.

Let $L = J \cap K$. Then L is a nonzero ideal of B. As L is an ideal of K, $L \in \mathrm{BW}(B)$ and L is scattered. As L is an ideal of J, it follows from (15.27) that

$$\mathcal{R}_{hc}(L) = \mathcal{R}_{cq}(L) = (0). \tag{15.28}$$

As L is a Berger–Wang algebra, it follows from (15.11) that $\mathrm{Rad}(L) \subseteq \mathcal{R}_{cq}(L)$. Therefore, L is a semisimple nonzero Banach algebra by (15.28). By Barnes' theorem [3], L has nonzero socle $\mathrm{soc}(L)$, that is, $\mathcal{R}_{hc}(L) \neq (0)$, a contradiction. □

Theorem 15.23. $\mathcal{R}_{hc} \vee \mathcal{R}_{cq} = \mathcal{R}_{bw} \circ \mathcal{R}_{sc} = \mathcal{R}_{bw} \wedge \mathcal{R}_{sc}$.

Proof. Let A be a Banach algebra and $I = \mathcal{R}_{sc}(A)$. Then I is a scattered algebra and $\mathcal{R}_{bw}(I) = (\mathcal{R}_{hc} \vee \mathcal{R}_{cq})(I)$ by Theorem 15.21. As $\mathcal{R}_{hc} \vee \mathcal{R}_{cq} \leq \mathcal{R}_{sc}$ and $\mathcal{R}_{hc} \vee \mathcal{R}_{cq} \leq \mathcal{R}_{bw}$, we obtain that

$$\begin{aligned}(\mathcal{R}_{hc} \vee \mathcal{R}_{cq})(A) &= (\mathcal{R}_{hc} \vee \mathcal{R}_{cq})((\mathcal{R}_{hc} \vee \mathcal{R}_{cq})(A)) \\ &\subseteq (\mathcal{R}_{hc} \vee \mathcal{R}_{cq})(\mathcal{R}_{sc}(A)) \subseteq \mathcal{R}_{bw}(\mathcal{R}_{sc}(A)) \\ &= \mathcal{R}_{bw}(I) = (\mathcal{R}_{hc} \vee \mathcal{R}_{cq})(I) \\ &\subseteq (\mathcal{R}_{hc} \vee \mathcal{R}_{cq})(A),\end{aligned}$$

that is, $\mathcal{R}_{hc} \vee \mathcal{R}_{cq} = \mathcal{R}_{bw} \circ \mathcal{R}_{sc}$.

It is clear that $\mathcal{R}_{bw} \circ \mathcal{R}_{sc} \leq \mathcal{R}_{bw}$ and $\mathcal{R}_{bw} \circ \mathcal{R}_{sc} \leq \mathcal{R}_{sc}$. As $\mathcal{R}_{bw} \circ \mathcal{R}_{sc}$ is a topological radical, then $\mathcal{R}_{bw} \circ \mathcal{R}_{sc} \leq \mathcal{R}_{bw} \wedge \mathcal{R}_{sc}$. By Lemma 15.20,

$$\mathcal{R}_{bw} \wedge \mathcal{R}_{sc} = (\mathcal{R}_{bw} \circ \mathcal{R}_{sc})^{\circ} \leq \mathcal{R}_{bw} \circ \mathcal{R}_{sc} \leq \mathcal{R}_{bw} \wedge \mathcal{R}_{sc}.$$

Therefore, $\mathcal{R}_{bw} \circ \mathcal{R}_{sc} = \mathcal{R}_{bw} \wedge \mathcal{R}_{sc}$. □

Let us call $\mathcal{R}_{sbw} := \mathcal{R}_{sc} \wedge \mathcal{R}_{bw}$ the *scattered* BW-*radical.*

15.4.6 The centralization procedure

Our next aim is to remove the frame of the class of scattered algebras by adding commutative algebras and forming transfinite extensions. For this in the theory of topological radicals, there exists a special procedure.

Let $\sum_a(A)$ be the sum of all commutative ideals of A, and let $\sum_\beta(A)$ be the sum of all nilpotent ideals. The maps $\sum_a$ and $\sum_\beta$ are preradicals on the class of Banach algebras.

Note that the ideals $\sum_a(A)$ and $\sum_\beta(A)$ can be nonclosed.

If A is semiprime, then $\sum_a(A)$ is the largest central ideal of A (see [25, Lemma 5.1]). Let $\mathcal{P}$ be a closed ideal map on the class of Banach algebras. Define an ideal map $\mathcal{P}^a$ by setting

$$\mathcal{P}^a = \sum\nolimits_a * \mathcal{P}.$$

Let $\sum_\beta \leq \mathcal{P}$. Then $\mathcal{P}^a(A)$ is the largest ideal of A commutative modulo $\mathcal{P}(A)$, and if $\mathcal{P}$ is a topological radical then, by [25, Theorem 5.3], $\mathcal{P}^a$ is a UTR.

Proposition 15.24. *Let $\mathcal{F}$ be a family of topological radicals, let $\sum_\beta \leq \mathcal{P} \in \mathcal{F}$ and $\mathcal{G} = \mathcal{F}\backslash\{\mathcal{P}\}$. Then $(\mathsf{H}_{\mathcal{F}})^a \leq \mathcal{P}^a * \mathsf{H}_{\mathcal{G}}$ and $(\mathsf{H}_{\mathcal{F}})^{a*} = (\mathcal{P}^a * \mathsf{H}_{\mathcal{G}})^*$.*

Proof. Let $\mathcal{T} = \mathsf{H}_{\mathcal{G}}$. Then $\mathcal{T}$ is a UTR. As the convolution operation is associative then

$$\mathcal{P}^a * \mathcal{T} = \left(\sum\nolimits_a * \mathcal{P}\right) * \mathcal{T} = \sum\nolimits_a * (\mathcal{P} * \mathcal{T}) = (\mathcal{P} * \mathcal{T})^a. \tag{15.29}$$

By Lemma 15.19, $\mathsf{H}_{\mathcal{F}} \leq \mathcal{P} * \mathcal{T}$. Then

$$(\mathsf{H}_{\mathcal{F}})^a \leq (\mathcal{P} * \mathcal{T})^a. \tag{15.30}$$

Let $\mathcal{R} = (\mathsf{H}_{\mathcal{F}})^{a*}$ and $\mathcal{S} = (\mathcal{P} * \mathcal{T})^{a*}$. It follows from (15.30) that $(\mathsf{H}_{\mathcal{F}})^a \leq (\mathcal{P} * \mathcal{T})^a \leq \mathcal{S}$ and, therefore,

$$\mathcal{R} = (\mathsf{H}_{\mathcal{F}})^{a*} \leq \mathcal{S}^* = \mathcal{S}.$$

On the other hand, $\mathsf{H}_{\mathcal{F}} \leq \mathcal{R}$ whence $\mathcal{P} * \mathcal{T} \leq \mathcal{R} * \mathcal{R} = \mathcal{R}$, $(\mathcal{P} * \mathcal{T})^a \leq \mathcal{R}^a = \mathcal{R}$ and

$$\mathcal{S} = (\mathcal{P} * \mathcal{T})^{a*} \leq \mathcal{R}^* = \mathcal{R}.$$ □

Sometimes $\mathcal{P}^a$ is a topological radical if $\mathcal{P}$ is a topological radical. We have the following.

Theorem 15.25 ([24, Theorem 5.13]). *$\mathcal{R}_{\mathrm{cq}}^a$ is a hereditary BW-radical.*

Corollary 15.26 ([24, Corollary 5.15]). *$\mathcal{R}_{\mathrm{hc}} \vee \mathcal{R}_{\mathrm{cq}}^a$ is a BW-radical.*

15.4.7 Centralization of *BW*-radicals and continuity of the joint spectral radius

Lemma 15.27. *$\mathcal{R}_{\mathrm{bw}}^a(A)$ and $\mathcal{R}_{\mathrm{sbw}}^a(A)$ are BW-ideals for every Banach algebra A.*

Proof. Indeed, $\rho(M) = \max\{\rho(M/\mathcal{R}_{bw}(A)), r(M)\}$ for every precompact set M in A by definition of $\mathcal{R}_{bw}$. Let $B = A/\mathcal{R}_{bw}(A)$ and $N = M/\mathcal{R}_{bw}(A)$. As $\sum_\beta \le \mathcal{R}_{cq} \le \mathcal{R}_{bw}$, B is semiprime and $\sum_a(B)$ is the largest central ideal of B. It is clear that $\sum_a(B)$ is closed. By [24, Lemma 5.5],

$$\rho(N) = \max\Big\{\rho\Big(N/\textstyle\sum_a(B)\Big), r_1(N)\Big\}$$

where $r_1(N) = \sup\{\rho(a) : a \in N\} \le r(N)$. Hence $\sum_a(B) = \mathcal{R}^a_{bw}(A)/\mathcal{R}_{bw}(A)$ and $\mathcal{R}_{bw}(A)$ are BW-ideals. By Proposition 15.6, BW-ideals are stable with respect to extensions. So $\mathcal{R}^a_{bw}(A)$ is a BW-ideal.

As $\mathcal{R}^a_{sbw}(A) \subseteq \mathcal{R}^a_{bw}(A)$, then $\mathcal{R}^a_{sbw}(A)$ is also a BW-ideal. □

Theorem 15.28. *$\mathcal{R}^{a*}_{sbw}$ is a BW-radical and $\mathcal{R}^a_{bw} = \mathcal{R}_{bw}$.*

Proof. Let $\mathcal{P}$ be $\mathcal{R}_{sbw}$ or $\mathcal{R}_{bw}$. Clearly, $\sum_\beta \le \mathcal{P}$. Let A be a Banach algebra, and let $(J_\alpha)_{\alpha\le\gamma+1}$ be an increasing transfinite chain of closed ideals of A such that $J_1 = \mathcal{P}^a(A)$ and $J_{\gamma+1} = J_\gamma$, and $J_{\alpha+1}/J_\alpha = \mathcal{P}^a(A/J_\alpha)$ for all $\alpha \le \gamma$. By Lemma 15.27, ideals J_1 and $J_{\alpha+1}/J_\alpha$ are BW-ideals. By Proposition 15.6, $\mathcal{P}^{a*}(A)$ is a BW-ideal for every Banach algebra A, that is, $\mathcal{P}^{a*}$ is a BW-radical.

As $\mathcal{R}_{bw}$ is the largest BW-radical, then $\mathcal{R}^{a*}_{bw} \le \mathcal{R}_{bw}$. We obtain that

$$\mathcal{R}_{bw} \le \mathcal{R}^a_{bw} \le \mathcal{R}^{a*}_{bw} \le \mathcal{R}_{bw}$$

whence $\mathcal{R}^a_{bw} = \mathcal{R}_{bw}$. □

This formally gives the following.

Corollary 15.29. *Any Banach algebra A commutative modulo the radical $\mathcal{R}_{bw}(A)$ is $\mathcal{R}_{bw}$-radical.*

Proof. Let $B = A/\mathcal{R}_{bw}(A)$. Clearly, $B = \mathcal{R}^a(B)$ for every topological radical $\mathcal{R}$. Therefore,

$$B = \mathcal{R}^a_{bw}(B) = \mathcal{R}_{bw}(B) = \mathcal{R}_{bw}\big(A/\mathcal{R}_{bw}(A)\big) = (0)$$

whence $A = \mathcal{R}_{bw}(A)$. □

Theorem 15.30. $\mathcal{R}^{a*}_{sbw} = \mathcal{R}_{hc} \vee \mathcal{R}^a_{cq}$.

Proof. Let $\mathcal{S} = ((\mathcal{R}_{cq} * \mathcal{R}_{hc})^a)^*$. By Lemma 15.20 and formula (15.29) applied to $\mathcal{P} = \mathcal{R}_{cq}$ and $\mathcal{T} = \mathcal{R}_{hc}$, we have that

$$\mathcal{R}_{sbw} = \mathcal{R}_{hc} \vee \mathcal{R}_{cq} \le \mathcal{R}_{hc} \vee \mathcal{R}^a_{cq} = \left(\mathcal{R}^a_{cq} * \mathcal{R}_{hc}\right)^* = \left((\mathcal{R}_{cq} * \mathcal{R}_{hc})^a\right)^* = \mathcal{S}.$$

Let A be a Banach algebra, and let $I = \mathcal{S}(A)$. By Lemma 15.20,

$$\mathcal{R}^a_{sbw}(A) \subseteq \mathcal{S}^a(A) = q_I^{-1}\Big(\textstyle\sum_a(A/I)\Big).$$

As $(\mathcal{R}_{\mathrm{cq}} * \mathcal{R}_{\mathrm{hc}})^a * \mathcal{S} = \mathcal{S}$ then $\mathcal{R}_{\mathrm{hc}}(A/I) = (0)$. Indeed, if $\mathcal{R}_{\mathrm{hc}}(A/I) \neq (0)$ then

$$\mathcal{R}_{\mathrm{hc}} * \mathcal{S}(A) = q_I^{-1}(\mathcal{R}_{\mathrm{hc}}(A/I))$$

differs from I and

$$\begin{aligned}(\mathcal{R}_{\mathrm{cq}} * \mathcal{R}_{\mathrm{hc}})^a * \mathcal{S}(A) &= (\mathcal{R}_{\mathrm{cq}}^a * \mathcal{R}_{\mathrm{hc}}) * \mathcal{S}(A) = \mathcal{R}_{\mathrm{cq}}^a * (\mathcal{R}_{\mathrm{hc}} * \mathcal{S})(A) \\ &\neq I = \mathcal{S}(A),\end{aligned}$$

a contradiction. Therefore, $\mathcal{R}_{\mathrm{hc}} * \mathcal{S} = \mathcal{S}$.

Similarly, we obtain that $\mathcal{R}_{\mathrm{cq}}(A/I) = (0)$ and $\mathcal{R}_{\mathrm{cq}} * \mathcal{S} = \mathcal{S}$. Then

$$\begin{aligned}\mathcal{R}_{\mathrm{sbw}}^a(A) \subseteq \mathcal{S}^a(A) &= \sum\nolimits_a * \mathcal{S}(A) = \sum\nolimits_a * (\mathcal{R}_{\mathrm{cq}} * (\mathcal{R}_{\mathrm{hc}} * \mathcal{S}))(A) \\ &= \Big(\sum\nolimits_a * (\mathcal{R}_{\mathrm{cq}} * \mathcal{R}_{\mathrm{hc}})\Big) * \mathcal{S}(A) = (\mathcal{R}_{\mathrm{cq}} * \mathcal{R}_{\mathrm{hc}})^a * \mathcal{S}(A) \\ &= \mathcal{S}(A),\end{aligned}$$

that is, $\mathcal{R}_{\mathrm{sbw}}^a \leq \mathcal{S}$. As $\mathcal{S}$ is a topological radical,

$$\mathcal{R}_{\mathrm{sbw}}^{a*} = (\mathcal{R}_{\mathrm{sbw}}^a)^* \leq \mathcal{S}^* = \mathcal{S} = \mathcal{R}_{\mathrm{hc}} \vee \mathcal{R}_{\mathrm{cq}}^a.$$

On the other hand, as $\mathcal{R}_{\mathrm{cq}} * \mathcal{R}_{\mathrm{hc}} \leq (\mathcal{R}_{\mathrm{hc}} \vee \mathcal{R}_{\mathrm{cq}}) * (\mathcal{R}_{\mathrm{hc}} \vee \mathcal{R}_{\mathrm{cq}}) = \mathcal{R}_{\mathrm{hc}} \vee \mathcal{R}_{\mathrm{cq}} = \mathcal{R}_{\mathrm{sbw}}$,

$$(\mathcal{R}_{\mathrm{cq}} * \mathcal{R}_{\mathrm{hc}})^a \leq \mathcal{R}_{\mathrm{sbw}}^a$$

and then

$$\mathcal{R}_{\mathrm{hc}} \vee \mathcal{R}_{\mathrm{cq}}^a = \mathcal{S} = ((\mathcal{R}_{\mathrm{cq}} * \mathcal{R}_{\mathrm{hc}})^a)^* \leq (\mathcal{R}_{\mathrm{sbw}}^a)^* = \mathcal{R}_{\mathrm{sbw}}^{a*}. \qquad \square$$

We will mention now an application of this result to the problem of continuity of joint spectral radius. Let us recall the required definitions. Consider the function $M \longmapsto \rho(M)$ for bounded sets M of a Banach algebra A. This function is upper continuous (see [18, Theorem 3.1]), that is,

$$\limsup \rho(M_n) \leq \rho(M) \tag{15.31}$$

when M_n converges to M in the Hausdorff metric. The set M is a *point of continuity of the joint spectral radius* if $\rho(M_n) \to \rho(M)$ for every sequence (M_n) convergent to M.

Corollary 15.31. *Let M be a precompact set in a Banach algebra A. If $\rho(M/\mathcal{R}_{\mathrm{sbw}}^{a*}(A)) < \rho(M)$, then M is a point of continuity of the joint spectral radius.*

Proof. By virtue of Theorem 15.30, it is sufficient to remark that M is a point of continuity of the joint spectral radius if $\rho(M/(\mathcal{R}_{\mathrm{hc}} \vee \mathcal{R}_{\mathrm{cq}}^a)(A)) < \rho(M)$ by [24, Theorem 6.3]. $\square$

The following corollary is a consequence of [24, Corollary 6.4] and Theorem 15.30.

Corollary 15.32. *Let A be a Banach algebra, and let G be a semigroup in $\mathcal{R}^{a*}_{\mathrm{sbw}}(A)$. If G consists of quasinilpotent elements of A then the closed subalgebra $\overline{A(G)}$ generated by G is compactly quasinilpotent.*

Proof. As G consists of quasinilpotent elements, $r(M) = 0$ for every precompact set M in G. As $\mathcal{R}^{a*}_{\mathrm{sbw}} \leq \mathcal{R}_{\mathrm{bw}}$, then $\rho(M) = r(M)$ for every precompact set M in $\mathcal{R}^{a*}_{\mathrm{sbw}}(A)$. Hence $\rho(M) = 0$ for every precompact set M in G. As it was described above (see, for instance, [28, Proposition 3.5]), $A(G)$ is finitely quasinilpotent. It follows from Corollary 15.31 that ρ is continuous at any precompact set in $\mathcal{R}^{a*}_{\mathrm{sbw}}(A)$. As the closure $\overline{A(G)}$ is contained in $\mathcal{R}^{a*}_{\mathrm{sbw}}(A)$, and each compact subset of $\overline{A(G)}$ is a limit of a net of finite subsets of $A(G)$, the algebra $\overline{A(G)}$ is compactly quasinilpotent. □

Bibliography

[1] S. A. Argyros and R. Haydon, *A hereditarily indecomposable L_∞-space that solves the scalar-plus-compact problem*, Acta Math. **206** (2011), 1–54.

[2] B. Aupetit, *A Primer on Spectral Theory*, Springer, New York, 1991.

[3] B. A. Barnes, *On the existence of minimal ideals in a Banach algebra*, Trans. Am. Math. Soc. **133** (1968), 511–517.

[4] M. A. Berger and Y. Wang, *Bounded semigroups of matrices*, Linear Algebra Appl. **166** (1992), 21–27.

[5] P. Cao and Yu. V. Turovskii, *Topological radicals, VI. Scattered elements in Banach Jordan and associative algebras*, Stud. Math. **235** (2) (2016), 171–208.

[6] P. G. Dixon, *Topologically irreducible representations and radicals in Banach algebras*, Proc. Lond. Math. Soc. (3) **74** (1997), 174–200.

[7] L. S. Goldenstein and A. S. Markus, On a measure of non-compactness of bounded sets and linear operators, in *Studies in Algebra and Mathematic Analysis*, 45–54, Izdat. Karta Moldovenjaski, Kishinev, 1965 (in Russian).

[8] R. Jungers, *Joint Spectral Radius, Theory and Applications*, Springer-Verlag, Berlin, 2009, 197 p.

[9] M. Kennedy, V. S. Shulman and Yu. V. Turovskii, *Invariant subspaces of subgraded Lie algebras of compact operators*, Integral Equ. Oper. Theory **63** (2009), 47–93.

[10] E. Kissin, V. S. Shulman and Yu. V. Turovskii, *Banach Lie algebras with Lie subalgebras of finite codimension have Lie ideals*, J. Lond. Math. Soc. **80** (2009), 603–626.

[11] E. Kissin, V. S. Shulman and Yu. V. Turovskii, *Topological radicals and Frattini theory of Banach Lie algebras*, Integral Equ. Oper. Theory **74** (2012), 51–121.

[12] V. I. Lomonosov, *Invariant subspaces for operators commuting with compact operators*, Funct. Anal. Appl. **7** (1973), 213–214.

[13] I. D. Morris, *The generalized Berger–Wang formula and the spectral radius of linear cocycles*, J. Funct. Anal. **262** (2012), 811–824.

[14] I. D. Morris, The generalized Berger–Wang formula and the spectral radius of linear cocycles, preprint: ArXiv:0906.2915v1 [math.DS] 16 Jun 2009.

[15] G.-K. Pedersen, *C*-algebras and Their Automorphism Groups*, Academic Press, London, 2018.

[16] G.-C. Rota and W. G. Strang, *A note on the joint spectral radius*, Indag. Math. **22** (1960), 379–381.
[17] V. S. Shulman, *On invariant subspaces of Volterra operators*, Funct. Anal. Appl. **18** (1984), 84–85 (in Russian).
[18] V. S. Shulman and Yu. V. Turovskii, *Joint spectral radius, operator semigroups and a problem of Wojtynskii*, J. Funct. Anal. **177** (2000), 383–441.
[19] V. S. Shulman and Yu. V. Turovskii, *Radicals in Banach algebras and some unsolved problems in the theory of radical Banach algebras*, Funct. Anal. Appl. **35** (2001), 312–314.
[20] V. S. Shulman and Yu. V. Turovskii, *Formulae for joint spectral radii of sets of operators*, Stud. Math. **149** (2002), 23–37.
[21] V. S. Shulman and Yu. V. Turovskii, *Invariant subspaces of operator Lie algebras and Lie algebras with compact adjoint action*, J. Funct. Anal. **223** (2005), 425–508.
[22] V. S. Shulman and Yu. V. Turovskii, *Topological radicals, I. Basic properties, tensor products and joint quasinilpotence*, Banach Cent. Publ. **67** (2005), 293–333.
[23] V. S. Shulman and Yu. V. Turovskii, *Topological radicals, II. Applications to the spectral theory of multiplication operators*, Oper. Theory, Adv. Appl. **212** (2010), 45–114.
[24] V. S. Shulman and Yu. V. Turovskii, *Topological radicals and joint spectral radius*, Funct. Anal. Appl. **46** (2012), 287–304.
[25] V. S. Shulman and Yu. V. Turovskii, *Topological radicals, V. From algebra to spectral theory*, Oper. Theory, Adv. Appl. **233**, Algebraic Methods in Functional Analysis (2014), 171–280.
[26] V. S. Shulman and Yu. V. Turovskii, Application of topological radicals to calculation of joint spectral radius, preprint: arXiv:0805.0209 [math.FA] 2 May 2008.
[27] P. Thieullen, *Fibres dynamiques asymptotiquement compacts. Exposants de Lyapunov. Entropie. Dimension*, Ann. Inst. H. Poincare Anal. Non Lineare **4** (1) (1987), 49–97.
[28] Yu. V. Turovskii, Spectral properties of certain Lie subalgebras and the spectral radius of subsets of a Banach algebra, in *Spectral Theory of Operators and Its Applications, vol. 6*, 144–181, Elm, Baku, 1985 (in Russian).
[29] Yu. V. Turovskii, *Volterra semigroups have invariant subspaces*, J. Funct. Anal. **182** (1999), 313–323.
[30] K. Vala, *On compact sets of compact operators*, Ann. Acad. Sci. Fenn. Ser. A I **351** (1964), 1–8.

Edward Kissin, Victor S. Shulman, and Yurii V. Turovskii

16 Pontryagin–Krein theorem: Lomonosov's proof and related results

To the memory of our dear friend and colleague Victor Lomonosov

Abstract: We discuss Lomonosov's proof of the Pontryagin–Krein theorem on invariant maximal non-positive subspaces, prove the refinement of one theorem from [23] on common fixed points for a group of fractional-linear maps of operator ball and deduce its consequences. Some Burnside-type counterparts of the Pontryagin–Krein theorem are also considered.

Keywords: Invariant subspace, Pontryagin space, Krein space, indefinite metric

MSC 2010: Primary 47A15, Secondary 47L10

16.1 Introduction and preliminaries

In 1944, L. S. Pontryagin, stimulated by actual problems of mechanics, published his famous paper [25] where it was proved that if an operator T is self-adjoint with respect to a scalar product with finite number k of negative squares then T has invariant nonpositive subspace of dimension k. The importance of results of this kind for stability of some mechanical problems was discovered by S. L. Sobolev in 1938, who proved the existence of nonpositive eigenvectors in the case $k = 1$.

Before giving precise formulations, we introduce some notation. By indefinite metric space, we mean a linear space H supplied with a semilinear form $[x, y]$ satisfying the following condition: H can be decomposed in a direct sum of two subspaces H_+, H_- ($x = x_+ + x_-$, for each $x \in H$) in such a way that H is a Hilbert space with respect to the form

$$(x, y) = [x_+, y_+] - [x_-, y_-].$$

The decomposition of this kind is not unique but the dimensions of the summands and the topology on H do not depend on the choice of the decomposition. We assume

Acknowledgement: This paper is in final form and no version of it will be submitted for publication elsewhere.
The authors are grateful to the referee for many helpful suggestions and remarks.

Edward Kissin, STORM, London Metropolitan University, 166-220 Holloway Road, London N7 8DB, Great Britain, e-mail: e.kissin@londonmet.ac.uk
Victor S. Shulman, Yurii V. Turovskii, Department of Mathematics, Vologda State University, Vologda, Russia, e-mails: shulman.victor80@gmail.com, yuri.turovskii@gmail.com

https://doi.org/10.1515/9783110656756-016

in what follows that $\dim H_+ \geq \dim H_-$. If $\dim H_- = k < \infty$, then one says that H is a *Pontryagin space* Π_k, otherwise H is called a *Krein space.*

A vector $x \in H$ is called *positive* (*nonnegative, negative, nonpositive, neutral*) if

$$[x,x] > 0 \quad (\text{resp., } [x,x] \geq 0,\ [x,x] < 0,\ [x,x] \leq 0,\ [x,x] = 0).$$

A subspace is *positive* (*nonnegative, nonpositive,negative, neutral*) if its nonzero elements are positive (resp. nonnegative, nonpositive, negative, neutral). For brevity, we write MNPS for maximal nonpositive subspaces.

Subspaces H_1, H_2 of H form *a dual pair* if H_1 is positive, H_2 is negative and $H = H_1 + H_2$.

Sometimes it is convenient to start with a Hilbert space H decomposed in the orthogonal sum of two subspaces $H = H_+ \oplus H_-$ and to set

$$[x,y] = (x_+, y_+) - (x_-, y_-).$$

Denoting by P_+ and P_- the projections onto H_+ and H_-, respectively, set $J = P_+ - P_-$. Then one can write the relation between two "scalar products" in the form

$$[x,y] = (Jx,y) \quad \text{and} \quad (x,y) = [Jx,y].$$

This notation determines the standard terminology. A space with indefinite metric is often called a J-space, a vector x is J-orthogonal to a vector y if $[x,y] = 0$. An operator B (we consider only bounded linear operators) on H is called *J-adjoint* to an operator A if $[Ax,y] = [x,By]$, for all $x,y \in H$; we write $B = A^\sharp$. If $A^\sharp = A$ then A is called J-self-adjoint; an equivalent condition is $[Ax,x] \in \mathbb{R}$, for all $x \in H$. If $\operatorname{Im}([Ax,x]) \geq 0$ for all x, then A is called J-dissipative.

Furthermore, A is *J-unitary* if $A^\sharp = A^{-1}$ (equivalently, A is surjective and $[Ax,Ay] = [x,y]$, for $x,y \in H$); A is *J-expanding* if $[Ax,Ax] \geq [x,x]$, for all $x \in H$.

In 1949, I. S. Iohvidov [10] constructed an analogue of Caley transform for indefinite metric spaces which allowed him to deduce from Pontryagin's theorem the existence of an invariant MNSP for J-unitary operators on Π_k-spaces. Then M. G. Krein [14], using absolutely different approach, proved that a J-unitary (and, more generally, J-expanding) operator U in arbitrary indefinite metric space has an invariant MNPS, if its "corner" P_-UP_+ is compact. Clearly, this condition holds in Π_k-spaces. In 1964, Ky Fan [6] extended Krein's theorem to operators on Banach spaces preserving indefinite norms $\nu(x) = \|(1-P)x\| - \|Px\|$ where P is a projection of finite rank.

Now we have the following *Pontryagin–Krein theorem* (hereafter *PK-theorem*).

Theorem 16.1. *Let an operator A on a Krein space H be J-dissipative and let P_+AP_- be compact. Then there exists an MNPS invariant for A.*

Note that the proof of Pontryagin's result in [25] was very complicated and long. The Krein's proof in [14] was short but far from elementary, because it was based on the Schauder–Tichonov fixed-point theorem. Moreover Ky Fan, to prove his version of the PK-theorem, previously obtained a more general fixed-point theorem. We add that to deduce the result for J-dissipative operators from the Krein's theorem about J-expanding operators, one needs to use Iohvidov's theory of Caley transformation for Krein spaces which is also very nontrivial.

In 1986, Victor Lomonosov in a talk at the Voronezh Winter School presented a proof of Theorem 16.1 which was extremely short and completely elementary; this proof was published in [18]. In Section 16.2 of our paper, we present the Lomonosov's proof in a complete form including the consideration of the finite-dimensional case. In Section 16.3, we consider the approach based on some fixed-point theorems and discuss several results obtained in this way. In Section 16.4, we prove Theorem 16.8 which refines a theorem of M. Ostrovskii, V. S. Shulman, and L. Turowska [23] about common fixed points for a group of fractional-linear maps of the operator ball. This allows us to estimate the similarity degree for a bounded representation of a group on a Hilbert space which preserves a quadratic form with finite number of negative squares. In Section 16.5, we prove by using Theorem 16.8 that any bounded quasi-positive definite function on a group is a difference of two positive definite functions (this was known earlier only for amenable groups). In the final section, we discuss Burnside-type counterparts of the PK-theorem.

16.2 Lomonosov's proof of the PK-theorem

As usual, $\mathcal{B}(H_1, H_2)$ is the space of all bounded linear operators from H_1 to H_2, and $\mathcal{B}(H) = \mathcal{B}(H, H)$ is the algebra of all bounded linear operators on H. To any operator $W : H_- \to H_+$ there corresponds the graph-subspace $L_W = \{x + Wx : x \in H_-\}$; it is easy to see that L_W is maximal nonpositive if and only if W is contractive, that is, $\|W\| \le 1$. Conversely, each MNPS is of the form L_W, for some contraction $W \in \mathcal{B}(H_-, H_+)$. It is not difficult to check that L_W is invariant under an operator $A \in \mathcal{B}(H)$ if and only if

$$WA_{11} + WA_{12}W - A_{21} - A_{22}W = 0, \tag{16.1}$$

where

$$A_{11} = P_-AP_-, \quad A_{12} = P_-AP_+, \quad A_{21} = P_+AP_-, \quad A_{22} = P_+AP_+. \tag{16.2}$$

Lomonosov in [18] introduced a "mixed" convergence (*M-convergence*) in $\mathcal{B}(H)$: a sequence $\{A^{(k)}\}_{k=1}^{\infty}$ of operators M-converges to an operator A, if $A_{11}^{(k)} \to A_{11}$ and $(A_{22}^{(k)})^* \to (A_{22})^*$ in the strong operator topology (SOT), $A_{21}^{(k)} \to A_{21}$ in the weak operator topology (WOT) and $A_{12}^{(k)} \to A_{12}$ in norm.

Theorem 16.2 ([18]). *Let a sequence $\{A^{(k)}\}_{k=1}^{\infty}$ of operators M-converge to an operator A. If each $A^{(k)}$ has an MNPS, then A has an MNPS.*

Proof. It follows from our assumptions, that for each k, there is a contraction $W_k \in \mathcal{B}(H_-, H_+)$ satisfying

$$W_k A_{11}^{(k)} + W_k A_{12}^{(k)} W_k - A_{21}^{(k)} - A_{22}^{(k)} W_k = 0. \tag{16.3}$$

Choosing a subsequence if necessary, one can assume that the sequence $\{W_k\}_{k=1}^{\infty}$ WOT-converges to some contraction $W \in \mathcal{B}(H_-, H_+)$. It follows easily from the definition of M-convergence that W satisfies (16.1). □

Deduction of Theorem 16.1 *from Theorem* 16.2. Denote by $(P_-^{(k)})_{k=1}^{\infty}$ and $(P_+^{(k)})_{k=1}^{\infty}$ increasing sequences of finite-dimensional projections such that $P_-^{(k)} \xrightarrow{\text{sot}} P_-$ and $P_+^{(k)} \xrightarrow{\text{sot}} P_+$, and set $P^{(k)} = P_-^{(k)} + P_+^{(k)}$. Then the operators $A^{(k)} = P^{(k)} A P^{(k)}$ are J-dissipative, finite-dimensional, and M-converge to A (the condition $\|A_{12}^{(k)} - A_{12}\| \to 0$ follows from the compactness of A_{12}). To see that each $A^{(k)}$ has an MNPS, it suffices to show that any J-dissipative operator in a finite-dimensional indefinite metric space has an MNPS. □

The proof of the PK-theorem in the finite-dimensional case was dropped in [18] as an easy one. In fact, the usual proof of this theorem for matrices (see, e. g., [7]) is not simple and is not direct: it goes via study of J-expanding operators and application of Caley transform. To present Lomonosov's result in the complete form, we add a short direct proof for the finite-dimensional case *which again uses Theorem* 16.2.

Completion of the proof of Theorem 16.1. Let A be a J-dissipative operator on a finite-dimensional indefinite metric space H. For each $t > 0$, the operator $B = A + tJ$ satisfies the condition of *strong J-dissipativity*:

$$\operatorname{Im}[Bx, x] > 0 \quad \text{if } x \neq 0.$$

Since $A + tJ \to A$ when $t \to 0$, Theorem 16.2 allows us to assume that A is strongly dissipative. In this case, A has no real eigenvalues: if $Ax = tx$, for some $t \in \mathbb{R}$ and $0 \neq x \in H$, then $[Ax, x] = t[x, x] \in \mathbb{R}$, a contradiction. Let us denote by H_+ and H_- the spectral subspaces of A corresponding to sets $\mathbb{C}_+ = \{z \in \mathbb{C} : \operatorname{Im} z > 0\}$ and $\mathbb{C}_- = \{z \in \mathbb{C} : \operatorname{Im} z < 0\}$, respectively. We will show that subspaces H_+ and H_- are positive and negative, respectively.

If an operator T is strongly J-dissipative, then also $-T^{-1}$ is strongly J-dissipative. Indeed,

$$-\operatorname{Im}\left[T^{-1}x, x\right] = \operatorname{Im}\left[x, T^{-1}x\right] = \operatorname{Im}\left[TT^{-1}x, T^{-1}x\right] > 0$$

if $x \neq 0$. Since $A - t\mathbb{1}$ is strongly J-dissipative, for each $t \in \mathbb{R}$, we get that $-(A - t\mathbb{1})^{-1}$ is strongly J-dissipative. Now, for each $0 \neq x \in H_+$, one has

$$x = \frac{i}{\pi} \int_{-\infty}^{\infty} (A - t\mathbb{1})^{-1} x \, dt$$

whence

$$[x,x] = \operatorname{Re}[x,x] = -\operatorname{Im}\left(\frac{1}{\pi}\int_{-\infty}^{\infty}[(A-t1)^{-1}x,x]dt\right) > 0.$$

Thus H_+ is positive. Similarly, H_- is negative.

So $H = H_- + H_+$ is the decomposition of H into the direct sum of a negative subspace and a positive subspace. It follows that H_- is an invariant MNPS. □

We add that

- In works of T. J. Azizov, H. Langer, A. A. Shkalikov, and other mathematicians, Theorem 16.1 was extended to various classes of unbounded operators (see, e. g., [30] and references therein);
- M. A. Naimark [20] (see also a much more general result in [21]) proved that any commutative family Q of J-self-adjoint operators in a Π_k-space has a common invariant MNPS. It follows that the result holds for any commutative family Q of operators which is *J-symmetric*: $T \in Q$ implies $T^\sharp \in Q$.

16.3 Fixed points

Let us return to Krein's proof of the existence of invariant MNPS for J-unitary operators. It is clear that any J-unitary operator U maps any MNPS onto an MNPS. Using the bijection $W \mapsto L_W$ between MNP subspaces and contractions, we see that U determines the map ϕ_U from the closed unit ball $\mathcal{B}_1(H_-,H_+)$ of the space $\mathcal{B}(H_-,H_+)$ into itself. It is easy to obtain the direct expression of ϕ_U in terms of U:

$$\phi_U(W) = (U_{21} + U_{22}W)(U_{11} + U_{12}W)^{-1} \tag{16.4}$$

(we use notation from (16.2)). It was shown in [14] that if U_{12} is compact then the map ϕ_U is WOT-continuous; since $\mathcal{B}_1(H_-,H_+)$ is WOT-compact, the fixed-point theorem implies the existence of a contraction W with $\phi_U(W) = W$. This means that L_W is invariant with respect to U. We get the following result.

Theorem 16.3 ([14]). *Let U be a J-unitary operator on a Krein space $H = H_+ + H_-$. If the "corner "U_{12} in the block-matrix of U with respect to the decomposition $H = H_+ + H_-$ is compact, then U has an invariant MNPS.*

This result can be reformulated independently of the choice of the decomposition $H = H_+ + H_-$ and without matrix terminology.

Theorem 16.4. *If J-unitary operator U on a Krein space H is a compact perturbation of an operator that preserves a maximal negative subspace, then it has an invariant MNPS.*

To prove this, let $U = R + K$, where K is compact, R preserves a maximal negative subspace $L \subset H$. Let $M = L^{\perp}$, and let P be the projection onto L along M. Then

$$(1 - P)UP = (1 - P)RP + (1 - P)KP = (1 - P)KP$$

is a compact operator. But $(1 - P)UP$ is the corner of the block-matrix U with respect to the decomposition $H = L + M$. So, by Krein's theorem, U has an invariant MNPS.

Note that for Π_k-spaces the assumption of compactness of U_{21} is automatically satisfied, so Krein's theorem implies that any J-unitary operator on a Π_k-space has an invariant MNPS.

The fractional-linear maps ϕ_U defined by (16.4) preserve the open unit ball $\mathfrak{B} = \{X \in \mathcal{B}(H_-, H_+) : \|X\| < 1\}$ and their restrictions to $\mathfrak{B}$ form the group of all biholomorphic automorphisms of $\mathfrak{B}$ (we refer to [1] or [13] for more information). So the existence of fixed points for such maps and families of such maps are of independent interest. After Naimark's result, it was natural to try to prove the existence of common fixed points for commutative sets of fractional-linear maps. Note that this does not follow directly from Naimark's theorem, because the maps ϕ_U and ϕ_V commute if and only if the operators U and V commute *up to a scalar multiple*: $UV = \lambda VU, \lambda \in \mathbb{C}$. The positive answer was obtained by J. W. Helton.

Theorem 16.5 ([9]). *Let H_1, H_2 be Hilbert spaces and* $\dim H_1 < \infty$. *Then any commutative family of fractional-linear maps of the closed unit ball in $\mathcal{B}(H_1, H_2)$ has a common fixed point.*

This result implies Naimark's theorem, but the proof uses it. Another result of Helton [8] based on the consideration of fractional-linear maps states that a commutative group of J-unitary operators on a Krein space $H_1 \oplus H_2$ has an invariant maximal positive subspace if it contains a compact perturbation of an operator $A \oplus B$ with $\sigma(A) \cap \sigma(B) = \emptyset$. This extends the Naimark theorem because the identity operator 1 in a Π_k-space is a compact perturbation of J.

The following result on fixed points of groups of fractional-linear maps was proved by M. Ostrovskii, V. S. Shulman, and L. Turowska [23, 22] (see also [32] where the case $k = 1$ was considered).

Theorem 16.6. *Let* $\dim H_2 = k < \infty$ *and let a group Γ of fractional-linear maps of the open unit ball $\mathfrak{B}$ in $\mathcal{B}(H_2, H_1)$ have an orbit separated from the boundary* ($\sup_{\phi \in \Gamma} \|\phi(K)\| < 1$, *for some $K \in \mathfrak{B}$). Then there is $K_0 \in \mathfrak{B}$ such that $\phi(K_0) = K_0$, for all $\phi \in \Gamma$.*

Corollary 16.7. *Any bounded group of J-unitary operators in a Π_k-space has an invariant dual pair of subspaces.*

We will obtain some related results in the next two sections.

16.4 Orthogonalization of bounded representations

In many situations (see the book [24] for examples and discussions), it is important to know if a given representation π of a group G in a Hilbert space is similar to a unitary representation:

$$\pi(g) = V^{-1}U(g)V, \quad \text{for all } g \in G,$$

the operators $U(g)$ are unitary, V is an invertible operator. The infimum $c(\pi)$ of values $\|V\|\|V^{-1}\|$ for all possible V's, is called the constant of similarity of π. It is obvious that a representation can be similar to a unitary one only if it is bounded:

$$\|\pi\| := \sup_{g\in G}\|\pi(g)\| < \infty;$$

clearly $\|\pi\| \leq c(\pi)$.

By a quadratic form, we mean a function $\Phi(x) = (Ax, x)$ on a Hilbert space H, where A is an invertible selfadjoint operator on H. Changing the scalar product if necessary, one can reduce the situation to the case that

$$\Phi(x) = (P_1x, x) - (P_2x, x), \tag{16.5}$$

where P_1 and P_2 are projections with $P_1 + P_2 = 1$ (if a form is given as above then P_1 and P_2 are spectral projections of A corresponding to the intervals $(-\infty, 0)$ and $(0, \infty)$). So we consider only forms given by (16.5). The number $\dim(P_2H)$ is called the number of negative squares of Φ.

A representation π is said to preserve the form (16.5) if $\Phi(\pi(g)x) = \Phi(x)$, for all $x \in H, g \in G$.

Theorem 16.8. *Any bounded representation π preserving a form with finite number of negative squares is similar to a unitary representation. Moreover,*

$$c(\pi) \leq 2\|\pi\|^2 + 1. \tag{16.6}$$

The first statement of the theorem was proved in [23]; to prove the inequality (16.6) we will repeat some steps of the proof in [23] adding necessary changes and estimations.

We begin with a general result on fixed points of groups of isometries.

Let us say that a metric space $(\mathcal{X}, d)$ is *ball-compact* if a family of balls

$$E_{a,r} = \{x \in X : d(a, x) \leq r\}$$

has nonvoid intersection provided each its finite subfamily has nonvoid intersection (see [33]).

A subset $M \subset \mathcal{X}$ is called *ball-convex* if it is the intersection of a family of balls. The compactness property extends from balls to ball-convex sets: if $(\mathcal{X}, d)$ is ball-compact, then a family $\{M_\lambda : \lambda \in \Lambda\}$ of ball-convex subsets of $\mathcal{X}$ has nonvoid intersection if each its finite subfamily has nonvoid intersection.

The *diameter* of a subset $M \subset \mathcal{X}$ is defined by

$$\operatorname{diam}(M) = \sup\{d(x,y) : x, y \in M\}. \tag{16.7}$$

A point $a \in M$ is called *diametral* if

$$\sup\{d(a,x) : x \in M\} = \operatorname{diam}(M).$$

A metric space $\mathcal{X}$ is said to have *normal structure* if every ball-convex subset of $\mathcal{X}$ with more than one element has a nondiametral point.

Lemma 16.9. *Suppose that a metric space $(\mathcal{X}, d)$ is ball-compact and has normal structure. If a group Γ of isometries of $(\mathcal{X}, d)$ has a bounded orbit O, then it has a fixed point x_0. Moreover, x_0 belongs to the intersection of all ball-convex subsets containing O.*

Proof. The family Φ of all balls containing O is nonvoid. Since O is invariant under Γ, the family Φ is also invariant: $g(E) \in \Phi$, for each $E \in \Phi$. Hence the intersection M_1 of all elements of Φ is a nonvoid Γ-invariant ball-convex set; moreover, it follows easily from the definition that M_1 is the intersection of all ball-convex subsets containing O.

Thus the family $\mathcal{M}$ of all nonvoid Γ-invariant ball-convex subsets of M_1 is nonvoid. Therefore, the intersection of a decreasing chain of sets in $\mathcal{M}$ belongs to $\mathcal{M}$ and, by Zorn Lemma, $\mathcal{M}$ has minimal elements. Our aim is to prove that any minimal element M of $\mathcal{M}$ consists of one point.

Assuming the contrary, let $\operatorname{diam}(M) = \alpha > 0$. Since $(\mathcal{X}, d)$ has normal structure, M contains a non-diametral point a. It follows that $M \subset \{x \in \mathcal{X} : d(a,x) \le \delta\}$ for some $\delta < \alpha$. Set

$$D = \bigcap_{b \in M} E_{b,\delta}.$$

The set D is nonvoid because $a \in D$. Furthermore, D is ball-convex by definition. To see that D is a proper subset of M, take $b, c \in M$ with $d(b,c) > \delta$, then $c \notin E_{b,\delta}$, hence $c \notin D$.

Since Γ is a group of isometric transformations and M is invariant under each element of Γ, D is Γ-invariant. We get a contradiction with the minimality of M.

Thus $M = \{x_0\}$, for some $x_0 \in M_1$. □

Let now H_1 and H_2 be Hilbert spaces, $\dim H_2 < \infty$. We denote by $\mathfrak{B}$ the open unit ball of the space $\mathcal{B}(H_2, H_1)$ of all linear operators from H_2 to H_1.

For each $A \in \mathfrak{B}$, we define a transformation μ_A of $\mathfrak{B}$ (*a Möbius transformation*) by setting

$$\mu_A(X) = (1 - AA^*)^{-1/2}(A + X)(1 + A^*X)^{-1}(1 - A^*A)^{1/2}. \tag{16.8}$$

It can be easily checked that $\mu_A(0) = A$ and $\mu_A^{-1} = \mu_{-A}$, for each $A \in \mathfrak{B}$.

We set

$$\rho(A,B) = \tanh^{-1}(\|\mu_{-A}(B)\|). \tag{16.9}$$

It was proved in [23, Theorem 6.1] that the space $(\mathfrak{B}, \rho)$ is ball-compact and has a normal structure. It can be also verified that ρ coincides with the Carathéodory distance $c_{\mathfrak{B}}$ in $\mathfrak{B}$. Therefore, all biholomorphic maps of $\mathfrak{B}$ preserve ρ. Applying Lemma 16.9, we get the following statement.

Lemma 16.10. *If a group of biholomorphic transformations of $\mathfrak{B}$ has an orbit contained in the ball $r\overline{\mathfrak{B}} = \{X \in \mathcal{B}(H_2, H_1) : \|X\| \le r\}$, where $r < 1$, then it has a fixed point $K \in r\overline{\mathfrak{B}}$.*

As we know, biholomorphic transformations of $\mathfrak{B}$ are just fractional-linear transformations corresponding to J-unitary operators in $H = H_1 + H_2$ with the indefinite scalar product $[x,y] = (P_1x,y) - (P_2x,y)$.

Let us denote by $\mathcal{T}$ the group of all fractional-linear transformations of $\mathfrak{B}$. Note that $\mathcal{T}$ contains all Möbius maps. Indeed it can be easily checked that $\mu_A = \phi_{M_A}$ where M_A is the J-unitary operator with the matrix

$$\begin{pmatrix} (1_H - A^*A)^{-1/2} & A^*(1_K - AA^*)^{-1/2} \\ A(1_H - A^*A)^{-1/2} & (1_K - AA^*)^{-1/2} \end{pmatrix}.$$

Since $\mu_A(0) = A$, we see that $\mathcal{T}$ acts transitively on $\mathfrak{B}$.

Lemma 16.11. *Let U be a J-unitary operator on a Π_k-space H, ϕ_U the corresponding fractional-linear map and $A = \phi_U(0)$. Let $C = \|U\|$ and $r = \|A\|$. Then*

$$C \le \sqrt{(1+r)(1-r)^{-1}}. \tag{16.10}$$

and

$$r \le \sqrt{(C^2-1)/(C^2+1)}. \tag{16.11}$$

Proof. Let $V = M_A^{-1}U$, then $\phi_V(0) = (\mu_A)^{-1}(A) = 0$, so that the J-unitary operator V preserves subspaces H_1 and H_2; it follows that V is a unitary operator on H. Thus $\|U\| = \|M_AV\| = \|M_A\|$, so it suffices to prove the inequalities (16.10) and (16.11) for $U = M_A$.

Let, for brevity, $S = (1 + A^*A)(1 - A^*A)^{-1}$ and $T = (1 + AA^*)(1 - AA^*)^{-1}$. For any $z = x_1 + x_2 \in H_1 + H_2$, a direct calculation gives

$$\|M_Az\|^2 = (Sx_1, x_1) + (Tx_2, x_2) + 4\,\mathrm{Re}\left((1 - AA^*)^{-1}Ax_1, x_2\right).$$

Recall that in our notation $\|A\| = r$, $\|M_A\| = C$. Since

$$\|S\| = \|T\| = (1 + r^2)(1 - r^2)^{-1}$$

and

$$\|(1-AA^*)^{-1}A\| = \|(1-AA^*)^{-1}AA^*(1-AA^*)^{-1}\|^{1/2}$$
$$= r(1-r^2)^{-1},$$

we get

$$\begin{aligned}\|M_Az\|^2 &\le (1+r^2)(1-r^2)^{-1}(\|x_1\|^2+\|x_2\|^2) + 4r(1-r^2)^{-1}\|x_1\|\|x_2\| \\ &\le (1+r^2)(1-r^2)^{-1}(\|x_1\|^2+\|x_2\|^2) + 2r(1-r^2)^{-1}(\|x_1\|^2+\|x_2\|^2) \\ &= (1+r)(1-r)^{-1}\|z\|^2,\end{aligned}$$

which proves (16.10).

On the other hand, for $x \in H_2$, we have

$$\begin{aligned}\|M_Ax\|^2 &= (AA^*(1-AA^*)^{-1}x, x) + ((1-AA^*)^{-1}x, x) \\ &= \|\sqrt{T}x\|^2\end{aligned}$$

whence

$$\sqrt{(1+r^2)/(1-r^2)} = \|\sqrt{T}\| \le \|M_A\| = C.$$

This shows that the inequality (16.11) holds. □

The proof of (16.6) in Theorem 16.8. Now recall that by the assumptions of theorem we have a bounded group $\{\pi(g) : g \in G\}$ of operators on a Hilbert space H preserving the form Φ given by (16.5). Introducing the indefinite scalar product $[x, y] = (P_1x, y) - (P_2x, y)$ on H, we convert H into a Π_k-space:

$$H = H_1 + H_2, \quad \text{where } H_i = P_iH.$$

Since $\Phi(x) = [x, x]$, all operators $\pi(g)$ are J-unitary. Let $\Gamma = \{\phi_{\pi(g)} : g \in G\}$ be the corresponding group of fractional-linear transformations of the open unit ball $\mathfrak{B}$ of $\mathcal{B}(H_2, H_1)$, and consider the Γ-orbit O of the point $0 \in \mathfrak{B}$.

For $g \in G$, the inequality $\|\pi(g)\| \le \|\pi\|$, Lemma 16.11 and monotonicity of the function $t \mapsto \sqrt{(t^2-1)/(t^2+1)}$ imply that

$$\|\phi_{\pi(g)}(0)\| \le R := \sqrt{(\|\pi\|^2-1)/(\|\pi\|^2+1)},$$

so $O \subset R\overline{\mathfrak{B}}$. By Lemma 16.10, there is an operator $K \in R\overline{\mathfrak{B}}$ such that $\phi_{\pi(g)}(K) = K$, for all $g \in G$.

Let $V = M_K$ and $U(g) = V\pi(g)V^{-1}$ for each $g \in G$. Then $U(g)$ is J-unitary and

$$\phi_{U(g)}(0) = \mu_K \circ \phi_{\pi(g)}\mu_{-K}(0) = \mu_K(\phi_{\pi(g)}(K)) = \mu_K(K) = 0.$$

Therefore, $U(g)$ preserves H_1 and H_2. Since $(x, y) = [x_1, y_1] - [x_2, y_2]$, where $x_i = P_i x \in H_i$, $y_i = P_i y \in H_i$, $i = 1, 2$, we see that

$$\begin{aligned}(U(g)x, U(g)y) &= [U(g)x_1, U(g)y_1] - [U(g)x_2, U(g)y_2] = [x_1, y_1] - [x_2, y_2] \\ &= (x, y),\end{aligned}$$

for all $x, y \in H$. Thus $U(g)$ is a unitary operator in H. We proved that π is similar to a unitary representation; moreover, by Lemma 16.11,

$$\begin{aligned}c(\pi) &\le \|V\|\|V^{-1}\| = \|M_K\|\|M_{-K}\| \le \sqrt{(1+R)(1-R)^{-1}}^{\,2} \\ &= (1+R)(1-R)^{-1}.\end{aligned}$$

Since $R = \sqrt{(\|\pi\|^2 - 1)/(\|\pi\|^2 + 1)}$, we get that

$$c(\pi) \le \|\pi\|^2 + 1 + \sqrt{\|\pi\|^4 - 1} < 2\|\pi\|^2 + 1,$$

which completes the proof. □

The fact that our estimate of the similarity degree does not depend on the number of negative squares leads to the conjecture that the result extends to representations preserving forms with infinite number of negative squares. We shall see now that this is not true.

It is known (see [24]) that for some groups there exist bounded representations which are not similar to unitary ones (there is a conjecture that all nonamenable groups have such representations). Let π be such a representation of a group G on a Hilbert space H. We define a representation τ of G on $\mathcal{H} = H \oplus H$ by setting

$$\tau(g) = \begin{pmatrix} \pi(g) & 0 \\ 0 & \pi(g^{-1})^* \end{pmatrix}.$$

Clearly, τ is bounded. Moreover, it is not similar to a unitary representation because otherwise π, being its restriction to an invariant subspace, would be similar to a restriction of a unitary representation, which is again unitary.

The space $\mathcal{H}$ is a Krein space with respect to the inner product $[x_1 \oplus y_1, x_2 \oplus y_2] = (x_1, y_2) + (y_1, x_2)$. Indeed, $\mathcal{H} = \mathcal{H}_+ + \mathcal{H}_-$, where the subspaces $\mathcal{H}_+ = \{x \oplus x : x \in H\}$ and $\mathcal{H}_- = \{x \oplus (-x) : x \in H\}$ are respectively positive and negative. It remains to check that the form $\Phi(x \oplus y) = [x \oplus y, x \oplus y]$ is preserved by operators $\tau(g)$:

$$\begin{aligned}[\tau(g)(x \oplus y), \tau(g)(x \oplus y)] &= (\pi(g)x, \pi(g^{-1})^* y) + (\pi(g^{-1})^* y, \pi(g)x) \\ &= (x, y) + (y, x) = [x \oplus y, x \oplus y].\end{aligned}$$

16.5 Quasi-positive definite functions

Recall that a function ϕ on a group G is *positive definite* (PD, for brevity) if $\phi(g^{-1}) = \overline{\phi(g)}$, for $g \in G$, and the matrices $A_n = (\phi(g_i^{-1}g_j))_{i,j=1}^n$ have no negative eigenvalues, for all $n \in \mathbb{N}$ and all n-tuples $g_1, \dots, g_n \in G$. In other words, the quadratic forms $\sum_{i,j=1}^n \phi(g_i^{-1}g_j)z_i\overline{z_j}$ are positive for all $n \in \mathbb{N}$. A famous theorem of Bochner [2] states that all such functions can be described as matrix elements of unitary representations:

$$\phi(g) = (\pi(g)x, x),$$

where π is a unitary representation of G in a Hilbert space H and $x \in H$.

We say that ϕ is PD *of finite type* if the corresponding representation is finite-dimensional. It could be proved that ϕ is PD of finite type if and only if it satisfies the condition

$$\phi(g^{-1}h) = \sum_{i=1}^{m} a_i(g)\overline{a_i(h)} \quad \text{for all } g, h \in G,$$

where a_i are some functions on G. For example, the function $\cos x$ is PD of finite type on $\mathbb{R}$.

A function ϕ on a group G is called *quasi-positive definite* (QPD hereafter) if $\phi(g^{-1}) = \overline{\phi(g)}$, for $g \in G$, and there is $k \in \mathbb{N}$ such that, for any $n \in \mathbb{N}$ and any n-tuple $g_1, \dots, g_n \in G$, the matrix $(\phi(g_i^{-1}g_j))_{i,j=1}^n$ has at most k negative eigenvalues. In other words, the quadratic form $\sum_{i,j=1}^n \phi(g_i^{-1}g_j)z_i\overline{z_j}$ should have at most k negative squares.

The study of QPD functions was initiated by M. G. Krein [15] motivated by applications to probability theory—in particular, to infinite divisible distributions and, more generally, to stochastic processes with stationary increments. Other applications of theory of QPD functions are related to moment problems, Toeplitz forms and other topics of functional analysis; see [28, 26] and references therein.

It is easy to see that the difference $a(g) - b(g)$ of two PD functions is a QPD function if b is of finite type. Clearly, such QPD functions are bounded. The following theorem shows that all bounded QPD functions are of this type.

Theorem 16.12. *Every bounded QPD function ϕ can be written in the form*

$$\phi(g) = \phi_1(g) - \phi_2(g),$$

where ϕ_1 is a PD function and ϕ_2 is a PD function of finite type.

Proof. There is a standard way to associate with ϕ a J-unitary representation of G on a Π_k-space. Let W be the linear space of all finitely supported functions on G; we define an indefinite scalar product $[\cdot,\cdot]$ on W by setting

$$[f_1, f_2] = \sum_{g,h \in G} f_1(g)\overline{f_2(h)}\phi(g^{-1}h). \tag{16.12}$$

For each $g \in G$, we define an operator T_g on W by setting $T_g f(h) = f(g^{-1}h)$. It is easy to check that the operators T_g preserve $[\cdot,\cdot]$, that is, $[T_g f_1, T_g f_2] = [f_1, f_2]$, for all f_1, f_2. Clearly, the map $g \mapsto T_g$ is a representation of G on W.

Defining by ε_g, for $g \in G$, the function on G equal 1 at g and 0 at other elements, we see that the matrix $(\phi(g_i^{-1}g_j))_{i,j=1}^n$ is the Gram matrix for the family $\varepsilon_{g_1}, \dots, \varepsilon_{g_n}$. Since the linear span of vectors ε_g coincides with W, the condition "ϕ is QPD" implies that the dimension of any negative subspace of W does not exceed k. It follows that $W = W_1 + H_-$, where W_1 is a positive subspace, H_- is negative, and $\dim H_- = k$. Denoting by H_+ the completion of W_1 with respect to the scalar product $[\cdot,\cdot]|_{W_1}$, we get a Π_k-space $H = H_+ + H_-$. It is not difficult to show that operators T_g extend to bounded J-unitary operators $U(g)$ on H. It follows easily from the definition that $\phi(g) = [U(g)f, f]$, where f is the image of ε_e in H.

Since ϕ is bounded, the representation U is bounded (see, e. g., [26, Theorem 3.2]). By Corollary 16.7, there is a decomposition $H = K_+ + K_-$ where K_+ is positive, K_- is negative, and both subspaces are invariant for operators $U(g)$. In other words, the operators $U(g)$ commute with the projection P on K_+. Setting $f_+ = Pf$, $f_- = (1-P)f$, we get

$$\begin{aligned}\phi(g) &= [U(g)f, f] = [U(g)f_+, f_+] + [U(g)f_-, f_-] = (U(g)f_+, f_+) - (U(g)f_-, f_-)\\ &= \phi_1(g) - \phi_2(g),\end{aligned}$$

which is what we need because the functions ϕ_1 and ϕ_2 are PD, and ϕ_2 is of finite type. □

For amenable groups, the result was proved by K. Sakai [26].

16.6 *J*-symmetric algebras and Burnside-type theorems

As in linear algebra, after proving the existence of a nontrivial invariant subspace (IS, for brevity) for a single operator, one looks for conditions under which a family of operators has a common IS. Since the lattice $\mathrm{Lat}(E)$ of invariant subspaces of a family $E \subset \mathcal{B}(H)$ coincides with $\mathrm{Lat}(\mathcal{A}(E))$, where $\mathcal{A}(E)$ is the algebra generated by E, it is reasonable to restrict ourself by study of nonpositive invariant subspaces for algebras (more precisely, for J-symmetric operator algebras in a Π_k-space H). Thus one may rewrite the Naimark's theorem in the form: all commutative J-symmetric algebras in H have invariant MNPS. What else?

For algebras of operators in a finite-dimensional space, the problem of existence of invariant subspaces was completely solved by W. Burnside [3]: the only algebra that has no IS is the algebra of all operators. For infinite-dimensional Hilbert spaces, the problem is unsolved: it is unknown if there exists an algebra $A \subset \mathcal{B}(H)$ which has no (closed) IS and is not WOT-dense in $\mathcal{B}(H)$. In the presence of compact operators, the answer was given by Victor Lomonosov [17]: if an algebra A contains at least one nonzero compact operator, then either A has an invariant subspace or it is WOT-dense in $\mathcal{B}(H)$. In fact, he proved much more: if an algebra A contains a nonzero compact operator and has no invariant subspaces then the *norm*-closure of A contains the algebra $\mathcal{K}(H)$ of all compact operators. These results were further extended in Lomonosov's work [19].

For $\star$-algebras of operators, von Neumann's *double commutant Theorem* [5] immediately implies a Burnside-type result: a $\star$-algebra of operators has an invariant subspace if and only if it is not WOT-dense in $\mathcal{B}(H)$.

Since Theorem 16.1 establishes that a J-symmetric operator has an invariant subspace, it leads to the traditional Burnside-type problem for J-symmetric algebras: which J-symmetric algebras of operators on a space of Π_k-type have no invariant subspaces?

The first answer was given by R. S. Ismagilov [11]: a J-symmetric WOT-closed algebra A in a Π_k-space either has an invariant subspace or coincides with $\mathcal{B}(H)$ (this work presents also another proof of Pontryagin's theorem, which is short but based on a deep result of J. Schwartz [29] about invariant subspaces of finite-rank perturbations of self-adjoint operators). Furthermore, A. I. Loginov and V. S. Shulman [16] (see [12] and [13] for a more transparent presentation) proved the corresponding result for *norm-closed* J-symmetric algebras in Π_k-spaces: a J-symmetric algebra $A \subset \mathcal{B}(H)$ has no invariant subspaces if and only if its norm-closure contains the algebra $\mathcal{K}(H)$. The proof is quite complicated and uses the striking theorem of J. Cuntz [4] about C^*-equivalent Banach $\star$-algebras.

The following Burnside-type result is more closely related to the Pontryagin–Krein theorem: it describes J-symmetric algebras that have no *nonpositive* invariant subspaces. To formulate it, let us consider a Hilbert space E and the direct sum $H = \bigoplus_{i=1}^{n} E_i$ of $n \leq \infty$ copies of E. Let $\mathcal{B}(E)^{(n)}$ be the algebra of all operators on H of the form $T \oplus T \oplus \cdots$, where $T \in \mathcal{B}(E)$. On each summand $E_i = E$ in H, we choose a projection P_i with $0 \leq \dim P_i E = k_i < \dim E$, assuming that $\sum_i k_i = k < \infty$, and set $P = P_1 \oplus P_2 \oplus \cdots$, $J = 1-2P$. Then H is a Π_k-space with respect to the inner product $[x, y] = (Jx, y)$. The algebra $\mathcal{B}(E)^{(n)}$ is clearly J-symmetric; J-symmetric algebras of this form are called *model algebras*.

Theorem 16.13. *A WOT closed J-symmetric algebra A on a Π_k-space H does not have nonpositive invariant subspaces if and only if it is a direct J-orthogonal sum of a W^*-algebra on a Hilbert space and a finite number of model algebras.*

The proof can be easily deduced from [13, Theorem 13.7] that gives a description of all algebras that have no neutral invariant subspaces. To describe *norm-closed* J-symmetric algebras without nonpositive invariant subspaces, one should replace in Theorem 16.13 a W*-algebra by a C*-algebra and model algebras $\mathcal{B}(E)^{(n)}$ by the algebras $A^{(n)}$, where $A \subset \mathcal{B}(E)$ is a C*-algebra containing $\mathcal{K}(E)$.

Another natural version of the problem is to describe Banach *-algebras with the property that all their J-symmetric representations in a Π_k-space have MNPS. It is shown in [13, Theorem 19.4] that this property is equivalent to the absence of irreducible Π_k-representations; let us denote by $(\mathcal{K})$ the class of all Banach *-algebras that possess it.

It follows from Naimark's theorem that $(\mathcal{K})$ contains all commutative algebras. On the other hand, Theorem 16.8 implies that any Banach algebra, generated by a bounded subgroup of unitary elements belongs to $(\mathcal{K})$. This implies that $(\mathcal{K})$ contains all C*-algebras (this was proved earlier in [31]).

Recall that a Banach *-algebra A is *Hermitian* if all its self-adjoint elements have real spectra. Let us say that A is *almost Hermitian* if the elements with real spectra are dense in the space of all self-adjoint elements. It is proved in [13, Corollary 20.6] that all almost Hermitian algebras belong to $(\mathcal{K})$; this result has applications to the study of unbounded derivations of C^*-algebras (see [13]).

It is known that the group algebras $L^1(G)$ of locally compact groups are not Hermitian for some G (the referee kindly informed us about a recent result of Samei and Wiersma [27] which states that $L^1(G)$ is not Hermitian if G is not amenable). It is not known if all algebras $L^1(G)$ are almost Hermitian. Nevertheless, all $L^1(G)$ belong to $(\mathcal{K})$; moreover, the following result holds.

Theorem 16.14. *If G is a locally compact group then any J-symmetric representation of $L^1(G)$ on a Π_k-space H has invariant dual pair of subspaces.*

We begin the proof of this theorem with a general statement which is undoubtedly known but it is difficult to give a precise reference.

Recall that the *essential subspace* for a representation D of an algebra A on a Banach space X is the closure of the linear span $D(A)X$ of all vectors $D(a)x$, where $a \in A$, $x \in X$. If the essential subspace for D coincides with X, then D is called *essential.*

Lemma 16.15. *Let L be an ideal of a Banach algebra A, and $D : L \to \mathcal{B}(X)$ be a bounded essential representation of L in a Banach space X. If L has a bounded approximate identity $\{u_n\}$, then D extends to a bounded representation $\widetilde{D}$ of A in X, and $\|\widetilde{D}\| \leq C\|D\|$ where $C = \sup_n \|u_n\|$.*

Proof. Let us show that

$$\left\|\sum_{i=1}^{n} D(ab_i)x_i\right\| \leq C\|D\|\|a\|\left\|\sum_{i=1}^{n} D(b_i)x_i\right\|,$$

for any $a \in A$, $b_i \in L$, $x_i \in X$. Indeed,

$$\left\|\sum_{i=1}^{n} D(au_n b_i)x_i\right\| = \left\|\sum_{i=1}^{n} D((au_n)b_i)x_i\right\|$$

$$= \left\| D(au_n)\left(\sum_{i=1}^{n} D(b_i)x_i\right)\right\|$$
$$\leq \|D(au_n)\| \left\|\sum_{i=1}^{n} D(b_i)x_i\right\|$$
$$\leq C\|D\|\|a\| \left\|\sum_{i=1}^{n} D(b_i)x_i\right\|,$$

and it remains to note that

$$\left\|\sum_{i=1}^{n} D(ab_i)x_i - \sum_{i=1}^{n} D(au_nb_i)x_i\right\| \to 0 \quad \text{when } n \to \infty.$$

Now we may define a map T_a on the space $D(L)X$ by setting

$$T_a\left(\sum_{i=1}^{n} D(b_i)x_i\right) = \sum_{i=1}^{n} D(ab_i)x_i \quad \text{for all } b_i \in L \text{ and } x_i \in X.$$

By the above, T_a is a well-defined linear operator on $D(L)X$ and

$$\|T_a\| \leq C\|D\|\|a\|.$$

Denoting by $\widetilde{D}(a)$ the closure of T_a, we obtain an operator on X with

$$\|\widetilde{D}(a)\| \leq \|C\|D\|\|a\|.$$

It is easy to see that the map $\widetilde{D} : a \mapsto \widetilde{D}(a)$ is a representation of A on X, extending D.

□

Now we need a result about J-symmetric representations of $*$-algebras. Recall that a closed subspace L of an indefinite metric space H is *nondegenerate* if $L \cap L^{\perp} = 0$.

Lemma 16.16. *Let a *-algebra L have a bounded approximate identity $\{u_n\}$, and let D be a J-symmetric representation of L on a Krein space H. Then the essential subspace $H_0 = \overline{D(L)H}$ of D is nondegenerate, and $H_0^{\perp} \subset \ker D(L)$.*

Proof. Let $K = H_0 \cap H_0^{\perp}$. For any $x \in H$, $y \in H_0^{\perp}$ and $a \in L$, we have $[x, D(a)y] = [D(a^*)x, y] = 0$ whence $D(a)y = 0$. We proved that $H_0^{\perp} \subset \ker D(L)$.

On the other hand, since $K \subset H_0$, then for each $y \in K$ and each $\varepsilon > 0$ there is $z \in D(A)H$ with $\|z - y\| < \varepsilon$. Note that $\|D(u_n)z - z\| \to 0$ when $n \to \infty$, because $D(u_n)D(a)x = D(u_na)x \to D(a)x$. Since $D(u_n)y = 0$, we get that

$$\|z\| = \lim\|D(u_n)(z - y)\| \leq C\|D\|\varepsilon,$$

where $C = \sup_n \|u_n\|$. Therefore,

$$\|y\| \leq \|z\| + \|y - z\| \leq \varepsilon(1 + C\|D\|).$$

Since ε can be arbitrary, we conclude that $y = 0$. Thus $K = 0$ and H_0 is nondegenerate.

□

The proof of Theorem 16.14. Let now H be a Π_k-space and $D : L^1(G) \to B(H)$ be a continuous J-symmetric representation. It is known that $L^1(G)$ has a bounded approximate identity $\{u_n\}$ (moreover $\|u_n\| = 1$, for all n). So, by Lemma 16.16, H decomposes in J-orthogonal sum of subspaces $H = H_0 + H_0^\perp$, where H_0 is the essential subspace for D.

The algebra $L^1(G)$ is an ideal of the $\star$-algebra $M(G)$ of all finite measures on G; we will denote the involution in $M(G)$ by $\mu \mapsto \mu^\flat$ and the product by $\mu * \nu$. Applying Lemma 16.15 to the restriction of D to H_0, we have that there is a representation $\widetilde{D}$ of $M(G)$ on H_0 extending D. To check that $\widetilde{D}$ is J-symmetric, it suffices to check the equality $[\widetilde{D}(\mu)x, y] = [x, \widetilde{D}(\mu^\flat)y]$, for x of the form $D(f)z$, where $f \in L^1(G)$, $z \in H_0$. In this case, we have

$$\begin{aligned}[\widetilde{D}(\mu)x, y] &= [\widetilde{D}(\mu)D(f)z, y] = [D(\mu * f)z, y] = [z, D(f^\flat * \mu^\flat)y] \\ &= [z, D(f^\flat)\widetilde{D}(\mu^\flat)y] = [D(f)z, \widetilde{D}(\mu^\flat)y] \\ &= [x, \widetilde{D}(\mu^\flat)y].\end{aligned}$$

For each $g \in G$, we denote by δ_g the point measure in g. Setting $\pi(g) = \widetilde{D}(\delta_g)$, we obtain a J-unitary representation of G. Indeed, since $(\delta_g)^{-1} = \delta_{g^{-1}}$, we have

$$\pi(gh) = \widetilde{D}(\delta_{gh}) = \widetilde{D}(\delta_g * \delta_h) = \widetilde{D}(\delta_g)\widetilde{D}(\delta_h) = \pi(g)\pi(h),$$

and

$$\pi(g)^\sharp = (\widetilde{D}(\delta_g))^\sharp = \widetilde{D}(\delta_{g^{-1}}) = \pi(g)^{-1}.$$

Since $\|\delta_g\| = 1$,

$$\|\pi(g)\| \le \|\widetilde{D}\|,$$

so π is bounded.

Let us check that the representation π is strongly continuous. Since π is bounded, it suffices to verify that the function $g \mapsto \pi(g)x$ is continuous for x in a dense subset of H_0. So we may take $x = D(f)y$, for some $f \in L^1(G)$, $y \in H_0$. Since the map $g \longmapsto (\delta_g * f)(h) = f(g^{-1}h)$ from G to $L^1(G)$ is continuous, we get that

$$\pi(g)x = \pi(g)D(f)y = D(\delta_g * f)y$$

continuously depends on g.

Applying Corollary 16.7, we find an invariant dual pair of subspaces K_+, K_- of H_0 invariant for all operators $\pi(g)$. To see that these subspaces are invariant for $D(L^1(G))$, let us denote by W the representation of $L^1(G)$ generated by π:

$$W(f) = \int_G f(g)\pi(g)dg.$$

Clearly, K_+ and K_- are invariant for all operators $W(f)$, and we have only to show that $W(f) = D(f)$, for all $f \in L^1(G)$.

Since $\pi(g)(D(f)x) = D(\delta_g * f)x$, for all $x \in H$ and $f \in L^1(G)$, we have

$$\begin{aligned} W(u)D(f)x &= \int_G u(g)\pi(g)D(f)xdg = \int_G u(g)D(\delta_g * f)xdg \\ &= D\Big(\int_G u(g)(\delta_g * f)dg\Big)x = D(u * f)x \\ &= D(u)D(f)x \end{aligned}$$

for each $u \in L^1(G)$. Since vectors of the form $D(f)x$ generate H, we conclude that $W(u) = D(u)$.

As we know, the restrictions of all operators $D(f)$, $f \in L^1(G)$, to $H_0^\perp$ are trivial. So we may choose any dual pair N_+, N_- of $H_0^\perp$ and, setting $H_+ = K_+ + N_+$, $H_- = K_- + N_-$, we will obtain a dual pair in H invariant for $D(L^1(G))$. □

Bibliography

[1] T. A. Azizov and I. S. Iohvidov, *Linear Operators in Spaces with Indefinite Metric*, John Wiley, Chichester, New York, 1989.

[2] S. Bochner, *Hilbert distances and positive definite functions*, Ann. Math. **42** (1941), 647–656.

[3] W. Burnside, *On the condition of reducibility of any group of linear substitution*, Proc. Lond. Math. Soc. **3** (1905), 430–434.

[4] J. Cuntz, *Locally C*-equivalent algebras*, J. Funct. Anal. **23** (1976), 95–106.

[5] P. G. Dixon, *Topologically irreducible representations and radicals in Banach algebras*, Proc. Lond. Math. Soc. (3) **74** (1997), 174–200.

[6] Ky Fan, *A generalization of Tichonoff's fixed point theorem*, Math. Anal. **142** (1961), 305–310.

[7] I. Gohberg, P. Lancaster and L. Rodman, *Idefinite Linear Algebra and Applications*, Birkhauser Verlag, 2005.

[8] J. W. Helton, *Unitary operators on a space with an indefinite inner product*, J. Funct. Anal. **6** (1970), 412–440.

[9] J. W. Helton, *Operators unitary in an indefinite metric and linear fractional transformations*, Acta Sci. Math. **32** (1971), 261–266.

[10] I. S. Iohvidov, *Unitary operators in spaces with indefinite metric*, Notes of NII of Math and Mech. Harkov University **21** (1949), 79–86.

[11] R. S. Ismagilov, *Rings of operators in a space with an indefinite metric*, Dokl. Akad. Nauk SSSR **171** (2) (1966), 269–271 (Soviet Math. Dokl. **7** (6) (1966), 1460–1462).

[12] E. Kissin, A. I. Loginov and V. S. Shulman, *Derivations of C*-algebras and almost Hermitian representations on Π_k-spaces*, Pac. J. Math. **174** (1996), 411–430.

[13] E. Kissin and V. S. Shulman, *Representations on Krein spaces and derivations of C*-algebras*, Pitman Monographs and Surveys in Pure and Applied mathematics, **89**, Addison Wesley Longman, 1997.

[14] M. G. Krein, *On an application of the fixed point principle in the theory of linear transformations of indefinite metric spaces*, Usp. Mat. Nauk **5** (1950), 180–190 (Am. Math. Transl. (2) **1** (1955), 27–35).

[15] M. G. Krein, *Integral representation of a continuous Hermitian-indefinite function with a finite number of negative squares*, Dokl. Akad. Nauk SSSR **125** (1958), 31–34.

[16] A. I. Loginov and V. S. Shulman, *Irreducible J-symmetric algebras of operators in spaces with an indefinite metric*, Dokl. Akad. Nauk SSSR **240** (1) (1978), 21–23 (Soviet Math. Dokl. **19** (3) (1978), 541–544).

[17] V. I. Lomonosov, *Invariant subspaces for the family of all operators commuting with completely continuous operators*, Funct. Anal. Appl. **7** (1973), 213–214.

[18] V. I. Lomonosov, *On stability of non-negative invariant subspaces*, New results on operator theory and its applications, Oper. Theory, Adv. Appl. **88** (1990), 186–189.

[19] V. I. Lomonosov, *An extension of Burnside's theorem to infinite-dimensional spaces*, Isr. J. Math. **75** (1991), 329–339.

[20] M. A. Naimark, *Commutative unitary operators on Π_k-spaces*, Dokl. Akad. Nauk SSSR **149** (1963), 1261–1263.

[21] M. A. Naimark, *On unitary representations of solvable groups in spaces with an indefinite metric*, Izv. Akad. Nauk SSSR **27** (1963), 1181–1185 (Am. Math. Transl. **49** (1966), 86–91).

[22] M. Ostrovskii, V. S. Shulman and L. Turowska, *Unitarizable representations and fixed points of groups of holomorphic transformations of operator balls*, J. Funct. Anal. **257** (2009), 2476–2496.

[23] M. Ostrovskii, V. S. Shulman and L. Turowska, *Fixed points of holomorphic transformations of operator balls*, Q. J. Math. Oxf. **62** (2011), 173–187.

[24] G. Pisier, *Similarity Problems and Completely Bounded Maps*, Lecture Notes of Math., **1618**, Springer, Berlin, 2001.

[25] L. S. Pontryagin, *Hermitian operators in spaces with indefinite metric*, Izv. Akad. Nauk SSSR **8** (1944), 243–280.

[26] K. Sakai, *On quasi-positive definite functions and unitary representations of groups in Pontryagin spaces*, J. Math. Kyoto Univ. **19** (1979), 71–90.

[27] E. Samei and M.Wiersma, *Quasi-Hermitian locally compact groups are amenable*, arXiv:1805.07908, July 6, 2019.

[28] Z. Sasvari, *On bounded functions with a finite number of negative squares*, Monatshefte Math. **99** (1985), 223–234.

[29] J. T. Schwartz, *Subdiagonalization of operators in Hilbert space with compact imaginary part*, Commun. Pure Appl. Math. **15** (1962), 159–172.

[30] A. A. Shkalikov, *On the existence of invariant subspaces for dissipative operators in spaces with indefinite metric*, Proc. Steklov Inst. Math. **248** (2005), 287–296.

[31] V. S. Shulman, *On representation of C*-algebras on indefinite metric spaces*, Matem. Zametki **22** (1977), 583–592.

[32] V. S. Shulman, *On fixed points of fractional-linear transformations*, Funct. Anal. Appl. **14** (1980), 93–94.

[33] W. Takahashi, *A convexity in metric space and non-expansive mappings, I*, Kodai Math. Semin. Rep. **22** (1970), 142–149.

Dong Li and Alexander Volberg

17 Poincaré type and spectral gap inequalities with fractional Laplacians on Hamming cube

Dedicated to Victor Lomonosov, sunny person, incisive mathematician, and a friend

Abstract: We prove here some dimension free Poincaré-type inequalities on Hamming cube for function with different spectral properties and for fractional Laplacians. In this note, the main attention is paid to estimates in L^1 norm on Hamming cube. We build the examples showing that our assumptions on spectral properties of functions cannot be dropped in general.

Keywords: Fractional Laplacian, Hamming cube, spectral gap, Poincaré-type inequalities for fractional Laplacian

MSC 2010: 42B20, 42B35, 47A30

17.1 Poincaré-type inequalities with Laplacian

Lemma 17.1. *Let $0 < \beta \le 2$. Let $(\Omega, d\mu)$ be a probability space. Then for any random variable $g : \Omega \to \mathbb{R}$ with $\mathbb{E}|g|^2 < \infty$, we have*

$$\mathbb{E}|g - \mathbb{E}g|^2 \ge c_1 \mathbb{E}|g|^2 - 2^{\frac{1}{\beta}} \cdot \left|\mathbb{E}[|g|^\beta \operatorname{sgn}(g)]\right|^{\frac{2}{\beta}},$$

where $c_1 > 0$ is an absolute constant.

Proof. Without the loss of generality, we assume $\mathbb{E}|g|^2 = 1$. Let $c_1 > 0$ be a sufficiently small absolute constant. If $\mathbb{E}|g-\mathbb{E}g|^2 \ge c_1$ we are done. Now assume $\mathbb{E}|g-\mathbb{E}g|^2 < c_1 \ll 1$. Together with the condition $\mathbb{E}|g|^2 = 1$, we infer that $0 \le 1 - |\mathbb{E}g| \ll 1$. Replacing g by $-g$, if necessary we may assume $|1 - \mathbb{E}g| \ll 1$.

Combining this with $\mathbb{E}|g - \mathbb{E}g|^2 < c_1$, we conclude that g is very close to 1 with probability very close to 1 (the closeness depends only on small absolute constant c_1).

Acknowledgement: Volberg is partially supported by the NSF DMS-1600065.

Dong Li, Department of Mathematics, Hong Kong University of Science and Technology, Clear Water Bay, Hong Kong, e-mail: madli@ust.hk
Alexander Volberg, Department of Mathematics, Michigan State University, East Lansing, MI 48823, USA, e-mail: volberg@math.msu.edu

https://doi.org/10.1515/9783110656756-017

Let $\eta = c_1^{\frac{1}{10}}$. Then for c_1 sufficiently small (below the smallness of c_1 is independent of β since $0 < \beta \le 2$), we have

$$\begin{aligned}\int |g|^\beta \operatorname{sgn}(g)d\mu &\ge \int_{|g-1|\le\eta} |g|^\beta \operatorname{sgn}(g)d\mu - \int_{|g-1|>\eta} |g|^\beta d\mu \\ &\ge \sqrt{\frac{3}{4}} - \int_{|g-1|>\eta} 4\cdot(|g-1|^\beta + 1)d\mu \\ &\ge \sqrt{\frac{3}{4}} - 4\int |g-1|^2 d\mu - 8\int_{|g-1|>\eta} d\mu \\ &\ge \frac{1}{\sqrt{2}}.\end{aligned}$$

In the above chain of inequalities, we used that g is very close to 1 with probability very close to 1, and also that $|g|^\beta \le 4\cdot(|g-1|^\beta + 1)$ for $\beta \in (0,2]$. The desired inequality then obviously follows. □

Let us recall that for function $g : \{-1,1\}^n \to \mathbb{R}$ its gradient is given by formula:

$$|\nabla g|^2(x) = \sum_{i=1}^n \left(\frac{g(x) - g(x^i)}{2}\right)^2,$$

where $x^i \in \{-1,1\}^n$ is the point of the Hamming cube such that $x_k = x_k^i$, $k \ne i$ and $x_i = -x_i^i$.

Proposition 17.2. *Let $0 < \beta \le 2$. Then for any $g : \{-1,1\}^n \to \mathbb{R}$, we have*

$$\mathbb{E}|\nabla g|^2 \ge c_1 \mathbb{E}|g|^2 - 2^{\frac{1}{\beta}}|\mathbb{E}(|g|^\beta \operatorname{sgn}(g))|^{\frac{2}{\beta}},$$

where $c_1 > 0$ is an absolute constant.

Proof. This follows from the Poincaré inequality with $p = 2$ on Hamming cube:

$$\mathbb{E}|\nabla g|^2 \ge \mathbb{E}|g - \mathbb{E}g|^2$$

and the previous lemma. □

Next is an elementary lemma.

Lemma 17.3. *Let $a, b \in \mathbb{R}$, $p > 1$. Then there exists $c_p > 0$ such that*

$$(a-b)(|a|^{p-1}\operatorname{sgn} a - |b|^{p-1}\operatorname{sgn} b) \ge c_p(|a|^{\frac{p}{2}}\operatorname{sgn} a - |b|^{\frac{p}{2}}\operatorname{sgn} b)^2.$$

Moreover,

$$c_p = \min_{0\le t\le 1} \frac{1 - t^{\frac{2}{p}}}{1-t}\cdot\frac{1 - t^{\frac{2}{p'}}}{1-t} = 2\min\left(\frac{1}{p}, \frac{1}{p'}\right). \tag{17.1}$$

Proof. Notice that by symmetry we can think that either a, b are both positive or that $a > 0 > b$. Then by homogeneity the case $a > 0 > b$ is reduced to the estimate

$$(1+x)(1+x^{p-1}) \geq (1+x^{\frac{p}{2}})^2, \quad x \geq 0,$$

which is the same as $2x^{\frac{p}{2}} \leq x + x^{p-1}$. The latter inequality is just $2AB \leq A^2 + B^2$.

The case when both a, b are positive becomes

$$(1-x)(1-x^{p-1}) \geq c_p(1-x^{\frac{p}{2}})^2, \quad 0 \leq x \leq 1.$$

Notice that this inequality is false for $p = 1$, but it holds for $p > 1$. This is just because after the change of variable $x = t^{\frac{2}{p}}$ one can observe that

$$\lim_{t\to 1-} \frac{1-t^{\frac{2}{p}}}{1-t} > 0, \quad \lim_{t\to 1-} \frac{1-t^{\frac{2}{p'}}}{1-t} > 0.$$

From this, one sees immediately that

$$c_p := \inf_{0\leq x\leq 1} \frac{(1-x)(1-x^{p-1})}{(1-x^{\frac{p}{2}})^2} > 0. \qquad \square$$

Theorem 17.4. *Let $1 < p < \infty$. Then for any $f : \{-1,1\}^n \to \mathbb{R}$, we have*

$$-\mathbb{E}(\Delta f|f|^{p-1}\operatorname{sgn}(f)) \geq C_1 \cdot c_p \cdot \mathbb{E}|f|^p - 2^{\frac{p}{2}} \cdot c_p \cdot |\mathbb{E}f|^p,$$

where $C_1 > 0$ is an absolute constant, and $c_p = 2\min(\frac{1}{p}, \frac{1}{p'})$.

Proof. By an explicit computation, we have

$$-\mathbb{E}\Delta f|f|^{p-1}\operatorname{sgn}(f) \geq c_p\mathbb{E}\Big(\sum_{y\sim x} \big||f(y)|^{\frac{p}{2}}\operatorname{sgn}(f(y)) - |f(x)|^{\frac{p}{2}}\operatorname{sgn}(f(x))\big|^2\Big), \tag{17.2}$$

where the expectation is taken with respect to the x-variable. In fact, let

$$L_i f = \frac{f(x) - f(x^i)}{2},$$

where x_i is the same as x but its ith coordinate is changed to the opposite one. Then

$$-\Delta = \sum_i L_i.$$

But it is easy to see that

$$(L_i f, g) = (L_i f, L_i g),$$

because by definition

$$L_i f = \sum_{S\subset[n], i\in S} \hat{f}(S)x^S,$$

and thus

$$-(\Delta f, g) = \sum_i (L_i f, L_i g) = \sum_{x\sim y} \frac{f(x)-f(y)}{2} \cdot \frac{g(x)-g(y)}{2}.$$

$$-\mathbb{E}\Delta f |f|^{p-1} \operatorname{sgn}(f) \asymp \sum_{x\sim y} (f(x)-f(y))(|f|^{p-1}\operatorname{sgn}(f)(x) - |f|^{p-1}\operatorname{sgn}(f)(y)). \tag{17.3}$$

Now we combine Lemma 17.3 and this inequality to get inequality (17.2). Notice that we need $p > 1$ for this to be true.

Now we make a change of variable and denote $g(x) = |f(x)|^{\frac{p}{2}} \operatorname{sgn}(f(x))$. Note that g and f have the same sign. Clearly,

$$\mathbb{E}f = \mathbb{E}[|g|^\beta \operatorname{sgn}(g)],$$

where $\beta = \frac{2}{p} \in (0,2)$ since $1 < p < \infty$. The desired inequality then clearly follows from the previous proposition. □

17.2 Fractional Laplacian on Hamming cube and its spectral gap estimates

For $0 < \gamma \le 1$, we introduce

$$\Delta_\gamma = -(-\Delta)^\gamma,$$

by the following formula:

$$\Delta_\gamma x_{i_1} \dots x_{i_k} = k^\gamma x_{i_1} \dots x_{i_k}.$$

We call it the fractional Laplacian operator on Hamming cube.

The first claim of the next theorem is very well known for $p = 2$; it is the claim that Laplacian on Hamming cube has a spectral gap. It is interesting that this "spectral gap" estimate can be extrapolated to $1 < p < \infty$, and even, as we will see later, for $p = 1$ sometimes.

In Section 17.4, we will see that with extra spectral assumptions on f it holds even for $p = 1$.

Theorem 17.5. *Let $1 < p < \infty$. Then for any $f : \{-1,1\}^n \to \mathbb{R}$, we have*

$$\|e^{t\Delta}(f - \mathbb{E}f)\|_p \le e^{-k_1 t}\|f - \mathbb{E}f\|_p, \quad \forall t > 0,$$

where $k_1 = C_1 \cdot c_p$, $C_1 > 0$ is an absolute constant and $c_p = 2\min(\frac{1}{p}, \frac{1}{p'})$. Similarly for Δ_γ,

$$\|e^{t\Delta_\gamma}(f - \mathbb{E}f)\|_p \le e^{-k_\gamma t}\|f - \mathbb{E}f\|_p, \quad \forall t > 0,$$

where the constant $k_\gamma = k_1^\gamma$.

Proof. Without loss of generality, we can assume $\mathbb{E}f = 0$. Denote $I(t) = \mathbb{E}(|e^{t\Delta}f|^p)$. Since μ is uniform counting measure, we can directly differentiate and this gives

$$\frac{d}{dt}I(t) = p\mathbb{E}(\Delta g|g|^{p-1}\operatorname{sgn}(g)),$$

where $g = e^{t\Delta}f$. Note that $\mathbb{E}g = 0$. Thus by Theorem 17.4, we have

$$\frac{d}{dt}I(t) \leq -p \cdot C_1 \cdot c_p I(t).$$

Integrating in time then yields the desired inequality with $k_1 = C_1 \cdot c_p$. For the fractional Laplacian case, we can use the subordination identity

$$e^{-\lambda^\gamma} = \int_0^\infty e^{-\tau\lambda} d\rho(\tau), \quad \lambda \geq 0,$$

where $d\rho(\tau)$ is a probability measure on $[0, \infty)$. Clearly, then

$$e^{-t\lambda^\gamma} = \int_0^\infty e^{-\tau t^{\frac{1}{\gamma}}\lambda} d\rho(\tau).$$

It follows that

$$\begin{aligned}\|e^{t\Delta_\gamma}f\|_p &\leq \int_0^\infty e^{-k_1\tau t^{\frac{1}{\gamma}}} d\rho(\tau)\|f\|_p \\ &= e^{-k_2 t}\|f\|_p, \qquad k_2 = k_1^\gamma.\end{aligned}$$

□

17.3 Counterexamples

17.3.1 Counterexample to $\|e^{t\Delta}f\|_1 < e^{-ct}\|f\|_1$, $\mathbb{E}f = 0$

One cannot get independent of n estimate of Theorem 17.5 for $p = 1$. In fact, let $f(1, \ldots, 1) = 2^{n-1}$, $f(-1, \ldots, -1) = -2^{n-1}$, and $f(x) = 0$ for all other points $x \in \{-1, 1\}^n$. Then $\mathbb{E}f = 0$, $\|f\|_1 = 1$, and

$$e^{t\Delta}f(x) = 2^{-1}\left(\prod_{i=1}^n (1 + e^{-t}x_i) - \prod_{i=1}^n (1 - e^{-t}x_i)\right).$$

Hence,

$$\begin{aligned}\|e^{t\Delta}f\|_1 &= \frac{1}{2^{n+1}}\sum_{k=0}^n \binom{n}{k}|(1+e^{-t})^{n-k}(1-e^{-t})^k - (1-e^{-t})^{n-k}(1+e^{-t})^k| \\ &= \frac{1}{2^n}\sum_{0\leq k\leq \frac{n}{2}} \binom{n}{k}((1+e^{-t})^{n-k}(1-e^{-t})^k - (1-e^{-t})^{n-k}(1+e^{-t})^k).\end{aligned} \tag{17.4}$$

Now let us assume that there exists a universal constant $\kappa < 1$ such that for all n and all functions $f \in L^1(\{-1,1\}^n)$, $\mathbb{E}f = 0$, there exists t_0 such that for all $t \geq t_0$

$$\|e^{t\Delta}f\|_1 \leq \kappa\|f\|_1, \quad \text{if } \mathbb{E}f = 0. \tag{17.5}$$

Then by semigroup property (17.5) would imply the universal $t_1 = 2t_0 \cdot \log 2/\log\frac{1}{\kappa}$ such that for all n simultaneously

$$\|e^{t_1\Delta}f\|_1 < \frac{1}{2}\|f\|_1, \quad \text{if } \mathbb{E}f = 0. \tag{17.6}$$

Proposition 17.6. *Let $0 < \epsilon \leq 1/2$. Then for n sufficiently large, we have*

$$\frac{1}{2^n}\sum_{0\leq k\leq n/2}\binom{n}{k}\cdot\left((1+\epsilon)^{n-k}(1-\epsilon)^k - (1+\epsilon)^k(1-\epsilon)^{n-k}\right) \geq \frac{1}{2}(1-(1-\epsilon^2)^{\frac{n}{2}}).$$

Proof. We have

$$\begin{aligned}
2\cdot\text{LHS} \geq & \sum_{0\leq k\leq\frac{n}{2}}\frac{1}{2^n}\binom{n}{k}\cdot(1+\epsilon)^{n-k}(1-\epsilon)^k + \sum_{k>n/2}\frac{1}{2^n}\binom{n}{k}\cdot(1+\epsilon)^{n-k}(1-\epsilon)^k \\
& - \sum_{0\leq k\leq\frac{n}{2}}\frac{1}{2^{n-1}}\binom{n}{k}\cdot(1+\epsilon)^{\frac{n}{2}}(1-\epsilon)^{\frac{n}{2}} \\
\geq & \, 1-(1-\epsilon^2)^{\frac{n}{2}},
\end{aligned}$$

where in the last inequality we may assume n is an odd integer so that $k = n/2$ cannot be obtained. If n is even, one can get a similar bound. □

Now we use (17.4) and the Proposition to come to contradiction with (17.6). Hence (17.5) is false, too.

17.3.2 Counterexample to $\|e^{t\Delta_\gamma}f\|_{L^1} \leq e^{-ct}\|f\|_{L^1}$ for f with $\mathbb{E}f = 0$

Fix $0 < \gamma < 1$. Again we shall argue by contradiction. Assume the desired estimate is true. Similar to the Laplacian case, this would imply that there exists universal $t_1 > 0$ independent of n, such that for all f with $\mathbb{E}f = 0$, we have

$$\|e^{t_1\Delta_\gamma}f\|_1 \leq \frac{1}{4}\|f\|_1.$$

Now take the same f as in the Laplacian case. By using the subordination formula,

$$e^{-t\lambda^\gamma} = \int_0^\infty e^{-\tau t^{\frac{1}{\gamma}}\lambda}d\rho(\tau),$$

we get

$$(e^{t\Delta_\gamma}f)(x) = \frac{1}{2}\int_0^\infty \left(\prod_{j=1}^n (1 + e^{-\tau t^{\frac{1}{\gamma}}} x_j) - \prod_{j=1}^n (1 - e^{-\tau t^{\frac{1}{\gamma}}} x_j)\right) d\rho(\tau).$$

Hence

$$\begin{aligned}\|e^{t\Delta_\gamma}f\|_1 &= \frac{1}{2^n}\sum_{0\le k\le \frac{n}{2}} \binom{n}{k} \int_0^\infty ((1 + e^{-\tau t^{\frac{1}{\gamma}}})^{n-k}(1 - e^{-\tau t^{\frac{1}{\gamma}}})^k - (1 - e^{-\tau t^{\frac{1}{\gamma}}})^{n-k}(1 + e^{-\tau t^{\frac{1}{\gamma}}})^k) d\rho(\tau)\\ &\ge \frac{1}{2}\int_0^\infty (1 - (1 - e^{-2\tau t^{\frac{1}{\gamma}}})^{\frac{n}{2}}) d\rho(\tau).\end{aligned}$$

Now take $t = t_1$ and send n to infinity. We clearly arrive at a contradiction!

17.3.3 Counterexample to $\|e^{t\Delta}f\|_1 \le e^{-ct}\|f\|_1$ for band-limited f with small t

Consider the Gaussian space case. Let $\rho(x) = e^{-\frac{x^2}{2}}$ and consider $f(x) = x^3 = \mathrm{He}_3(x) + 3\,\mathrm{He}_1(x)$. Denote $\Delta_{\mathrm{ou}}f = f'' - xf'$. Then one can verify that

$$\int_{f\neq 0} (-\Delta_{\mathrm{ou}}f)\,\mathrm{sgn}(f)\rho(x)dx = 0.$$

This in turn implies that

$$\|e^{t\Delta}f\|_1 \ge \|f\|_1 - O(t^2),$$

for small t, which of course contradicts $\|e^{t\Delta}f\|_1 \le e^{-c_0 t}\|f\|_1 \le (1 - c_0 t + O(t^2))\|f\|_1$, $c_0 > 0$.

17.4 Band spectrum and $p = 1$

We first prove a certain Poincaré-type inequality involving $\Delta_\gamma f$, $0 < \gamma < 1$ in $L^1(\{-1,1\}^n)$. It will work for functions with band spectrum. Then we derive from it the inequality of "spectrum gap type" for functions in $L^1(\{-1,1\}^n)$ having band spectrum. Namely, we get the following.

Theorem 17.7. *For every $\gamma \in (0,1)$ there exits $c_\gamma > 0$ independent of n such that for every n and every $f \in L^1(\{-1,1\}^n)$ with band spectrum (meaning that it has only, say, 1-mode and 2-mode only), or, more generally, finite number of modes and $\mathbb{E}f = 0$, we have*

$$\|e^{t\Delta_\gamma}f\|_1 \le e^{-c_\gamma t}\|f\|_1. \tag{17.7}$$

This result will be proved, in fact, by two different methods. The second method shows, in particular, that the L^1-norm can be changed to any shift invariant norm (as $\{-1,1\}^n$ is isomorphic to F_2^n and shift can be understood on this group).

However, the Poincaré inequality in $L^1(\{-1,1\}^n)$ from the Subsection 17.4.2 below seems to have an independent interest and it looks slightly unusual.

But first we need a known result on hypercontractivity.

17.4.1 Hypercontractivity helps

Theorem 17.8. *Let f be Fourier localized to only* 1*-mode and* 2*-mode. Then for large t*

$$\|e^{t\Delta}f\|_1 \le e^{-ct}\|f\|_1. \tag{17.8}$$

Proof. This follows easily from Theorem 9.22 of [8]. We will repeat the reasoning for the sake of convenience of the reader. Let f_1 be the 1-mode of f, f_2 be its 2-mode. We first want to find the universal constant K such that

$$\|f_1 + \rho f_2\|_1 \le K\|f\|_1, \quad \forall \rho \in [0,1],\ \forall n. \tag{17.9}$$

Obviously, this holds for $\|\cdot\|_2$-norm. So the only thing we need to prove (17.9) is

$$\|f_1 + f_2\|_2 \le K\|f_1 + f_2\|_1, \tag{17.10}$$

which we will now deduce by repeating the proof of Theorem 9.22 of [8]. Let $q = 2+\varepsilon$, $t = \frac{1}{2}\log(1+\varepsilon)$. Then the application of the operator $e^{t\Delta}$ to f multiplies f_1 by $\frac{1}{\sqrt{q-1}}$ and multiplies f_2 by $\frac{1}{q-1}$. So by the well-known real hypercontractivity result, it maps L^2 to L^q with norm at most 1.

Therefore,

$$\|f_1 + f_2\|_q = \|f\|_q = \|e^{t\Delta}e^{-t\Delta}f\|_q \le \|e^{-t\Delta}f\|_2 \le (q-1)\|f\|_2$$

as the application of the operator $e^{-t\Delta}$ to f multiplies f_1 by $\sqrt{q-1}$ and multiplies f_2 by $q-1$. Now interpolate L^2-norm between $L^q = L^{2+\varepsilon}$-norm and L^1-norm, namely, let $\theta = \frac{1}{2}\frac{\varepsilon}{1+\varepsilon}$. Then

$$\|f\|_2 \le \|f\|_q^{1-\theta}\|f\|_1^{\theta} \le (q-1)^{1-\theta}\|f\|_2^{1-\theta}\|f\|_1^{\theta}.$$

Or,

$$\|f\|_2 \le (q-1)^{\frac{1-\theta}{\theta}}\|f\|_1 \le (1+\varepsilon)^{\frac{2}{\varepsilon}}\|f\|_1 \le e^2\|f\|_1.$$

Then (17.8) follows easily:

$$\|e^{t\Delta}(f_1+f_2)\|_1 = e^{-t}\|f_1 + e^{-t}f_2\|_1 \le e^{-t}\|f_1 + e^{-t}f_2\|_2 \le e^2e^{-t}\|f\|_1.$$

This gives (17.8) for $t \ge 4$ and $c = \frac{1}{2}$. □

Remark 17.9. The inequality

$$\|e^{t\Delta}(f_1 + f_2)\|_1 \le e^{-ct} \|f_1 + f_2\|_1 \tag{17.11}$$

is not true for small t. The counterexample in Subsection 17.3.3 shows that.

17.4.2 Poincaré inequality with Δ_γ in L^1

Recall that $\Delta_\gamma = -(-\Delta)^\gamma$.

Theorem 17.10. *For every $\gamma \in (0,1)$, there exits $b_\gamma > 0$ independent of n such that for every n and every $f \in L^1(\{-1,1\}^n)$ with band spectrum (meaning that it has only, say, 1-mode and 2-mode only), or, more generally, finite number of modes and $\mathbb{E}f = 0$, we have*

$$b_\gamma \|f\|_1 \le \mathbb{E}[(-\Delta_\gamma f) \cdot \operatorname{sgn} f] - \mathbb{E}[|\Delta_\gamma f| \cdot \mathbf{1}_{f=0}]. \tag{17.12}$$

Proof. Let $\gamma \in (0,1)$, put

$$C_\gamma := \int_0^\infty (1 - e^{-u}) \frac{du}{u^{1+\gamma}} < \infty.$$

It is then obvious that for any test function f such that $\mathbb{E}f = 0$ one has

$$-\Delta_\gamma f = C_\gamma^{-1} \int_0^\infty (\mathrm{Id} - e^{t\Delta}) f \frac{dt}{t^{1+\gamma}} = C_\gamma^{-1} \int_\delta^\infty (\mathrm{Id} - e^{t\Delta}) f \frac{dt}{t^{1+\gamma}} + \frac{1}{1-\gamma} O(\delta^{1-\gamma}) \|f\|_1, \tag{17.13}$$

where $O(\delta^{1-\gamma})$ depends on the number of nonzero modes of f. We use here Theorem 9.22 of [8] again. Hence,

$$\begin{aligned}
\mathbb{E}[\operatorname{sgn} f \cdot (-\Delta_\gamma f) \cdot \mathbf{1}_{f \ne 0}] &\ge C_\gamma^{-1} \int_\delta^\infty \frac{dt}{t^{1+\gamma}} \mathbb{E}((f - e^{t\Delta} f) \cdot \operatorname{sgn} f \cdot \mathbf{1}_{f \ne 0}) + \frac{1}{1-\gamma} O(\delta^{1-\gamma}) \|f\|_1 \\
&\ge C_\gamma^{-1} \int_\delta^\infty \frac{dt}{t^{1+\gamma}} \mathbb{E}|f| - C_\gamma^{-1} \int_\delta^\infty \frac{dt}{t^{1+\gamma}} \mathbb{E}|e^{t\Delta} f| + \frac{1}{1-\gamma} O(\delta^{1-\gamma}) \|f\|_1 \\
&\ge C_\gamma^{-1} \int_\delta^T \frac{dt}{t^{1+\gamma}} (\|f\|_1 - \|e^{t\Delta} f\|_1) + C_\gamma^{-1} \int_T^\infty \frac{dt}{t^{1+\gamma}} (\|f\|_1 - \|e^{t\Delta} f\|_1) + \frac{1}{1-\gamma} O(\delta^{1-\gamma}) \|f\|_1 \\
&\ge C_\gamma^{-1} \int_T^\infty \frac{dt}{t^{1+\gamma}} (\|f\|_1 - \|e^{t\Delta} f\|_1) + \frac{1}{1-\gamma} O(\delta^{1-\gamma}) \|f\|_1,
\end{aligned}$$

because fortunately $\|e^{t\Delta}$ contracts in L^1 as well. Now we choose T from Theorem 17.8 to have $\|e^{t\Delta}f\| \le \frac{1}{2}\|f\|_1$ for all $t \ge T$. It is an absolute constant bigger than 1 say. So we get

$$\mathbb{E}[\operatorname{sgn} f \cdot (-\Delta_\gamma f)] \ge A\gamma^{-1}C_\gamma^{-1}\|f\|_1 - \mathbb{E}[\mathbf{1}_{f=0}|\Delta_\gamma f|] + \frac{1}{1-\gamma}O(\delta^{1-\gamma})\|f\|_1.$$

Now we can tend δ to zero. We are done as after that we get the claim of Theorem 17.10. □

17.4.3 The first proof of Theorem 17.7 via Poincaré inequality in L^1

Denote

$$I(t) = \mathbb{E}\big|e^{t\Delta_\gamma}f\big|.$$

We want to estimate $\frac{d}{dt}I(t)$ for a test function f. Let $F := F_t := e^{t\Delta_\gamma}f$. Then for every test function f we have

$$\big|e^{-\varepsilon(-\Delta_\gamma)}F\big| - |F| = \begin{cases} \varepsilon \operatorname{sgn} F \cdot (-\Delta_\gamma F) + O(\varepsilon^2), & \text{if } F(x) \ne 0; \\ \varepsilon|\Delta_\gamma F| + O(\varepsilon^2), & \text{if } F(x) = 0. \end{cases} \tag{17.14}$$

Let us think that f is a test function with only finitely many nonzero Fourier–Walsh coefficients. If we look at $\frac{d}{dt}I(t)$ as the expression,

$$\frac{d}{dt}I(t) := \lim_{\varepsilon\to 0}\frac{I(t+\varepsilon) - I(t)}{\varepsilon},$$

we notice that the limit exists and that we can go to the limit under the sign of $\mathbb{E}$. So we get from (17.14) that

$$\frac{d}{dt}I(t) = \mathbb{E}(\operatorname{sgn} F_t \cdot (-\Delta_\gamma F_t)\cdot \mathbf{1}_{F_t\ne 0}) - \mathbb{E}(|\Delta_\gamma F_t| \cdot \mathbf{1}_{F_t=0}) \le -b_\gamma \mathbb{E}|F_t|.$$

The last inequality follows from Theorem 17.10. Hence,

$$\frac{d}{dt}I(t) \le -b_\gamma I(t), \quad I(0) = \|f\|_1.$$

Therefore, (17.7) is proved for test functions f with universal constant, and so Theorem 17.7 is proved just by density argument.

17.4.4 The second proof of Theorem 17.7 via the modification of the kernel of $e^{t\Delta_\gamma}$

We start with $\gamma = \frac{1}{2}$. See [3] for the formula

$$e^{-\xi^{1/2}} = \frac{1}{\sqrt{2\pi}}\int_0^\infty e^{-\frac{1}{2\sigma}}e^{-\frac{\sigma}{2}\xi}\frac{d\sigma}{\sigma^{3/2}}, \quad \xi \ge 0. \tag{17.15}$$

Let S be an arbitrary subset of the set $\{1,\dots,n\}$, that is $S \subset [n]$, and let $|S|$ denote its cardinality. Let $t \in [0,1]$. Then (17.15) shows

$$e^{-t|S|^{1/2}} = \frac{1}{\sqrt{2\pi}} \int_0^\infty e^{-\frac{1}{2\sigma}} e^{-\frac{\sigma}{2}t^2|S|} \frac{d\sigma}{\sigma^{3/2}}. \tag{17.16}$$

On the other hand, let $x = (x_1,\dots,x_n) \in \{-1,1\}^n =: C_n$ be a point in the Hamming cube C_n, we put

$$K_t^P(x) = \sum_{S\subset[n]} e^{-t|S|^{1/2}} x^S,$$

where $x^S := x_1^{s_1} \cdot \ldots \cdot x_n^{s_n}$, and we use the convention to associate a subset S with the string of 0 and 1, $S = (s_1,\dots,s_n)$, where $s_j = 1$ if and only if $j \in S$.

The function $K_P(x\cdot y)$ is called the Poisson kernel on Hamming cube C_n. Here, $x\cdot y$ is the usual product in the group $\{-1,1\}$. We also need a heat kernel. Put

$$K_r^H(x) = \sum_{S\subset[n]} e^{-r|S|} x^S = \prod_{i=1}^n (1 + e^{-r} x_i), \quad r \geq 0.$$

Now use (17.16) to write

$$K_t^P(x) = \int_0^\infty K^H_{t^2\frac{\sigma}{2}} e^{-\frac{1}{2\sigma}} \frac{d\sigma}{\sigma^{3/2}}. \tag{17.17}$$

We rewrite this as follows:

$$K_t^P(x) = \frac{1}{\sqrt{2\pi}} \int_0^\infty \prod_{i=1}^n (1 + e^{-t^2\frac{\sigma}{2}} x_i) e^{-\frac{1}{2\sigma}} \frac{d\sigma}{\sigma^{3/2}}. \tag{17.18}$$

More generally, we have

$$K_t^\alpha(x) = \sum_{S\subset[n]} e^{-t|S|^\alpha} x^S, \quad 0 < \alpha < 1.$$

It is known that with a positive kernel P_α one has (see [9], Proposition 1.2.12)

$$K_t^\alpha(x) = \int_0^\infty K^H_{t^2\frac{\sigma}{2}}(x) P_\alpha(\sigma) d\sigma. \tag{17.19}$$

Moreover, the asymptotic of P_α is as follows:

$$P_\alpha(\sigma) \approx \frac{c}{\sigma^{1+\alpha}}, \quad \sigma \to \infty. \tag{17.20}$$

In fact, consider function $p_\alpha(\sigma)$ is given by the relationship

$$\int_0^\infty e^{-\sigma\zeta} p_\alpha(\sigma)d\sigma = e^{-\zeta^\alpha}, \quad \zeta = \eta + i\xi, \quad \eta \geq 0. \tag{17.21}$$

To see that such a formula should exist, consider function $F_\alpha := e^{-\zeta^\alpha}$, $\zeta = \eta + i\xi$ in the right half-plane Π_+. If $\varphi = \arg\zeta \in [-\pi/2, \pi/2]$, then

$$|e^{-\zeta^\alpha}| = e^{-\Re\zeta^\alpha} = e^{-|\zeta|^\alpha \cos(\alpha\varphi)},$$

and $\cos(\alpha\varphi) > \cos(\alpha\frac{\pi}{2}) =: d_\alpha > 0$ if $\alpha \in (0,1)$. So function $F_\alpha = e^{-\zeta^\alpha}$ is a bounded analytic function in the right half plane, and on the imaginary axis $i\mathbb{R} = \{i\xi : \xi \in \mathbb{R}\}$ it is $e^{-d_\alpha|\xi|^\alpha}$, in particular it is in $L^2(\mathbb{R}, d\xi)$. So $F_\alpha \in H^2(\Pi_+) \cap H^\infty(\Pi_+)$. Therefore, it can be represented in the form (17.21) with $p_\alpha \in L^2(\mathbb{R}_+)$. Moreover, $p_\alpha(\sigma)$ is actually in $L^1(\mathbb{R}_+, d\sigma)$ and we even know its behavior for large σ. It is listed in the following lemma from [2]; see also [7].

Lemma 17.11. *Let $\alpha \in (0,2)$, then*

$$\frac{1}{2\pi}\int_{\mathbb{R}} e^{-|\xi|^\alpha} e^{i\xi\sigma} d\xi \asymp \frac{1}{(1+|\sigma|)^{1+\alpha}}.$$

Coming back to (17.21), we apply this formula to $\zeta = \eta > 0$. Then we get

$$2\int_0^\infty e^{-\frac{\sigma}{2}\eta} p_\alpha(\sigma/2)d\sigma = e^{-\eta^\alpha}, \quad \eta > 0. \tag{17.22}$$

Hence, using Lemma 17.11 we get (17.19) with asymptotic (17.20) for function $P_\alpha(\sigma)$. By famous theorem of Bernstein, p_α is positive function as the right-hand side is a completely positive function.

We denote $P_\alpha(\sigma) := 2p_\alpha(\sigma/2)$, and then formula (17.22) means

$$e^{-t|S|^\alpha} = \frac{1}{\sqrt{2\pi}}\int_0^\infty e^{-\frac{\sigma}{2}t^2|S|} P_\alpha(\sigma)d\sigma. \tag{17.23}$$

Now let us make a notational convention that whenever we have a product as above, we can decompose it to the sum, whose terms are polynomials in x_i variables, and of degree $0, 1, 2, \ldots$. We call corresponding polynomials "modes": 0-mode, 1-mode, 2-mode, et cetera. Then

$$\begin{aligned} K_t^P(x) &= \text{0-mode} + \text{1-mode} + \text{2-mode} + \cdots \\ &= 1 + \sum_i x_i \int_0^\infty e^{-t^2\frac{\sigma}{2}} e^{-\frac{1}{2\sigma}} \frac{d\sigma}{\sigma^{3/2}} + \sum_{i\neq j} x_i x_j \int_0^\infty e^{-t^2\sigma} e^{-\frac{1}{2\sigma}} \frac{d\sigma}{\sigma^{3/2}} + \cdots. \end{aligned} \tag{17.24}$$

Now let us first think about $t = 1$ and use the modification of the kernel idea of [7] to modify $K_1^P(x)$. Let us choose a function $\varphi(\sigma)$ supported on $[1,2]$ and orthogonal to two function $e^{-\frac{\sigma}{2}}$ and $e^{-\sigma}$:

$$\int_0^\infty \varphi(\sigma)e^{-\frac{\sigma}{2}}\,d\sigma = 0, \quad \int_0^\infty \varphi(\sigma)e^{-\sigma}\,d\sigma = 0. \tag{17.25}$$

Then

$$\int_0^\infty \varphi(t^2\sigma)e^{-t^2\frac{\sigma}{2}}\,d\sigma = 0, \quad \int_0^\infty \varphi(t^2\sigma)e^{-t^2\sigma}\,d\sigma = 0. \tag{17.26}$$

Consider now modified K_t^P:

$$\tilde{K}_t^P(x) = \frac{1}{\sqrt{2\pi}}\int_0^\infty \prod_{i=1}^n (1 + e^{-t^2\frac{\sigma}{2}}x_i)\Big(e^{-\frac{1}{2\sigma}}\frac{d\sigma}{\sigma^{3/2}} - \kappa t^3\varphi(t^2\sigma)\Big). \tag{17.27}$$

We will choose κ momentarily. First, notice that if $|\kappa|$ is small (absolutely), then the expression in bracket is positive. In fact, as φ is supported in $[1,2]$ the relevant σ (remember that t is small, $0 \le t \le 1$) is only such that $\sigma \approx \frac{1}{t^2}(\ge 1)$. Otherwise, the modification does not exist. But for such σ the term $e^{-\frac{1}{2\sigma}}\frac{d\sigma}{\sigma^{3/2}}$ dominates the term $\kappa t^3\varphi(t^2\sigma)$ if κ is small in **absolute value.**

By (17.24) and (17.26), this modification does not change 1-mode and 2-mode at all. Namely, the 1-mode and the 2-mode of

$$\tilde{K}_t^P(x) = \frac{1}{\sqrt{2\pi}}\int_0^\infty \prod_{i=1}^n (1 + e^{-t^2\frac{\sigma}{2}}x_i)\Big(e^{-\frac{1}{2\sigma}}\frac{d\sigma}{\sigma^{3/2}} - \kappa t^3\varphi(t^2\sigma)\Big)$$

are exactly the same as in (17.24). This is because of orthogonality (17.26).

But we saw that modified kernel is still positive if absolute value of κ is absolutely small. What about the 0-mode, how it changed in transition from K_t^P to $\tilde{K}_t^P$?

$$\text{0-mode of } \tilde{K}_t^P = \text{0-mode of } K_t^P - \kappa\int_0^\infty t^3\varphi(t^2\sigma) = 1 - \kappa t\int_0^\infty \varphi(\sigma)d\sigma. \tag{17.28}$$

Function φ can be chosen **not orthogonal** to **1.** So the integral is some nonzero number, say 1/2. Now choose κ to be absolutely small but positive and get

$$\text{0-mode of } \tilde{K}_t^P = 1 - c_0 t \approx e^{-c_0 t}. \tag{17.29}$$

At the same time $\tilde{K}_t^P$, $t \in [0,1]$, is positive and has the same 1-mode and 2-mode as Poisson K_t^P.

Therefore, on any function $f(x)$ which is the sum of polynomials of say degree 1 and 2 (no constant term and no higher degree polynomials), that is on any function f with correct band Fourier localization we obtain

$$\|e^{t\Delta_{1/2}}f\|_p \le e^{-c_0 t}\|f\|_p, \quad t \in [0,1],\ 1 \le p < \infty.$$

This is just because

$$e^{t\Delta_{1/2}}f(x) = \mathbb{E}[K_t^P(x \cdot y)f(y)] = \mathbb{E}[\tilde{K}_t^P(x \cdot y)f(y)].$$

Of course, we used here the positivity of the modified kernel: that $\tilde{K}_t^P \ge 0$, and hence

$$\|\tilde{K}_t^P\|_1 = \mathbb{E}\tilde{K}_t^P(x) = 0\text{-mode } \tilde{K}_t^p = 1 - c_0 t.$$

We also use the convolution nature of operator with kernel $\tilde{K}_t^P(x \cdot y)$.

Moreover, by using (17.19) and (17.20), we obtain verbatim as for $\alpha = \frac{1}{2}$–the following more general inequality for band localized functions f for some universal $c_\alpha > 0$ (if $0 < \alpha < 1$) independent of band localized f.

Theorem 17.12. *Let function $f : C_n \to \mathbb{R}$ is band localized to, say, the first and the second mode only, then independent of n and for all such f we have*

$$\|e^{t\Delta_\alpha}f\|_p \le e^{-c_\alpha t}\|f\|_p, \quad t \in [0,1],\ 1 \le p < \infty,\ 0 < \alpha < 1. \tag{17.30}$$

Moreover, the norm $\|\cdot\|_p, 1 \le p < \infty$ can be replaced here by the norm of any shift invariant Banach space on Hamming cube.

Remark 17.13. For $p > 1$ and $\alpha = 1$, we have even stronger Theorem 17.5. It is stronger because it can be formulated as

$$\|e^{t\Delta}f\|_p \le e^{-c_1 t}\|f\|_p, \quad t \in [0,1],\ 1 < p < \infty, \tag{17.31}$$

independently of n for all functions f that are very weakly spectral localized, namely, for f such that only 0-mode vanishes: $\mathbb{E}f = 0$.

Remark 17.14. Also for $p = 1$, $\alpha = 1$ one has the estimate (17.30)—but only for large t; see Theorem 17.8. As to the case $p = 1$, $\alpha = 1$, t is small, and f is band localized, Subsection 17.3.3 shows that such drop of norm can be false. So this is the case when even for band localized functions we do not have "spectral gap" type inequality. But as soon as either 1) $p > 1$ and any $\alpha \le 1$ or 2) $\alpha < 1$, $p = 1$ we have "spectral gap" inequality

$$\|e^{t\Delta_\alpha}f\| \le e^{-ct}\|f\|, \quad c > 0.$$

In case (1), we just need very weak spectral localization, namely, just $\mathbb{E}f = 0$. In case (2), we used that f is band localized. This condition cannot be dropped as counterexample in Subsection 17.3.2 shows.

17.5 Hypercontractivity and two Log-Sobolev inequality inequalities

We give here a certain proof of Log-Sobolev inequality on Hamming cube, which easily generalizes to several other interesting inequalities. We are not sure, it might be that the proof below was already in the literature on Log-Sobolev inequality. This literature is huge, the latest proof of Log-Sobolev inequality on Hamming cube can be found in [5], the first one is in Leonard's Gross [4].

We first consider the classical (Gaussian) case on $\mathbb{R}^n$ with $\Delta_{ou}f = \Delta f - x\cdot\nabla f$. Suppose $f > 0$ and let $u = e^{t\Delta_{ou}}f$. We use $\mathbb{E}$ to denote the expectation with respect to the standard Gaussian density.

Lemma 17.15. *Fix any index $k \in \{1,\dots,n\}$. Denote $v = \sqrt{u}$. Then*

$$\partial_t((\partial_k v)^2) = \Delta_{ou}((\partial_k v)^2) - 2(\partial_k v)^2 - \left|\nabla\partial_k v - \frac{\partial_k v \nabla v}{v}\right|^2.$$

Consequently,

$$\int_0^\infty \mathbb{E}(\partial_k v)^2 dt \le \frac{1}{2}\mathbb{E}(\partial_k(\sqrt{f}))^2, \tag{17.32}$$

or in the usual form, upon summation in k,

$$\int_0^\infty \mathbb{E}\frac{|\nabla u|^2}{u} dt \le \frac{1}{2}\mathbb{E}\frac{|\nabla f|^2}{f}. \tag{17.33}$$

Proof. Let us prove the first equality of the lemma. We denoted $u := P_t f := e^{t\Delta_{ou}}f$, $v = \sqrt{u}$. Then

$$\partial_k v = \frac{1}{2}\frac{\partial_k P_t f}{\sqrt{P_t f}}, \quad (\partial_k v)^2 = \frac{1}{4}\frac{[\partial_k(P_t f)]^2}{P_t f}. \tag{17.34}$$

Just because $\partial_k P_t f = e^{-t}P_t(\partial_k f)$, we have

$$\begin{aligned}
I := (\partial_t - \Delta_{ou})(\partial_k v)^2 &= \frac{1}{4}(\partial_t - \Delta_{ou})\frac{(\partial_k P_t f)^2}{P_t f} = \frac{1}{4}(\partial_t - \Delta_{ou})\frac{e^{-2t}[P_t(\partial_k f)]^2}{P_t f}\\
&= \frac{1}{4}e^{-2t}(\partial_t - \Delta_{ou})\frac{[P_t(\partial_k f)]^2}{P_t f} - 2\frac{1}{4}e^{-2t}\frac{[P_t(\partial_k f)]^2}{P_t f}\\
&= \frac{1}{4}e^{-2t}(\partial_t - \Delta_{ou})\frac{[P_t(\partial_k f)]^2}{P_t f} - 2\frac{1}{4}\frac{[e^{-t}P_t(\partial_k f)]^2}{P_t f}\\
&= \frac{1}{4}e^{-2t}(\partial_t - \Delta_{ou})\frac{[P_t(\partial_k f)]^2}{P_t f} - 2(\partial_k v)^2.
\end{aligned}$$

Now denote temporarily

$$g = \partial_k f, \quad F(x,y) := \frac{y^2}{x},$$

then we see that we want to calculate

$$I = \frac{1}{4} e^{-2t} (\partial_t - \Delta_{\mathrm{ou}}) F(P_t g, P_t f) - 2(\partial_k v)^2.$$

Now we use Lemma 17.16 below. The Hessian

$$\operatorname{Hess} F = \begin{bmatrix} \frac{2}{x}, & -\frac{2y}{x^2} \\ -\frac{2y}{x^2}, & \frac{2y^2}{x^3} \end{bmatrix} = \frac{2}{x} \begin{bmatrix} 1, & -\frac{y}{x} \\ -\frac{y}{x}, & \frac{y^2}{x^2} \end{bmatrix}.$$

So

$$\begin{aligned} I = -\frac{1}{4} e^{-2t} (\partial_t - \Delta_{\mathrm{ou}}) F(P_t g, P_t f) &= -\frac{1}{4} \frac{2e^{-2t}}{P_t f} \sum_j \Big(\partial_j P_t g - \frac{P_t g}{P_t f} \partial_j P_t f \Big)^2 - 2(\partial_k v)^2 \\ &= -\frac{1}{2} e^{-2t} \sum_j \Big(\frac{\partial_j P_t g}{v} - \frac{(P_t g) \partial_j P_t f}{v^3} \Big)^2 - 2(\partial_k v)^2. \end{aligned} \tag{17.35}$$

Now this can be expressed via $\partial_k v$ and its derivatives. In fact, looking at (17.34) we notice that

$$\begin{aligned} \nabla(\partial_k v) &= \frac{1}{2} \frac{\nabla(\partial_k P_t f)}{v} - \frac{1}{4} \frac{(\partial_k P_t f) \nabla P_t f}{v^3} \\ &= e^{-t} \Big(\frac{1}{2} \frac{\nabla(P_t g)}{v} - \frac{1}{4} \frac{(P_t g) \nabla P_t f}{v^3} \Big). \end{aligned}$$

Also $\nabla v = \nabla \sqrt{P_t f} = \frac{1}{2} \frac{\nabla P_t f}{v}$, and so from this and (17.34) we get

$$\frac{\nabla v}{v} \partial_k v = \frac{1}{4} \frac{\partial_k P_t f \nabla P_t f}{v^3} = e^{-t} \frac{1}{4} \frac{(P_t g) \nabla P_t f}{v^3}.$$

Hence, from two last display formulas we get

$$e^{-t} \frac{1}{2} \frac{\nabla(P_t g)}{v} = \nabla(\partial_k v) + \frac{\nabla v}{v} \partial_k v \tag{17.36}$$

$$e^{-t} \frac{1}{4} \frac{(P_t g) \nabla P_t f}{v^3} = \frac{\nabla v}{v} \partial_k v. \tag{17.37}$$

Therefore, from this and 17.35 we get

$$\begin{aligned} (\partial_t - \Delta_{\mathrm{ou}})(\partial_k v)^2 = I &= -\frac{1}{2} \Big| 2\Big(\nabla(\partial_k v) + \frac{\nabla v}{v} \partial_k v \Big) - 4\Big(\frac{\nabla v}{v} \partial_k v \Big) \Big|^2 - 2(\partial_k v)^2 \\ &= -2 \Big| \nabla(\partial_k v) - \frac{\nabla v}{v} \partial_k v \Big|^2 - 2(\partial_k v)^2. \end{aligned} \tag{17.38}$$

We think that f is a test function. This is just to avoid the problems of convergence and of interchange the signs of integral and differential.We hit the last inequality by $\mathbb{E}$, which makes Δ_{ou} term disappear.

We also throw away a negative term $-2|\nabla(\partial_k v) - \frac{\nabla v}{v}\partial_k v|^2$. Then we get

$$\mathbb{E}(\partial_k v)^2 \leq -\frac{1}{2}\partial_t \mathbb{E}(\partial_k v)^2,$$

or

$$\int_0^\infty \mathbb{E}(\partial_k v)^2 \leq \frac{1}{2}\mathbb{E}(\partial_k \sqrt{f})^2.$$

This is (17.32), so the second inequality of lemma is proved. Then the last inequality (17.33) of Lemma follows from the latter inequality by summing over k. Lemma is proved. □

Lemma 17.16.

$$(\partial_t - \Delta_{\mathrm{ou}})F(P_t g, P_t f) = -\operatorname{tr}[\operatorname{Hess} F \cdot \Gamma],$$

where

$$\Gamma := \begin{bmatrix} \nabla P_t g \cdot \nabla P_g f, & \nabla P_t g \cdot \nabla P_t f \\ \nabla P_t g \cdot \nabla P_t f, & \nabla P_t f \cdot \nabla P_t f \end{bmatrix}.$$

Proof. When we calculate $\partial_t F(\cdot,\cdot)$ we get $\nabla F \cdot (\partial_t P_t g, \partial_t P_t f)$. When we calculate $\Delta_{\mathrm{ou}} F(\cdot,\cdot)$, we first of all get $\nabla F \cdot (\Delta_{\mathrm{ou}} P_t g, \Delta_{\mathrm{ou}} P_t f)$ (and the difference of these two terms vanishes), we also get the second derivatives of F: exactly in the form $-\operatorname{tr}[\operatorname{Hess} F \cdot \Gamma]$. See also [6]. □

Corollary 17.17 (Usual entropy inequality). *For any $f > 0$, we have*

$$\mathbb{E}\left(f \log \frac{f}{\mathbb{E}f}\right) \leq \frac{1}{2}\mathbb{E}\frac{|\nabla f|^2}{f},$$

Proof. Write $u = e^{t\Delta_{\mathrm{ou}}} f$. Then

$$\begin{aligned} \mathbb{E}\left(f \log \frac{f}{\mathbb{E}f}\right) &= -\int_0^\infty \mathbb{E}(\log u \Delta_{\mathrm{ou}} u)dt \\ &= \int_0^\infty \mathbb{E}\left(\frac{|\nabla u|^2}{u}\right)dt \leq \frac{1}{2}\mathbb{E}\frac{|\nabla f|^2}{f}. \end{aligned}$$ □

Now we turn to the case on the cube. Denote

$$\begin{aligned} \partial_j f &= \frac{1}{2}(f(x_j = 1) - f(x_j = -1)); \\ \mathbb{E}_j f &= \frac{1}{2}(f(x_j = 1) + f(x_j = -1)). \end{aligned}$$

Lemma 17.18 (Calculus on the cube). *For any $f, g\colon \{-1,1\}^n \to \mathbb{R}$, we have*

$$\partial_j(fg) = (\partial_j f)\mathbb{E}_j g + (\mathbb{E}_j f)\partial_j g;$$
$$\Delta f = \sum_{j=1}^{n}(\mathbb{E}_j f - f) = \sum_{j=1}^{n}(-x_j\partial_j f);$$
$$\Delta(fg) = f\Delta g + g\Delta f + 2\sum_{j=1}^{n}\partial_j f\partial_j g;$$
$$\partial_j\Delta f = \Delta\partial_j f - \partial_j f.$$

Lemma 17.19. *For any $v > 0$ on the cube, we have*

$$(\partial_l\partial_k v)^2 - \partial_k\left(\frac{1}{v}(\partial_l v)^2\right)\partial_k v \geq 0,$$

where k, l are any fixed index.

Proof. Denote $A = \partial_l v$. Then

$$\text{LHS} = (\partial_k A)^2 - \partial_k\left(\frac{1}{v}\right)\mathbb{E}_k A^2\partial_k v - \mathbb{E}_k\left(\frac{1}{v}\right)\partial_k(A^2)\partial_k v.$$

The desired inequality then easily follows from the elementary inequality

$$\left(\frac{A_1 - A_2}{2}\right)^2 + \frac{A_1^2 + A_2^2}{2}\cdot\frac{1}{2}\cdot\left(\frac{1}{a} - \frac{1}{b}\right)\cdot\frac{1}{2}(b-a) \geq \frac{1}{2}\left(\frac{1}{a} + \frac{1}{b}\right)\cdot\frac{1}{2}|A_1^2 - A_2^2|\cdot\frac{b-a}{2},$$

where $b \geq a \geq 0$, $A_1, A_2 \in \mathbb{R}$. □

Now take $f > 0$ on the cube and denote $u = e^{t\Delta}f$.

Lemma 17.20. *Fix any index $k \in \{1,\dots,n\}$. Denote $v = \sqrt{u}$. Then*

$$\partial_t((\partial_k v)^2) = \Delta((\partial_k v)^2) - 2(\partial_k v)^2 - F(t),$$

where $F(t) \geq 0$. Consequently,

$$\int_0^\infty \mathbb{E}(\partial_k v)^2 dt \leq \frac{1}{2}\mathbb{E}(\partial_k(\sqrt{f}))^2,$$

and upon summation in k,

$$\int_0^\infty \mathbb{E}|\nabla v|^2 dt \leq \frac{1}{2}\mathbb{E}|\nabla\sqrt{f}|^2.$$

Remark 17.21. One should note that on the cube, the quantities $|\nabla\sqrt{u}|^2$ and $\frac{|\nabla u|^2}{4u}$ are different!

Proof. First, observe that

$$\partial_t v = \Delta v + \frac{1}{v}(\partial v)^2,$$

where we denote $(\partial v)^2 = \sum_{j=1}^n (\partial_j v)^2$. Now note that

$$\begin{aligned}\partial_t \partial_k v &= \partial_k \Delta v + \partial_k\left(\frac{1}{v}(\partial v)^2\right) \\ &= \Delta \partial_k v - \partial_k v + \partial_k\left(\frac{1}{v}(\partial v)^2\right).\end{aligned}$$

Then

$$\frac{1}{2}\partial_t((\partial_k v)^2) = \frac{1}{2}\Delta((\partial_k v)^2) - (\partial_k v)^2 - (\partial\partial_k v)^2 + \partial_k\left(\frac{1}{v}(\partial v)^2\right)\partial_k v.$$

□

The next lemma is for showing the monotonicity of the quantity $(\partial_k u)^2/u$ for $u = e^{t\Delta}f, f > 0$ on the cube.

Lemma 17.22. *Fix any $k \in \{1, \dots, n\}$. For any $l \in \{1, \dots, n\}$, we have*

$$\left(\frac{\Delta_l u}{u^2} + \Delta_l\left(\frac{1}{u}\right)\right) \cdot (\partial_k u)^2 + \frac{2}{u}(\partial_l \partial_k u)^2 - 2\partial_l\left(\frac{1}{u}\right) \cdot \partial_l((\partial_k u)^2) \geq 0,$$

where $\Delta_l = -x_l \partial_l$. It follows that

$$\left(\frac{\Delta u}{u^2} + \Delta\left(\frac{1}{u}\right)\right)(\partial_k u)^2 + \frac{2}{u}\sum_{l=1}^n (\partial_l \partial_k u)^2 - 2\sum_{l=1}^n \partial_l\left(\frac{1}{u}\right)\partial_l((\partial_k u)^2) \geq 0.$$

Proof. This follows from the elementary inequality:

$$\frac{(\alpha_1 - \alpha_2)^2}{2\alpha_1^2 \alpha_2} + \frac{2}{\alpha_1}\left(\frac{1-x}{2}\right)^2 - \left|\frac{1}{2} \cdot \frac{\alpha_1 - \alpha_2}{\alpha_1 \alpha_2}(1 - x^2)\right| \geq 0,$$

where $\alpha_1 > 0$, $\alpha_2 > 0$, $|x| \leq 1$. □

Lemma 17.23. *Fix any $k \in \{1, \dots, n\}$. Denote $h = (\partial_k u)^2/u$. Then*

$$\partial_t h = \Delta h - 2h - F(t),$$

where $F(t) \geq 0$. In particular,

$$\int_0^\infty \mathbb{E}h(t)dt \leq \frac{1}{2}\mathbb{E}\frac{(\partial_k f)^2}{f},$$

and

$$\int_0^\infty \mathbb{E}\frac{|\nabla u|^2}{u} \leq \frac{1}{2}\mathbb{E}\frac{|\nabla f|^2}{f}.$$

Proof. We have

$$\begin{aligned}\partial_t h &= -\frac{\Delta u}{u^2}(\partial_k u)^2 + \frac{2\partial_k u}{u}\partial_k(\Delta u)\\ &= -2h - \frac{\Delta u}{u^2}(\partial_k u)^2 \frac{2\partial_k u \Delta(\partial_k u)}{u}\\ &= -2h - \frac{\Delta u}{u^2}(\partial_k u)^2 - \frac{2}{u}\sum_{l=1}^{n}(\partial_l\partial_k u)^2 + \Delta h - 2\sum_{l=1}^{n}\partial_l\left(\frac{1}{u}\right)\partial_l((\partial_k u)^2) - \Delta\left(\frac{1}{u}\right)(\partial_k u)^2,\end{aligned}$$

where in the last equality, we used the fact that

$$\begin{aligned}\Delta h &= \Delta\left(\frac{1}{u}\cdot(\partial_k u)^2\right)\\ &= \frac{1}{u}\Delta(\partial_k u)^2 + 2\sum_{l=1}^{n}\partial_l\left(\frac{1}{u}\right)\partial_l((\partial_k u)^2) + \Delta\left(\frac{1}{u}\right)\cdot(\partial_k u)^2.\end{aligned}$$

One can then use the previous lemma to get the positivity of F. □

Lemma 17.24. *For any $a > 0$, $b > 0$, we have*

$$\left(\log\frac{b}{a}\right)(b-a) \le \frac{1}{2}\left(\frac{1}{a}+\frac{1}{b}\right)(b-a)^2.$$

Proof. By scaling, we only need to show for $0 < x \le 1$,

$$-x\log x \le \frac{1}{2}(1-x^2).$$

But this is easy to check. □

Lemma 17.25. *For any $u > 0$ on the cube, we have*

$$\mathbb{E}(-\log u\Delta u) \le \mathbb{E}\frac{|\nabla u|^2}{u}.$$

Proof. By using the previous lemma, we have

$$\begin{aligned}\mathrm{LHS} &= \mathbb{E}^x\left(\sum_{y\sim x}\frac{\log u(x)-\log u(y)}{2}\cdot\frac{u(x)-u(y)}{2}\right)\\ &\le \frac{1}{4}\mathbb{E}^x\sum_{y\sim x}(u(x)-u(y))^2\cdot\frac{1}{2}\left(\frac{1}{u(x)}+\frac{1}{u(y)}\right)\\ &= \mathbb{E}\frac{|\nabla u|^2}{u}.\end{aligned}$$

□

We now prove the entropy inequality on the cube for $f > 0$.

Theorem 17.26. *For any $f > 0$ on the cube, we have*

$$\mathbb{E}\left(f\log\frac{f}{\mathbb{E}f}\right) \le \frac{1}{2}\mathbb{E}\frac{|\nabla f|^2}{f}.$$

Proof. Denote $u = e^{t\Delta}f$. Then we have

$$\mathbb{E}\left(f \log \frac{f}{\mathbb{E}f}\right) = -\int_0^\infty \mathbb{E}(\log u \Delta u)dt$$
$$\leq \int_0^\infty \mathbb{E}\left(\frac{|\nabla u|^2}{u}\right)dt$$
$$\leq \frac{1}{2}\mathbb{E}\frac{|\nabla f|^2}{f}. \qquad \square$$

17.5.1 How to get the full inequality for Hamming cube case

Consider again $u = e^{t\Delta}f$ with $f > 0$. Set $v = \sqrt{u} > 0$. Then

$$\partial_t v = \Delta v + \frac{|\nabla v|^2}{v}.$$

Now observe

$$\partial_t(\partial_k v) = -(\partial_k v) + \Delta(\partial_k v) + \partial_k\left(\frac{|\nabla v|^2}{v}\right).$$

We then obtain

$$\frac{1}{2}\partial_t(\|\nabla v\|_2^2) = -\|\nabla v\|_2^2 - \sum_{k,l} \mathbb{E}\left((\partial_l \partial_k v)^2 - \partial_k\left(\frac{(\partial_l v)^2}{v}\right)\partial_k v\right).$$

Observe that

$$\sum_{k=1}^n \mathbb{E}\left(-\partial_k\left(\frac{(\partial_k v)^2}{v}\right)\partial_k v\right) \gtrsim \sum_{k=1}^n \mathbb{E}\left(\frac{(b_k - a_k)^4}{a_k b_k}\right)$$
$$\gtrsim \sum_{k=1}^n \mathbb{E}\left(|b_k^2 - a_k^2| \cdot \left|\log\left(\frac{b_k}{a_k}\right)\right|, \; a_k \ll b_k \text{ or } b_k \ll a_k\right),$$

where $a_k = v(x_k = 1)$, $b_k = v(x_k = -1)$. Thus now

$$\mathbb{E}\left(v^2 \log \frac{v^2}{\mathbb{E}v^2}\right) = -\int_0^\infty \mathbb{E}(\log(v^2)\Delta(v^2))dt$$
$$\lesssim \int_0^\infty \sum_{k=1}^n \mathbb{E}\left((b_k^2 - a_k^2)\log\left(\frac{b_k}{a_k}\right), \; a_k \sim b_k\right)$$
$$+ \mathbb{E}\left(|b_k^2 - a_k^2| \cdot \left|\log\left(\frac{b_k}{a_k}\right)\right|, \; a_k \ll b_k \text{ or } b_k \ll a_k\right)dt$$

$$\lesssim \int_0^\infty \left(\text{const}\, \|\nabla v\|_2^2 + \frac{1}{2}\frac{d}{dt}(\|\nabla v\|_2^2) \right) dt$$
$$\lesssim \|\nabla \sqrt{f}\|_2^2.$$

Now to get the sharp version, we use the following inequality.

Lemma 17.27. *For any* $0 < B < A$*, we have*

$$\frac{(A-B)^4}{4AB} + (A-B)^2 \geq \frac{1}{2} \cdot (A^2 - B^2) \log\left(\frac{A}{B}\right).$$

Proof. By scaling it reduces to verifying for $0 < x < 1$:

$$\frac{(1-x)^3}{4x} + 1 - x \geq \frac{1}{2}(1+x) \log\left(\frac{1}{x}\right).$$ □

Lemma 17.28.

$$-\mathbb{E}(\log(v^2)\Delta(v^2)) \leq 4\left(\|\nabla v\|_2^2 + \sum_k \mathbb{E}\left(-\partial_k\left(\frac{(\partial_k v)^2}{v} \right)\partial_k v \right)\right) \leq -2\frac{d}{dt}\|\nabla v\|_2^2.$$

Proof. Note that

$$\begin{aligned} \text{LHS} &= \sum_{k=1}^n \mathbb{E}\partial_k(\log v^2)\partial_k(v^2) \\ &= \sum_{k=1}^n \mathbb{E}\frac{1}{4} \cdot (\log A_k^2 - \log B_k^2)(A_k^2 - B_k^2). \end{aligned}$$

On the other hand,

$$\begin{aligned} &4\left(\|\nabla v\|_2^2 + \sum_k \mathbb{E}\left(-\partial_k\left(\frac{(\partial_k v)^2}{v} \right)\partial_k v \right)\right) \\ &\quad \geq 4 \sum_{k=1}^n \mathbb{E}\left(\frac{(A_k - B_k)^2}{4} + \frac{(A_k - B_k)^4}{16 A_k B_k} \right). \end{aligned}$$

One can then use the previous inequality to get the result. □

With the previous lemma in hand, it is then easy to see that

$$\mathbb{E}(f \log(f/\mathbb{E}(f))) \leq -2 \int_0^\infty \frac{d}{dt}\|\nabla v\|_2^2 dt = 2\|\nabla \sqrt{f}\|_2^2.$$

17.5.2 Beckner-type inequalities for general p

Consider first the Gaussian case on $\mathbb{R}$ with standard Gaussian density $dy = \frac{1}{\sqrt{2\pi}}e^{-\frac{x^2}{2}}dx$. We shall denote $\mathbb{E}_\gamma$ simply as $\mathbb{E}$.

Theorem 17.29. *Let $1 \le p \le 2$. For any $f > 0$, we have*

$$\mathbb{E}f^p - (\mathbb{E}f)^p \le C_p \mathbb{E}(|\nabla f|^2 f^{p-2}),$$

where $C_p = \frac{p(p-1)}{2}$.

Proof. We only need to consider $1 < p < 2$. We may rewrite the inequality as

$$\mathbb{E}f^p - (\mathbb{E}f)^p \le C_p \cdot \frac{4}{p^2}\mathbb{E}(|\nabla(f^{\frac{p}{2}})|^2) = \frac{2(p-1)}{p}\mathbb{E}(|\nabla(f^{\frac{p}{2}})|^2).$$

Let $u = e^{t\Delta_{\mathrm{ou}}}f$. Easy to check that

$$\mathbb{E}f^p - (\mathbb{E}f)^p = \int_0^\infty \mathbb{E}(-pu^{p-1}\Delta_{\mathrm{ou}}u)dt.$$

We can simplify the integrand as

$$\begin{aligned}\mathbb{E}(-pu^{p-1}\Delta_{\mathrm{ou}}u) &= \mathbb{E}(p(p-1)|\nabla u|^2 u^{p-2})\\ &= \frac{4(p-1)}{p}\mathbb{E}(|\nabla(u^{\frac{p}{2}})|^2).\end{aligned}$$

Denote $\alpha = \frac{p}{2} \in (\frac{1}{2}, 1)$. Introduce $v = u^\alpha$. Then it is not difficult to check that

$$\partial_t v = v'' - xv' + \left(\frac{1}{\alpha} - 1\right) \cdot \frac{(v')^2}{v}.$$

Denote $Lv = v'' - xv'$. Then

$$\partial_t v' = Lv' - v' + \left(\frac{1}{\alpha} - 1\right) \cdot \left(\frac{(v')^2}{v}\right)'.$$

Then

$$\begin{aligned}\frac{1}{2}\frac{d}{dt}(\|v'\|_2^2) &= \mathbb{E}(Lv'v') - \|v'\|_2^2 + \left(\frac{1}{\alpha} - 1\right)\mathbb{E}\left(\left(\frac{(v')^2}{v}\right)'v'\right)\\ &= -\|v''\|_2^2 - \|v'\|_2^2 + \left(\frac{1}{\alpha} - 1\right)\mathbb{E}\left(\left(\frac{(v')^2}{v}\right)'v'\right).\end{aligned}$$

It is clear that we now only need to check the inequality

$$\|v'\|_2^2 \le -\frac{1}{2}\frac{d}{dt}(\|v'\|_2^2).$$

This amounts to checking

$$\|v''\|_2^2 - \left(\frac{1}{\alpha} - 1\right)\mathbb{E}\left(\left(\frac{(v')^2}{v}\right)' v'\right) \geq 0.$$

Now denote $v = g^2$. Then

$$\|v''\|_2^2 = 4\|gg'' + (g')^2\|_2^2,$$

and

$$-\mathbb{E}\left(\left(\frac{(v')^2}{v}\right)' v'\right) = -16\mathbb{E}(gg''(g')^2) \geq -4\mathbb{E}(gg'' + (g')^2)^2.$$

The desired inequality is then obvious since $0 < \frac{1}{\alpha} - 1 < 1$. □

17.5.3 Log-Sobolev for fractional operators

Theorem 17.30 (Log-Sobolev for fractional operators). *Let* $0 < \gamma < 1$. *Then for any* f *with* $\mathbb{E}f = 0$, *we have*

$$\|(-\Delta_\gamma)^{\frac{1}{2}} f\|_2^2 \gtrsim \mathbb{E}\left(f^2 \log_+^\gamma \frac{f^2}{\mathbb{E}f^2}\right),$$

where $\log_+^\gamma x = \log^\gamma x$ *if* $x \geq 1$ *and* $\log_+^\gamma x = 0$ *if* $0 < x < 1$.

Proof. WLOG we assume $\mathbb{E}f^2 = 1$. Also by Poincaré, we have

$$\|(-\Delta_\gamma)^{\frac{1}{2}} f\|_2 \gtrsim \|f\|_2 = 1.$$

So in the computation below, we can afford any loss of $O(1)$. Easy to check that

$$\|(-\Delta_\gamma)^{\frac{1}{2}} f\|_2^2 = \text{const} \cdot \int_0^\infty \frac{\|f\|_2^2 - \|e^{\frac{t}{2}\Delta} f\|_2^2}{t^{1+\gamma}} dt.$$

By using hypercontractivity, we have

$$\|e^{\frac{t}{2}\Delta} f\|_2^2 \leq \|f\|_{1+e^{-t}}^2 \leq \|f\|_{1+e^{-t}}^{1+e^{-t}},$$

where in the last inequality we used the fact that $\mathbb{E}f^2 = 1$ so that $\mathbb{E}|f|^p \leq 1$ for $p < 2$. One may then bound

$$\int_0^1 \frac{\|f\|_2^2 - \|f\|_{1+e^{-t}}^{1+e^{-t}}}{t^{1+\gamma}} dt$$

$$\geq \mathbb{E}\left(|f|^2 \chi_{|f|>3} \int_0^1 \frac{1 - |f|^{e^{-t}-1}}{t^{1+\gamma}} dt\right) - O(\|f\|_2^2).$$

Note that in the above for the term involving the piece $|f| \le 3$, we used the inequality

$$|f|^{1+e^{-t}} \cdot 1 \le |f|^2 \cdot \frac{1+e^{-t}}{2} + \frac{1-e^{-t}}{2}.$$

On the other hand, for the piece involving $|f| > 3$, we can just use the elementary inequality for $a = \log|f| > 1$,

$$\int_0^1 \frac{1-e^{(e^{-t}-1)a}}{t^{1+\gamma}} dt \gtrsim a^{\gamma}.$$

□

17.6 Dependence on p

17.6.1 Exponential localization

For f such that $\mathbb{E}f = 0$ the inequality of Theorem 17.4 combined with (17.1) reads as follows:

$$\mathbb{E}|f|^p \le -2p\mathbb{E}(\Delta f|f|^{p-1}\operatorname{sgn}(f)), \quad p \ge 2. \tag{17.39}$$

It is not difficult to check that for $p > 2$, and any $a, b \in \mathbb{R}$:

$$(a-b)(|a|^{p-1}\operatorname{sgn}(a) - |b|^{p-1}\operatorname{sgn}(b)) \lesssim (p-1)(a-b)^2(|a|^{p-2} + |b|^{p-2}).$$

By using the above inequality, we then get a sort of Poincaré or Beckner inequality for $p \ge 2$ and f such that $\mathbb{E}f = 0$:

$$\mathbb{E}|f|^p \lesssim \frac{p(p-1)}{2}\mathbb{E}(|\nabla f|^2|f|^{p-2}), \quad p \ge 2. \tag{17.40}$$

Hence,

$$\mathbb{E}|f|^p \lesssim \frac{p(p-1)}{2}(\mathbb{E}(|\nabla f|^p))^{\frac{2}{p}}(\mathbb{E}|f|^p)^{1-\frac{2}{p}}, \quad p \ge 2.$$

Or,

$$\mathbb{E}f = 0 \Rightarrow \|f\|_p^p \lesssim p^p\|\nabla f\|_p^p, \quad p \ge 2. \tag{17.41}$$

Thus for Lipschitz f on the Hamming cube with $\||\nabla f|\|_\infty \le 1$, we get

$$\mathbb{E}|f|^p \lesssim p^p, \quad p \ge 2.$$

This in turn yields the following (weak) exponential localization for Lipschitz f on the Hamming cube.

Proposition 17.31. *For any Lipschitz f on the Hamming cube, we have*

$$\mathbb{E}\exp\left(c_1 \cdot \frac{|f - \mathbb{E}f|}{|||\nabla f|||_\infty}\right) < C_1,$$

where $c_1 > 0$, $C_1 > 0$ are absolute constants. Here we tacitly assume $|||\nabla f|||_\infty > 0$, that is, f is not identically a constant.

17.6.2 Gaussian localization

We first use a direct argument to show a nonsharp Gaussian localization. The sharp version will be given in the next subsection by using the Herbst argument [8].

17.6.2.1 The nonsharp argument

Proposition 17.32. *For any Lipschitz f on $\mathbb{R}^n$, and let $\mathbb{E}$ is the integration with respect to Gaussian measure, then we have*

$$\mathbb{E}\exp\left(c_2 \cdot \frac{|f - \mathbb{E}f|^2}{|||\nabla f|||_\infty^2}\right) < C_2,$$

where $c_2 > 0$, $C_2 > 0$ are absolute constants. Here, we assume f is not identically a constant.

Proof. WLOG we assume $\|\nabla f\|_\infty \le 1$ and $\mathbb{E}f = 0$. It suffices for us to show that for $p > 2$:

$$\mathbb{E}|f|^p \le A_1 \cdot p^{\frac{p}{2}} \cdot A_2^p,$$

where $A_1 \ge 1$, $A_2 \ge 1$ are constants independent of p.

Observe that

$$\begin{aligned}\mathbb{E}(|\nabla f|^2 |f|^{p-2}) &\lesssim \mathbb{E}(|f|^{p-2}) \\ &\lesssim (\mathbb{E}|f|^p)^{\frac{p-2}{p}}.\end{aligned}$$

On the other hand, by using Log-Sobolev, we get

$$\begin{aligned}p^2 \mathbb{E}|\nabla f|^2 |f|^{p-2} &\gtrsim \mathbb{E}|\nabla(|f|^{\frac{p}{2}})|^2 \\ &\gtrsim (\mathbb{E}|f|^p \log|f|^p - \mathbb{E}|f|^p \log(\mathbb{E}|f|^p)).\end{aligned}$$

We then obtain the basic inequality:

$$\mathbb{E}|f|^p \log|f|^p - \mathbb{E}|f|^p \log(\mathbb{E}|f|^p) \le C_0 p^2 (\mathbb{E}|f|^p)^{\frac{p-2}{p}},$$

where $C_0 > 0$ is an absolute constant. Denote $a_p = \mathbb{E}|f|^p$. Note that a_p is a smooth function of p since we are on the Hamming cube. Clearly, $\frac{d}{dp}a_p = \mathbb{E}|f|^p \log|f|$. We then get

$$\frac{d}{dp}a_p \le \frac{1}{p}a_p \log a_p + C_0 p a_p^{1-\frac{2}{p}}.$$

Now consider $b_p = e^{\frac{1}{2}p\log p + \beta p}$. It is easy to check that

$$\frac{d}{dp}b_p = \frac{1}{p}b_p \log b_p + \frac{1}{2}b_p = \frac{1}{p}b_p \log b_p + C_0 \cdot p \cdot b_p^{1-\frac{2}{p}} \cdot \frac{e^{2\beta}}{C_0}.$$

By choosing β sufficiently large (note that $a_2 \lesssim 1$) and a simple ODE comparison argument, it is not difficult to show that $a_p \le b_p$ for all p. This then concludes the proof. □

17.6.2.2 Sharp version on Hamming cube

Lemma 17.33. *For any $u > 0$ on the Hamming cube, we have*

$$\frac{1}{4}\mathbb{E}(|\nabla(\log u)|^2 u) \ge \mathbb{E}|\nabla(\sqrt{u})|^2 \ge \frac{1}{2}\mathbb{E}\Big(u \log \frac{u}{\mathbb{E}u}\Big).$$

Proof. The second inequality is already proved in the previous section. For the first inequality, one can just use the elementary inequality:

$$\frac{1}{4}\cdot(\log(b/a))^2 \cdot \frac{a+b}{2} \ge (\sqrt{b}-\sqrt{a})^2, \quad \text{for any } b > a > 0.$$

By scaling and a change of variable $x \to x^2$, the above inequality reduces to the inequality:

$$(\log x)^2(1+x^2) - 2(1-x)^2 \ge 0, \quad \forall\, 0 < x < 1.$$

This can be easily checked. □

Proposition 17.34 (Gaussian localization). *For any Lipschitz f on the Hamming cube with $\||\nabla f|\|_\infty \le 1$, we have*

$$\mathbb{P}(f - \mathbb{E}f \ge a) \le e^{-\frac{1}{2}a^2}, \quad \forall\, a \ge 0.$$

Proof. WLOG assume $\mathbb{E}f = 0$. We follow (by now standard) Herbst argument. By applying the previous lemma to the function $u = e^{\lambda f}$ with $\lambda \ge 0$, we get

$$\mathbb{E}\Big(e^{\lambda f}\log\Big(\frac{e^{\lambda f}}{\mathbb{E}e^{\lambda f}}\Big)\Big) \le \frac{1}{2}\lambda^2\mathbb{E}(|\nabla f|^2 e^{\lambda f}) \le \frac{1}{2}\lambda^2\mathbb{E}(e^{\lambda f}).$$

Denote $g(\lambda) = \mathbb{E}e^{\lambda f}$. One can then obtain an ODE for g as

$$\frac{d}{d\lambda}\left(\frac{\log g}{\lambda}\right) \le \frac{1}{2}.$$

Solving this ODE then easily yields $g(\lambda) \le e^{\frac{1}{2}\lambda^2}$. Then Chebyshev inequality implies the desired inequality. See also [1]. □

17.7 Nonlocal derivatives on Hamming cube

Consider again the operator $\Delta_\gamma = -(-\Delta)^\gamma$ on $C_n := \{-1,1\}^n$. We want to understand its kernel representation. In the "flat case" of $\mathbb{R}^d$ this representation is given by

$$\Delta_\gamma f(x) = c_{d,\gamma} p.v. \int \frac{f(x) - f(y)}{|x-y|^{d+2\gamma}} dy$$

if $\gamma \in (0, 1/2)$.

On Hamming cube, we have

$$K_t^H(x) - 1 = \sum_{S\subset[n], S\neq\emptyset} e^{-t|S|}x^S = \prod_{i=1}^n (1 + e^{-t}x_i) - 1.$$

Now just multiply the latter expression by $t^{\alpha-1}$ and integrate from 0 to ∞. Having

$$\int_0^\infty t^{\alpha-1}e^{-st}dt = \frac{\Gamma(\alpha)}{s^\alpha}, \quad \alpha > 0,$$

we get what gives us fractional integration

$$\begin{aligned} I_\alpha(x) &= \sum_{S\subset[n]} \frac{1}{|S|^\alpha} x^S = 1 + \left(\sum x_i\right) + \left(\sum_{i_1<i_2} x_{i_1}x_{i_2}\right) 2! \frac{1}{2^\alpha} + \cdots \\ &\quad + \left(\sum_{i_1<i_2<\cdots<i_k} x_{i_1}x_{i_2}\ldots x_{i_k}\right) k! \frac{1}{k^\alpha} + \cdots \\ &= 1 + \frac{1}{\Gamma(\alpha)} \int_0^\infty t^{\alpha-1}(K_t^H - 1)dt \\ &= 1 + \frac{1}{\Gamma(\alpha)} \int_0^\infty t^{\alpha-1}\left(\prod_{i=1}^n (1 + e^{-t}x_i) - 1\right)dt. \end{aligned}$$

The kernel $I_\alpha(x \cdot y)$ is the kernel of fractional integration. It does not look like this ζ-type expression can have a closed form.

So operator of fractional differentiation Δ_γ should have the kernel $D_\gamma(x \cdot y)$, where

$$D_\gamma(x) = 1 + \left(\left(\sum x_i\right) + \left(\sum_{i_1<i_2} x_{i_1}x_{i_2}\right)\frac{2^{2\gamma}}{2!} + \cdots + \left(\sum_{i_1<i_2<\cdots<i_k} x_{i_1}x_{i_2}\ldots x_{i_k}\right)\frac{k^{2\gamma}}{k!}\ldots\right).$$

Another way of writing it is

$$D_\gamma(x) = 1 + \left(\left(\sum x_i\right) + \left(\sum_{i_1,i_2} x_{i_1}x_{i_2}\right)2^{2\gamma} + \cdots + \left(\sum_{i_1,i_2,\ldots,i_k} x_{i_1}x_{i_2}\ldots x_{i_k}\right)k^{2\gamma} + \cdots + x_1\ldots x_n n^{2\gamma}\right).$$

It does not look like this ζ-type expression can have a closed form either.

Bibliography

[1] L. Ben Efraim and F. Lust-Piquard, Poincaré type inequalities on the discrete cube and in the CAR algebra, Probab. Theory Relat. Fields **141** (3–4) (2008), 569–602.

[2] R. M. Blumenthal and R. K. Getoor, *Some theorems on stable processes*, Trans. Am. Math. Soc. **95** (1960), 263–273.

[3] D. Chamorro and P. G. Lemarié-Rieusset, *Quasi-geostrophic equations, nonlinear Bernstein inequalities and α-stable processes*, Rev. Mat. Iberoam. **28** (4) (2012), 1109–1122.

[4] L. Gross, *Logarithmic Sobolev inequalities*, Am. J. Math. **97** (4) (1975), 1061–1083.

[5] P. Ivanisvili, F. Nazarov and A. Volberg, *Square function and the Hamming cube: Duality*, Discrete Anal. (2018)

[6] Paata Ivanisvili and Alexander Volberg, *Isoperimetric functional inequalities via the maximum principle: the exterior differential systems approach*, arxiv: 1511.06895, Oper. Theory, Adv. Appl. **261**, 279–303. Birkhauser volume dedicated to V. P. Khavin.

[7] D. Li, *On a frequency localized Bernstein inequality and some generalized Poincaré-type inequalities*, Math. Res. Lett. **20** (5) (2013), 933–945.

[8] R. O'Donnell, Analysis of Boolean Functions, Cambridge University Press, 2014.

[9] G. Samorodnitskyand M. Taqqu, *Stable Non-Gaussian Random Processes*, Chapman and Hall, New York, London, 1994, 632 pp.

Pedro J. Miana and Jesús Oliva-Maza

18 Spectra of generalized Poisson integral operators on $L^p(\mathbb{R}^+)$

Abstract: For $\alpha, \beta, \mu > 0$, the following integral operators, that generalize the Poisson transform,

$$\mathcal{P}_{\alpha,\beta,\mu} f(t) := t^{\alpha\mu-\beta} \int_0^\infty \frac{s^{\beta-1}}{(s^\alpha + t^\alpha)^\mu} f(s)ds, \quad t > 0,$$

are studied in detail on Lebesgue spaces $L^p(\mathbb{R}^+)$ for $1 \le p < \infty$. If $0 < \beta - 1/p < \alpha\mu$, then these operators $\mathcal{P}_{\alpha,\beta,\mu}$ are bounded (and we compute their operator norms which depend on p); and commute on their range. We calculate and represent explicitly their spectra $\sigma(\mathcal{P}_{\alpha,\beta,\mu})$. The main technique is to subordinate these operators in terms of C_0 groups of isometries $T_{t,p} f(s) := e^{-\frac{t}{p}} f(e^{-t}s)$ for $f \in L^p(\mathbb{R}^+)$ (which is isometrically isomorphic to the C_0 group of translations on $L^p(\mathbb{R})$) and transfer properties from some special functions. As consequences, we show that these integral operators are noncompact, have thin spectrum and nontrivial invariant subspaces on $L^p(\mathbb{R}^+)$ for $1 \le p < \infty$.

Keywords: Integral operators, Lebesgue spaces, beta function, spectrum, non-trivial invariant subspaces

MSC 2010: Primary 44A15, 47A10, Secondary 44A35, 47A15

18.1 Introduction

After the works of Enflo [8] and Read [15], it is known that there are bounded operators acting on separable Banach spaces without nontrivial closed invariant subspaces. In [1], two following conjectures regarding the invariant subspace problem are listed:

Acknowledgement: The author has been partially supported by Project MTM2016-77710-P, DGI-FEDER, of the MCYTS and Project E26-17R, D. G. Aragón, Spain.
The authors thank the referee for several interesting ideas, suggestions, and the proof of Theorem 18.10, which have led to the final improvement of the article. We also thank E. A. Gallardo-Gutiérrez, P. Tradacete, and A. Siskakis for several comments and additional information, in particular several references.

Pedro J. Miana, Jesús Oliva-Maza, Departamento de Matemáticas, Instituto Universitario de Matemáticas y Aplicaciones, Universidad de Zaragoza, 50009 Zaragoza, Spain, e-mails: pjmiana@unizar.es, joliva@unizar.es

https://doi.org/10.1515/9783110656756-018

Conjecture 1. Every positive operator on a separable Banach lattice has a nontrivial closed invariant subspace.

Conjecture 2. Every adjoint operator has a nontrivial closed invariant subspace.

The second conjecture was posed by Lomonosov in [13]. Our main aim is to study a family of three parameter integral operators $\mathcal{P}_{\alpha,\beta,\mu}$ which are positive, adjoint, and act on a separable Banach lattice, $L^p(\mathbb{R}^+)$. For $1 \le p < \infty$, these operators have nontrivial invariant subspaces in $L^p(\mathbb{R}^+)$; see Theorem 18.10. In particular, these integral operators are normal in $L^2(\mathbb{R}^+)$ and have nontrivial invariant subspaces as a direct consequence of the spectral theorem; see, for example, [17, Theorem 12.27].

Let $(L^p(\mathbb{R}^+), \| \ \|_p)$ be the classical Lebesgue space formed by measurable functions f on $\mathbb{R}^+$ such that

$$\|f\|_p := \left(\int_0^\infty |f(t)|^p dt\right)^{\frac{1}{p}} < \infty,$$

for $1 < p < \infty$. The integral operator $\mathcal{S}$, sometimes called Carleman or Stieltjes operator, where

$$\mathcal{S}f(t) := \int_0^\infty \frac{f(s)}{s+t} ds, \quad t > 0,$$

is the origin of different theories in many fields of (real and complex) mathematical analysis and differential equations. In particular, this operator is bounded on $L^p(\mathbb{R}^+)$,

$$\|\mathcal{S}\| = \frac{\pi}{\sin(\frac{\pi}{p})}, \quad 1 < p < \infty,$$

and $\sigma(\mathcal{S}) = [0, \pi]$ for $p = 2$, which was originally proved by T. Carleman in [6, p. 169].

The point of view of Carleman was followed in [11, Theorem 319], to show that the integral operator

$$f \mapsto \int_0^\infty K(s, \cdot) f(s) ds, \quad f \in L^p(\mathbb{R}^+),$$

is bounded on $L^p(\mathbb{R}^+)$ for $p > 1$ where the kernel $K(\cdot,\cdot)$ is nonnegative and homogeneous of degree -1. Here, we consider the kernel $K(s,t) := \frac{t^{\alpha\mu-\beta} s^{\beta-1}}{(s^\alpha+t^\alpha)^\mu}$ for $\alpha, \beta, \mu > 0$ to study in detail the integral operator $\mathcal{P}_{\alpha,\beta,\mu}$ where

$$\mathcal{P}_{\alpha,\beta,\mu} f(t) := t^{\alpha\mu-\beta} \int_0^\infty \frac{s^{\beta-1}}{(s^\alpha + t^\alpha)^\mu} f(s) ds, \quad t \ge 0. \tag{18.1}$$

Note that $\mathcal{P}_{1,1,1} = \mathcal{S}$ and $\mathcal{P}_{2,2,1} = \mathcal{P}$ is the classical Poisson transform,

$$\mathcal{P}(f)(t) := \int_0^\infty \frac{s}{s^2+t^2} f(s)ds, \quad t \geq 0.$$

From now on, we label this three-parameter family of integral operators $\mathcal{P}_{\alpha,\beta,\mu}$ as generalized Poisson operators, on behalf of the Poisson transform.

In this paper, the main idea is to subordinate these generalized Poisson operators $\mathcal{P}_{\alpha,\beta,\mu}$ in terms of the C_0 group of invertible isometries $\{T_{t,p}\}_{t\in\mathbb{R}}$ defined on $L^p(\mathbb{R}^+)$ by

$$T_{t,p}f(s) := e^{-\frac{t}{p}} f(e^{-t}s), \quad f \in L^p(\mathbb{R}^+). \tag{18.2}$$

In fact, this C_0 group $\{T_{t,p}\}_{t\in\mathbb{R}}$ defined on $L^p(\mathbb{R}^+)$ is linearly and isometrically isomorphic to the C_0 group of translations on $L^p(\mathbb{R})$; see (18.8).

This strategy of subordination has been also followed by other authors. To study the property of subnormality of the Cesàro operator on $L^2(\mathbb{R}^+)$, Cowen considered this C_0 group of isometries in [7]; later in [4], the Cesàro operator on the Hardy spaces of the half-plane was studied using this subordination. Finally, the generalized Cesàro operator $\mathcal{C}_\beta$ where

$$\mathcal{C}_\beta f(t) := \frac{\beta}{t^\beta} \int_0^t (t-s)^{\beta-1} f(s)ds, \quad t > 0,$$

with $\beta > 0$ was also treated on some subspaces contained on $L^p(\mathbb{R}^+)$ ([12]). Recently, in [14], authors have applied this C_0 group to the generalized Stieltjes operator on the half and whole real line.

The subordination process is a useful tool and a natural extension of the Fourier transform in abstract Banach spaces. Let X be a Banach spaces and $(T(t))_{t\in\mathbb{R}} \subset \mathcal{B}(X)$ a C_0 group of uniformly bounded operators on X, that is, $T(t+s) = T(s)T(t)$, for $t, s \in \mathbb{R}$; $\lim_{t\to 0} T(t)x = x$ for $x \in X$ and $M := \sup_{t\in\mathbb{R}} \|T(t)\| < \infty$ ([3, Definition 3.1.19]). Let θ denote the map: $\theta : L^1(\mathbb{R}) \longrightarrow \mathcal{B}(X)$ such that

$$\theta(g)x := \int_{-\infty}^\infty g(t)T(t)x dt, \quad x \in X,\ g \in L^1(\mathbb{R}). \tag{18.3}$$

Then the map θ is actually well-defined, and is a linear and bounded operator, $\|\theta(g)\| \leq M\|g\|_1$ for $g \in L^1(\mathbb{R})$, and $\|\theta\| \leq M$ ([9, Lemma IV.3.17]). As a consequence, θ is commutative in its range $\theta(L^1(\mathbb{R})) \subset \mathcal{B}(X)$, that is, $\theta(f)\theta(g) = \theta(f*g) = \theta(g)\theta(f)$ for $f, g \in L^1(\mathbb{R})$. By the spectral mapping theorem,

$$\sigma(\theta(g)) = \overline{\widehat{g}(\sigma(iA))}$$

where $\hat{g}$ is the Fourier transform of g and A is the infinitesimal generator of the C_0 group (see, e. g., [18, Theorem 3.1]).

The outline of this paper has been set as follows. In Section 18.2, a three parameter family of exponential functions $\varphi_{\alpha,\beta,\mu}$ that belong to $L^1(\mathbb{R})$, is introduced. Note that their Fourier transform is given by

$$\widehat{\varphi_{\alpha,\beta,\mu}}(\xi) = \frac{1}{\alpha}B\left(\frac{\beta}{\alpha} - i\frac{\xi}{\alpha}, \mu - \frac{\beta}{\alpha} + i\frac{\xi}{\alpha}\right), \quad \xi \in \mathbb{R},$$

(Theorem 18.1), where $B(\cdot,\cdot)$ is the Beta function, $B(z,w) := \int_0^1 t^{z-1}(1-t)^{w-1}dt$, for $\Re z, \Re w > 0$.

Generalized Poisson operators $\mathcal{P}_{\alpha,\beta,\mu}$ acting on $L^p(\mathbb{R}^+)$ are analyzed in Section 18.3. We are able to subordinate them in terms of the C_0 of isometries $(T_{t,p})_{t\in\mathbb{R}}$ (18.2) and the family of $L^1(\mathbb{R})$ functions $\varphi_{\alpha,\beta,\mu}$, that is,

$$\mathcal{P}_{\alpha,\beta,\mu}f = \int_{-\infty}^{\infty} \varphi_{\alpha,\alpha\mu-\beta+1/p,\mu}(r)T_{r,p}f dr = \theta(\varphi_{\alpha,\alpha\mu-\beta+1/p,\mu})f, \quad f \in L^p(\mathbb{R}^+),$$

for $0 < \beta - 1/p < \alpha\mu$ and $1 \le p < \infty$. Then these operators are bounded on $L^p(\mathbb{R}^+)$ and

$$\|\mathcal{P}_{\alpha,\beta,\mu}\| = \frac{1}{\alpha}B\left(\mu - \frac{1}{\alpha}\left(\beta - \frac{1}{p}\right), \frac{1}{\alpha}\left(\beta - \frac{1}{p}\right)\right),$$

for $0 < \beta - 1/p < \alpha\mu$ and $1 \le p < \infty$ (Theorem 18.3). Also, by the spectral mapping theorem, we have that

$$\sigma(\mathcal{P}_{\alpha,\beta,\mu}) = \left\{\frac{1}{\alpha}B\left(\frac{1}{\alpha}\left(\beta - \frac{1}{p}\right) + it, \mu - \frac{1}{\alpha}\left(\beta - \frac{1}{p}\right) - it\right) : t \in \mathbb{R}\right\} \cup \{0\}.$$

(Theorem 18.5); in particular, $\sigma(\mathcal{S}) = [0,\pi]$ on $L^2(\mathbb{R}^+)$ and $\sigma(\mathcal{P}) = [0, \frac{\pi}{2}]$ on $L^1(\mathbb{R}^+)$.

For $p > 1$ and $0 < \beta - 1/p < \alpha\mu$, the dual to the generalized Poisson operator $\mathcal{P}_{\alpha,\beta,\mu}$ on $L^p(\mathbb{R}^+)$, is the generalized Poisson operator $\mathcal{P}_{\alpha,\alpha\mu-\beta+1,\mu}$, acting on $L^{p'}(\mathbb{R})$, that is, $(\mathcal{P}_{\alpha,\beta,\mu})' = \mathcal{P}_{\alpha,\alpha\mu-\beta+1,\mu}$, where $1/p + 1/p' = 1$; see Theorem 18.8. In $L^2(\mathbb{R}^+)$, these integral operators are normal and $\mathcal{P}_{\alpha,\frac{\alpha\mu+1}{2},\mu}$ are self-adjoint.

Finally, in the last section, we use the software Mathematica to visualize spectrum, $\sigma(\mathcal{P}_{\alpha,\beta,\mu})$, on $L^p(\mathbb{R}^+)$ in some particular cases. We also conclude that generalized Poisson operators $\mathcal{P}_{\alpha,\beta,\mu}$ on $L^p(\mathbb{R}^+)$ has nontrivial invariant subspaces (Theorem 18.10).

18.2 Three parametric exponential functions on $L^1(\mathbb{R})$

We define the set of functions $(\varphi_{\alpha,\beta,\mu})_{\alpha,\beta,\mu\in\mathbb{R}}$ by

$$\varphi_{\alpha,\beta,\mu}(t) := \frac{e^{\beta t}}{(1+e^{\alpha t})^{\mu}}, \quad t \in \mathbb{R}. \tag{18.4}$$

Note that $\varphi_{\alpha,\beta,\mu}(-t) = \varphi_{\alpha,\alpha\mu-\beta,\mu}(t)$ for $t \in \mathbb{R}$ and $\varphi_{\alpha,\beta,\mu}\varphi_{\alpha,\gamma,\nu} = \varphi_{\alpha,\beta+\gamma,\mu+\nu}$ for $\alpha, \beta, \gamma, \mu, \nu \in \mathbb{R}$. It is direct to check that

$$\varphi'_{\alpha,\beta,\mu} = \varphi_{\alpha,\beta,\mu}(\beta - \mu\alpha\varphi_{\alpha,\alpha,1}) = \varphi_{\alpha,\beta,\mu}(\beta + \mu\alpha(\varphi_{\alpha,0,1} - 1)).$$

Theorem 18.1. *Fixed $\alpha, \beta, \mu > 0$.*
(i) *For $1 \le p < \infty$, $\varphi_{\alpha,\beta,\mu} \in L^p(\mathbb{R})$ if and only if $0 < \beta < \alpha\mu$ and*

$$\|\varphi_{\alpha,\beta,\mu}\|_p = \left(\frac{1}{\alpha}B\left(\frac{p\beta}{\alpha}, p\left(\mu - \frac{\beta}{\alpha}\right)\right)\right)^{\frac{1}{p}}.$$

(ii) *For $0 < \beta < \alpha\mu$, we have that*

$$\widehat{\varphi_{\alpha,\beta,\mu}}(\xi) = \frac{1}{\alpha}B\left(\frac{\beta}{\alpha} - i\frac{\xi}{\alpha}, \mu - \frac{\beta}{\alpha} + i\frac{\xi}{\alpha}\right), \quad \xi \in \mathbb{R}.$$

Proof. (i) Note that $\varphi_{\alpha,\beta,\mu} \in L^1(\mathbb{R})$ when $0 < \beta < \alpha\mu$; moreover,

$$\|\varphi_{\alpha,\beta,\mu}\|_1 = \int_0^\infty \frac{s^{\beta-1}}{(1+s^\alpha)^\mu}ds = \frac{1}{\alpha}\int_0^\infty \frac{r^{\frac{\beta}{\alpha}-1}}{(1+r)^\mu}dr = \frac{1}{\alpha}B\left(\frac{\beta}{\alpha}, \mu - \frac{\beta}{\alpha}\right).$$

For $1 \le p < \infty$, $\varphi_{\alpha,\beta,\mu} \in L^p(\mathbb{R})$ if and only if $\varphi_{\alpha,\beta p,\mu p} \in L^1(\mathbb{R})$, and in this case, $0 < \beta < \alpha\mu$ and

$$\|\varphi_{\alpha,\beta,\mu}\|_p = \left(\frac{1}{\alpha}B\left(\frac{p\beta}{\alpha}, p\left(\mu - \frac{\beta}{\alpha}\right)\right)\right)^{\frac{1}{p}}.$$

(ii) For $\xi \in \mathbb{R}$, we have that

$$\widehat{\varphi_{\alpha,\beta,\mu}}(\xi) = \int_0^\infty \frac{s^{\beta-i\xi-1}}{(1+s^\alpha)^\mu}ds = \frac{1}{\alpha}B\left(\frac{\beta}{\alpha} - i\frac{\xi}{\alpha}, \mu - \frac{\beta}{\alpha} + i\frac{\xi}{\alpha}\right), \quad \xi \in \mathbb{R},$$

and we conclude the proof. □

18.3 Generalized Poisson operators on $L^p(\mathbb{R}^+)$

For $\alpha, \beta, \mu > 0$, the generalized Poisson operator $\mathcal{P}_{\alpha,\beta,\mu}$ on $\mathbb{R}^+$ is defined by

$$\mathcal{P}_{\alpha,\beta,\mu}f(t) := t^{\mu\alpha-\beta}\int_0^\infty \frac{s^{\beta-1}}{(s^\alpha + t^\alpha)^\mu}f(s)ds = \int_0^\infty \frac{u^{\beta-1}}{(1+u^\alpha)^\mu}f(tu)du, \quad t > 0, \tag{18.5}$$

for functions f defined on $\mathbb{R}^+$. It is direct to get

$$\mathcal{P}_{\alpha,\beta,\mu}f(t) = \frac{1}{\alpha\Gamma(\mu)}t^{\mu\alpha-\beta}\mathcal{L}(x^{\mu-1}\mathcal{L}(s^{\frac{\beta}{\alpha}-1}f(s^{\frac{1}{\alpha}}))(x))(t^{\alpha}), \quad f \in C_c(\mathbb{R}^+), \tag{18.6}$$

for $t > 0$, where $\mathcal{L}$ is the usual Laplace transform and $C_c(\mathbb{R}^+)$ the set of continuous functions of compact support.

Now we check how these operators act on some particular functions.

Example 18.2.

(i) For $\gamma > 0$, we define g_γ by $g_\gamma(t) := \frac{t^{\gamma-1}}{\Gamma(\gamma)}$, for $t > 0$. Then

$$\mathcal{P}_{\alpha,\beta,\mu}(g_\gamma) = \frac{1}{\alpha}B\Big(\frac{1}{\alpha}(\beta+\gamma-1), \mu - \frac{1}{\alpha}(\beta+\gamma-1)\Big)g_\gamma,$$

for $\alpha\mu > \gamma + \beta - 1 > 0$. Under these conditions, functions g_γ are eigenfunctions of $\mathcal{P}_{\alpha,\beta,\mu}$, in particular $\mathcal{P}_{\alpha,\beta,\mu}(\chi_{(0,\infty)}) = \frac{1}{\alpha}B(\frac{\beta}{\alpha}, \mu - \frac{\beta}{\alpha})\chi_{(0,\infty)}$ for $0 < \beta < \alpha\mu$.

(ii) Take $e_\lambda(s) := e^{-\lambda s}$ for $s > 0$ and $\lambda \in \mathbb{C}^+ := \{z \in \mathbb{C} : \Re z > 0\}$. For $\alpha, \beta, \mu > 0$, we have that

$$\mathcal{P}_{\alpha,\beta,\mu}(e_\lambda)(t) = \mathcal{L}\Big(\frac{u^{\beta-1}}{(1+u^\alpha)^\mu}\Big)(\lambda t), \quad t > 0.$$

As we have commented in the Introduction, the operator $\mathcal{P}_{\alpha,\beta,\mu}$ defines a bounded operator on $L^p(\mathbb{R}^+)$ for $0 < \beta - \frac{1}{p} < \alpha\mu$ and $1 \le p < \infty$.

The first result in this section is the following theorem. Although the first part of the proof can be found in [11, Theorem 319], we include it here for the sake of completeness.

Theorem 18.3. *The operator $\mathcal{P}_{\alpha,\beta,\mu}$ is bounded on $L^p(\mathbb{R}^+)$ and*

$$\|\mathcal{P}_{\alpha,\beta,\mu}\| = \frac{1}{\alpha}B\Big(\mu - \frac{1}{\alpha}\Big(\beta - \frac{1}{p}\Big), \frac{1}{\alpha}\Big(\beta - \frac{1}{p}\Big)\Big),$$

for $0 < \beta - 1/p < \alpha\mu$ and $1 \le p < \infty$.

Proof. Take $f \in L^p(\mathbb{R}^+)$ with $1 \le p$. By (18.5) and the Minkowski inequality, we have that

$$\begin{aligned}\|\mathcal{P}_{\alpha,\beta,\mu}f\|_p &= \Big(\int_0^\infty\Big|\int_0^\infty \frac{u^{\beta-1}}{(1+u^\alpha)^\mu}f(tu)du\Big|^p dt\Big)^{\frac{1}{p}} \\ &\le \int_0^\infty \frac{u^{\beta-1}}{(1+u^\alpha)^\mu}\Big(\int_0^\infty |f(tu)|^p dt\Big)^{\frac{1}{p}} du = \|f\|_p \int_0^\infty \frac{u^{\beta-\frac{1}{p}-1}}{(1+u^\alpha)^\mu}du \\ &= \frac{1}{\alpha}B\Big(\mu - \frac{1}{\alpha}(\beta - 1/p), \frac{1}{\alpha}(\beta - 1/p)\Big)\|f\|_p.\end{aligned}$$

Now we prove that this upper bound is optimal for $p > 1$. For $\varepsilon > 0$, take $f_\varepsilon(t) := t^{-\frac{1+\varepsilon}{p}}\chi_{[1,\infty)}(t)$ and $g_\varepsilon(t) := t^{-\frac{1+\varepsilon}{p'}}\chi_{[1,\infty)}(t)$. Note that $f_\varepsilon \in L^p(\mathbb{R}^+)$, $g_\varepsilon \in L^{p'}(\mathbb{R}^+)$, and $\|f_\varepsilon\|_p\|g_\varepsilon\|_{p'} = 1/\varepsilon$, where $1/p + 1/p' = 1$. Then it follows that

$$\begin{aligned}\frac{\langle g_\varepsilon, \mathcal{P}_{\alpha,\beta,\mu} f_\varepsilon\rangle}{\|f_\varepsilon\|_p\|g_\varepsilon\|_{p'}} &= \varepsilon\int_1^\infty t^{-\frac{1+\varepsilon}{p'}}\int_{1/t}^\infty \frac{u^{\beta-1}}{(1+u^\alpha)^\mu}(ut)^{-\frac{1+\varepsilon}{p}}\,dudt\\ &= \varepsilon\int_1^\infty t^{-(1+\varepsilon)}\int_0^\infty \frac{u^{\beta-1}}{(1+u^\alpha)^\mu}u^{-\frac{1+\varepsilon}{p}}\,dudt - \varepsilon h_{\alpha,\beta,\mu}(\varepsilon)\\ &= \frac{1}{\alpha}B\left(\frac{1}{\alpha}\left(\beta - \frac{1+\varepsilon}{p}\right), \mu - \frac{1}{\alpha}\left(\beta - \frac{1+\varepsilon}{p}\right)\right) - \varepsilon h_{\alpha,\beta,\mu}(\varepsilon),\end{aligned}$$

where we restrict on $\varepsilon \in (0, p\beta - 1)$ and the (positive) function $h_{\alpha,\beta,\mu}$ is given by

$$\begin{aligned}h_{\alpha,\beta,\mu}(\varepsilon) &:= \int_1^\infty t^{-(1+\varepsilon)}\int_0^{1/t} \frac{u^{\beta-1}}{(1+u^\alpha)^\mu}u^{-\frac{1+\varepsilon}{p}}\,dudt \le \int_1^\infty t^{-(1+\varepsilon)}\int_0^{1/t} u^{\beta-1-\frac{1+\varepsilon}{p}}\,dudt\\ &= \frac{1}{\beta - \frac{1+\varepsilon}{p}}\int_1^\infty \frac{dt}{t^{1+\varepsilon+\beta-\frac{1+\varepsilon}{p}}} = \frac{1}{(\beta - \frac{1+\varepsilon}{p})(\varepsilon + \beta - \frac{1+\varepsilon}{p})},\end{aligned}$$

so that $\lim_{\varepsilon\to 0^+} \varepsilon h_{\alpha,\beta,\mu}(\varepsilon) = 0$. We conclude the proof taking the limit $\varepsilon \to 0^+$ of $\frac{1}{\|f_\varepsilon\|_p\|g_\varepsilon\|_{p'}}\langle g_\varepsilon, \mathcal{P}_{\alpha,\beta,\mu} f_\varepsilon\rangle$.

For $p = 1$, one can take $f_\varepsilon(t) := t^{-(1+\varepsilon)}\chi_{[1,\infty)}(t) \in L^1(\mathbb{R}^+)$ and $g_\varepsilon := \chi_{[1,\infty)} \in L^\infty(\mathbb{R}^+)$, and apply same reasoning as above. □

Remark 18.4. For $\mu = \beta = \alpha = 1$ and $1 < p < \infty$, we have that $\|\mathcal{S}\| = \frac{\pi}{\sin(\pi p)}$ ([11, Section 9.5, p. 232]). It is clear that for $p = 1$ the Stieltjes operator $\mathcal{S}$ does not take $L^1(\mathbb{R}^+)$ into $L^1(\mathbb{R}^+)$. Indeed, the function h_2, given by $h_2(t) := (1+t)^{-2}$ for $t > 0$, belongs to $L^1(\mathbb{R}^+)$ and

$$\mathcal{S}h_2(t) = \frac{t - \log(t) - 1}{(t-1)^2}, \quad t > 0,$$

which does not belong to $L^1(\mathbb{R}^+)$.

For $\alpha = \beta = 2$, $\mu = 1$ and $1 \le p < \infty$, we get that the Poisson transform $\mathcal{P}$ verifies that

$$\|\mathcal{P}\| = \frac{\pi}{2\sin(\frac{\pi}{2p})},$$

where we apply the Euler's reflection formula $\Gamma(z)\Gamma(1-z) = \frac{\pi}{\sin(\pi z)}$, $z \notin \mathbb{Z}$.

In these Lebesgue spaces, the family of operators $(T_{t,p})_{t\in\mathbb{R}}$ defined by

$$T_{t,p}f(s) := e^{-\frac{t}{p}}f(e^{-t}s), \quad s \ge 0, \tag{18.7}$$

is a C_0 group of isometries on $L^p(\mathbb{R}^+)$ for $1 \le p < \infty$ and on $C_0(\mathbb{R}^+)$, taking $p = \infty$ (i. e., $T_{t,\infty}f(s) := f(e^{-t}s)$). The isometric property and the group law are fairly simple to check. As for the strong continuity, it is also part of folklore: for $1 \le p < \infty$, $h \in C_c(\mathbb{R}^+)$, and $s, t \in \mathbb{R}$,

$$\|T_{t,p}h - T_{s,p}h\|_p^p = \int_0^\infty |e^{\frac{-t}{p}}h(e^{-t}r) - e^{\frac{-s}{p}}h(e^{-s}r)|^p\, dr,$$

and so $\|T_{t,p}h - T_{s,p}h\|_p^p \to 0$ as $t \to s$, by the dominated converge theorem, for example, since supp(h) is compact. For arbitrary $f \in L^p(\mathbb{R}^+)$, one obtains $\|T_{t,p}f - T_{s,p}f\|_p^p \to 0$ when $t \to s$, using the density of $C_c(\mathbb{R}^+)$ and the fact that the operator $T_p(t)$ is an isometry for each $t \in \mathbb{R}$. The case of $C_0(\mathbb{R}^+)$ is even simpler, and is left as an exercise to the reader.

The infinitesimal generator Λ of the C_0 group $\{T_{t,p}\}_{t\in\mathbb{R}}$ is given by

$$(\Lambda f)(s) := -sf'(s) - \frac{1}{p}f(s), \quad s \ge 0,$$

with domain $D(\Lambda) = \{f \in L^p(\mathbb{R}^+) \; : \; tf' \in L^p(\mathbb{R}^+)\}$; the point spectrum $\sigma_p(\Lambda) = \emptyset$; and the usual spectrum $\sigma(\Lambda) = \sigma_a(\Lambda) = i\mathbb{R}$ (where $\sigma_a(\Lambda)$ is the approximate spectrum); see similar ideas in [4, Proposition 2.3]. Also, the groups $(T_{t,p})_{t\in\mathbb{R}}$ and $(T_{-t,p'})_{t\in\mathbb{R}}$ are adjoint operators of each other acting on $L^p(\mathbb{R}^+)$ and $L^{p'}(\mathbb{R}^+)$ with $\frac{1}{p} + \frac{1}{p'} = 1$.

For a fixed $1 \le p < \infty$, consider the isometric isomorphism $U_p : L^p(\mathbb{R}^+) \to L^p(\mathbb{R})$, defined by

$$U_pf(s) := e^{\frac{s}{p}}f(e^s), \quad s \in \mathbb{R}.$$

Observe that its inverse is given by $U_p^{-1}g(s) = s^{-\frac{1}{p}}g(\ln s)$. Then it is straightforward to check that the C_0 group of isometries on $L^p(\mathbb{R}^+)$, $\{T_{t,p}\}_{t\in\mathbb{R}}$, is linearly and isometrically isomorphic to the C_0 group of isometries on $L^p(\mathbb{R})$ given by $\{U_pT_{t,p}U_p^{-1}\}_{t\in\mathbb{R}}$, which is simply the group of translations on $\mathbb{R}$, that is,

$$U_pT_{t,p}U_p^{-1}f(s) = f(s - t), \quad s, t \in \mathbb{R}. \tag{18.8}$$

We will make use of this group isomorphism to find nontrivial invariant subspaces of the generalized Poisson operator $\mathcal{P}_{\alpha,\beta,\mu}$ in Theorem 18.10.

It is known the C_0 group $\{T_{t,p}\}_{t\in\mathbb{R}}$ subordinates several operators: generalized Cesàro operator ([12]); continuous Hilbert transform ([2]) or generalized Stieltjes operators ([14]). Moreover, the classical Cesàro operator $\mathcal{C}$ is equal to $(\lambda_p - \Lambda)^{-1}$, for $\lambda_p = 1 - 1/p > 0$, where

$$\mathcal{C}f(t) := \frac{1}{t}\int_0^t f(s)ds, \quad f \in L^p(\mathbb{R}^+), \; t > 0,$$

([4, 12]). In the next result, we describe this subordination for generalized Poisson operators $\mathcal{P}_{\alpha,\beta,\mu}$ and identify $\sigma(\mathcal{P}_{\alpha,\beta,\mu})$ for suitable $\alpha, \beta, \mu > 0$.

Theorem 18.5. *Let $1 \le p < \infty$ and let $\mathcal{P}_{\alpha,\beta,\mu}$ be the generalized Poisson operator given by (18.5) with $0 < \beta - 1/p < \alpha\mu$.*

(i) *If $f \in L^p(\mathbb{R}^+)$, then*

$$\mathcal{P}_{\alpha,\beta,\mu} f(t) = \int_{-\infty}^{\infty} \varphi_{\alpha,\alpha\mu-\beta+1/p,\mu}(r) T_{r,p} f(t)\,dr, \quad t \ge 0.$$

(ii)

$$\sigma(\mathcal{P}_{\alpha,\beta,\mu}) = \left\{ \frac{1}{\alpha} B\left(\frac{1}{\alpha}\left(\beta - \frac{1}{p}\right) + it, \mu - \frac{1}{\alpha}\left(\beta - \frac{1}{p}\right) - it \right) \; : \; t \in \mathbb{R} \right\} \cup \{0\}.$$

In particular the operator $\mathcal{P}_{\alpha,\beta,\mu}$ is not compact on $L^p(\mathbb{R}^+)$.

Proof. (i) Let $0 < \beta - 1/p < \alpha\mu$ be, $1 \le p < \infty$ and take $f \in L^p(\mathbb{R}^+)$. We apply the change of variable $s = te^{-r}$ to get that

$$\mathcal{P}_{\alpha,\beta,\mu} f(t) = t^{\alpha\mu-\beta} \int_0^{\infty} \frac{s^{\beta-1}}{(s^\alpha + t^\alpha)^\mu} f(s)\,ds = \int_{-\infty}^{\infty} \frac{e^{(\alpha\mu-\beta+1/p)r}}{(1+e^{\alpha r})^\mu} e^{-r/p} f(te^{-r})\,dr,$$

and the equality is proved. Observe that by this equality, the operator $\mathcal{P}_{\alpha,\beta,\mu}$ is a bounded operator on $L^p(\mathbb{R}^+)$ for $1 \le p < \infty$ and

$$\|\mathcal{P}_{\alpha,\beta,\mu}\| \le \|\varphi_{\alpha,\alpha\mu-\beta+1/p,\mu}\|_1 = \frac{1}{\alpha} B\left(\frac{1}{\alpha}\left(\beta - \frac{1}{p}\right), \mu - \frac{1}{\alpha}\left(\beta - \frac{1}{p}\right) \right),$$

where we have applied Theorem 18.1(i).

(ii) Since $(T_{t,p})_{t\in\mathbb{R}}$ is a C_0 group of isometries whose infinitesimal generator is $(\Lambda, D(\Lambda))$ and $\mathcal{P}_{\alpha,\beta,\mu} = \theta(\varphi_{\alpha,\alpha\mu-\beta+1/p,\mu})$, (see part (i)) we apply [18, Theorem 3.1] to obtain

$$\sigma(\mathcal{P}_{\alpha,\beta,\mu}) = \overline{\widehat{\varphi_{\alpha,\alpha\mu-\beta+1/p,\mu}}(\sigma(i\Lambda))} = \overline{\widehat{\varphi_{\alpha,\alpha\mu-\beta+1/p,\mu}}(\mathbb{R})}.$$

Now we apply Theorem 18.1(ii) to conclude

$$\sigma(\mathcal{P}_{\alpha,\beta,\mu}) = \left\{ \frac{1}{\alpha} B\left(\frac{1}{\alpha}\left(\beta - \frac{1}{p}\right) + it, \mu - \frac{1}{\alpha}\left(\beta - \frac{1}{p}\right) - it \right) \; : \; t \in \mathbb{R} \right\} \cup \{0\},$$

where we have applied that $\lim_{t\to\pm\infty} \Gamma(a + it) = 0$ for $a > 0$. Finally, we deduce that the operator $\mathcal{P}_{\alpha,\beta,\mu}$ is not compact on $L^p(\mathbb{R}^+)$: in the opposite case, its spectrum must contain only a countable number of eigenvalues and 0, which is not the case. □

Remark 18.6. In the case that $\mu = 1$ and $\alpha > \beta - \frac{1}{p} > 0$, we obtain that

$$\sigma(\mathcal{P}_{\alpha,\beta,1}) = \frac{1}{\alpha}\left\{\frac{\pi}{\sin(\frac{\pi}{\alpha}(\beta - \frac{1}{p}) + it)} : t \in \mathbb{R}\right\} \cup \{0\}.$$

For $\alpha = 2$, $\beta = 1 + \frac{1}{p}$; we have that

$$\sigma(\mathcal{P}_{2,1+\frac{1}{p},1}) = \frac{1}{2}\left\{\frac{\pi}{\cosh(t)} : t \in \mathbb{R}\right\} \cup \{0\} = \left[0, \frac{\pi}{2}\right].$$

In particular, it applies to $\mathcal{P}_{2,\frac{3}{2},1}$ in $L^2(\mathbb{R}^+)$, or $\mathcal{P} = \mathcal{P}_{2,2,1}$ in $L^1(\mathbb{R}^+)$, where $\mathcal{P}$ is the classical Poisson transform (see the Introduction). In the last section, we draw some of these families of spectra.

Corollary 18.7. *Let $1 \le p < \infty$, and let $\mathcal{P}_{\alpha,\beta,\mu}$ and $\mathcal{P}_{\alpha',\beta',\mu'}$ be generalized Poisson operators on $L^p(\mathbb{R}^+)$ with $0 < \beta - 1/p < \alpha\mu$ and $0 < \beta' - 1/p < \alpha'\mu'$. Then these operators commute*

$$\mathcal{P}_{\alpha,\beta,\mu}\mathcal{P}_{\alpha',\beta',\mu'} = \mathcal{P}_{\alpha',\beta',\mu'}\mathcal{P}_{\alpha,\beta,\mu}.$$

Proof. By Theorem 18.5(i), $\mathcal{P}_{\alpha,\beta,\mu} = \theta(\varphi_{\alpha,\alpha\mu-\beta+1/p,\mu})$ and $\mathcal{P}_{\alpha',\beta',\mu'} = \theta(\varphi_{\alpha',\alpha'\mu'-\beta'+1/p,\mu'})$ where the algebra homomorphism θ is defined in (18.3). Then

$$\begin{aligned}\mathcal{P}_{\alpha,\beta,\mu}\mathcal{P}_{\alpha',\beta',\mu'} &= \theta(\varphi_{\alpha,\alpha\mu-\beta+1/p,\mu})\theta(\varphi_{\alpha',\alpha'\mu'-\beta'+1/p,\mu'}) = \theta(\varphi_{\alpha,\alpha\mu-\beta+1/p,\mu} * \varphi_{\alpha',\alpha'\mu'-\beta'+1/p,\mu'})\\ &= \theta(\varphi_{\alpha',\alpha'\mu'-\beta'+1/p,\mu'} * \varphi_{\alpha,\alpha\mu-\beta+1/p,\mu}) = \mathcal{P}_{\alpha',\beta',\mu'}\mathcal{P}_{\alpha,\beta,\mu},\end{aligned}$$

and we conclude the proof. □

Now we identify the dual to the generalized Poisson operator $(\mathcal{P}_{\alpha,\beta,\mu})'$ on $L^{p'}(\mathbb{R}^+)$.

Theorem 18.8. *For $p > 1$ and $0 < \beta - 1/p < \alpha\mu$, the dual to the generalized Poisson operator $\mathcal{P}_{\alpha,\beta,\mu}$ on $L^p(\mathbb{R}^+)$ is the general Poisson operator $\mathcal{P}_{\alpha,\alpha\mu-\beta+1,\mu}$, acting on $L^{p'}(\mathbb{R}^+)$, that is,*

$$\langle \mathcal{P}_{\alpha,\beta,\mu}f, g\rangle = \langle f, \mathcal{P}_{\alpha,\alpha\mu-\beta+1,\mu}g\rangle, \quad f \in L^p(\mathbb{R}^+),\ g \in L^{p'}(\mathbb{R}^+),$$

where $\frac{1}{p} + \frac{1}{p'} = 1$. As a consequence, $\mathcal{P}_{\alpha,\beta,\mu}$ is an injective, nonsurjective and of dense range on $L^p(\mathbb{R}^+)$.

In the case of $L^2(\mathbb{R}^+)$, and $0 < \beta - \frac{1}{2} < \alpha\mu$, the operator $\mathcal{P}_{\alpha,\beta,\mu}$ is normal and has nontrivial invariant subspaces; moreover, $\mathcal{P}_{\alpha,\frac{\alpha\mu+1}{2},\mu}$ is self-adjoint.

Proof. We apply the Fubini theorem to get that

$$\langle \mathcal{P}_{\alpha,\beta,\mu}f, g\rangle = \int_0^\infty\int_0^\infty \frac{t^{\alpha\mu-\beta}s^{\beta-1}}{(s^\alpha + t^\alpha)^\mu} g(t)f(s)\,ds\,dt$$

$$= \int_0^\infty f(s)\, s^{\beta-1} \int_0^\infty \frac{t^{\alpha\mu-\beta}}{(t^\alpha + s^\alpha)^\mu} g(t)\,dt\,ds = \langle f, \mathcal{P}_{\alpha,\alpha\mu-\beta+1,\mu} g\rangle.$$

The injectivity of the Laplace transform $\mathcal{L}$ and (18.6) imply the injectivity of $\mathcal{P}_{\alpha,\beta,\mu}$ on $L^p(\mathbb{R}^+)$. By Theorem 18.5, $0 \in \sigma(\mathcal{P}_{\alpha,\beta,\mu})$ and $\mathcal{P}_{\alpha,\beta,\mu}$ is not invertible so, by the open mapping theorem, $\mathcal{P}_{\alpha,\beta,\mu}$ cannot be surjective. Then, as the dual of operator $\mathcal{P}_{\alpha,\beta,\mu}$ is injective (which is $\mathcal{P}_{\alpha,\alpha\mu-\beta+1,\mu}$, as we have just shown), we conclude that $\mathcal{P}_{\alpha,\beta,\mu}$ is of dense range on $L^p(\mathbb{R}^+)$.

Now we consider the case $p = 2$, the operator $\mathcal{P}_{\alpha,\beta,\mu}$ is normal, that is,

$$\mathcal{P}_{\alpha,\beta,\mu}(\mathcal{P}_{\alpha,\beta,\mu})' = \mathcal{P}_{\alpha,\beta,\mu}\mathcal{P}_{\alpha,\alpha\mu-\beta+1,\mu} = \mathcal{P}_{\alpha,\alpha\mu-\beta+1,\mu}\mathcal{P}_{\alpha,\beta,\mu} = (\mathcal{P}_{\alpha,\beta,\mu})'\mathcal{P}_{\alpha,\beta,\mu}.$$

By the spectral theorem for normal operators (see, e. g., [17, Theorem 12.27]), the operator $\mathcal{P}_{\alpha,\beta,\mu}$ has nontrivial invariant subspaces. Finally, we get that $\mathcal{P}_{\alpha,\beta,\mu} = (\mathcal{P}_{\alpha,\beta,\mu})' = \mathcal{P}_{\alpha,\alpha\mu-\beta+1,\mu}$ if and only if $2\beta = \alpha\mu + 1$. □

Remark 18.9. As the operator $\mathcal{P}_{\alpha,\frac{\alpha\mu+1}{2},\mu}$ is self-adjoint on $L^2(\mathbb{R}^+)$ for $\alpha, \mu > 0$, the spectrum $\sigma(\mathcal{P}_{\alpha,\frac{\alpha\mu+1}{2},\mu})$ is a subset of real numbers,

$$\begin{aligned}\sigma(\mathcal{P}_{\alpha,\frac{\alpha\mu+1}{2},\mu}) &= \frac{1}{\alpha\Gamma(\mu)}\left\{\Gamma\left(\frac{\mu}{2} - it\right)\Gamma\left(\frac{\mu}{2} + it\right) : t \in \mathbb{R}\right\} \cup \{0\} \\ &= \left[0, \frac{1}{\alpha}B\left(\frac{\mu}{2}, \frac{\mu}{2}\right)\right],\end{aligned}$$

where we have used that $\Gamma(z)\Gamma(\overline{z}) \in \mathbb{R}$. This result was proved for the operator $\mathcal{S}$ in [6, p. 169] and finally for $\mathcal{P}_{1,\beta,2\beta-1}$ for $\beta > \frac{1}{2}$ in [10, Proposition 1.1].

18.4 Spectral pictures and nontrivial invariant subspaces

The main aim of this last section is that the reader visualizes the spectra of some generalized Poisson operators, $\sigma(\mathcal{P}_{\alpha,\beta,\mu})$, on $L^p(\mathbb{R}^+)$. We will manage the software Mathematica in order to represent these spectra. We also proved that these operators $\mathcal{P}_{\alpha,\beta,\mu}$ have nontrivial invariant subspaces on $L^p(\mathbb{R}^+)$ for $1 \le p < \infty$ (Theorem 18.10).

By Theorem 18.5(ii), we have that

$$\sigma(\mathcal{P}_{\alpha,\beta,\mu}) = \left\{\frac{1}{\alpha}B\left(\frac{1}{\alpha}\left(\beta - \frac{1}{p}\right) + it, \mu - \frac{1}{\alpha}\left(\beta - \frac{1}{p}\right) - it\right) : t \in \mathbb{R}\right\} \cup \{0\}.$$

For $0 < \beta - 1/p < \alpha\mu$, the closed curve $\sigma(\mathcal{P}_{\alpha,\beta,\mu})$ is symmetrical with respect to the OX axis and takes the point $\frac{1}{\alpha}B(\frac{1}{\alpha}(\beta - \frac{1}{p}), \mu - \frac{1}{\alpha}(\beta - \frac{1}{p}))$ on the complex plane (at $t = 0$).

Moreover, note that the curve $\sigma(\mathcal{P}_{\alpha,\beta,\mu})$ is contained in the circle of center $(0,0)$ and radius $\frac{1}{\alpha}B(\frac{1}{\alpha}(\beta-\frac{1}{p}),\mu-\frac{1}{\alpha}(\beta-\frac{1}{p}))$, due to the following inequality holds for $t\in\mathbb{R}$:

$$\left|B\left(\frac{1}{\alpha}\left(\beta-\frac{1}{p}\right)+it,\mu-\frac{1}{\alpha}\left(\beta-\frac{1}{p}\right)-it\right)\right|\leq B\left(\frac{1}{\alpha}\left(\beta-\frac{1}{p}\right),\mu-\frac{1}{\alpha}\left(\beta-\frac{1}{p}\right)\right).$$

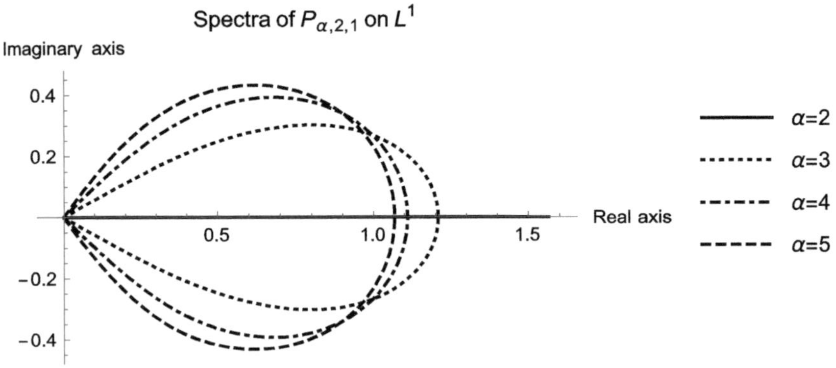

Figure 18.1: Fixed $\beta=2$, $\mu=1$ and $p=1$.

The special case $\mu\alpha=2(\beta-\frac{1}{p})$ has remarkable properties. Since $\Gamma(\overline{z})=\overline{\Gamma(z)}$, then $B(\frac{1}{\alpha}(\beta-\frac{1}{p})+it,\frac{1}{\alpha}(\beta-\frac{1}{p})-it)\geq 0$ for $t\in\mathbb{R}$ and

$$\sigma(\mathcal{P}_{\alpha,\frac{\alpha\mu}{2}+\frac{1}{p},\mu})=\left[0,\frac{1}{\alpha}B\left(\frac{\mu}{2},\frac{\mu}{2}\right)\right].$$

The case $p=2$ is considered in Remark 18.9. The spectrum of the self-adjoint operator $\mathcal{P}_{\alpha,\frac{\alpha\mu+1}{2},\mu}$ on $L^2(\mathbb{R}^+)$ is $[0,\frac{1}{\alpha}B(\frac{\mu}{2},\frac{\mu}{2})]$; in particular $\sigma(\mathcal{S})=[0,\pi]$ and $\sigma(\mathcal{P}_{2,\frac{3}{2},1})=[0,\frac{\pi}{2}]$.

For $\mu=1$, and $0<\beta-\frac{1}{p}<\alpha$, we define $\gamma=\frac{1}{\alpha}(\beta-\frac{1}{p})\in(0,1)$ and

$$\begin{aligned}B(\gamma+it,1-\gamma-it)&=\Gamma(\gamma+it)\Gamma(1-\gamma-it)=\frac{\pi}{\sin(\pi(\gamma+it))}\\&=\frac{\pi}{\sin^2(\pi\gamma)+\sinh^2(\pi t)}(\sin(\pi\gamma)\cosh(\pi t)-i\cos(\pi\gamma)\sinh(\pi t)),\end{aligned}$$

where we conclude that $\sigma(\mathcal{P}_{\alpha,\beta,1})\subset\mathbb{C}^+$; see Figure 18.1. In this figure, we fix $\beta=2$, $\mu=1$ and $p=1$. Then we consider $\alpha>1$. Note that for $\alpha=2$, (i. e., $\gamma=\frac{1}{2}$), $\sigma(\mathcal{P}_{2,2,1})=[0,\frac{\pi}{2}]$; and for $\alpha=4$ (i. e., $\gamma=\frac{1}{4}$), we have that

$$\sigma(\mathcal{P}_{4,2,1})=\left\{\frac{\pi}{2\sqrt{2}\cosh(2\pi t)}(\cosh(\pi t)-i\sinh(\pi t))\;:\;t\in\mathbb{R}\right\}\cup\{0\}.$$

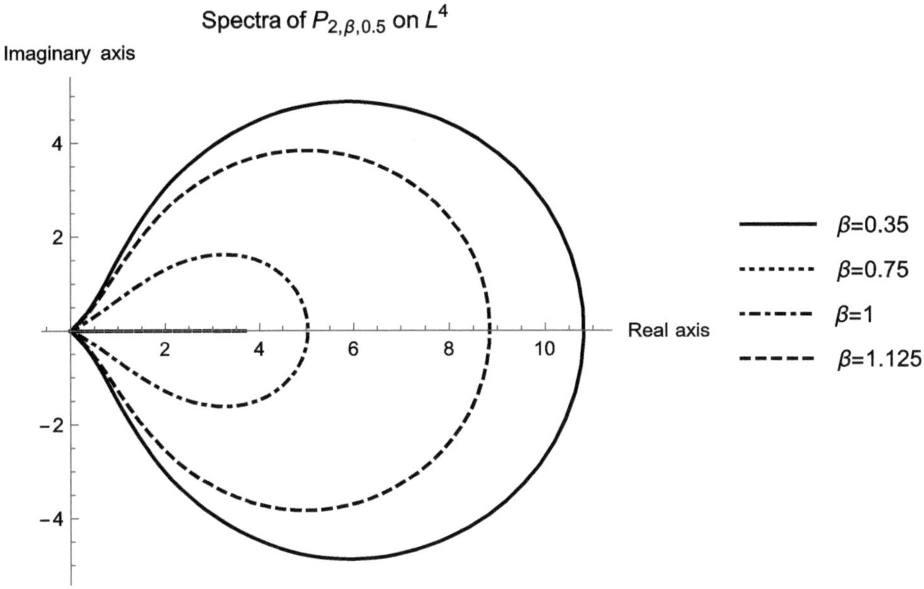

Figure 18.2: Fixed $\alpha = 2$, $\mu = 0.5$ and $p = 4$.

In Figure 18.2, we fix $\alpha = 2$, $\mu = \frac{1}{2}$ and $p = 4$. Then we may consider $\frac{1}{4} < \beta < \frac{5}{4}$. For $\beta = \frac{3}{4}$, we obtain that

$$\sigma(\mathcal{P}_{2,\frac{3}{4},\frac{1}{2}}) = \left[0, \frac{(\Gamma(\frac{1}{4}))^2}{2\sqrt{\pi}}\right].$$

In Figure 18.3, we fix $\alpha = 1$, $\beta = 1$ and $p = 2$. Then we may consider $\mu > \frac{1}{2}$. As we have commented, we get $\sigma(\mathcal{P}_{1,1,1}) = [0, \pi]$ for $\mu = 1$. For $\mu = 2$, note that

$$B\left(\frac{1}{2} + it, \frac{3}{2} - it\right) = \frac{\pi(\frac{1}{2} - it)}{\cosh(\pi t)}, \quad t \in \mathbb{R},$$

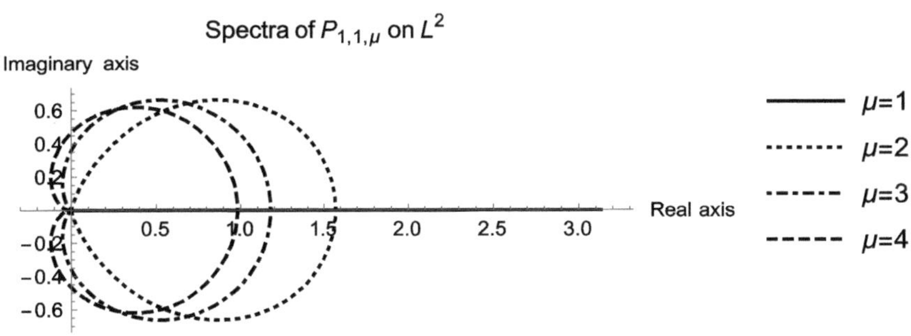

Figure 18.3: Fixed $\alpha = 1$, $\beta = 1$ and $p = 2$.

and we conclude that $\sigma(\mathcal{P}_{1,1,2}) \subset \mathbb{C}^+$. However for $\mu = 3$, and $\mu = 4$ we obtain that

$$B\left(\frac{1}{2} + it, \frac{5}{2} - it\right) = \frac{\pi(3 - 4t^2 - 8ti)}{8\cosh(\pi t)},$$
$$B\left(\frac{1}{2} + it, \frac{7}{2} - it\right) = \frac{\pi(15 - 36t^2 + (8t^3 - 46t)i)}{48\cosh(\pi t)}, \qquad t \in \mathbb{R},$$

and $\sigma(\mathcal{P}_{1,1,3}), \sigma(\mathcal{P}_{1,1,4}) \not\subset \mathbb{C}^+$.

Nontrivial invariant subspaces for $\mathcal{P}_{\alpha,\beta,\mu}$

As we have commented, operators $\mathcal{P}_{\alpha,\beta,\mu}$ are normal in $L^2(\mathbb{R}^+)$ for $0 < \beta - \frac{1}{2} < \alpha\mu$, and they have nontrivial invariant subspaces (Theorem 18.8). For $L^p(\mathbb{R}^+)$ with $1 \le p \ne 2 < \infty$, it seems natural that the same assertion holds; see conjectures in the Introduction and [1]. For $0 < \beta - \frac{1}{p} < \alpha\mu$, note that $\mathcal{P}_{\alpha,\beta,\mu}$ is a positive and adjoint operator in a Banach lattice whose thin spectrum has empty interior.

Note that the C_0 group of translations on $L^p(\mathbb{R})$ has natural nontrivial invariant subspaces. For example, let $\mathcal{W}$ be any proper closed subset of the real line with non-empty interior, and let $\mathcal{F} : L^p(\mathbb{R}) \to \mathcal{S}'$ be the Fourier transform where $\mathcal{S}'$ is the set of tempered distributions on $\mathbb{R}$. Then the set $\{f \in L^p(\mathbb{R}) : \operatorname{supp} \mathcal{F}f \subset \mathcal{W}\}$ is a nontrivial translation invariant subspace due to $\mathcal{F}$ is a continuous operator from $L^p(\mathbb{R})$ to $\mathcal{S}'$. Other nontrivial translation invariant subspaces may be found in [5, 16].

Theorem 18.10. *Let $1 \le p < \infty$ and let $0 < \beta - \frac{1}{p} < \alpha\mu$. Then the generalized Poisson operator $\mathcal{P}_{\alpha,\beta,\mu}$ on $L^p(\mathbb{R}^+)$ has nontrivial invariant subspaces.*

Proof. By the subordination of $\mathcal{P}_{\alpha,\beta,\mu}$ in terms of the C_0 group $\{T_{t,p}\}_{t\in\mathbb{R}}$ given in Theorem 18.5, one can deduce that if $Z \subset L^p(\mathbb{R}^+)$ is a nontrivial invariant subspace for every operator of the C_0 group $\{T_{t,p}\}_{t\in\mathbb{R}}$, then it is also a nontrivial invariant subspace of the generalized Poisson operator $\mathcal{P}_{\alpha,\beta,\mu}$ on $L^p(\mathbb{R}^+)$.

By (18.8), the C_0 group $\{T_{t,p}\}_{t\in\mathbb{R}}$ on $L^p(\mathbb{R}^+)$ is isomorphic to the C_0 group of translations on $L^p(\mathbb{R})$, via the isometric isomorphism $U_p : L^p(\mathbb{R}^+) \to L^p(\mathbb{R})$. Then it follows that $Z \subset L^p(\mathbb{R}^+)$ is a nontrivial invariant subspace for $\{T_{t,p}\}_{t\in\mathbb{R}}$, if and only if $U_p(Z) \subset L^p(\mathbb{R})$ is a nontrivial translation invariant subspace on $L^p(\mathbb{R})$. As these subspaces exist for any $1 \le p < \infty$ (see beginning of this subsection and also [5] and [16]), the image by $U_p^{-1} : L^p(\mathbb{R}) \to L^p(\mathbb{R}^+)$ of these nontrivial translation invariant subspaces in $L^p(\mathbb{R})$ are nontrivial invariant subspaces of the generalized Poisson operator $\mathcal{P}_{\alpha,\beta,\mu}$ in $L^p(\mathbb{R}^+)$. □

Remark 18.11. By Example 18.2(i),

$$\mathcal{P}_{\alpha,\beta,\mu}(g_\gamma) = \frac{1}{\alpha} B\left(\frac{1}{\alpha}(\beta + \gamma - 1), \mu - \frac{1}{\alpha}(\beta + \gamma - 1)\right) g_\gamma$$

where $g_\gamma(t) = \frac{t^{\gamma-1}}{\Gamma(\gamma)}$, for $t > 0$ and $\alpha\mu > \gamma + \beta - 1 > 0$. For $1 < p < \infty$, we take $\gamma = \frac{1}{p'} + i\alpha\xi$ where $\frac{1}{p} + \frac{1}{p'} = 1$ and $\xi \in \mathbb{R}$. Similarly, we compute that

$$\mathcal{P}_{\alpha,\beta,\mu}(g_{\frac{1}{p'}+i\alpha\xi}) = \frac{1}{\alpha}B\left(\frac{1}{\alpha}\left(\beta - \frac{1}{p}\right) + i\xi, \mu - \frac{1}{\alpha}\left(\beta - \frac{1}{p}\right) - i\xi\right)g_{\frac{1}{p'}+i\alpha\xi}$$

whose eigenvalues describe $\sigma(\mathcal{P}_{\alpha,\beta,\mu})\setminus\{0\}$. Although functions $g_{\frac{1}{p'}+i\alpha\xi} \notin L^p(\mathbb{R}^+)$, it seems natural to conjecture that these functions are useful to construct explicitly non-trivial invariant subspaces of $\mathcal{P}_{\alpha,\beta,\mu}$ in $L^p(\mathbb{R}^+)$.

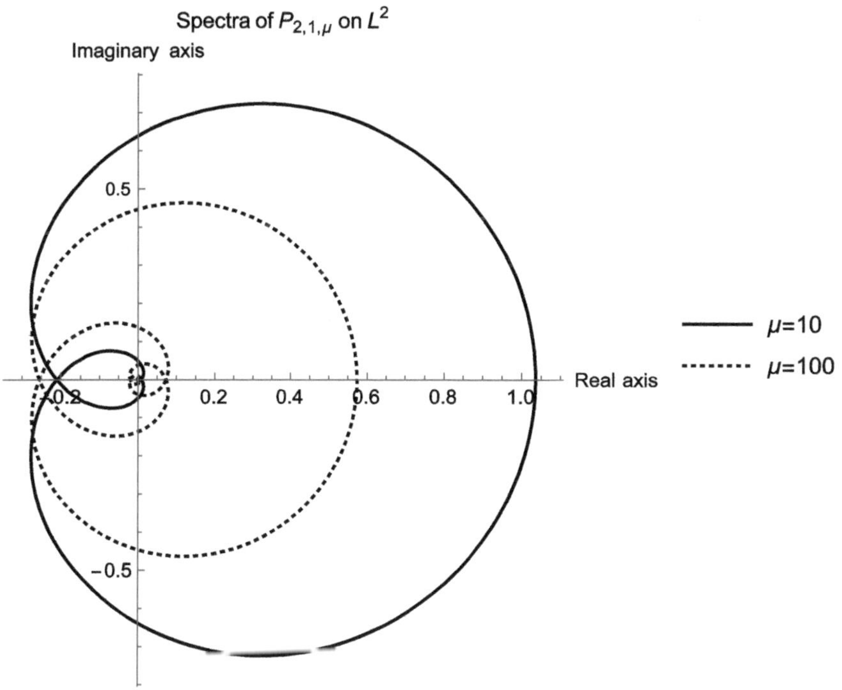

Figure 18.4: Fixed $\alpha = 2$, $\beta = 1$ and $p = 1$.

The spectrum of $\mathcal{P}_{\alpha,\beta,\mu}$ is not always a Jordan curve, as Figure 18.4 shows. In the case that the spectrum of $\mathcal{P}_{\alpha,\beta,\mu}$ is a Jordan curve, one may consider the resolvent operator $(\lambda - \mathcal{P}_{\alpha,\beta,\mu})^{-1}$ for $\lambda \notin \sigma(\mathcal{P}_{\alpha,\beta,\mu})$ and study their nontrivial invariant subspaces, that is, rationally invariant subspaces of $\mathcal{P}_{\alpha,\beta,\mu}$. One might use the theory of C_0 groups to identify nontrivial rationally invariant subspaces and to relate with nontrivial translation invariant subspaces. We hope to address this research in the near future.

Bibliography

[1] Y. A. Abramovich, C. D. Aliprantis, G. Sirotkin and V. G. Troitsky, *Some open problems and conjectures associated with the invariant subspace problem*, Positivity **9** (2005), 273–286.

[2] A. Aleman, A. G. Siskakis and D. Vukotic, *On the Hilbert matrix* (tentative title), work in progress.

[3] W. Arendt, C. J. K. Batty, M. Hieber and F. Neubrander, *Vector-valued Laplace Transforms and Cauchy Problems*, Monographs in Math., **96**, Birkhäuser, 2001.

[4] A. G. Arvanitidis and A. G. Siskakis, *Cesàro operators on the Hardy spaces of the half-plane*, Can. Math. Bull. **56** (2013), 229–240.

[5] A. Atzmon, *Translation invariant subspaces of $L^p(G)$*, Stud. Math. **48** (1973), 245–250. Addition in: Studia Math. **52** (1974/75), 291–292.

[6] T. Carleman, *Sur les équations intégrales singulières à noyau réel et symétrique*, Almqvist and Wiksell, Uppsala, Sweden, 1923.

[7] C. C. Cowen, *Subnormality of the Cesàro operator and a semigroup of composition operators*, Indiana Univ. Math. J. **33** (1984), 305–318.

[8] P. Enflo, *On the invariant subspace problem for Banach spaces*, Seminaire Maurey–Schwarz (1975–1976), Acta Math. **158** (1987), 213–313.

[9] K.-J. Engel and R. Nagel, *One-parameter Semigroups for Linear Evolution Equations*, Springer, New York, 2000.

[10] E. Fedele and A. Pushnitski, *Weighted integral Hankel operators with continuous spectrum*, Concr. Oper. **4** (2017), 121–129.

[11] G. Hardy, J. Littlewood and G. Pólya, *Inequalities*, Cambridge University Press, 1934.

[12] C. Lizama, P. J. Miana, R. Ponce and L. Sánchez-Lajusticia, *On the boundedness of generalized Cesàro operators on Sobolev spaces*, J. Math. Anal. Appl. **419** (2014), 373–394.

[13] V. I. Lomonosov, *An extension of Burnside's theorem to infinite-dimensional spaces*, Isr. J. Math. **75** (1991), 329–339.

[14] P. J. Miana and J. Oliva-Maza, *Generalized Stieltjes operators on Sobolev–Lebesgue spaces*, Preprint, (2019), 1–43.

[15] C. J. Read, *A solution to the invariant subspace problem on the space ℓ^1*, Bull. Lond. Math. Soc. **17** (1985), 305–317.

[16] J. M. Rosenblatt and K. L. Shuman, *Cyclic functions in $L^p(\mathbb{R})$, $1 \le p < \infty$*, J. Fourier Anal. Appl. **9** (3) (2003), 289–300.

[17] W. Rudin, *Functional Analysis*, McGraw-Hill, Singapore, 1973.

[18] H. Seferoĝlu, *A spectral mapping theorem for representations of one-parameter groups*, Proc. Am. Math. Soc. **134** (8), (2006), 2457–2463.

T. Oikhberg and M. A. Tursi

19 Order extreme points and solid convex hulls

To the memory of Victor Lomonosov

Abstract: We consider the “order” analogues of some classical notions of Banach space geometry: extreme points and convex hulls. A Hahn–Banach-type separation result is obtained, which allows us to establish an “order” Krein–Milman theorem. We show that the unit ball of any infinite dimensional reflexive space contains uncountably many order extreme points, and investigate the set of positive norm-attaining functionals. Finally, we introduce the “solid” version of the Krein–Milman property, and show it is equivalent to the Radon–Nikodým property.

Keywords: Banach lattice, extreme point, convex hull, Radon–Nikodým property

MSC 2010: 46B22, 46B42

19.1 Introduction

At the very heart of Banach space geometry lies the study of three interrelated subjects: (i) separation results (starting from the Hahn–Banach theorem), (ii) the structure of extreme points, and (iii) convex hulls (for instance, the Krein–Milman theorem on convex hulls of extreme points). Certain counterparts of these notions exist in the theory of Banach lattices as well. For instance, there are positive separation/extension results; see, for example, [1, Section 1.2]. One can view solid convex hulls as lattice analogues of convex hulls; these objects have been studied, and we mention some of their properties in the paper. However, no unified treatment of all three phenomena listed above has been attempted.

In the present paper, we endeavor to investigate the lattice versions of (i), (ii), and (iii) above. We introduce the order version of the classical notion of an extreme point: if A is a subset of a Banach lattice X, then $a \in A$ is called an *order extreme point* of A if for all $x_0, x_1 \in A$ and $t \in (0,1)$ the inequality $a \leq (1-t)x_0 + tx_1$ implies $x_0 = a = x_1$. Note that, in this case, if $x \geq a$ and $x \in A$, then $x = a$ (write $a \leq (x+a)/2$).

Acknowledgement: We would like to thank the anonymous referee for reading the paper carefully, and providing numerous helpful suggestions. We are also grateful to Professor Anton Schep for finding an error in an earlier version of Proposition 19.6.

T. Oikhberg, M. A. Tursi, Dept. of Mathematics, University of Illinois, Urbana, IL 61801, USA, e-mails: oikhberg@illinois.edu, gramcko2@illinois.edu

https://doi.org/10.1515/9783110656756-019

Throughout, we work with real spaces. We will be using the standard Banach lattice results and terminology (found in, for instance, [1], [19], or [22]). We also say that a subset of a Banach lattice is *bounded* when it is norm bounded, as opposed to order bounded.

Some special notation is introduced in Section 19.2. In the same section, we establish some basic facts about order extreme points and solid hulls. In particular, we note a connection between order and "canonical" extreme points (Theorem 19.2).

In Section 19.3, we prove a "Hahn–Banach-"type result (Proposition 19.7), involving separation by positive functionals. This result is used in Section 19.4 to establish a "solid" analogue of the Krein–Milman theorem. We prove that solid compact sets are solid convex hulls of their order extreme points (see Theorem 19.10). A "solid" Milman theorem is also proved (Theorem 19.13).

In Section 19.5, we study order extreme points in *AM*-spaces. For instance, we show that, for an AM-space X, the following three statements are equivalent: (i) X is a $C(K)$ space; (ii) the unit ball of X is the solid convex hull of finitely many of its elements; (iii) the unit ball of X has an order extreme point (Propositions 19.24 and 19.25).

Further in Section 19.5, we investigate norm-attaining positive functionals. Functionals attaining their maximum on certain sets have been investigated since the early days of functional analysis; here, we must mention V. Lomonosov's papers on the subject (see, e. g., the excellent summary [3], and the references contained there). In this paper, we show that a separable AM-space is a $C(K)$ space iff any positive functional on it attains its norm (Proposition 19.26). On the other hand, an order continuous lattice is reflexive iff every positive operator on it attains its norm (Proposition 19.27).

In Section 19.6, we show that the unit ball of any reflexive infinite-dimensional Banach lattice has uncountably many order extreme points (Theorem 19.28).

Finally, in Section 19.7 we define the "solid" version of the Krein–Milman property, and show that it is equivalent to the Radon–Nikodým property (Theorem 19.30).

To close this Introduction, we would like to mention that related ideas have been explored before, in other branches of functional analysis. In the theory of C^* algebras, and, later, operator spaces, the notions of "matrix" or "C^*" extreme points and convex hulls have been used. The reader is referred to, for example, [11], [12], [14], [23] for more information; for a recent operator-valued separation theorem, see [18].

19.2 Preliminaries

In this section, we introduce the notation commonly used in the paper, and mention some basic facts.

The closed unit ball (sphere) of a Banach space X is denoted by $\mathbf{B}(X)$ (resp., $\mathbf{S}(X)$). If X is a Banach lattice, and $C \subset X$, write $C_+ = C \cap X_+$, where X_+ stands for the positive cone of X. Further, we say that $C \subset X$ is *solid* if, for $x \in X$ and $z \in C$, the inequality

$|x| \le |z|$ implies the inclusion $x \in C$. In particular, $x \in X$ belongs to C if and only if $|x|$ does. Note that any solid set is automatically *balanced*; that is, $C = -C$.

Restricting our attention to the positive cone X_+, we say that $C \subset X_+$ is *positive-solid* if for any $x \in X_+$, the existence of $z \in C$ satisfying $x \le z$ implies the inclusion $x \in C$.

We will denote the set of order extreme points of C (defined in Section 19.1) by $\mathrm{OEP}(C)$; the set of "classical" extreme points is denoted by $\mathrm{EP}(C)$.

Remark 19.1. It is easy to see that the set of all extreme points of a compact metrizable set is G_δ. The same can be said for the set of order extreme points of A, whenever A is a closed solid bounded subset of a separable reflexive Banach lattice. Indeed, then the weak topology is induced by a metric d. For each n let F_n be the set of all $x \in A$ for which there exist $x_1, x_2, \in A$ with $x \le (x_1 + x_2)/2$, and $d(x_1, x_2) \ge 1/n$. By compactness, F_n is closed. Now observe that $\bigcup_n F_n$ is the complement of the set of all order extreme points.

Note that every order extreme point is an extreme point in the usual sense, but the converse is not true: for instance, $\mathbf{1}_{(0,1)}$ is an extreme point of $\mathbf{B}(L_\infty(0,2))_+$, but not its order extreme point. However, a connection between "classical" and order extreme points exists:

Theorem 19.2. *Suppose A is a solid subset of a Banach lattice X. Then a is an extreme point of A if and only if $|a|$ is its order extreme point.*

The proof of Theorem 19.2 uses the notion of a quasi-unit. Recall [19, Definition 1.2.6] that for $e, v \in X_+$, v is a *quasi-unit* of e if $v \wedge (e - v) = 0$. This terminology is not universally accepted: the same objects can be referred to as *components* [1], or *fragments* [20].

Proof. Suppose $|a|$ is order extreme. Let $0 < t < 1$ be such that $a = tx + (1-t)y$. Then since A is solid and $|a| \le t|x| + (1-t)|y|$, one has $|x| = |y| = |a|$. Thus the latter inequality is in fact equality. Thus $|a| + a - 2a_+ = 2tx_+ + 2(1-t)y_+$, so $a_+ = tx_+ + (1-t)y_+$. Similarly, $a_- = tx_- + (1-t)y_-$. It follows that $x_+ \perp y_-$ and $x_- \perp y_+$. Since $x_+ + x_- = |x| = |y| = y_+ + y_-$, we have that $x_+ = x_+ \wedge (y_+ + y_-) = x_+ \wedge y_+ + x_+ \wedge y_-$ (since y_+, y_- are disjoint). Now since $x_+ \perp y_-$, the latter is just $x_+ \wedge y_+$, hence $x_+ \le y_+$. By similar argument, one can show the opposite inequality to conclude that $x_+ = y_+$, and likewise $x_- = y_-$, so $x = y = a$.

Now suppose a is extreme. It is sufficient to show that $|a|$ is order extreme for A_+. Indeed, if $|a| \le tx + (1-t)y$ (with $0 \le t \le 1$ and $x, y \in A$), then $|a| \le t|x| + (1-t)|y|$. As $|a|$ is an order extreme point of A_+, we conclude that $|x| = |y| = |a|$, so $|a| = tx + (1-t)y = t|x| + (1-t)|y|$. The latter implies that $x_- = y_- = 0$, hence $x = |x| = |a| = |y| = y$.

Therefore, suppose $|a| \le tx + (1-t)y$ with $0 \le t \le 1$, and $x, y \in A_+$. First, show that $|a|$ is a quasi-unit of x (and by similar argument of y). To this end, note that $a_+ - tx \wedge a_+ \le (1-t)y \wedge a_+$. Since A is solid,

$$A \ni z_+ := \frac{1}{1-t}(a_+ - tx \wedge a_+)$$

and similarly, since $a_- - tx \wedge a_- \leq (1-t)y \wedge a_-$,

$$A \ni z_- := \frac{1}{1-t}(a_- - tx \wedge a_-).$$

These inequalities imply that $z_+ \perp z_-$, so they correspond to the positive and negative parts of some $z = z_+ - z_-$. Also, $z \in A$ since $|z| \leq |a|$. Now $a_+ = t(x \wedge \frac{a_+}{t}) + (1-t)z_+$ and $a_- = t(x \wedge \frac{a_-}{t}) + (1-t)z_-$. In addition, $|x \wedge \frac{a_+}{t} - x \wedge \frac{a_-}{t}| \leq x$, so since A is solid,

$$z' := x \wedge \frac{a_+}{t} - x \wedge \frac{a_-}{t} \in A.$$

Therefore, $a = a_+ - a_- = tz' + (1-t)z$. Since a is an extreme point, $a = z$, hence

$$(1-t)z_+ = (1-t)a_+ = a_+ - tx \wedge a_+$$

so $tx \wedge a_+ = ta_+$ which implies that $(t(x-a_+)) \wedge ((1-t)a_+) = 0$. As $0 < t < 1$, we have that a_+ (and likewise a_-) is a quasi-unit of x (and similarly of y). Thus $|a|$ is a quasi-unit of x and of y.

Now let $s = x - |a|$. Then $a + s, a - s \in A$, since $|a \pm s| = x$. We have

$$a = \frac{a-s}{2} + \frac{a+s}{2},$$

but since a is extreme, s must be 0. Hence $x = |a|$, and similarly $y = |a|$. □

The situation is different if A is a positive-solid set: the paragraph preceding Theorem 19.2 shows that A can have extreme points which are not order extreme. If, however, a positive-solid set satisfies certain compactness conditions, then some connections between extreme and order extreme points can be established; see Proposition 19.20, and the remark following it.

If C is a subset of a Banach lattice X, denote by $\mathrm{S}(C)$ the *solid hull* of C, which is the smallest solid set containing C. It is easy to see that $\mathrm{S}(C)$ is the set of all $z \in X$ for which there exists $x \in C$ satisfying $|z| \leq |x|$. Clearly, $\mathrm{S}(C) = \mathrm{S}(|C|)$, where $|C| = \{|x| : x \in C\}$. Further, we denote by $\mathrm{CH}(C)$ the *convex hull* of C. For future reference, observe the following.

Proposition 19.3. *If X is a Banach lattice, then* $\mathrm{S}(\mathrm{CH}(|C|)) = \mathrm{CH}(\mathrm{S}(C))$ *for any $C \subset X$.*

Proof. Let $x \in \mathrm{CH}(\mathrm{S}(C))$. Then $x = \sum a_i y_i$, where $\sum a_i = 1$, $a_i > 0$, and $|y_i| \leq |k_i|$ for some $k_i \in C$. Then

$$|x| \leq \sum a_i |y_i| \leq \sum a_i |k_i| \in \mathrm{CH}(|C|),$$

so $x \in \mathrm{S}(\mathrm{CH}(|C|))$. If $x \in \mathrm{S}(\mathrm{CH}(|C|))$, then

$$|x| \leq \sum_1^n a_i y_i, \quad y_i \in |C|, \quad 0 < a_i, \quad \sum a_i = 1.$$

We use induction on n to prove that $x \in \mathrm{CH}(\mathrm{S}(C))$. If $n = 1$, $x \in \mathrm{S}(C)$ and we are done. Now, suppose we have shown that if $|x| \le \sum_1^{n-1} a_i y_i$ then there are $z_1, \dots, z_{n-1} \in \mathrm{S}(C)_+$ such that $|x| = \sum_1^{n-1} a_i z_i$. From there, we have that

$$|x| = \left(\sum_1^n a_i y_i\right) \wedge |x| \le \left(\sum_1^{n-1} a_i y_i\right) \wedge |x| + (a_n y_n) \wedge |x|.$$

Now

$$0 \le |x| - \left(\sum_1^{n-1} a_i y_i\right) \wedge |x| \le a_n\left(y_n \wedge \frac{|x|}{a_n}\right).$$

Let $z_n := \frac{1}{a_n}(|x| - (\sum_1^{n-1} a_i y_i) \wedge |x|)$. By the above, $z_n \in \mathrm{S}(C)_+$. Furthermore,

$$\frac{1}{1-a_n}\left(|x| \wedge \sum_1^{n-1} a_i y_i\right) \le \sum_1^{n-1} \frac{a_i}{1-a_n} y_i \in \mathrm{CH}(|C|),$$

so by induction there exist $z_1, \dots, z_{n-1} \in \mathrm{S}(C)_+$ such that

$$|x| \wedge \left(\sum_1^{n-1} a_i y_i\right) = \sum_1^{n-1} \frac{a_i}{1-a_n} z_i.$$

Therefore, $|x| = \sum_1^n a_i z_i$. Now for each n, $a_i z_i \le |x|$, so $|x| = \sum((a_i z_i) \wedge |x|)$, and

$$a_i z_i = a_i z_i \wedge x_+ + a_i z_i \wedge x_- = a_i\left(z_i \wedge \left(\frac{x_+}{a_i}\right) + z_i \wedge \left(\frac{x_-}{a_i}\right)\right).$$

Let $w_i = z_i \wedge (\frac{x_+}{a_i}) - z_i \wedge (\frac{x_-}{a_i})$. Note that $|w_i| = z_i$, so $w_i \in \mathrm{S}(C)$. It follows that $x = \sum a_i w_i \in \mathrm{CH}(\mathrm{S}(C))$. □

For $C \subset X$ (as before, X is a Banach lattice) we define the *solid convex hull* of C to be the smallest convex, solid set containing C, and denote it by $\mathrm{SCH}(C)$; the norm (equivalently, weak) closure of the latter set is denoted by $\mathrm{CSCH}(C)$, and referred to as the *closed solid convex hull* of C.

Corollary 19.4. *Let $C \subseteq X$. Then:*
(1) $\mathrm{SCH}(C) = \mathrm{CH}(\mathrm{S}(C)) = \mathrm{SCH}(|C|)$, *and consequently,* $\mathrm{CSCH}(C) = \mathrm{CSCH}(|C|)$.
(2) *If $C \subseteq X_+$, then* $\mathrm{SCH}(C) = \mathrm{S}(\mathrm{CH}(C))$.

Proof. (1) Suppose $C \subseteq D$, where D is convex and solid. Then $\mathrm{CH}(\mathrm{S}(C)) \subseteq D$. Consequently, $\mathrm{CH}(\mathrm{S}(C)) \subset \mathrm{SCH}(C)$. On the other hand, by Proposition 19.3, $\mathrm{CH}(\mathrm{S}(C))$ is also solid, so $\mathrm{SCH}(C) \subseteq \mathrm{CH}(\mathrm{S}(C))$. Thus, $\mathrm{SCH}(C) = \mathrm{CH}(\mathrm{S}(C)) = \mathrm{CH}(\mathrm{S}(|C|)) = \mathrm{SCH}(|C|)$.

(2) This follows from (1) and the equality in Proposition 19.3. □

Remark 19.5. The two examples below show that $\mathrm{S}(C)$ need not be closed, even if C itself is. Example (1) exhibits an unbounded closed set C with $\mathrm{S}(C)$ not closed; in Example (2), C is closed and bounded, but the ambient Banach lattice needs to be infinite dimensional.

(1) Let X be a Banach lattice of dimension at least two, and consider disjoint norm one $e_1, e_2 \in \mathbf{B}(X)_+$. Let $C = \{x_n : n \in \mathbb{N}\}$, where $x_n = \frac{n}{n+1}e_1 + ne_2$. Now, C is norm-closed: if $m > n$, then $\|x_m - x_n\| \geq \|e_2\| = 1$. However, $\mathrm{S}(C)$ is not closed: it contains re_1 for any $r \in (0,1)$, but not e_1.

(2) If X is infinite dimensional, then there exists a closed *bounded* $C \subset X_+$, for which $\mathrm{S}(C)$ is not closed. Indeed, find disjoint norm one elements $e_1, e_2, \ldots \in X_+$. For $n \in \mathbb{N}$, let $y_n = \sum_{k=1}^n 2^{-k} e_k$ and $x_n = y_n + e_n$. Then clearly $\|x_n\| \leq 2$ for any n; further, $\|x_n - x_m\| \geq 1$ for any $n \neq m$, hence $C = \{x_1, x_2, \ldots\}$ is closed. However, $y_n \in \mathrm{S}(C)$ for any n, and the sequence (y_n) converges to $\sum_{k=1}^\infty 2^{-k} e_k \notin \mathrm{S}(C)$.

However, under certain conditions we can show that the solid hull of a closed set is closed.

Proposition 19.6. *A Banach lattice X is reflexive if and only if, for any norm closed, bounded convex $C \subset X_+$, $\mathrm{S}(C)$ is norm closed.*

Proof. Support first X is reflexive, and C is a norm closed bounded convex subset of X_+. Suppose (x_n) is a sequence in $\mathrm{S}(C)$, which converges to some x in norm; show that x belongs to $\mathrm{S}(C)$ as well. Clearly, $|x_n| \to |x|$ in norm. For each n find $y_n \in C$ so that $|x_n| \leq y_n$. By passing to a subsequence if necessary, we assume that the sequence (y_n) converges to some $y \in X$ in the weak topology. For convex sets, norm and weak closures coincide, hence y belongs to C. For each n, $\pm x_n \leq y_n$; passing to the weak limit gives $\pm x \leq y$, hence $|x| \leq y$.

Now suppose X is not reflexive. By [1, Theorem 4.71], there exists a sequence of disjoint elements $e_i \in \mathbf{S}(X)_+$, equivalent to the natural basis of either c_0 or ℓ_1.

First, consider the c_0 case. Let C be the closed convex hull of

$$x_1 = \frac{e_1}{2}, \quad x_n = (1 - 2^{-n})e_1 + \sum_{j=2}^n e_j \quad (n \geq 2).$$

We shall show that any element of C can be written as $ce_1 + \sum_{i=2}^\infty c_i e_i$, with $c < 1$. This will imply that $\mathrm{S}(C)$ is not closed: clearly, $e_1 \in \overline{\mathrm{S}(C)} \backslash \mathrm{S}(C)$.

The elements of $\mathrm{CH}(x_1, x_2, \ldots)$ are of the form $\sum_{i=1}^\infty t_i x_i = ce_1 + \sum_{i=2}^\infty c_i e_i$; here, $t_i \geq 0$, $t_i \neq 0$ for finitely many values of i only, and $\sum_i t_i = 1$. Note that $c_i = \sum_{j=i}^\infty t_j$ for $i \geq 2$ (so $c_i = 0$ eventually); for convenience, let $c_1 = \sum_{j=1}^\infty t_i = 1$. Then $t_i = c_i - c_{i+1}$; Abel's summation technique gives

$$c = \sum_{i=1}^\infty (1 - 2^{-i}) t_i = 1 - \sum_{i=1}^\infty 2^{-i}(c_i - c_{i+1}) = \frac{1}{2} + \sum_{j=2}^\infty 2^{-j} c_j.$$

Now consider $x \in C$. Then x is the norm limit of the sequence

$$x^{(m)} = c^{(m)} e_1 + \sum_{i=2}^{\infty} c_i^{(m)} e_i \in \mathrm{CH}(x_1, x_2, \ldots);$$

for each m, the sequence $(c_i^{(m)})$ has only finitely many nonzero terms, $c^{(m)} = \frac{1}{2} + \sum_{j=2}^{\infty} 2^{-j} c_j^{(m)}$, and for all $m, n \in \mathbb{N}$, $|c_i^{(m)} - c_i^{(n)}| \leq \|x^{(m)} - x^{(n)}\|$. Thus, $x = ce_1 + \sum_{i=2}^{\infty} c_i e_i$, with $c = \frac{1}{2} + \sum_{j=2}^{\infty} 2^{-j} c_j$. As $0 \leq c_j \leq 1$, and $\lim_j c_j = 0$, we conclude that $c < 1$, as claimed.

Now suppose (e_i) are equivalent to the natural basis of ℓ_1. Let C be the closed convex hull of the vectors

$$x_n = (1 - 2^{-n}) e_1 + e_n \quad (n \geq 2),$$

and show that $e_1 \in \overline{\mathrm{S}(C)} \backslash \mathrm{S}(C)$. Note that

$$C = \left\{ \left(\sum_{i=2}^{\infty} (1 - 2^{-n}) t_i \right) e_1 + \sum_{i=2}^{\infty} t_i e_i : t_2, t_3, \ldots \geq 0, \sum_{i=2}^{\infty} t_i = 1 \right\}.$$

Clearly, e_1 belongs to $\overline{\mathrm{S}(C)}$, but not to $\mathrm{S}(C)$. □

19.3 Separation by positive functionals

Throughout the section, X is a Banach lattice, equipped with a locally convex Hausdorff topology τ. This topology is called *sufficiently rich* if the following conditions are satisfied:

(i) The space X^τ of τ-continuous functionals on X is a Banach lattice (with lattice operations defined by Riesz–Kantorovich formulas).
(ii) X_+ is τ-closed.

Note that (i) and (ii) together imply that positive τ-continuous functionals separate points. That is, for every $x \in X \backslash \{0\}$ there exists $f \in X_+^\tau$ so that $f(x) \neq 0$. Indeed, without loss of generality, $x_+ \neq 0$. Then $-x_+ \notin X_+$, hence there exists $f \in X_+^\tau$ so that $f(x_+) > 0$. By [19, Proposition 1.4.13], there exists $g \in X_+^\tau$ so that $g(x_+) > f(x_+)/2$ and $g(x_-) < f(x_+)/2$. Then $g(x) > 0$.

Clearly, the norm and weak topologies are sufficiently rich; in this case, $X^\tau = X^*$. The weak* topology on X, induced by the predual Banach lattice X_*, is sufficiently rich as well; then $X^\tau = X_*$.

Proposition 19.7 (Separation). *Suppose τ is a sufficiently rich topology on a Banach lattice X, and $A \subset X_+$ is a τ-closed positive-solid bounded subset of X_+. Suppose, furthermore, $x \in X_+$ does not belong to A. Then there exists $f \in X_+^\tau$ so that $f(x) > \sup_{a \in A} f(a)$.*

Lemma 19.8. *Suppose A and X are as above, and $f \in X^\tau$. Then*

$$\sup_{a\in A} f(a) = \sup_{a\in A} f_+(a).$$

Proof. Clearly, $\sup_{a\in A} f(a) \le \sup_{a\in A} f_+(a)$. To prove the reverse inequality, write $f = f_+ - f_-$, with $f_+ \wedge f_- = 0$. Fix $a \in A$; then

$$0 = [f_+ \wedge f_-](a) = \inf_{0\le x\le a} (f_+(a-x) + f_-(x)).$$

For any $\varepsilon > 0$, we can find $x \in A$ so that $f_+(a-x), f_-(x) < \varepsilon$. Then $f_+(x) = f_+(a) - f_+(a-x) > f_+(a) - \varepsilon$ and, therefore, $f(x) = f_+(x) - f_-(x) > f_+(a) - 2\varepsilon$. Now recall that $\varepsilon > 0$ and $a \in A$ are arbitrary. □

Proof of Proposition 19.7. Use Hahn–Banach theorem to find f strictly separating x from A. By Lemma 19.8, f_+ achieves the separation as well. □

Remark 19.9. In this paper, we do not consider separation results on general ordered spaces. Our reasoning will fail without lattice structure. For instance, Lemma 19.8 is false when X is not a lattice, but merely an ordered space. Indeed, consider $X = M_2$ (the space of real 2×2 matrices), $f = \begin{pmatrix} 1 & 0 \\ 0 & -1 \end{pmatrix}$, and $A = \{ta_0 : 0 \le t \le 1\}$, where $a_0 = \begin{pmatrix} 1 & 1 \\ 1 & 1 \end{pmatrix}$; one can check that $A = \{x \in M_2 : 0 \le x \le a_0\}$. Then $f|_A = 0$, while $\sup_{x\in A} f_+(x) = 1$.

The reader interested in the separation results in the nonlattice ordered setting is referred to an interesting result of [15], recently reproved in [2].

19.4 Solid convex hulls: theorems of Krein-Milman and Milman

Throughout this section, the topology τ is assumed to be sufficiently rich (defined in the beginning of Section 19.3).

Theorem 19.10 (“Solid” Krein–Milman). *Any τ-compact positive-solid subset A of X_+ coincides with the τ-closed positive-solid convex hull of its order extreme points.*

Proof. Let A be a τ-compact positive-solid subset of X_+. Denote the τ-closed positive convex hull of OEP(A) by B; then clearly $B \subset A$. The proof of the reverse inclusion is similar to that of the “usual” Krein–Milman.

Suppose C is a τ-compact subset of X. We say that a nonvoid closed $F \subset C$ is an *order extreme subset* of C if, whenever $x \in F$ and $a_1, a_2 \in C$ satisfy $x \le (a_1 + a_2)/2$, then necessarily $a_1, a_2 \in F$. The set $\mathcal{F}(C)$ of order extreme subsets of C can be ordered by reverse inclusion (this makes C the minimal order extreme subset of itself). By compactness, each chain has an upper bound; therefore, by Zorn’s lemma, $\mathcal{F}(C)$ has a maximal element. We claim that these maximal elements are singletons, and they are the order extreme points of C.

We need to show that if $F \in \mathcal{F}(C)$ is not a singleton, then there exists $G \subsetneq F$ which is also an order extreme set. To this end, find distinct $a_1, a_2 \in F$, and $f \in X_+^\tau$ which separates them – say $f(a_1) > f(a_2)$. Let $\alpha = \max_{x \in F} f(x)$, then $G = F \cap f^{-1}(\alpha)$ is a proper, order extreme subset of F.

Suppose, for the sake of contradiction, that there exists $x \in A \backslash B$. Use Proposition 19.7 to find $f \in X_+^\tau$ so that $f(x) > \max_{y \in B} f(y)$. Let $\alpha = \max_{x \in A} f(x)$, then $A \cap f^{-1}(\alpha)$ is an order extreme subset of A, disjoint from B. As noted above, this subset contains at least one extreme point. This yields a contradiction, as we started out assuming all order extreme points lie in B. □

Corollary 19.11. *Any τ-compact solid subset of X coincides with the τ-closed solid convex hull of its order extreme points.*

Of course, there exist Banach lattices whose unit ball has no order extreme points at all – $L_1(0,1)$, for instance. However, an order analogue of [16, Lemma 1] holds.

Proposition 19.12. *For a Banach lattice X, the following two statements are equivalent:*
(1) *Every bounded closed solid convex subset of X has an order extreme point.*
(2) *Every bounded closed solid convex subset of X is the closed solid convex hull of its order extreme points.*

Proof. (2) ⇒ (1) is evident; we shall prove (1) ⇒ (2). Suppose $A \subset X$ is closed, bounded, convex, and solid. Let $B = \mathrm{CSCH}(\mathrm{OEP}(A))$ (which is not empty, by (1)). Suppose, for the sake of contradiction, that B is a proper subset of A. Let $a \in A \backslash B$. Since B and A are solid, $|a| \in A \backslash B$ as well, so without loss of generality we assume that $a \geq 0$. Then there exists $f \in \mathbf{S}(X^*)_+$ which strictly separates a from B; consequently,

$$\sup_{x \in A} f(x) \geq f(a) > \sup_{x \in B} f(x).$$

Fix $\varepsilon > 0$ so that

$$2\sqrt{2\varepsilon}\alpha < \sup_{x \in A} f(x) - \sup_{x \in B} f(x), \quad \text{where } \alpha = \sup_{x \in A} \|x\|.$$

By Bishop–Phelps–Bollobás theorem (see, e. g., [4] or [9]), there exists $f' \in \mathbf{S}(X^*)$, attaining its maximum on A, so that $\|f - f'\| \leq \sqrt{2\varepsilon}$.

Let $g = |f'|$, then $\|f - g\| \leq \|f - f'\| \leq \sqrt{2\varepsilon}$. Further, g attains its maximum on A_+, and $\max_{g \in A} g(x) > \sup_{x \in B} g(x)$. Indeed, the first statement follows immediately from the definition of g. To establish the second one, note that the triangle inequality gives us

$$\sup_{x \in B} g(x) \leq \sqrt{2\varepsilon}\alpha + \sup_{x \in B} f(x), \quad \sup_{x \in A} g(x) \geq \sup_{x \in A} f(x) - \sqrt{2\varepsilon}\alpha.$$

Our assumption on ε gives us $\max_{g \in A} g(x) > \sup_{x \in B} g(x)$.

Let $D = \{a \in A : g(a) = \sup_{x \in A} g(x)\}$. Due to (1), D has an order extreme point which is also an order extreme point of A; this point lies inside of B, leading to the desired contradiction. □

Milman's theorem [21, 3.25] states that, if both K and $\overline{\mathrm{CH}(K)}^{\tau}$ are compact, then $\mathrm{EP}(\overline{\mathrm{CH}(K)}^{\tau}) \subset K$. An order analogue of Milman's theorem exists.

Theorem 19.13. *Suppose X is a Banach lattice.*
(1) *If $K \subset X_+$ and $\overline{\mathrm{CH}(K)}^{\tau}$ are τ-compact, then* $\mathrm{OEP}(\overline{\mathrm{SCH}(K)}^{\tau}) \subseteq K$.
(2) *If $K \subset X_+$ is weakly compact, then* $\mathrm{OEP}(\mathrm{CSCH}(K)) \subseteq K$.
(3) *If $K \subset X$ is norm compact, then* $\mathrm{OEP}(\mathrm{CSCH}(K)) \subseteq |K|$.

The following lemma describes the solid hull of a τ-compact set.

Lemma 19.14. *Suppose a Banach lattice X is equipped with a sufficiently rich topology τ. If $C \subset X_+$ is τ-compact, then* $\mathrm{S}(C)$ *is τ-closed.*

Proof. Suppose a net $(y_i) \subset \mathrm{S}(C)$ τ-converges to $y \in X$. For each i find $x_i \in C$ so that $|y_i| \leq x_i$ – or equivalently, $y_i \leq x_i$ and $-y_i \leq x_i$. Passing to a subnet if necessary, we assume that $x_i \to x \in C$ in the topology τ. Then $\pm y \leq x$, which is equivalent to $|y| \leq x$. □

Proof of Theorem 19.13. (1) We first consider a τ-compact $K \subseteq X_+$. Milman's traditional theorem holds that $\mathrm{EP}(\overline{\mathrm{CH}(K)}^{\tau}) \subseteq K$. Every order extreme point of a set is extreme, hence the order extreme points of $\overline{\mathrm{CH}(K)}^{\tau}$ are in K. Therefore, by Lemma 19.14 and Corollary 19.4,

$$\overline{\mathrm{SCH}(K)}^{\tau} = \overline{\mathrm{S}(\mathrm{CH}(K))}^{\tau} \subseteq \mathrm{S}(\overline{\mathrm{CH}(K)}^{\tau}) = \{x : |x| \leq y \in \overline{\mathrm{CH}(K)}^{\tau}\}.$$

Thus, the points of $\overline{\mathrm{SCH}(K)}^{\tau} \backslash \overline{\mathrm{CH}(K)}^{\tau}$ cannot be order extreme due to being dominated by $\overline{\mathrm{CH}(K)}^{\tau}$. Therefore, $\mathrm{OEP}(\overline{\mathrm{SCH}(K)}^{\tau}) \subseteq \mathrm{OEP}(\overline{\mathrm{CH}(K)}^{\tau}) \subseteq K$.

(2) Combine (1) with Krein's theorem (see, e. g., [13, Theorem 3.133]), which states that $\overline{\mathrm{CH}(K)}^{W} = \overline{\mathrm{CH}(K)}$ is weakly compact.

(3) Finally, suppose $K \subseteq X$ is norm compact. By Corollary 19.4, $\mathrm{CSCH}(K) = \mathrm{CSCH}(|K|)$. $|K|$ is norm compact, hence by [21, Theorem 3.20], so is $\overline{\mathrm{CH}(|K|)}$. By the proof of part (1), $\mathrm{OEP}(\mathrm{CSCH}(K)) \subseteq |K|$. □

We turn our attention to interchanging "solidification" and norm closure. We work with the norm topology, unless specified otherwise.

Lemma 19.15. *Let $C \subseteq X$, where X is a Banach lattice, and suppose that* $\mathrm{S}(\overline{|C|})$ *is closed. Then* $\overline{\mathrm{S}(C)} = \mathrm{S}(\overline{|C|})$.

Proof. One direction is easy: $\mathrm{S}(C) = \mathrm{S}(|C|) \subseteq \mathrm{S}(\overline{|C|})$, hence $\overline{\mathrm{S}(C)} \subseteq \overline{\mathrm{S}(\overline{|C|})} = \mathrm{S}(\overline{|C|})$.

Now consider $x \in \mathrm{S}(\overline{|C|})$. Then by definition, $|x| \leq y$ for some $y \in \overline{|C|}$. Take $y_n \in |C|$ such that $y_n \to y$. Then $|x| \wedge y_n \in \mathrm{S}(|C|) = \mathrm{S}(C)$ for all n. Furthermore,

$$|x_+ \wedge y_n - x_- \wedge y_n| = |x| \wedge y_n,$$

so, $x_+ \wedge y_n - x_- \wedge y_n \in \mathrm{S}(C)$. By norm continuity of $\wedge$,

$$x_+ \wedge y_n - x_- \wedge y_n \to x_+ \wedge y - x_- \wedge y = x,$$

hence $x \in \overline{\mathrm{S}(C)}$. □

Remark 19.16. The assumption of $\mathrm{S}(\overline{|C|})$ being closed is necessary: Remark 19.5 shows that, for a closed $C \subset X_+$, $\mathrm{S}(C)$ need not be closed.

Corollary 19.17. *Suppose $C \subseteq X$ is relatively compact in the norm topology. Then $\overline{\mathrm{S}(C)} = \mathrm{S}(\overline{C})$.*

Proof. The set $\overline{C}$ is compact, hence, by the continuity of $|\cdot|$, the same is true for $|\overline{C}|$. Consequently, $|\overline{C}| \subseteq \overline{|C|} \subseteq \overline{|\overline{C}|} = |\overline{C}|$, hence $|\overline{C}| = \overline{|C|}$. By Lemmas 19.14 and 19.15, $\mathrm{S}(\overline{C}) = \mathrm{S}(|\overline{C}|) = \mathrm{S}(\overline{|C|}) = \overline{\mathrm{S}(C)}$. □

Remark 19.18. In the weak topology, the equality $|\overline{C}| = \overline{|C|}$ may fail. Indeed, equip the Cantor set $\Delta = \{0,1\}^{\mathbb{N}}$ with its uniform probability measure μ. Define $x_i \in L_2(\mu)$ by setting, for $t = (t_1, t_2, \ldots) \in \Delta$, $x_i(t) = t_i - 1/4$ (i. e., x_i equals to either $3/4$ or $-1/4$, depending on whether t_i is 1 or 0). Then $C = \{x_i : i \in \mathbb{N}\}$ belongs to the unit ball of $L_2(\mu)$, hence it is relatively compact. It is clear that $\overline{C}$ contains $\mathbf{1}/4$ (here and below, $\mathbf{1}$ denotes the constant 1 function). On the other hand, $\overline{C}$ does not contain $\mathbf{1}/2$, which can be witnessed by applying the integration functional. Conversely, $\overline{|C|}$ contains $\mathbf{1}/2$, but not $\mathbf{1}/4$.

Remark 19.19. Relative weak compactness of solid hulls have been studied before. If X is a Banach lattice, then, by [1, Theorem 4.39], it is order continuous iff the solid hull of any weakly compact subset of X_+ is relatively weakly compact. Further, by [8], the following three statements are equivalent:

(1) The solid hull of any relatively weakly compact set is relatively weakly compact.
(2) If $C \subset X$ is relatively weakly compact, then so is $|C|$.
(3) X is a direct sum of a KB-space and a purely atomic order continuous Banach lattice (a Banach lattice is called purely atomic if its atoms generate it, as a band).

Finally, we return to the connections between extreme points and order extreme points. As noted in the paragraph preceding Theorem 19.2, a nonzero extreme point of a positive-solid set need not be order extreme. However, we have the following.

Proposition 19.20. *Suppose τ is a sufficiently rich topology, and A is a τ-compact positive-solid convex subset of X_+. Then for any extreme point $a \in A$ there exists an order extreme point $b \in A$ so that $a \leq b$.*

Remark 19.21. The compactness assumption is essential. Consider, for instance, the closed set $A \subset C[-1,1]$, consisting of all functions f so that $0 \leq f \leq \mathbf{1}$, and $f(x) \leq x$ for $x \geq 0$. Then $g(x) = x \vee 0$ is an extreme point of A; however, A has no order extreme points.

Proof. If a is not an order extreme point, then we can find distinct $x_1, x_2 \in A$ so that $2a \le x_1 + x_2$. Then $2a \le (x_1 + x_2) \wedge (2a) \le x_1 \wedge (2a) + x_2 \wedge (2a) \le x_1 + x_2$. Write $2a = x_1 \wedge (2a) + (2a - x_1 \wedge (2a))$. Both summands are positive, and both belong to A (for the second summand, note that $2a - x_1 \wedge (2a) \le x_2$). Therefore, $x_1 \wedge (2a) = a = 2a - x_1 \wedge (2a)$, hence in particular $x_1 \wedge (2a) = a$. Similarly, $x_2 \wedge (2a) = a$. Therefore, we can write x_1 as a disjoint sum $x_1 = x_1' + a$ (a, x_1' are quasi-units of x_1). In the same way, $x_2 = x_2' + a$ (disjoint sum).

Now consider the τ-closed set $B = \{x \in A : x \ge a\}$. As in the proof of Theorem 19.10, we show that the family of τ-closed extreme subsets of B has a maximal element; moreover, such an element is a singleton $\{b\}$. It remains to prove that b is an order extreme point of A. Indeed, suppose $x_1, x_2 \in A$ satisfy $2b \le x_1 + x_2$. A fortiori, $2a \le x_1 + x_2$, hence, by the preceding paragraph, $x_1, x_2 \in B$. Thus, $x_1 = b = x_2$. □

19.5 Examples: AM-spaces and their relatives

The following example shows that, in some cases, $\mathbf{B}(X)$ is much larger than the closed convex hull of its extreme points, yet is equal to the closed solid convex hull of its order extreme points.

Proposition 19.22. *For a Banach lattice X, $\mathbf{B}(X)$ is the (closed) solid convex hull of n disjoint nonzero elements if and only if X is lattice isometric to $C(K_1) \oplus_1 \cdots \oplus_1 C(K_n)$ for suitable nontrivial Hausdorff compact topological spaces $K_1, \dots, K_n$.*

Proof. Clearly, the only order extreme points of $\mathbf{B}(C(K_1) \oplus_1 \cdots \oplus_1 C(K_n))$ are $\mathbf{1}_{K_i}$, with $1 \le i \le n$.

Conversely, suppose $\mathbf{B}(X) = \mathrm{CSCH}(x_1, \dots, x_n)$, where $x_1, \dots, x_n \in \mathbf{B}(X)_+$ are disjoint. It is easy to see that, in this case, $\mathbf{B}(X) = \mathrm{SCH}(x_1, \dots, x_n)$. Moreover, $x_i \in \mathbf{S}(X)_+$ for each i. Indeed, otherwise there exists $i \in \{1, \dots, n\}$ and $\lambda > 1$ so that $\lambda x_i \in \mathrm{SCH}(x_1, \dots, x_n)$, or in other words, $\lambda x_i \le \sum_{j=1}^n t_j x_j$, with $t_j \ge 0$ and $\sum_j t_j \le 1$. Consequently, due to the disjointness of x_j's,

$$\lambda x_i = (\lambda x_i) \wedge (\lambda x_i) \le \left(\sum_{j=1}^n t_j x_j \right) \wedge (\lambda x_i) \le \sum_{j=1}^n (t_j x_j) \wedge (\lambda x_i) \le t_i x_i,$$

which yields the desired contradiction.

Let E_i be the ideal of X generated by x_i, meaning the set of all $x \in X$ for which there exists $c > 0$ so that $|x| \le c|x_i|$. Note that, for such x, $\|x\|$ is the infimum of all c's with the above property. Indeed, if $|x| \le |x_i|$, then clearly $x \in \mathbf{B}(X)$. Conversely, suppose $x \in \mathbf{B}(X) \cap E_i$. In other words, $|x| \le c x_i$ for some c, and also $|x| \le \sum_j t_j x_j$, with $t_j \ge 0$, and $\sum_j t_j = 1$. Then $|x| \le (c x_i) \wedge (\sum_j t_j x_j) = (c \wedge t_i) x_i$. Consequently, E_i (with the norm inherited from X) is an AM-space, whose strong unit is x_i. By [19, Theorem 2.1.3], E_i can be identified with $C(K_i)$, for some Hausdorff compact K_i.

Further, Proposition 19.3 shows that X is the direct sum of the ideals E_i: any $y \in X$ has a unique disjoint decomposition $y = \sum_{i=1}^n y_i$, with $y_i \in E_i$. We have to show that $\|y\| = \sum_i \|y_i\|$. Indeed, suppose $\|y\| \le 1$. Then $|y| = \sum_i |y_i| \le \sum_j t_j x_j$, with $t_j \ge 0$, and $\sum_j t_j = 1$. Note that $\|y_i\| \le 1$ for every i, or equivalently, $|y_i| \le x_i$. Therefore,

$$|y_i| = |y| \wedge x_i = \Big(\sum_j t_j x_j\Big) \wedge x_i = t_i,$$

which leads to $\|y_i\| \le t_i$; consequently, $\|y\| \le \sum_i t_i \le 1$. □

Example 19.23. For $X = (C(K_1) \oplus_1 C(K_2)) \oplus_\infty C(K_3)$, order extreme points of $\mathbf{B}(X)$ are $\mathbf{1}_{K_1} \oplus_\infty \mathbf{1}_{K_3}$ and $\mathbf{1}_{K_2} \oplus_\infty \mathbf{1}_{K_3}$; $\mathbf{B}(X)$ is the solid convex hull of these points. Thus, the word "disjoint" in the statement of Proposition 19.22 cannot be omitted.

Note that $\mathbf{B}(C(K))$ is the closed solid convex hull of its only order extreme point – namely, $\mathbf{1}_K$. This is the only type of AM-spaces with this property.

Proposition 19.24. *Suppose X is an AM-space, and $\mathbf{B}(X)$ is the closed solid convex hull of finitely many of its elements. Then $X = C(K)$ for some Hausdorff compact K.*

Proof. Suppose $\mathbf{B}(X)$ is the closed solid convex hull of $x_1, \dots, x_n \in \mathbf{B}(X)_+$. Then $x_0 := x_1 \vee \cdots \vee x_n \in \mathbf{B}(X)_+$ (due to X being an AM-space), hence $x \in \mathbf{B}(X)$ iff $|x| \le x_0$. Thus, x_0 is the strong unit of X. □

Proposition 19.25. *If X is an AM-space, and $\mathbf{B}(X)$ has an order extreme point, then X is lattice isometric to $C(K)$, for some Hausdorff compact K.*

Proof. Suppose a is order extreme point of $\mathbf{B}(X)$. We claim that a is a strong unit, which means that $a \ge x$ for any $x \in \mathbf{B}(X)_+$. Suppose, for the sake of contradiction, that the inequality $a \ge x$ fails for some $x \in \mathbf{B}(X)_+$. Then $b = a \vee x \in \mathbf{B}(X)_+$ (due to the definition of an AM-space), and $a \le (a+b)/2$, contradicting the definition of an order extreme point. □

We next consider norm-attaining functionals. It is known that, for a Banach space X, any element of X^* attains its norm iff X is reflexive. If we restrict ourself to positive functionals on a Banach lattice, the situation is different: clearly every positive functional on $C(K)$ attains its norm at $\mathbf{1}$. Below we show that, among separable AM-spaces, only $C(K)$ has this property.

Proposition 19.26. *Suppose X is a separable AM-space, so that every positive linear functional attains its norm. Then X is lattice isometric to $C(K)$.*

Proof. Let $(x_i)_{i=1}^\infty$ be a dense sequence in $\mathbf{S}(X)_+$. For each i find $x_i^* \in \mathbf{B}(X_+^*)$ so that $x_i^*(x_i) = 1$. Let $x^* = \sum_{i=1}^\infty 2^{-i} x_i^*$. We shall show that $\|x^*\| = 1$. Indeed, $\|x^*\| \le \sum_i 2^{-i} = 1$ by the triangle inequality. For the opposite inequality, fix $N \in \mathbb{N}$, and let $x = x_1 \vee \cdots \vee x_N$.

Then $x \in \mathbf{S}(X)_+$, and

$$\|x^*\| \geq x^*(x) \geq \sum_{i=1}^{N} 2^{-i} x_i^*(x) \geq \sum_{i=1}^{N} 2^{-i} x_i^*(x_i) = \sum_{i=1}^{N} 2^{-i} = 1 - 2^{-N}.$$

As N can be arbitrarily large, we obtain the desired estimate on $\|x^*\|$.

Now suppose x^* attains its norm on $a \in \mathbf{S}(X)_+$. We claim that a is the strong unit for X. Suppose otherwise; then there exists $y \in \mathbf{B}(X)_+$ so that $a \geq y$ fails. Let $b = a \vee y$, then $z = b - y$ belongs to $X_+ \backslash \{0\}$. Then $1 \geq x^*(b) \geq x^*(a) = 1$, hence $x^*(z) = 0$. However, x^* cannot vanish at z. Indeed, find i so that $\|z/\|z\| - x_i\| < 1/2$. Then $x_i^*(z) \geq \|z\|/2$, hence $x^*(z) > 2^{-i-1}\|z\| > 0$. This gives the desired contradiction. □

In connection to this, we also mention a result about norm-attaining functionals on order continuous Banach lattices.

Proposition 19.27. *An order continuous Banach lattice X is reflexive if and only if every positive linear functional on it attains its norm.*

Proof. If an order continuous Banach lattice X is reflexive, then clearly every linear functional is norm-attaining. If X is not reflexive, then by the classical result of James, there exists $x^* \in X^*$ which does not attain its norm. We show that $|x^*|$ does not either.

Let $B_+ = \{x \in X : x_+^*(|x|) = 0\}$, and define B_- similarly. As all linear functionals on X are order continuous [19, Section 2.4], B_+ and B_- are bands [19, Section 1.4]. Due to the order continuity of X [19, Section 2.4], $B_\pm$ are ranges of band projections $P_\pm$. Let B be the range of $P = P_+P_-$; let B_+^o be the range of $P_+^o = P_+P_-^\perp = P_+ - P$ (where we set $Q^\perp = I_X - Q$), and similarly for B_-^o and P_-^o. Note that $P_+^o + P_-^o = P^\perp$.

Suppose for the sake of contradiction that $x \in \mathbf{S}(X)_+$ satisfies $|x^*|(x) = \|x^*\|$. Replacing x by $P^\perp x$ if necessary, we assume that $Px = 0$, so $x = P_+^o x + P_-^o x$. Then $\|P_+^o x - P_-^o x\| = 1$, and

$$\begin{aligned} x^*(P_-^o x - P_+^o x) &= x_+^*(P_-^o x) - x_+^*(P_+^o x) - x_-^*(P_-^o x) + x_-^*(P_+^o x) \\ &= x_+^*(P_-^o x) + x_-^*(P_+^o x) = |x^*|(x) = \|x^*\|, \end{aligned}$$

which contradicts our assumption that x^* does not attain its norm. □

19.6 On the number of order extreme points

It is shown in [17] that, if a Banach space X is reflexive and infinite-dimensional Banach lattice, then $\mathbf{B}(X)$ has uncountably many extreme points. Here, we establish a similar lattice result.

Theorem 19.28. *If X is a reflexive infinite-dimensional Banach lattice, then $\mathbf{B}(X)$ has uncountably many order extreme points.*

Note that if X is a reflexive infinite-dimensional Banach lattice, then Theorems 19.2 and 19.28 imply that $\mathbf{B}(X)$ has uncountably many extreme points, reproving the result of [17] in this case.

Proof. Suppose, for the sake of contradiction, that there were only countably many such points $\{x_n\}$. For each such x_n, we define $F_n = \{f \in \mathbf{B}(X^*)_+ : f(x_n) = \|f\|\}$. Clearly, F_n is weak* (= weakly) compact.

By the reflexivity of X, any $f \in \mathbf{B}(X^*)$ attains its norm at some $x \in \mathrm{EP}(\mathbf{B}(X))$. Since $f(x) \leq |f|(|x|)$ we assume that any positive functional attains its norm at a positive extreme point in $\mathbf{B}(X)$. By Theorem 19.2, these are precisely the order extreme points. Therefore, $\bigcup F_n = \mathbf{B}(X^*)_+$. By the Baire category theorem, one of these sets F_n must have nonempty interior in $\mathbf{B}(X^*)_+$.

Assume it is F_1. Pick $f_0 \in F_1$, and $y_1, \ldots, y_k \in X$, such that if $f \in \mathbf{B}(X^*)_+$ and for each y_i, $|f(y_i) - f_0(y_i)| < 1$, then $f \in F_1$. Without loss of generality, we assume that $\|f_0\| < 1$, and also that each $y_i \geq 0$.

Further, we can and do assume that there exist mutually disjoint $u_1, u_2, \ldots \in \mathbf{S}(X)_+$ which are disjoint from $y = \vee_i y_i$. Indeed, find mutually disjoint $z_1, z_2, \ldots \in \mathbf{S}(X)_+$. Denote the corresponding band projections by $P_1, P_2, \ldots$ (such projections exist, due to the σ-Dedekind completeness of X). Then the vectors $P_n y$ are mutually disjoint, and dominated by y. As X is reflexive, it must be order continuous and, therefore, $\lim_n \|P_n y\| = 0$. Find $n_1 < n_2 < \cdots$ so that $\sum_j \|P_{n_j} y\| < 1/2$. Let $w_i = \sum_j P_{n_j} y_i$ and $y_i' = 2(y_i - w_i)$. Then if $|(f_0 - g)(y_i')| < 1$, with $g \geq 0$, $\|g\| \leq 1$, it follows that

$$
\begin{aligned}
|(f_0 - g)(y_i)| &\leq \frac{1}{2}\big(|(f_0 - g)(y_i')| + |(f_0 - g)(w_i)|\big) \\
&\leq \frac{1}{2}(1 + \|f_0 - g\|\|w_i\|) < \frac{1}{2}\left(1 + 2 \cdot \frac{1}{2}\right) = 1.
\end{aligned}
$$

We can therefore replace y_i with y_i' to ensure sufficient conditions for being in F_1. Then the vectors $u_j = z_{n_j}$ have the desired properties. Let P be the band projection complementary to $\sum_j P_{n_j}$ (in other words, complementary to the band projection of $\sum_j 2^{-j} u_j$); then $P y_i = y_i$ for any i.

By [19, Lemma 1.4.3 and its proof], there exist linear functionals $g_j \in \mathbf{S}(X^*)_+$ so that $g_j(u_j) = 1$, and $g_j = P_{n_j}^* g_j$. Consequently, the functionals g_j are mutually disjoint, and $g_j|_{\mathrm{ran}\, P} = 0$. For $j \in \mathbb{N}$, find $\alpha_j \in [1 - \|P^* f_0\|, 1]$ so that $\|f_j\| = 1$, where $f_j = P^* f_0 + \alpha_j g_j$. Then, for $1 \leq i \leq k$, $f_j(y_i) = (P^* f_0)(y_i) + \alpha_j g_j(y_i) = f_0(y_i)$, which implies that, for every j, f_j belongs to F_1, hence attains its norm at x_1.

On the other hand, note that $\lim_j g_j(x_1) = 0$. Indeed, otherwise, there exist $\gamma > 0$ and a sequence (j_k) so that $g_{j_k}(x_1) \geq \gamma$ for every k. For any finite sequence of positive numbers (β_k), we have

$$
\sum_k |\beta_k| \geq \Big\|\sum_k \beta_k g_{j_k}\Big\| \geq \sum_k \beta_k g_{j_k}(x_1) \geq \gamma \sum_k |\beta_k|.
$$

As the functionals g_{j_k} are mutually disjoint, the inequalities

$$\sum_k |\beta_k| \geq \left\| \sum_k \beta_k g_{j_k} \right\| \geq \gamma \sum_k |\beta_k|$$

hold for every finite sequence (β_k). We conclude that $\overline{\text{span}}[g_{j_k} : k \in \mathbb{N}]$ is isomorphic to ℓ_1, which contradicts the reflexivity of X. Thus, $\lim_j g_j(x_1) = 0$, hence $\lim_j f_j(x_1) = f_0(Px_1) \leq \|f_0\| < 1$. □

Corollary 19.29. *Suppose C is a closed, bounded, solid, convex subset of a reflexive Banach lattice, having nonempty interior. Then C contains uncountably many order extreme points.*

Proof. We assume without loss of generality that $\sup_{x \in C} \|x\| = 1$. Note that 0 is an interior point of C. Indeed, suppose x is an interior point. Pick $\varepsilon > 0$ such that $x + \varepsilon \mathbf{B}(X) \subset C$. For any k such that $\|k\| < \varepsilon$, we have $\frac{k}{2} = \frac{-x}{2} + \frac{x+k}{2} \in C$, since C is solid and convex. Hence $\frac{\varepsilon}{2}\mathbf{B}(X) \subseteq C$. Since C is bounded, we can then define an equivalent norm, with $\|y\|_C = \inf\{\lambda > 0 : y \in \lambda C\}$. Since C is solid, $\|y\|_C = \| |y| \|_C$, and the norm is consistent with the order. Finally, $\|\cdot\|_C$ is equivalent to $\|\cdot\|$, since for all $y \in X$, we have that $\frac{\varepsilon}{2}\|y\|_C \leq \|y\| \leq \|y\|_C$. The conclusion follows by Theorem 19.28. □

19.7 The solid Krein–Milman Property and the RNP

We say that a Banach lattice (or, more generally, an ordered Banach space) X has the *Solid Krein–Milman Property* (*SKMP*) if every solid closed bounded subset of X is the closed solid convex hull of its order extreme points. This is analogous to the canonical Krein–Milman Property (KMP) in Banach spaces, which is defined in the similar manner, but without any references to order. It follows from Theorem 19.2 that the KMP implies the SKMP.

These geometric properties turn out to be related to the Radon–Nikodým Property (RNP). It is known that the RNP implies the KMP, and, for Banach lattices, the converse is also true (see [7] for a simple proof). For more information about the RNP in Banach lattices, see [19, Section 5.4]; a good source of information about the RNP in general is [6] or [10].

One of the equivalent definitions of the RNP of a Banach space X involves integral representations of operators $T : L_1 \to X$. If X is a Banach lattice, then, by [22, Theorem IV.1.5], any such operator is regular (can be expressed as a difference of two positive ones); so positivity comes naturally into the picture.

Theorem 19.30. *For a Banach lattice X, the SKMP, KMP, and RNP are equivalent.*

Proof. The implications RNP $\Leftrightarrow$ KMP $\Rightarrow$ SKMP are noted above. Now suppose X fails the RNP (equivalently, the KMP). We shall establish the failure of the SKMP in two different ways, depending on whether X is a KB-space, or not.

(1) If X is not a KB-space, then [19, Theorem 2.4.12] there exist disjoint $e_1, e_2, \ldots \in \mathbf{S}(X)_+$, equivalent to the canonical basis of c_0. Then the set

$$C = \overline{S\left(\left\{\sum_i \alpha_i e_i : \max_i |\alpha_i| = 1, \lim_i \alpha_i = 0\right\}\right)}$$

is solid, bounded, and closed. To give a more intuitive description of C, for $x \in X$ we let $x_i = |x| \wedge e_i$. It is easy to see that $x \in C$ if and only if $\lim_i \|x_i\| = 0$, and $|x| = \sum_i x_i$. Finally, show that $x \in C_+$ cannot be an order extreme point. Find i so that $\|x_i\| < 1/2$, and consider $x' = \sum_{j \neq i} x_j + e_i$. Then clearly $x' \in C$, and $x' - x \in X_+ \backslash \{0\}$.

(2) If X is a KB-space failing the RNP, then by [19, Proposition 5.4.9], X contains a separable sublattice Y failing the RNP. Find a quasi-interior point $u \in Y$ – that is, $u \in Y_+$ so that $y = \lim_n y \wedge (nu)$ for any $y \in Y_+$ (for properties of quasi-interior points and their existence in separable Banach lattices, see [1, pp. 266–267]). By [19, Corollary 5.4.20], Y is not order dentable, that is, Y_+ contains a nonempty, convex, bounded subset A so that for every $n \in \mathbb{N}$, $A = \overline{\mathrm{CH}(A \backslash H_n)}$, where $H_n = \{y \in Y_+ : \|u \wedge y\| \geq \frac{1}{n}\}$.

Any KB-space is order continuous, hence by [19, Theorem 2.4.2], its order intervals are weakly compact. This permits us to use the techniques (and notation) of [5] to construct a set C witnessing the failure of the SKMP. For $f \in Y^*$, let $M(A, f) = \sup_{x \in A} |f(x)|$. For $\alpha > 0$, define the *slice* $T(A, f, \alpha) = \{x \in A : f(x) > M(A, f) - \alpha\}$. By [5] (proof of the main result – p. 96), we can construct increasing measure spaces Σ_n on $[0, 1]$ with $|\Sigma_n|$ finite, as well as Σ_n-measurable functions $Y_n : [0, 1] \to A$, $f_n : [0, 1] \to Y^*$, and $\alpha_n : [0, 1] \to (0, \infty)$ such that:

(1) For any n and t, $Y_n(t) \in \overline{T(A, f_n(t), \alpha_n(t))}$.
(2) (Y_n) is a martingale, that is, $Y_n(t) = \mathbb{E}^{\Sigma_n}(Y_{n+1}(t))$, for any t and n ($\mathbb{E}$ stands for the conditional expectation).
(3) For any n and t, $H_n \cap T(A, f_n(t), \alpha_n(t)) = \emptyset$.
(4) For any n and t, $T(A, f_{n+1}(t), \alpha_{n+1}(t)) \subseteq T(A, f_n(t), \alpha_n(t))$.

Now let $C' = \overline{\mathrm{CH}(\{Y_n(t), n \in \mathbb{N}, t \in [0, 1]\})}$, then the set $C = \overline{S(C')}$ (the solid hull is in X) is closed, bounded, convex, and solid. We will show that C has no order extreme points. By Theorem 19.2, it suffices to show that no $x \in C_+ \backslash \{0\}$ can be an extreme point of C, or equivalently, of $C_+ = C \cap X_+$.

From now on, fix $x \in C_+ \backslash \{0\}$. Note that $x \wedge u \neq 0$. Indeed, suppose, for the sake of contradiction, that $x \wedge u = 0$. Find $y' \in C' \subset Y_+$, so that $x \leq y'$. For any n, we have $y' \wedge (nu) = (y' - x) \wedge (nu) \leq y' - x$. Thus, $\|y' - y' \wedge (nu)\| \geq \|x\|$. However, u is a quasi-interior point of Y, hence $y' = \lim_n y' \wedge (nu)$. This is the desired contradiction.

Find $n \in \mathbb{N}$ so that $\|x \wedge u\| > \frac{1}{n}$. Let $I_1, \ldots, I_m$ be the atoms of Σ_n. For $i \leq m$, define $C_i' = \overline{\mathrm{CH}(\{Y_m(t) : m \geq n, t \in I_i\})}$, and let $C_i = \overline{S(C_i')}_+$.

The sequence (Y_k) is a martingale, hence $C' = \overline{\mathrm{CH}(\bigcup_{i=1}^m C_i')}$. Thus, by Proposition 19.3,

$$C = \overline{\mathrm{S}(C')} = \overline{\mathrm{S}\Big(\overline{\mathrm{CH}\Big(\bigcup_{i=1}^m C_i'\Big)}\Big)} = \overline{\mathrm{S}\Big(\mathrm{CH}\Big(\bigcup_{i=1}^m C_i\Big)\Big)}.$$

By [5, Lemma 3], $\mathrm{CH}(\bigcup_{i=1}^m C_i)$ is closed. This set is clearly positive-solid, so by norm continuity of $|\cdot|$, $\mathrm{S}(\mathrm{CH}(\bigcup_1^m C_i))$ is closed, hence equal to C. In particular, $C_+ = \mathrm{CH}(\bigcup_{i=1}^m C_i)$. Therefore, if x is an extreme point of C_+, then it must belong to C_i, for some i. We show this cannot happen.

If $y \in \mathrm{S}(C_i')_+$, then we can find $y' \in C_i'$ with $y \leq y'$. By parts (1) and (4), $C_i' \subseteq \overline{T(A, f_n(t), \alpha_n(t))}$ for $t \in I_i$. By (3), $\|z \wedge u\| < \frac{1}{n}$ for any $z \in T(A, f_n(t), \alpha_n(t))$, hence, by the norm continuity of lattice operations, $\|y' \wedge u\| \leq \frac{1}{n}$. This implies $\|y \wedge u\| \leq \frac{1}{n}$. By the triangle inequality,

$$\|x \wedge u\| \leq \|y \wedge u\| + \|x - y\| \leq \frac{1}{n} + \|x - y\|.$$

Hence $\|x - y\| \geq \|x \wedge u\| - \frac{1}{n}$. Recall that n is selected in such a way that $\|x \wedge u\| > \frac{1}{n}$. As $C_i = \overline{\mathrm{S}(C_i')_+}$, it cannot contain x. Thus, C witnesses the failure of the SKMP. □

Bibliography

[1] C. Aliprantis and O. Burkinshaw, *Positive Operators*, Springer, Berlin, 2006.

[2] M. Amarante, *The Sandwich theorem via Pataraia's fixed point theorem*, Positivity **23** (2019), 97–100.

[3] R. Aron and V. Lomonosov, *After the Bishop–Phelps theorem*, Acta Comment. Univ. Tartu Math. **18** (2014), 39–49.

[4] B. Bollobás, *An extension to the theorem of Bishop and Phelps*, Bull. Lond. Math. Soc. **2** (1970), 181–182.

[5] J. Bourgain and M. Talagrand, *Dans un espace de Banach réticulé solide, la propriété de Radon–Nikodym et celle de Krein–Milman sont equivalentes*, Proc. Am. Math. Soc. **81** (1981), 93–96.

[6] R. Bourgin, *Geometric Aspects of Convex Sets with the Radon–Nikodým Property*, Springer, Berlin, 1983.

[7] V. Caselles, *A short proof of the equivalence of KMP and RNP in Banach lattices and preduals of von Neumann algebras*, Proc. Am. Math. Soc. **102** (1988), 973–974.

[8] Z. L. Chen and A. Wickstead, *Relative weak compactness of solid hulls in Banach lattices*, Indag. Math. (N.S.) **9** (1998), 187–196.

[9] M. Chica, V. Kadets, M. Martín, S. Moreno-Pulido and F. Rambla-Barreno, *Bishop–Phelps–Bollobás moduli of a Banach space*, J. Math. Anal. Appl. **412** (2014), 697–719.

[10] J. Diestel and J. Uhl, *Vector Measures*, American Mathematical Society, Providence, RI, 1977.

[11] E. Effros and C. Webster, Operator analogues of locally convex spaces, in: *Operator algebras and applications*, 163–207, NATO Adv. Sci. Inst. Ser. C Math. Phys. Sci., **495**, Kluwer Acad. Publ., Dordrecht, 1997.

[12] E. Effros and S. Winkler, *Matrix convexity: operator analogues of the bipolar and Hahn–Banach theorems*, J. Funct. Anal. **144** (1997), 117–152.

[13] M. Fabian, P. Hajek, P. Habala, V. Montesinos and V. Zizler, *Banach Space Theory. The Basis for Linear and Nonlinear Analysis*, Springer, New York, 2011.

[14] D. Farenick and P. Morenz, *C^*-extreme points in the generalised state spaces of a C^*-algebra*, Trans. Am. Math. Soc. **349** (1997), 1725–1748.

[15] B. Fuchssteiner and J. D. Maitland Wright, *Representing isotone operators on cones*, Q. J. Math. **28** (1977), 155–162.

[16] J. Lindenstrauss, *On extreme points in ℓ_1*, Isr. J. Math. **4** (1966), 59–61.

[17] J. Lindenstrauss and R. Phelps, *Extreme point properties of convex bodies in reflexive Banach spaces*, Isr. J. Math. **6** (1968), 39–48.

[18] B. Magajna, *C^*-convex sets and completely positive maps*, Integral Equ. Oper. Theory **85** (2016), 37–62.

[19] P. Meyer-Nieberg, *Banach Lattices*, Springer, Berlin, 1991.

[20] M. Popov and B. Randrianantoanina, *Narrow Operators on Function Spaces and Vector Lattices*, Walter de Gruyter, Berlin, 2013.

[21] W. Rudin, *Functional Analysis*, McGraw-Hill, Singapore, 1991.

[22] H. H. Schaefer, *Banach Lattices and Positive Operators*, Springer, New York, Heidelberg, 1974.

[23] C. Webster and S. Winkler, *The Krein–Milman theorem in operator convexity*, Trans. Am. Math. Soc. **351** (1999), 307–322.

Sasmita Patnaik, Srdjan Petrovic, and Gary Weiss

20 Universal block tridiagonalization in $\mathcal{B}(\mathcal{H})$ and beyond

In memoriam of Victor Lomonosov

Abstract: For $\mathcal{H}$, a separable infinite dimensional complex Hilbert space, we prove that every $\mathcal{B}(\mathcal{H})$ operator has a basis with respect to which its matrix representation has a universal block tridiagonal form with block sizes given by a simple exponential formula independent of the operator. From this, such a matrix representation can be further sparsified to slightly sparser forms; it can lead to a direct sum of even sparser forms reflecting in part some of its reducing subspace structure; and in the case of operators without invariant subspaces (if any exists), it gives a plethora of sparser block tridiagonal representations. An extension to unbounded operators occurs for a certain domain of definition condition. Moreover, this process gives rise to many different choices of block sizes.

Keywords: Hilbert space, orthonormal basis, block tridiagonal matrices, sparse matrices

MSC 2010: Primary 47A65, 47A67, 47A99, 15A21, 47A08, Secondary 47B47, 47B02

20.1 Introduction

How sparse can a matrix of an operator be? By this we mean the following: If T is a bounded linear operator on infinite dimensional, separable, complex Hilbert space $\mathcal{H}$, can we find an orthonormal basis with respect to which the matrix of T has as many zero entries as possible? A change of basis corresponds to a unitary operator U which yields the matrix representation of T in the new basis $\{Ue_n\}$, or equivalently, the matrix of $U^{-1}TU$ in the original basis $\{e_n\}$. Thus our question can be phrased as: How sparse can $U^{-1}TU$ be made?

An extreme example is the spectral theorem for normal compact operators yielding diagonal operators. It is well known that every self-adjoint operator (even when

Acknowledgement: Partially supported by Simons Foundation grant number 245014.

Sasmita Patnaik, Department of Mathematics and Statistics, Indian Institute of Technology, Kanpur 208016, India, e-mail: sasmita@iitk.ac.in
Srdjan Petrovic, Department of Mathematics, Western Michigan University, Kalamazoo, MI 49008, USA, e-mail: srdjan.petrovic@wmich.edu
Gary Weiss, Department of Mathematics, University of Cincinnati, Cincinnati, OH 45221-0025, USA, e-mail: gary.weiss@uc.edu

https://doi.org/10.1515/9783110656756-020

not compact) that possesses a cyclic vector can be represented as a tridiagonal matrix, and if it does not possess a cyclic vector then it can at least be represented as a direct sum (finite or infinite) of tridiagonal matrices. We found a way to universally generalize these tridiagonal phenomena but with block tridiagonal matrices. That is, we will extend the tridiagonal form idea to each $\mathcal{B}(\mathcal{H})$ operator but it will be a block tridiagonal matrix with universal block sizes independent of the operator. Moreover, our methods will hold more generally, that is, for all $\mathcal{B}(\mathcal{H})$ operators and for unbounded operators with a certain constraint.

Block tridiagonal matrices have been useful in establishing various results related to [4], the Pearcy–Topping compact commutator problem: What operators are commutators of compact operators, that is, operators of the form $[A, B] = AB - BA$, with A, B compact? An outstanding result in this direction is Anderson's construction in [1], where he employed block tridiagonal matrices with a particular arithmetic mean growth to prove that every rank one projection operator is a commutator of compact operators. From here, he proved the important consequence: Every compact operator is a commutator of a compact operator with a bounded operator. Moreover, in [4] Pearcy–Topping asked whether every trace class operator with zero trace is a commutator of Hilbert–Schmidt operators, or at least a finite sum of such commutators. In the same period the third author answered these questions in the negative in [5]. This work introduced the study of matrix sparsification in terms of staircase form representations.

Historically, [5] introduced staircase forms for general operators (Theorem 20.1) which herein leads us to universal block tridiagonal forms (Theorem 20.4). In this article, we will obtain a general matrix sparsification via a special unitary operator by showing how the staircase forms can be reorganized into block tridiagonal forms. As in Anderson's model [1], we hope block tridiagonal forms are more manageable for computations. (For our recent analysis of [1] using results herein, see [3].) Then we provide two independent further sparsifications, potentially even more manageable for computations.

20.2 Sparsifying arbitrary matrices

The following general staircase form is obtained (from a slight modification of [5, Lemma]) by considering the free semigroup on the two generators T, T^* with any basis $\{e_n\}$ to generate a new basis via applying Gram–Schmidt to e_1, Te_1, T^*e_1, e_2, T^2e_1, T^*Te_1, e_3, TT^*e_1, $T^{*2}e_1$, e_4, Te_2, T^*e_2, e_5, …. The latter list consists of $\mathcal{W}(T, T^*)$ (all words in T, T^*, including the empty word I) applied to all elements of $\{e_n\}$ and arranged in this special order: first list e_1 followed by T, T^* applied to e_1 to obtain the first three vectors; then list e_2, followed by T, T^* applied to the second vector in this list, namely Te_1; and so on.

An alternate way to view this is to start an induction with the first three terms e_1, Te_1, T^*e_1, followed by e_2, e_3, ..., and then use Gram–Schmidt to obtain f_1, f_2, f_3 (the first 3 vectors of the new basis). Inductively, assume that $f_1, f_2, \ldots, f_{3n}$ have been chosen orthonormal with $e_1, e_2, \ldots, e_n$ in their span and $e_i, Tf_i, T^*f_i \in \bigvee_{j=1}^{3n} f_j$, for $1 \le i \le n$; then we extend this list to $f_{3n+1}, f_{3n+2}, f_{3(n+1)}$ by continuing the Gram–Schmidt process with the first three new linearly independent vectors from the list

$$f_1, f_2, \ldots, f_{3n}, e_{n+1}, Tf_{n+1}, T^*f_{n+1}, e_{n+2}, e_{n+3}, \ldots.$$

Clearly, because $\{e_n\}$ spans $\mathcal{H}$, $\{f_n\}$ forms a basis. And moreover, this basis together with this inductive condition yield the following matrix form with respect to $\{f_n\}$ by using the condition $Tf_n, T^*f_n \in \bigvee_{j=1}^{3n} f_j$ for all n.

Theorem 20.1 ([5, Theorem 2]). *For every $T \in B(H)$, there is a basis whose implementing operator U fixes an arbitrary e_1 and with respect to which basis $\{e_n\}$ the matrix of $U^{-1}TU$ takes the staircase form*

$$U^{-1}TU = \begin{pmatrix} * & * & * & 0 & \cdots & & & & \\ * & * & * & * & * & * & 0 & \cdots & \\ * & * & \cdots & & & & & & \\ 0 & * & & & & & & & \\ 0 & * & & & & & & & \\ 0 & * & & & & & & & \\ \vdots & 0 & & & & & & & \end{pmatrix}, \tag{20.1}$$

with each $e_n \in \bigvee_{k=1}^{3n} Ue_k$ and where this (20.1) matrix has row and column support lengths $3, 6, 9, \ldots$.

In addition, if $S_1, \ldots, S_N$ is any finite collection of self-adjoint operators, then there is a unitary operator U that fixes e_1 for which each operator $U^{-1}S_kU$ has the form (20.1) with $3, 6, 9, \ldots$ replaced by $N+1, 2(N+1), 3(N+1), \ldots$, with each $e_n \in \bigvee_{k=1}^{1+(n-1)(N+1)} Ue_k$. When $S_1, \ldots, S_N$ are not necessarily selfadjoint, we have $3, 6, 9, \ldots$ replaced by $2N+1, 2(2N+1), 3(2N+1), \ldots$, with each $e_n \in \bigvee_{k=1}^{1+(n-1)(2N+1)} Ue_k$.

Definition 20.2. We call the $*$-entries in (20.1) the support entries. Albeit some can also be zero as we shall see in Theorems 20.7 and 20.8.

Remark 20.3. (i) In fact, we have a little more. The proof for a single operator yields $2, 5, 8, \ldots$ for the columns and $3, 6, 9, \ldots$ for the rows, that is, Tf_n is a linear combination of at most $f_1, f_2, \ldots, f_{3n-1}$ vectors and T^*f_n is a linear combination of at most $f_1, f_2, \ldots, f_{3n}$ vectors, for $n \ge 1$. From here, an important question is: Is there a more substantial sparsification of the form in (20.1)? Theorems 20.7 and 20.8 achieve this and imply the obvious question: Can we sparsify these forms even further? One goal as mentioned earlier is to improve their computation potential.

When T possesses a cyclic vector v (i. e., $v, Tv, T^2v, \dots$ spans $\mathcal{H}$), (20.1) would have instead column support sizes $2, 3, 4, \dots$ obtaining an upper Hessenberg form [2, Problem 44]. Or if one preferred also simultaneous sparsification of the rows, and had a joint cyclic vector in that $v, Tv, T^*v, T^2v, T^*Tv, TT^*v, T^{*2}v, \dots$ span $\mathcal{H}$ (the free semigroup on the two generators T, T^* but arranged in this specific order), then applying Gram–Schmidt to this sequence obtains the matrix pattern $2, 4, 6, \dots$ columns and $3, 5, 7, \dots$ rows. So in particular, for an operator T with no invariant subspaces (if indeed one exists), every nonzero vector is cyclic (otherwise for noncyclic v, the span of $v, Tv, T^2v, \dots$ is a nontrivial invariant subspace). Or, if T has no proper reducing subspaces (which can occur) then every vector is jointly cyclic, that is, $v, Tv, T^*v, T^2v, T^*Tv, TT^*v, T^{*2}v, \dots$ spans $\mathcal{H}$. In either case, we obtain for T this above $2, 4, 6, \dots / 3, 5, 7, \dots$ sparser staircase pattern.

(ii) The same applies to those unbounded operators T whose free semigroup on these two generators has a basis on which all words $w(T, T^*)$ are defined. Albeit, we do not know how to test for this.

From staircase to block tridiagonal matrix forms

Theorem 20.1 gives a striking $3, 6, 9, \dots$ staircase universal form for an arbitrary operator and universal simultaneous staircase forms with larger stairs for arbitrary finite collections.

Staircase form (20.1) will allow us to represent a matrix in universal block tridiagonal form

$$T = \begin{pmatrix} C_1 & A_1 & 0 & & \\ B_1 & C_2 & A_2 & & \\ 0 & B_2 & C_3 & \ddots & \\ & & \ddots & \ddots & \end{pmatrix}, \tag{20.2}$$

which we believe may be fundamental and of broad general interest. Even for our orthonormal basis $\{e_n\}_{n=1}^{\infty}$ in which T is given by (20.2), the sizes of these blocks we will determine and see they are not unique (see Theorem 20.4). We believe it could be of further general interest to make the (20.2) matrix blocks as sparse as possible, that is, obtain further universal zeros. Two ways to accomplish this are demonstrated in Theorems 20.7–20.8. We begin with the following theorem which gives canonical dimensions for the blocks in (20.2) in order that they cover all the support entries. And since there will be no change of basis, we retain the Theorem 20.1 condition that each $e_n \in \bigvee_{k=1}^{3n} Ue_k$.

Theorem 20.4. *For all $T \in B(H)$, the block tridiagonal partition of the matrix of T induced by (20.1) is given by (20.2), where C_1, A_1, and B_1 are 1×1, 1×2 and 2×1 matrices,* *respectively; and for $k \geq 2$,*

$$\begin{aligned} &C_k : 2(3^{k-2}) \times 2(3^{k-2}) \text{ square matrix (i. e., } 2\times 2, 6\times 6, 18\times 18, \dots) \\ &A_k : 2(3^{k-2}) \times 2(3^{k-1}) \text{ rectangular wide matrix} \\ &B_k : 2(3^{k-1}) \times 2(3^{k-2}) \text{ rectangular tall matrix.} \end{aligned} \tag{20.3}$$

Alternatively, one can choose C_1, A_1, B_1 to be $n_1 \times n_1$, $n_1 \times 2n_1$, $2n_1 \times n_1$ matrices, respectively; and for $k \geq 2$, C_k, A_k, B_k, respectively of sizes $2(3^{k-2})n_1 \times 2(3^{k-2})n_1$, $2(3^{k-2})n_1 \times 2(3^{k-1})n_1$, $2(3^{k-1})n_1 \times 2(3^{k-2})n_1$. More generally, necessary and sufficient conditions that (20.2) cover the (20.1) staircase support entries are: n_1 is chosen arbitrarily and $n_{k+1} \geq 2(n_1 + n_2 + \cdots + n_k)$.

Proof. The partition of the matrix for T to blocks goes row by row. We select the sizes for C_1 and A_1 to be 1×1 and 1×2, respectively. This forces B_1 to be 2×1. For any $k \in \mathbb{N}$, if the sizes of C_k, A_k and B_k are $n_k \times n_k$, $n_k \times n_{k+1}$ and $n_{k+1} \times n_k$, respectively, then the members of the sequence $\{n_k\}$ must satisfy the condition $n_{k+1} \geq 2(n_1 + n_2 + \cdots + n_k)$. Indeed, the width of A_k has to be sufficient to cover all support entries on the right of C_k and these lie in rows $n_1 + n_2 + \cdots + n_{k-1} + 1$ through $n_1 + n_2 + \cdots + n_k$. The last of these support entries stretch out to the position $3(n_1 + n_2 + \cdots + n_k)$. However, block A_k starts with the column $n_1 + n_2 + \cdots + n_{k-1} + 1$, so by considering its last (20.1)-staircase row it needs to cover at least $3(n_1 + n_2 + \cdots + n_k) - (n_1 + n_2 + \cdots + n_k) = 2(n_1 + n_2 + \cdots + n_k)$ more support entries. Consequently, the C_{k+1} size $n_{k+1} \geq 2(n_1 + n_2 + \cdots + n_k)$. Taking equality for every k and $n_1 = 1$ yields (20.3).

The necessary and sufficient conditions that (20.2) cover the (20.1) staircase support entries: n_1 is chosen arbitrarily and $n_{k+1} \geq 2(n_1 + n_2 + \cdots + n_k)$, follows by the same argument. □

Remark 20.5. Although our canonical choice in Theorem 20.4 uses the minimal value of each n_k for $k \geq 1$, that is, $n_1 = 1$ and $n_k = 2(n_1 + \cdots + n_{k-1})$, the covering of support entries in (20.1) is *not* minimal in the block tridiagonal sense. That is, if we choose a different sequence $\{n'_k\}$ and denote the corresponding blocks in (20.2) by $\{A'_n\}$, $\{B'_n\}$, $\{C'_n\}$, they need not completely cover the full canonical blocks. For example: if we select $n'_1 = 4$, and $n'_k = 2(n'_1 + \cdots + n'_{k-1})$ for $k \geq 2$, the $(4, 27)$ entry in the matrix of T does not lie in any of the $\{A'_n\}$, $\{B'_n\}$, $\{C'_n\}$ blocks, yet it belongs to A_3. Since this example has $n_1 = 4$ one might ask whether the canonical blocks are minimal among those that have $n_1 = 1$. Once again, the answer is no: take $n'_1 = 1$, $n'_2 = 3$, and $n'_k = 2(n'_1 + \cdots + n'_{k-1})$ for $k \geq 3$. Again, the $(4, 27)$ entry in the matrix of T does not lie in any of the associated blocks, yet it belongs to A_3.

Cyclic vector consequences. In case T and T^* have a joint cyclic vector v, that is, the collection $\mathcal{W}(T, T^*)v$ spans $\mathcal{H}$, so if in particular the operator T has a cyclic vector v ($\mathcal{W}(T)v$ spans $\mathcal{H}$), then the (20.2) diagonal square blocks have smaller sizes 1×1, 2×2, 4×4, $8\times 8, \dots$ and with off-diagonal block sizes forced accordingly. Indeed, the list e_1,

$Te_1, T^*e_1, e_2, T^2e_1, T^*Te_1, e_3, TT^*e_1, T^{*2}e_1, e_4, Te_2, T^*e_2, e_5, \dots$, that was used in the proof of Theorem 20.1 is now substantially reduced. That is, the vectors e_n, for $n \geq 2$, can be deleted, their purpose being to ensure that the new orthogonal set spans $\mathcal{H}$. It follows that instead of the $3, 6, 9, \dots$ pattern we get $2, 4, 6, 8, \dots$. The same reasoning as in the proof of Theorem 20.4 now yields the inequality $n_{k+1} \geq n_1 + n_2 + \cdots + n_k$ and equality for all k leads to the blocks of smaller sizes than in (20.3). Namely from $n_{k+1} = n_1 + n_2 + \cdots + n_k$ and $n_k = n_1 + n_2 + \cdots + n_{k-1}$, we obtain $n_{k+1} - n_k = n_k$, $n_{k+1} = 2n_k$, and finally $n_k = 2^{k-1}n_1$. Then in general, $\mathcal{H}$ can be represented as an orthogonal direct sum of reducing subspaces on which T and T^* have a joint cyclic vector, and thus we have the following.

Theorem 20.6. *Every $\mathcal{B}(\mathcal{H})$ operator is a direct sum of operators of the form (20.2) where the sizes of diagonal blocks in each summand are 1×1, 2×2, 4×4, 8×8, $\dots$ and the sizes of the off-diagonal blocks are forced accordingly.*

It is natural to ask whether Theorem 20.4 can be further improved, that is, obtain more universal zeros in the blocks. Here, we are attempting to preserve the structure and block sizes as in Theorem 20.4, while ensuring that some additional entries in these blocks are universally zeros. The following theorem presents one way to achieve this. It applies more generally only requiring that $\{n_k\}$ be nondecreasing (so that A_k has width no less than its height).

Theorem 20.7. *Every block tridiagonal matrix with diagonal block sizes $\{n_k\}$ nondecreasing is unitarily equivalent to a block tridiagonal matrix with the same block sizes but also with A_n of the form $(A'_n \mid 0)$ with A'_n a positive square matrix. Alternatively, the same but with B_n of the form $(B'_n \mid 0)^T$ with B'_n a positive square matrix.*

Proof. Consider the block tridiagonal matrix form of T as in (20.2) with $\{n_k\}$ nondecreasing. We will define recursively a sequence of unitary matrices $\{U_n\}$ with the size of U_n same as the size of C_n (i. e., $n_k \times n_k$), and U their direct sum. The matrix for U^*TU becomes

$$\begin{pmatrix} U_1^*C_1U_1 & U_1^*A_1U_2 & 0 & \cdots \\ U_2^*B_1U_1 & U_2^*C_2U_2 & U_2^*A_2U_3 & \ddots \\ 0 & U_3^*B_2U_2 & U_3^*C_3U_3 & \ddots \\ \vdots & \ddots & \ddots & \ddots \end{pmatrix}.$$

Let $U_1 = I$ and suppose that matrices $\{U_i\}_{i=1}^k$ have been selected. Consider the square matrix

$$X = \begin{pmatrix} U_k^*A_k \\ 0 \end{pmatrix},$$

with the zero matrix of height $n_{k+1} - n_k$, and let $X^* = UP$ be the polar decomposition of X^*. Then $P = U^*X^* = XU$, because P is self-adjoint. Further,

$$XU = \begin{pmatrix} U_k^* A_k \\ 0 \end{pmatrix} U = \begin{pmatrix} U_k^* A_k U \\ 0 \end{pmatrix}.$$

On the other hand, $P^2 = PP^* = (XU)(XU)^* = XUU^*X^* = XX^*$, so

$$P = \left[\begin{pmatrix} U_k^* A_k \\ 0 \end{pmatrix} \begin{pmatrix} A_k^* U_k & 0 \end{pmatrix} \right]^{1/2} = \begin{pmatrix} [U_k^* A_k A_k^* U_k]^{1/2} & 0 \\ 0 & 0 \end{pmatrix},$$

and we define $U_{k+1} = U$. This implies that $U_k^* A_k U_{k+1} = ([U_k^* A_k A_k^* U_k]^{1/2} \mid 0)$, hence T is unitarily equivalent to the block tridiagonal form

$$\begin{pmatrix} \tilde{C}_1 & \tilde{A}_1 & 0 & \dots \\ \tilde{B}_1 & \tilde{C}_2 & \tilde{A}_2 & \ddots \\ 0 & \tilde{B}_2 & \tilde{C}_3 & \ddots \\ \vdots & \ddots & \ddots & \ddots \end{pmatrix} \tag{20.4}$$

where each $\tilde{A}_n$ has the form $(A_n' \mid 0)$ with A_n' a positive square matrix.

The "alternatively" part of Theorem 20.7 follows by applying to T^* the first part. □

At this point, it is natural to ask whether this form can be sparsified further. That is, is there a choice of an orthonormal basis in which (20.2) can be sparsified beyond the above matrix (20.4)? Recently, we have been able to prove such a result in the Theorems 20.1, 20.4 setting. However, we here get the weaker spanning condition: $e_n \in \bigvee_{k=1}^{3^n} Ue_k$, than that of Theorem 20.1.

Theorem 20.8. *For arbitrary $T \in \mathcal{B}(\mathcal{H})$ and any orthonormal basis $\{e_n\}$ of $\mathcal{H}$, there exists an orthonormal basis $\{f_n\}$ in which T has a block tridiagonal form as in (20.2) with the block sizes as in Theorem 20.4 (with $n_1 - 1$) and:*
(a) *each block $B_n = (B_n' \mid 0 \mid 0)^T$ where all three blocks are square and B_n' is upper triangular, that is, $B_n'(i,j) = 0$ if $i > j$;*
(b) *each block A_n is of the form $(A_n' \mid A_n'' \mid 0)$ where all three blocks are square and A_n'' is lower triangular, that is, $A_n''(i,j) = 0$ if $i < j$.*
(c) *$e_1 = f_1$ and $e_n \in \bigvee_{k=1}^{3^n} f_k$ for all $n \in \mathbb{N}$.*

Alternatively, T is unitarily equivalent to another matrix of the form (20.2) with the block sizes as in Theorem 20.4 (with $n_1 = 1$), where each $A_n = (A_n' \mid 0 \mid 0)$, all three blocks are square, and A_n' is lower triangular, while each B_n has the form $(B_n' \mid B_n'' \mid 0)^T$, all three blocks are square, and B_n'' is an upper triangular matrix, and (c) holds.

Proof. Our first step is to define recursively a sequence $\{g_n\}$ that contains $\{e_n\}$ dispersed more sparsely than the sequence described in the first paragraph of Section 20.2. We will use the notation $n_1 = 1$, $n_k = 2(3^{k-2})$, for $k \geq 2$, and $s_k = n_1 + n_2 + \cdots + n_k$, for $k \geq 1$ and $s_0 = 0$. It is clear that every positive integer n can be written in a unique way as

$$n = s_k + r, \quad \text{where } \leq r \leq n_{k+1}.$$

Then we define $g_1 = e_1, g_2 = Te_1, g_3 = T^* e_1$ and for $n \geq 4$ we use the following formulas:

$$g_n = Tg_{s_{k-1}+r}, \quad \text{when } s_k + 1 \leq n \leq s_k + n_k. \tag{20.5}$$

$$g_n = T^* g_{r+1-k} \quad \text{when } s_k + n_k + 1 \leq n \leq s_{k+1} - 1. \tag{20.6}$$

$$g_n = e_k \quad \text{when } n = s_{k+1}. \tag{20.7}$$

It is not hard to verify that $\{g_n\}$ is a well-defined sequence that contains $\{e_k\}$. Let us assume for a moment that $\{g_n\}$ is linearly independent. (The case of linear dependence we deal with at the end of the proof.) By applying the Gram–Schmidt process to $\{g_n\}$, we obtain an orthonormal basis $\{f_n\}$. An easy calculation yields $s_{k+1} = 3^k$ so the spanning condition follows from (20.7).

It follows from (20.5) that the length of the nonzero portion of the mth column, where $m = s_{k-1} + r$, does not exceed $n = s_k + r$. The inequalities in (20.5) imply that $1 \leq r \leq n_k$, hence $s_{k-1} + 1 \leq m \leq s_{k-1} + n_k = s_k$ and these characterize the columns that go through the block B_k. Since the length of the nonzero portions of these columns does not exceed $n = s_k + r = m + n_k$ it follows that B_k is upper triangular and all the blocks below B_k in (20.2) are zeros.

Similarly, the inequalities in (20.6) imply that $n_k + 1 \leq r \leq n_{k+1} - 1$, whence

$$s_{k-1} \leq n_k + 2 - k \leq r + 1 - k \leq n_{k+1} - k \leq s_{k+1}.$$

This shows that row $r + 1 - k$ goes through either the block A_k or A_{k+1}. More precisely, it goes through A_k if

$$n_k + 2 - k \leq r + 1 - k \leq s_k, \tag{20.8}$$

and through A_{k+1} if

$$s_k + 1 \leq r + 1 - k \leq n_{k+1} - k. \tag{20.9}$$

If we replace k by $k + 1$ in (20.8), we obtain rows numbered $n_{k+1} - k + 1$ through s_{k+1}. Together with (20.9), it shows that as k takes positive integer values all rows of the matrix for T appear in (20.6).

For rows that go through A_k, (resp., A_{k+1}) condition in (b) means that the length of the nonzero portion of row m should not exceed $m + 2n_{k+1}/3$ (resp., $m + 2n_{k+2}/3$). By (20.6), the length of the row $r+1-k$ does not exceed $n = s_k + r = (r+1-k) + (s_k + k - 1)$. A calculation shows that $s_k + k - 1 \leq 2n_{k+1}/3 \leq 2n_{k+2}/3$ for $k \geq 2$, so (b) is proved together with the fact that all blocks in (20.2) to the right of A_k are indeed zero blocks.

In the case that the sequence $\{g_n\}$ constructed above is not linearly independent, there exists $n_0 \in \mathbb{N}$ for which $g_{n_0} \in \bigvee_{k=1}^{n_0-1} g_k$. In that case, we will delete the equation that has g_{n_0} on the left side and in the subsequent equations g_{n_0+1} will be replaced by g_{n_0}, g_{n_0+2} by g_{n_0+1}, etc. Since in each equation of the form $g_n = Tg_i$ (resp., $g_n = T^*g_i$), n determines the maximum length of the ith row (resp., column), this will decrease the said maximum by one so the conclusions of the theorem will hold all the more. Of course, if there is a next such number, we apply the same procedure, and so on.

The "alternatively" part of Theorem 20.8 follows by applying to T^* the first part. □

Remark 20.9. Both Theorem 20.7 and Theorem 20.8 exhibit a lack of symmetry regarding the role of blocks $\{A_n\}$ and $\{B_n\}$, but we do not know if each of these further sparsifications A_n, B_n forms (Theorems 20.7 and 20.8) can be achieved symmetrically. We suspect not in general.

Remark 20.10. The results of this section share the same method of finding the desired orthonormal basis $\{f_n\}$. An arbitrary orthonormal basis $\{e_n\}$ is augmented by adding all vectors of the form $w(T, T^*)e_k$, (all words in T, T^*, that is, the free semigroup on two generators, applied to all e_k), arranged in a certain order, to which the Gram–Schmidt orthogonalization is applied. It follows that the same results hold even when T is an unbounded operator, as long as all words $w(T, T^*)e_k$ are defined. One condition that achieves this is when each $e_k \in (\text{domain } T) \cap (\text{domain } T^*)$ and $(\text{range } T) \cup (\text{range } T^*) \subset (\text{domain } T) \cap (\text{domain } T^*)$.

Both Theorems 20.7 and 20.8 show that each block A_n can be sparsified to $(A_n' \mid 0)$ with A_n' a positive square matrix in the former and a lower triangular matrix in the latter. (In both cases, the same number of variables have been eliminated.) Then one may ask whether there is a further sparsification in which every A_n' is a diagonal matrix.

Problem 20.11. *Given an operator T, is there an orthonormal basis in which T is of the form (20.2), where each A_n has the form $(A_n' \mid 0)$ and A_n' is a diagonal matrix?*

We considered the following 5×5 test matrix (as in Remark 20.3):

$$T = \begin{pmatrix} 1 & 1 & 1 & 0 & 0 \\ 1 & 1 & 1 & 1 & 1 \\ 0 & 1 & 1 & 1 & 1 \\ 0 & 1 & 1 & 1 & 1 \\ 0 & 1 & 1 & 1 & 1 \end{pmatrix}. \tag{20.10}$$

Note: replacing $t_{31} = 1$ makes T self-adjoint, hence diagonalizable and not an appropriate test case. Also, Remark 20.3 shows that one can obtain the $(3, 1)$ entry equal to 0. This question was answered affirmatively by Zack Cramer, University of Waterloo, who produced the following unitary matrix:

$$U = \frac{1}{\sqrt{2}} \begin{pmatrix} 0 & 0 & \sqrt{2} & 0 & 0 \\ 0 & 1 & 0 & -1 & 0 \\ 0 & 1 & 0 & 1 & 0 \\ 1 & 0 & 0 & 0 & 1 \\ 1 & 0 & 0 & 0 & -1 \end{pmatrix}.$$

It is easy to verify that

$$U^* TU = \frac{1}{2} \begin{pmatrix} 4 & 4 & 0 & 0 & 0 \\ 4 & 4 & \sqrt{2} & 0 & 0 \\ 0 & 2\sqrt{2} & 2 & 0 & 0 \\ 0 & 0 & -\sqrt{2} & 0 & 0 \\ 0 & 0 & 0 & 0 & 0 \end{pmatrix},$$

so T is unitarily equivalent to a matrix with the entries $(1,3)$, $(2,5)$ and $(3,4)$ equal to 0, and the five zero entries of T remaining zeros. Nevertheless, Problem 20.11 remains open even in the general 5×5 test case.

Bibliography

[1] J. Anderson, *Commutators of compact operators*, J. Reine Angew. Math. **291** (1977), 128–132, 10.1515/crll.1977.291.128, MR0442742.

[2] P. R. Halmos, *A Hilbert Space Problem Book*, 2nd edition, Graduate Texts in Mathematics, **19**, Springer, 1982. Encyclopedia of Mathematics and its Applications, **17**, MR675952.

[3] S. Patnaik, S. Petrovic and G. Weiss, *Commutators of Compact Operators and Sparsifying Arbitrary Matrices*, preprint.

[4] C. Pearcy and D. Topping, *On commutators in ideals of compact operators*, Mich. Math. J. **18** (1971), 247–252, MR0284853.

[5] G. Weiss, *Commutators of Hilbert–Schmidt operators. II*, Integral Equ. Oper. Theory, **3** (4) 1980, 574–600, 10.1007/BF01702316.

Mikhail Popov

21 Rademacher-type independence in Boolean algebras

To the memory of a brilliant mathematician, Victor Lomonosov

Abstract: We find necessary and sufficient conditions on a family $\mathcal{R} = (r_i)_{i\in I}$ in a σ-complete Boolean algebra $\mathcal{B}$ under which there exists a unique positive σ-additive measure μ on $\mathcal{B}$ such that $\mu(\bigwedge_{k=1}^{n} \theta_k r_{i_k}) = 2^{-n}$ for all distinct $i_1, \dots, i_n \in I$ and all signs $\theta_1, \dots, \theta_n \in \{-1, 1\}$, where the product θx of a sign θ by an element $x \in \mathcal{B}$ is defined by setting $1x = x$ and $-1x = -x = \mathbf{1} - x$. Such a family we call a σ-generating Rademacher family. We prove that σ-complete Boolean algebras admitting σ-generating Rademacher systems of the same cardinality are σ-isomorphic. As a consequence, we obtain that a σ-complete Boolean algebra is Maharam homogeneous measurable if and only if it admits a σ-generating Rademacher family. This axiomatic definition of a σ-generating Rademacher family gives an alternative approach to define a measure.

Keywords: Boolean algebra, measurable algebra, Rademacher system

MSC 2010: Primary 46G12, Secondary 28A60

21.1 Introduction

21.1.1 Why do we need Rademacher families in Boolean algebras?

The classical Rademacher system constitute an important tool for the investigation of the isomorphic structure of symmetric (rearrangement invariant) spaces [1], Köthe function spaces on measure spaces [9], as well as in probability theory. So, it would be natural to generalize a Rademacher system to the setting of Riesz spaces. However, the definition of a Rademacher-type system uses a measure which is not generally defined if we consider an arbitrary Riesz space. We want to define a Rademacher-type system on a general Riesz space without a measure. A Rademacher system, which is defined

Acknowledgement: The author is grateful to O. Maslyuchenko for valuable discussions during the preparation of the manuscript. Especially I thank the anonymous referee for numerous improvements of the text, which can be compared with a co-author's contribution. In particular, Example 21.30, which is an essential point in the idea of Rademacher families, is due to the referee.

Mikhail Popov, Institute of Mathematics, Pomeranian University in Słupsk, ul. Arciszewskiego 22d, PL-76-200 Słupsk, Poland; and Vasyl Stefanyk Precarpathian National University, Ivano-Frankivsk, Ukraine, e-mail: misham.popov@gmail.com

https://doi.org/10.1515/9783110656756-021

axiomatically, generates a probability measure by an obvious way. Since any element of a Rademacher-type system in a Riesz space E has to be of the form $r = a - b$, where $a, b \in E^+$, $a \perp b$ and the element $e = |r| = a \sqcup b$ (by $a \sqcup b$ we denote a disjoint sum, that is the sum $a+b$ of disjoint elements $a \perp b = 0$) is some fixed element of E considered as the "support" of the Rademacher system, to define a Rademacher system on e means to find a sequence of two-element partitions of e. Observe that elements a, b of any partition $e = a \sqcup b$ are fragments of e, that is, $a \perp (e - a)$ and the same with b. It is well known that the set $\mathfrak{F}_e$ of all fragments of e is a Boolean algebra with unity e. Thus, we deal with an arbitrary Boolean algebra $\mathcal{B}$ to define a Rademacher system, which becomes a sequence (or, more generally, a family) of two-element partitions of unity $\mathbf{1}$ of $\mathcal{B}$. For convenience of notation, instead of a sequence of partitions $\mathbf{1} = r_n \sqcup s_n$ we consider a sequence (r_n) of representatives of each partition, no matter which ones. To distinguish Rademacher systems in Riesz spaces and Boolean algebras; the later ones we call Rademacher families.

21.1.2 Terminology and notation

We find algebraic conditions on a family $(r_i)_{i \in I}$ of elements of a Boolean algebra $\mathcal{B}$ under which one can consider it as a Rademacher family. More precisely, we find conditions under which there is a unique countably additive positive (i. e., strictly positive at every nonzero element) measure on $\mathcal{B}$ possessing the equality

$$\mu\left(\bigwedge_{k=1}^{n} \theta_k r_{i_k}\right) = \frac{1}{2^n} \tag{21.1}$$

for all finite collections of distinct indices $i_1, \ldots, i_n \in I$ and all collections of signs $\theta_1, \ldots, \theta_n \in \{-1, 1\}$, where the product θx of a sign $\theta \in \{-1, 1\}$ by an element $x \in \mathcal{B}$ is defined by setting $1x = x$ and $-1x = -x = \mathbf{1} - x$.

To formulate the exact result, we need some definitions. Our terminology is standard; see, for example, [4] or [5]. *Zero* $\mathbf{0}$ and *unit* $\mathbf{1}$ of a Boolean algebra $\mathcal{B}$ we write in bold to distinguish them from the corresponding numbers.

Definition 21.1. Let $\mathcal{A}$ and $\mathcal{B}$ be Boolean algebras. A Boolean isomorphism $S : \mathcal{A} \to \mathcal{B}$ is called a *Boolean σ-isomorphism* (or *Boolean τ-isomorphism*) provided S and S^{-1} are order σ-continuous (resp., *order τ-continuous*).

Definition 21.2. Sequences $\mathcal{X} = (x_n)_{n=1}^{\infty}$ and $\mathcal{Y} = (y_n)_{n=1}^{\infty}$ (or transfinite sequences $\mathcal{X} = (x_\alpha)_{\alpha<\omega_\delta}$ and $\mathcal{Y} = (y_\alpha)_{\alpha<\omega_\delta}$) in Boolean algebras $\mathcal{A}$ and $\mathcal{B}$, respectively, are said to be *σ-equivalent* if there is a Boolean σ-isomorphism $S : \mathcal{A}_{\mathcal{X}} \to \mathcal{B}_{\mathcal{Y}}$ between the minimal σ-complete subalgebras $\mathcal{A}_{\mathcal{X}}$ and $\mathcal{B}_{\mathcal{Y}}$ of $\mathcal{A}$ and $\mathcal{B}$ containing $\mathcal{X}$ and $\mathcal{Y}$, respectively, such that $S(x_n) = y_n$ for all $n \in \mathbb{N}$ (or $S(x_\alpha) = y_\alpha$ for all $\alpha < \omega_\delta$).

Next, we recall the well-known definition of an independent family and introduce a new notion of a vanishing family.

Definition 21.3. An infinite family $\mathcal{R} = (r_i)_{i\in I}$ in a Boolean algebra $\mathcal{B}$ is called:
(1) *independent* if $\bigwedge_{j\in J} \theta_j r_j \neq \mathbf{0}$ for any finite subset $J \subseteq I$ and any collection of signs $\theta_j = \pm 1$, $j \in J$;
(2) *vanishing* if $\bigwedge_{j\in J} \theta_j r_j = \mathbf{0}$ for any infinite subset $J \subseteq I$ and any collection of signs $\theta_j = \pm 1$, $j \in J$.

Finite meets $\bigwedge_{j\in J} \theta_j r_j$ presenting in (1) are called *particles*[1] of $\mathcal{R}$.

So, a family $\mathcal{R}$ is independent provided all its particles are nonzero. Evidently, every subfamily of an independent family is an independent family.

Remark 21.4. If transfinite sequences $\mathcal{X} = (x_\alpha)_{\alpha<\omega_\delta}$ and $\mathcal{Y} = (y_\alpha)_{\alpha<\omega_\delta}$, at least one of which is independent (or vanishing), are σ-equivalent then the other one is independent (resp., vanishing) as well, and there is a unique Boolean σ-isomorphism $S : \mathcal{A}_\mathcal{X} \to \mathcal{B}_\mathcal{Y}$ such that $S(x_\alpha) = y_\alpha$ for all $\alpha < \omega_\delta$.

Throughout the chapter, we reserve the notation $\widehat{\mathcal{B}}$ for the σ-complete Boolean algebra of all equivalence classes of Borel subsets of $[0,1]$ *with respect to the Lebesgue measure, which will be frequently used in different contexts.*

Let $I_n^k = [\frac{k-1}{2^n}, \frac{k}{2^n})$ be the dyadic intervals, $n = 0, 1, 2, \ldots$, $k = 1, \ldots, 2^n$. We set

$$\boldsymbol{r}_n = \bigsqcup_{j=1}^{2^{n-1}} I_n^{2j-1}, \quad n = 1, 2, \ldots. \tag{21.2}$$

By the *usual Rademacher family* on $[0,1)$, we mean the sequence $\widehat{\mathcal{R}} = (\widehat{\boldsymbol{r}}_n)_{n\in\mathbb{N}}$ of the co-sets of $\boldsymbol{r}_n$ in the Boolean algebra $\widehat{\mathcal{B}}$. The equivalence classes in $\widehat{\mathcal{B}}$ containing I_n^k will be denoted by $\widehat{I}_n^k$.

A subalgebra $\mathcal{A}$ of a Boolean algebra $\mathcal{B}$ is said to be *order closed* (resp., *σ-order closed*) if for every (resp., countable) subset $C \subseteq \mathcal{A}$ the existence of $\sup C \in \mathcal{B}$ implies $\sup C \in \mathcal{A}$. Another equivalent definition contains an additional assumption on C to be upwards directed (for this and other equivalences see [4, 313E]).

For any $\mathcal{A} \subseteq \mathcal{B}$, we denote
- $\mathcal{B}(\mathcal{A})$ the smallest subalgebra of $\mathcal{B}$ including $\mathcal{A}$;
- $\mathcal{B}_\sigma(\mathcal{A})$ the smallest σ-order closed subalgebra of $\mathcal{B}$ including $\mathcal{A}$;
- $\mathcal{B}_\tau(\mathcal{A})$ the smallest order closed subalgebra of $\mathcal{B}$ including $\mathcal{A}$.

Anyway, $\mathcal{B}(\mathcal{A}) \subseteq \mathcal{B}_\sigma(\mathcal{A}) \subseteq \mathcal{B}_\tau(\mathcal{A})$, and if $\mathcal{B}$ possesses the countably chain condition (see preliminaries for the definition) then $\mathcal{B}_\sigma(\mathcal{A}) = \mathcal{B}_\tau(\mathcal{A})$ for all subsets $\mathcal{A} \subseteq \mathcal{B}$ [4, 331G].

---

1 *$\mathcal{R}$-atomic elements* in standard terminology, which is inconvenient for our purpose; see, for example, Theorem 21.17.

The following folklore theorem has a standard proof (it is stated as Exercise 6 of Section 9, [6, p. 139]).

Theorem 21.5. *Let $\mathcal{R} = (r_i)_{i\in I}$ be an independent family in a Boolean algebra $\mathcal{B}$. Then there is a unique finitely additive measure $\mu^* : \mathcal{B}(\mathcal{R}) \to [0,1]$ satisfying* (21.1).

Definition 21.6. Let $\mathcal{R} = (r_i)_{i\in I}$ be an independent family in a Boolean algebra $\mathcal{B}$. The finitely additive measure $\mu^* : \mathcal{B}(\mathcal{R}) \to [0,1]$ satisfying (21.1) is called the *dyadic measure* on the subalgebra $\mathcal{B}(\mathcal{R})$ generated by $\mathcal{R}$.

At first glance, it is a striking fact that in the most natural cases the dyadic measure on $\mathcal{B}(\mathcal{R})$ is not σ-additive. Indeed, we show that the restriction $\mu_0 = \mu|_{\widehat{\mathcal{B}}(\widehat{\mathcal{R}})}$ of the Lebesgue measure μ, which is σ-additive on $\widehat{\mathcal{B}}$, to the subalgebra $\widehat{\mathcal{B}}(\widehat{\mathcal{R}})$ generated by the usual Rademacher family $\widehat{\mathcal{R}}$, is not σ-additive. To do this, we provide an example of a sequence $(x_n)_{n=1}^{\infty}$ in $\widehat{\mathcal{B}}(\widehat{\mathcal{R}})$ with $x_{n+1} \le x_n$ for all n, $\inf_n x_n = \mathbf{0}$ in $\widehat{\mathcal{B}}(\widehat{\mathcal{R}})$ (but not in $\widehat{\mathcal{B}}$) with $\mu(x_n) \ge 1/2$ for all n, which is enough by the well known and easily proved fact [4, 326 F(c)]. Let $(I_n)_{n=1}^{\infty}$ be any numeration of the dyadic intervals. For all $n \in \mathbb{N}$, choose a dyadic interval $I_{k_n} \subseteq I_n$ of measure $\mu(I_{k_n}) \le 2^{-n-1}$ and set $x_m = [0,1) \setminus \bigcup_{n=1}^{m} I_{k_n}$. Then $x_{n+1} \le x_n$ for all n and

$$\mu(x_m) \ge 1 - \sum_{n=1}^{m} \mu(I_{k_n}) \ge 1 - \sum_{n=1}^{m} \frac{1}{2^{n+1}} > 1 - \frac{1}{2} = \frac{1}{2}.$$

Prove that $\inf_n x_n = \mathbf{0}$ in $\widehat{\mathcal{B}}(\widehat{\mathcal{R}})$. Let $z \in \widehat{\mathcal{B}}(\widehat{\mathcal{R}})$ be a lower bound for $\{x_n : n \in \mathbb{N}\}$. Assume on the contrary that $z > 0$. Then by Corollary 21.14 below (which is a well-known fact), there exists a dyadic interval I_m such that $I_m \le z$, and hence $I_{k_m} \subseteq I_m \le z \le x_m$, which contradicts the choice of x_m.

Definition 21.7. An independent family $\mathcal{R}$ in a Boolean algebra $\mathcal{B}$ is said to be *injective* if for every disjoint sequence of particles (p_n) of $\mathcal{R}$ the following implication holds:

$$\bigvee_{n=1}^{\infty} p_n = \mathbf{1} \quad \Longrightarrow \quad \sum_{n=1}^{\infty} \mu^*(p_n) = 1,$$

where μ^* is the dyadic measure on $\mathcal{B}(\mathcal{R})$.

21.1.3 The main results

Next is our first main result.

Theorem A. *Let (r_n) be a σ-generating sequence of elements of a σ-complete Boolean algebra $\mathcal{B}$. Then the following assertions are equivalent:*

(1) *There is a positive σ-additive probability measure on $\mathcal{B}$ possessing* (21.1).
(2) *There is a Boolean σ-isomorphism $J : \widehat{\mathcal{B}} \to \mathcal{B}$ such that $J(\widehat{r}_n) = r_n$ for all $n \in \mathbb{N}$.*

(3) (r_n) *is a vanishing injective independent family (i. e., a Rademacher family, see Definition* 21.27 *below).*

Here, our contribution is item (3), and the core of the proof is Theorem 21.23. The difficulty lies in the fact that (3) only provides information on certain supremum/infimum operation on particles, and one needs to extend that to the rest. The key property of the standard algebra $\widehat{\mathcal{B}}$ (as well as any other measurable algebra) which makes it possible is that every element of $\widehat{\mathcal{B}}$ is the infimum of suprema of particles (in general, σ-generation could require more iterations of countable suprema and infima of the generators).

Then we easily generalize Theorem A to transfinite sequences of any cardinality in Theorem 21.31. Corollary 21.32 asserts that any two Rademacher families of the same cardinality are σ-equivalent in any of their permutations.

Since a σ-generating Rademacher family defines a positive measure on a σ-complete Boolean algebra by (21.1), a σ-complete Boolean algebra admitting a σ-generating Rademacher family must be a measurable algebra. Moreover, the existence of a σ-generating Rademacher family is a necessary and sufficient condition on a Boolean algebra to be measurable and Maharam homogeneous.

For convenience of the reader, we include special terms, introduced in the chapter, to the Subject index.

Preliminaries

The order $x \leq y$ on a Boolean algebra $\mathcal{B}$ is defined to be equivalent to the equality $x \wedge y = x$, which in turn is equivalent to $x \vee y = y$. So, $x \vee y = \sup\{x, y\}$ and $x \wedge y = \inf\{x, y\}$ with respect to this order. The relation $\subseteq$ is used for subsets and $\leq$ is used for elements of a Boolean algebra. The join and the meet of an infinite subset $\mathcal{A} \subseteq \mathcal{B}$ is defined by $\bigvee \mathcal{A} = \sup \mathcal{A}$ and $\bigwedge \mathcal{A} = \inf \mathcal{A}$ with respect to the order $\leq$ only if the corresponding supremum or infimum exists. By a *subalgebra* of $\mathcal{B}$, we mean any subset of $\mathcal{B}$ containing $\mathbf{1}$ which is itself a Boolean algebra with the induced Boolean algebra structure. A subset $\mathcal{A}$ of a Boolean algebra $\mathcal{B}$ is said to be *disjoint* provided $x \wedge y = \mathbf{0}$ for all distinct $x, y \in \mathcal{A}$. By a *partition* (of unity) in a Boolean algebra $\mathcal{B}$ we mean a maximal disjoint subset $\mathcal{A} \subseteq \mathcal{B}$, that is, $(\forall x \in \mathcal{B})\,((\forall a \in \mathcal{A}\;\; a \wedge x = \mathbf{0}) \Rightarrow (x = \mathbf{0}))$. A *disjoint join* $\bigvee \mathcal{A}$ (i. e., the join of a disjoint system $\mathcal{A} \subseteq \mathcal{B}$), if exists, is denoted by $\bigsqcup \mathcal{A}$. Although in some cases an infinite join in a Boolean algebra does not exist, it is immediate that if $\mathcal{A}$ is a partition then $\bigsqcup \mathcal{A} = \mathbf{1}$ exists. Conversely, if $\bigsqcup \mathcal{A} = \mathbf{1}$ then $\mathcal{A}$ is a partition. A Boolean algebra $\mathcal{B}$ is said to have the *countable chain condition* (ccc, in short) if any disjoint subset $\mathcal{A} \subseteq \mathcal{B}$ is, at most, countable. A Boolean algebra $\mathcal{B}$ is called *measurable* if $\mathcal{B}$ is a σ-complete Boolean algebra and there is a finite positive σ-additive measure on $\mathcal{B}$ (by a *positive* measure we mean a strictly positive measure μ, that is, $\mu(x) > 0$ for all $x \in \mathcal{B} \setminus \{\mathbf{0}\}$). Obviously, every measurable Boolean algebra has the ccc.

A Boolean algebra $\mathcal{B}$ is said to be complete (resp., σ-complete) if every nonempty (resp., every nonempty countable) subset of $\mathcal{B}$ has a supremum (equivalently, infimum).

Given any $\mathcal{A} \subseteq \mathcal{C} \subseteq \mathcal{B}$, we say that
- $\mathcal{A}$ σ-generates $\mathcal{C}$ if $\mathcal{C} = \mathcal{B}_\sigma(\mathcal{A})$;
- $\mathcal{A}$ τ-generates $\mathcal{C}$ if $\mathcal{C} = \mathcal{B}_\tau(\mathcal{A})$.

We define the *density* $\operatorname{dens} \mathcal{B}$ of a Boolean algebra $\mathcal{B}$ to be the smallest cardinality of subsets $\mathcal{A} \subseteq \mathcal{B}$ that τ-generate $\mathcal{B}$. The *density* $\operatorname{dens} e$ of a nonzero element $e \in \mathcal{B}$ is defined to be the density of the Boolean algebra $\mathcal{B}_e = \{x \in \mathcal{B} : x \le e\}$ with operations induces by $\mathcal{B}$ and unit e. In particular, $\operatorname{dens} \mathbf{1} = \operatorname{dens} \mathcal{B}$. We say that a Boolean algebra $\mathcal{B}$ is *Maharam homogeneous* if for every $e \in \mathcal{B} \setminus \{0\}$ we have $\operatorname{dens} e = \operatorname{dens} \mathcal{B}$.

Let ω_δ be an arbitrary infinite cardinal, μ_{ω_δ} the Haar measure on the σ-algebra Σ_{ω_δ} of subsets of $\{-1, 1\}^{\omega_\delta}$ considered as a compact Abelian group, $\widehat{\Sigma}_{\omega_\delta}$ the quotient Boolean algebra of Σ_{ω_δ} modulo μ_{ω_δ}-null sets. The quotient map from Σ_{ω_δ} to $\widehat{\Sigma}_{\omega_\delta}$ we denote by Co. By the *generalized Rademacher family* $(\overline{r}_\alpha)_{\alpha<\omega_\delta}$ in $\widehat{\Sigma}_{\omega_\delta}$, we mean the co-sets of the following sets: $\overline{r}_\alpha = \operatorname{Co}\{x \in \{-1, 1\}^{\omega_\delta} : x(\alpha) = 1\}$.

By a *semialgebra* in a Boolean algebra $\mathcal{B}$, we mean a subset $P \subseteq \mathcal{B}$ possessing the following properties:
(1) $\mathbf{0}, \mathbf{1} \in P$;
(2) if $a, b \in P$ then $a \wedge b \in P$;
(3) if $a_1, b \in P$ with $a_1 \subset b$ then there are $n \in \mathbb{N}$ with $n > 1$ and $a_2, \dots, a_n \in P$ such that $b = \bigsqcup_{m=1}^n a_m$.

We will use the following description of the order closed subalgebra generated by a semialgebra.

Proposition 21.8. *Let $\mathcal{A}$ be a semialgebra in a Boolean algebra $\mathcal{B}$. Then:*
(1) *$\mathcal{B}(\mathcal{A})$ equals the set of all finite disjoint unions of elements of $\mathcal{A}$;*
(2) *$\mathcal{B}_\sigma(\mathcal{A})$ equals the set of all order limits of sequences from $\mathcal{B}(\mathcal{A})$*
(3) *$\mathcal{B}_\tau(\mathcal{A})$ equals the set of all order limits of nets from $\mathcal{B}(\mathcal{A})$.*

To prove item (1) of Proposition 21.8 is a standard technical exercise (see [2, Lemma 1.2.14]). Items (2) and (3) follow from (1) and the fact that the order closure of a subalgebra is a subalgebra [4, 313F(c)].

21.2 Auxiliary properties of independent families

The independence is an important property of a Rademacher family. However, as Theorem 21.21 shows, maximal independent families in the very natural Boolean algebras are not actually independent in the natural sense, because one of its elements belongs to the smallest order closed subalgebra generated by the rest of elements. In the present section, we analyze some properties of independent families that will be needed in the sequel. Perhaps, some of them are known.

21.2.1 The usual Rademacher family

Observe that the usual Rademacher family $\widehat{\mathcal{R}} = (\widehat{r}_n)_{n\in\mathbb{N}}$ possesses the following properties:

(R1) **Independence:** $\bigwedge_{j\in J} \theta_j \widehat{r}_j \neq \mathbf{0}$ for any finite subset $J \subset \mathbb{N}$ and any collection of signs $\theta_j = \pm 1$, $j \in J$;

(R2) **Vanishing:** $\bigwedge_{j\in J} \theta_j \widehat{r}_j = \mathbf{0}$ for any infinite subset $J \subseteq \mathbb{N}$ and any collection of signs $\theta_j = \pm 1$, $j \in J$;

(R3) **Irredundance:** for any $n_0 \in \mathbb{N}$ one has $\widehat{\mathcal{B}}_\tau((\widehat{r}_n)_{n\in\mathbb{N}\setminus\{n_0\}}) \neq \widehat{\mathcal{B}}_\tau(\widehat{\mathcal{R}})$;

(R4) **Injectivity:** For every disjoint sequence $(\widehat{I}_{n_j}^{k_j})_{j=1}^{\infty}$ the condition $\sup_j \widehat{I}_{n_j}^{k_j} = [0,1)$ in $\widehat{\mathcal{B}}$ implies $\sum_{j=1}^{\infty} 2^{-n_j} = 1$;

(R5) **σ-generation:** $\widehat{\mathcal{B}}_\sigma(\widehat{\mathcal{R}}) = \widehat{\mathcal{B}}$.

We introduce a Rademacher family in a Boolean algebra as a family of two-point partitions of unity to satisfy some of these properties. More precisely,

- (R1)&(R2)&(R3) determine a weak Rademacher family;
- (R1)&(R2)&(R4) determine a Rademacher family;
- (R1)&(R2)&(R4)&(R5) determine a σ-generating Rademacher family

We prove that a Rademacher family is a weak Rademacher family; however, the converse is not true.

We define the *usual Rademacher family supported on a fixed dyadic interval* $\widehat{I}_m^j$, $m \in \mathbb{N}$, $j \in \{1, \ldots, 2^m\}$, to be the sequence $\widehat{r}_n{}' = \widehat{r}_{m+n} \wedge \widehat{I}_m^j$, $n = 1, 2, \ldots$. Note that the usual Rademacher family supported on any dyadic interval has properties (R1), (R2), and (R3) and does not have (R5) if the dyadic interval is not $[0,1)$.

We provide below with examples showing that none of properties (R1)–(R3) follows from the rest ones even for the Boolean algebra $\widehat{\mathcal{B}}$.

Example 21.9. There is a sequence $\mathcal{R} = (r_n)_{n=1}^{\infty}$ in $\widehat{\mathcal{B}}$ satisfying (R2), (R3), and (R5) and failing (R1).

Proof. Let $(\widehat{r}_n{}')_{n\in\mathbb{N}}$ and $(\widehat{r}_n{}'')_{n\in\mathbb{N}}$ be the usual Rademacher families on $[0, \frac{1}{2})$ and $[\frac{1}{2}, 1)$, respectively. We define a sequence $(r_n)_{n\in\mathbb{N}}$ in $\widehat{\mathcal{B}}$ by setting $r_{2k-1} = \widehat{r}_k{}'$ and $r_{2k} = \widehat{r}_k{}''$ for $k = 1, 2, \ldots$. Then $\mathcal{R} = (r_n)_{n=1}^{\infty}$ has the desired properties. □

Example 21.10. There is a sequence $\mathcal{R} = (r_n)_{n=1}^{\infty}$ in $\widehat{\mathcal{B}}$ satisfying (R1), (R3), and (R5) and failing (R2).

Proof. Let $(\widehat{r}_n{}')_{n\in\mathbb{N}}$ and $(\widehat{r}_n{}'')_{n\in\mathbb{N}}$ be the usual Rademacher families on $[0,\frac{1}{2})$ and $[\frac{1}{2},1)$ respectively. We define a sequence $(r_n)_{n\in\mathbb{N}}$ in $\widehat{\mathcal{B}}$ by setting $r_{2k-1} = \widehat{[0,\frac{1}{2})} \sqcup \widehat{r}_{2k-1}{}''$ and $r_{2k} = \widehat{r}_k{}' \sqcup \widehat{r}_{2k}{}''$ for $k = 1,2,\ldots$. Then $\mathcal{R} = (r_n)_{n\in\mathbb{N}}$ satisfies (R1). Indeed, for any $J \subseteq \mathbb{N}$ and any collection of signs $\theta_j = \pm 1$, $j \in J$, one has

$$\bigwedge_{j\in J} \theta_j r_j \geq \bigwedge_{j\in J} \Big(\widehat{\Big[\frac{1}{2},1\Big)} \wedge \theta_j r_j \Big) = \widehat{\Big[\frac{1}{2},1\Big)} \wedge \bigwedge_{j\in J} \theta_j \widehat{r}_j{}'' \neq \mathbf{0}.$$

$\mathcal{R}$ does not satisfy (R2) because $\bigwedge_{k=1}^{\infty} r_{2k-1} = \widehat{[1,\frac{1}{2})} \neq \mathbf{0}$.
$\mathcal{R}$ satisfies (R3) because $(\widehat{r}_n{}'')_{n\in\mathbb{N}}$ satisfies (R3).
(R4) for $\mathcal{R}$ follows from the observation that $\widehat{\mathcal{B}}_\tau(\mathcal{R})$ contains every dyadic interval. □

Example 21.11. There is a sequence $\mathcal{R} = (r_n)_{n=1}^{\infty}$ in $\widehat{\mathcal{B}}$ satisfying (R1), (R2), and (R5) and failing (R3).

The existence of a family satisfying the claims of Example 21.11 is not so obvious and follows from Theorem 21.21 below.

21.2.2 Particle semialgebra

The following proposition has a standard proof.

Proposition 21.12. *Let $\mathcal{R}$ be an independent family in a Boolean algebra $\mathcal{B}$ and $\mathcal{P}$ the set of all particles of $\mathcal{R}$. Then $\overline{\mathcal{P}} = \mathcal{P} \cup \{\mathbf{0}\}$ is a semialgebra.*

Definition 21.13. Let $\mathcal{R}$ be an independent family in a Boolean algebra $\mathcal{B}$ with the set $\mathcal{P}$ of all particles. The semialgebra $\overline{\mathcal{P}} = \mathcal{P} \cup \{\mathbf{0}\}$ is called the *particle semialgebra* of the family $\mathcal{R}$.

Item (i) of Proposition 21.8 and Proposition 21.12 imply the following simple but very useful statement (see also [5, p. 81, Theorem 2]).

Corollary 21.14. *Let $\mathcal{R}$ be an independent family in a Boolean algebra $\mathcal{B}$. Then the subalgebra $\mathcal{B}(\mathcal{R})$ of $\mathcal{B}$ generated by $\mathcal{R}$ equals the set of all disjoint joins of particles of $\mathcal{R}$.*

21.2.3 Atoms of the order closed subalgebra generated by an independent family

Recall that a nonzero element a of a Boolean algebra $\mathcal{B}$ is called an *atom* if for any $x \in \mathcal{B}$ the inclusion $x \subseteq a$ implies that either $x = \mathbf{0}$ or $x = a$.

Definition 21.15. An independent family $\mathcal{R}$ in a Boolean algebra $\mathcal{B}$ is said to be:
- *σ-atomless* if $\mathcal{B}_\sigma(\mathcal{R})$ is atomless;
- *τ-atomless* if $\mathcal{B}_\tau(\mathcal{R})$ is atomless.

If, in addition, $\mathcal{B}$ possesses the ccc or is σ-complete, we say *atomless* for both σ- and τ-versions.[2]

Proofs of the following observations are straightforward.

Remark 21.16.
(1) The generalized Rademacher family $(\overline{r}_\gamma)_{\gamma<\omega_\alpha}$ in $\widehat{\Sigma}_{\omega_\alpha}$ is an atomless σ-generating independent family.
(2) The above properties of an independent family are preserved under a Boolean σ-isomorphism or Boolean τ-isomorphism, depending on each case (see Definition 21.1).

As we will see later, a subsequence of an atomless countable independent family need not be atomless (see item (1) of Remark 21.18).

Theorem 21.17. *Let $\mathcal{R} = (r_i)_{i\in I}$ be an infinite independent family in a Boolean algebra $\mathcal{B}$. Then:*
(1) *for every nonzero element a of $\mathcal{B}_\tau(\mathcal{R})$ the following assertions are equivalent*
   (a) *a is an atom in $\mathcal{B}_\tau(\mathcal{R})$;*
   (b) *there is a collection of signs $(\overline{\theta}_i)_{i\in I}$ such that $a = \bigwedge_{i\in I} \overline{\theta}_i r_i$ in $\mathcal{B}_\tau(\mathcal{R})$.*
(2) *the following assertions are equivalent*
   (a) *$\mathcal{B}_\tau(\mathcal{R})$ is atomless;*
   (b) *for every collection of signs $\theta_i = \pm 1$ one has that either $\bigwedge_{i\in I} \theta_i r_i = \mathbf{0}$ or $\bigwedge_{i\in I} \theta_i r_i$ does not exist.*

Proof. Observe that (2) is a direct consequence of (1). So, we prove (1). Let $\mathbf{0} < a \in \mathcal{B}_\tau(\mathcal{P})$.

(b) $\Rightarrow$ (a). Suppose $a = \bigwedge_{i\in I} \overline{\theta}_i r_i$ in $\mathcal{B}_\tau(\mathcal{R})$. Observe that $\mathcal{A} = \{z \in \mathcal{B} : a \le z \text{ or } a \le -z\}$ is a τ-closed subalgebra of $\mathcal{B}$ containing $\mathcal{R}$, and so, $\mathcal{B}_\tau(\mathcal{R}) \subseteq \mathcal{A}$. Hence, for every $x \in \mathcal{B}_\tau(\mathcal{R})$ with $x \le a$ one has that either $x \ge a$ (and so, $x = a$) or $x \le -a$ (and so, $x = \mathbf{0}$). Thus, a is an atom in $\mathcal{B}_\tau(\mathcal{R})$.

(a) $\Rightarrow$ (b). Let a be an atom in $\mathcal{B}_\tau(\mathcal{R})$. Fix any $i \in I$. Since $(r_i, -r_i)$ is a partition, either $a \wedge r_i \ne \mathbf{0}$ or $a \wedge -r_i \ne \mathbf{0}$. Therefore, since a is an atom, either $a \le r_i$ or $a \le -r_i$. Set $\overline{\theta}_i = 1$ if $a \le r_i$ and $\overline{\theta}_i = -1$ if $a \le -r_i$. Thus, signs $(\overline{\theta}_i)_{i\in I}$ are chosen so that $(\forall i \in I)\ \ a \le \overline{\theta}_i r_i$, that is, a is a lower bound for $\{\overline{\theta}_i r_i : i \in I\}$. Show that $a = \bigwedge_{i\in I} \overline{\theta}_i r_i$ (in particular, we show that the meet exists). Assume $x \in B_\tau(\mathcal{R})$ is any lower bound for $\{\overline{\theta}_i r_i : i \in I\}$.

2 remind that for ccc algebras one has $\mathcal{B}_\sigma(\mathcal{R}) = \mathcal{B}_\tau(\mathcal{R})$.

Our goal is to prove that $x \le a$. Observe that $\mathcal{A}' = \{z \in \mathcal{B} : a \vee x \le z \text{ or } a \vee x \le -z\}$ is a τ-closed subalgebra of $\mathcal{B}$ containing $\mathcal{R}$, and so, $\mathcal{B}_\tau(\mathcal{R}) \subseteq \mathcal{A}'$. Hence, $a \in \mathcal{A}'$. Since $a \vee x \le -a$ is false, we obtain that $a \vee x \le a$, which yields $x \le a$. □

Remark 21.18. Example 21.10 shows that:
(1) a subsequence of an atomless countable independent family need not be atomless;
(2) one cannot equivalently extend the claim of item (2)(b) in Theorem 21.17 to any infinite intersection $\bigwedge_{j\in J} \theta_j r_j = \mathbf{0}$ with $J \subsetneq I$ as far as in (R2).

The following example shows that the last possibility in item (b) of (2) in Theorem 21.17 that $\bigwedge_{i\in I} \theta_i r_i$ does not exist, can sometimes happen.

Example 21.19. There exist a Boolean algebra $\mathcal{B}_0$ and an independent family $(s_n)_{n=1}^\infty$ in $\mathcal{B}$ with the following properties:
(i) every subsequence of $(s_n)_{n=1}^\infty$ is an atomless independent family;
(ii) the meet $\bigwedge_{n\in M} s_n$ does not exist for every infinite subset $M \subseteq \mathbb{N}$.

Proof. Let $\mathcal{B}_0$ be the subalgebra of $\widehat{\mathcal{B}}$ generated by the usual Rademacher family $(\widehat{r}_n)_{n=1}^\infty$. Fix any irrational number $\alpha \in (0,1)$ and choose a sequence $(D_n)_{n=1}^\infty$ of intervals $[a,b) \subseteq [\alpha,1)$, $a < b$ such that:
(1) $[\alpha,1) = \bigsqcup_{k=0}^{2^n-1} D_{2^n+k}$ for all $n = 0,1,\ldots$;
(2) $D_{2^n+k} = D_{2^{n+1}+2k-1} \sqcup D_{2^{n+1}+2k}$ for all $n = 0,1,\ldots$ and $k = 0,\ldots,2^n-1$;
(3) the endpoints of D_n are dyadic numbers or α, ordered in such a way that $\alpha = \min D_{2^n}$, $\sup D_{2^n+k} = \min D_{2^n+k+1}$ and $\sup D_{2^{n+1}-1} = 1$ for every $n = 0,1,\ldots$ and $0 \le k \le 2^n - 2$;
(4) $\lim_n \max_{0\le k<2^n} \mu(D_{2^n+k}) = 0$.

Then set

$$s_n = \widehat{[0,\alpha)} \sqcup \bigsqcup_{j=0}^{2^{n-1}-1} \widehat{D}_{2^n+2j}, \quad n = 1,2,\ldots.$$

Since $\alpha = \min D_{2^n}$, the union $\widehat{[0,\alpha)} \sqcup \widehat{D}_{2^n}$ is a dyadic interval, and so $s_n \in \mathcal{B}_0$ for all $n \in \mathbb{N}$. By (1)–(2), $(s_n)_{n=1}^\infty$ is an independent family. First, we prove (ii). Let $M \subseteq \mathbb{N}$ be an infinite subset. Let $0 \le z \in \mathcal{B}_0$ be any lower bound for $\{s_n : n \in M\}$. Since $z \in \mathcal{B}_0$, one has that $z = \bigsqcup_{j\in J} \widehat{I}_n^j$ for suitable $n \in \mathbb{N}$ and $J \subseteq \{0,\ldots,2^n-1\}$. By (4), $I_n^j \subseteq [0,\alpha)$ for all $j \in J$. Since α is irrational and J is finite, there exists a dyadic number $k/2^m$ with

$$\max_{j\in J} \max I_n^j < \frac{k}{2^m} < \alpha.$$

Thus, $z < \widehat{[0, k/2^m)}$ and $\widehat{[0, k/2^m)}$ is a lower bound for $\{s_n : n \in M\}$ in $\mathcal{B}_0$ which is greater than z, and so (ii) is proved. Finally, (i) follows from (ii) and item (2) of Theorem 21.17.

□

21.2.4 Maximal independent families

Definition 21.20. An independent family $\mathcal{R}$ in a Boolean algebra $\mathcal{B}$ is said to be *maximal* if there is no independent family in $\mathcal{B}$ including $\mathcal{R}$, but $\mathcal{R}$ itself.

Using Zorn's lemma, one can easily prove that every independent family can be extended to a maximal independent family. However, the maximality is a bad property if one wants to define a measure by an independent family. To show this, we need the following theorem, mainly due to Rudin [10] (see also [3, 134J(b)]).

Theorem 21.21. *The usual Rademacher family* $(\widehat{r}_n)_{n=1}^{\infty}$ *is not maximal. Moreover, for every* $\gamma \in (0, 1)$ *there exists an element* $\widehat{r}_0 \in \widehat{\mathcal{B}}$ *of measure* $\mu(\widehat{r}_0) = \gamma$ *such that* $(\widehat{r}_n)_{n=0}^{\infty}$ *is an independent family in* $\widehat{\mathcal{B}}$.

Actually, it is proved in the cited literature the existence of a measurable subset $A \subseteq [0, 1]$ such that for every open interval $I \subseteq [0, 1]$ one has $\mu(A \cap I) > 0$ and $\mu(I \setminus A) > 0$. However, there is a direct argument to get such a set A_γ with $\mu(A_\gamma) = \gamma$. For every $t \in (0, 1]$, consider the function $\phi_t : [0, 1] \to [0, 1]$ given by $\phi_t(x) = x^t$. By the Lebesgue dominated convergence theorem, the function $f(t) = \mu(\phi_t(A)) = \int_A tx^{t-1} dt$ is continuous and satisfies $\lim_{t \to 0} f(t) = 0$ and $f(1) = \mu(A)$. It follows that $f(t)$ takes all values $\gamma \in (0, \mu(A)]$, and hence the set $A_\gamma = \phi_t(A)$ is as desired. Applying the same argument to the complement $B = [0, 1] \setminus A$, we also get sets with any measure from the interval $[\mu(A), 1)$.

We remark that the constructed above extended independent family $(\widehat{r}_n)_{n=0}^{\infty}$ cannot define a countably additive measure on the Borel σ-algebra $\widehat{\mathcal{B}}$ by (21.1) if $\gamma \neq 1/2$. Indeed, if such a measure $\widehat{\mu}$ existed, on the one hand, (21.1) would imply that $\widehat{\mu}(\widehat{r}_0) = 1/2$. But on the other hand, $\widehat{\mu}$ must coincide with the Lebesgue measure on $\widehat{\mathcal{B}}$ because both measures have the same values at dyadic intervals. Hence, $\widehat{\mu}(\widehat{r}_0) = \gamma$, a contradiction.

21.3 Sequences σ-equivalent to the dyadic tree of intervals

In this section, we find necessary and sufficient conditions on a sequence in a σ-complete Boolean algebra to be σ-equivalent to the dyadic tree of intervals on the real line. This gives an essential step in the proof of the main result.

Recall some notation: $\widehat{\mathcal{B}}$ is the quotient algebra modulo measure null sets of the Borel σ-algebra $\mathcal{B}$ on $[0, 1)$; $I_n^k = [\frac{k-1}{2^n}, \frac{k}{2^n})$ are the dyadic intervals, $n = 0, 1, 2, \ldots$, $k =$

$1, \ldots, 2^n$, and $\widehat{I}_n^k$ the element of $\widehat{\mathcal{B}}$ containing I_n^k; more general, $\widehat{I}$ denotes the element of $\widehat{\mathcal{B}}$ containing $I \in \mathcal{B}$.

For convenience, we introduce a new notation: $\widehat{b}_n = \widehat{I}_k^\ell$, where $n = 2^k + \ell$ with $k \in \mathbb{N}$, $\ell \in \{0, \ldots, 2^k - 1\}$, so the dyadic intervals $(\widehat{b}_n)_{n=1}^\infty$ possess the following property: $\widehat{b}_n = \widehat{b}_{2n} \sqcup \widehat{b}_{2n+1}$ for all $n \in \mathbb{N}$. Denote by $\widehat{P}$ the semialgebra of $\widehat{\mathcal{B}}$ consisting of zero and all elements of the sequence $(\widehat{b}_n)_{n=1}^\infty$, and by $\widehat{\mathcal{B}}(\widehat{P})$ the smallest subalgebra of $\widehat{\mathcal{B}}$ containing all elements of $(\widehat{b}_n)_{n=1}^\infty$, that is, the set of all finite disjoint joins of elements of $(\widehat{b}_n)_{n=1}^\infty$.

Definition 21.22. Let $\mathcal{B}$ be a Boolean algebra. A sequence $(b_n)_{n=1}^\infty$ in $\mathcal{B}^+ := \{x \in \mathcal{B} : x \neq \mathbf{0}\}$ satisfying $b_n = b_{2n} \sqcup b_{2n+1}$ for all $n \in \mathbb{N}$, is said to be a *regular tree* in $\mathcal{B}$. A regular tree $(b_n)_{n=1}^\infty$ in $\mathcal{B}$ is said to be

- *vanishing* if for every subsequence $(b_{n_k})_{k=1}^\infty$ one has $\bigwedge_{k=1}^\infty b_{n_k} = \mathbf{0}$;
- *injective* provided that for every disjoint subsequence $(b_{n_k})_{k=1}^\infty$ the condition $\bigvee_{k=1}^\infty b_{n_k} = b_1$ implies $\sum_{k=1}^\infty 2^{-[\log_2 n_k]} = 1$.

Observe that the sequence $(\widehat{b}_n)_{n=1}^\infty$ of dyadic intervals is a vanishing injective regular tree by (R2) and (R4).

The following theorem is the main result of the section.

Theorem 21.23. *A regular tree $(b_n)_{n=1}^\infty$ in a σ-complete Boolean algebra $\mathcal{B}$ is σ-equivalent to the tree of dyadic intervals $(\widehat{b}_n)_{n=1}^\infty$ if and only if $(b_n)_{n=1}^\infty$ is vanishing and injective.*

Before we start the proof, observe that the injectivity of a sequence $(b_n)_{n=1}^\infty$ in a Boolean algebra $\mathcal{B}$ can be equivalently reformulated as follows: *for every disjoint subsequence $(b_{n_k})_{k=1}^\infty$, the condition* $\sup_k b_{n_k} = b_1$ *implies* $\sup_k \widehat{b}_{n_k} = \widehat{[0,1)}$, because the sequence $(\widehat{b}_{n_k})_{k=1}^\infty$ is disjoint as well, and for the disjoint dyadic intervals $(\widehat{b}_{n_k})_{k=1}^\infty$ the conditions $\sum_{k=1}^\infty 2^{-[\log_2 n_k]} = 1$ and $\sup_k \widehat{b}_{n_k} = \widehat{[0,1)}$ are equivalent (remark that the Lebesgue measure of $\widehat{b}_m$ equals $2^{-[\log_2 m]}$).

Proof. The "only if" part of the proof is clear from the definitions. We prove the "if" part. Our goal is to construct a function $J : \widehat{\mathcal{B}} \to \mathcal{B}$ such that the following conditions hold:

(1) $(\forall n \in \mathbb{N})\, J(\widehat{b}_n) = b_n$;
(2) $(\forall x, y \in \widehat{\mathcal{B}})\quad x \le y \;\to\; J(x) \le J(y)$;
(3) $(\forall x, y \in \widehat{\mathcal{B}})\quad J(x) \le J(y) \;\to\; x \le y$.

This is enough to prove the theorem, because if these were true then J would be an order preserving bijection with order preserving inverse, which is a Boolean isomorphism by [4, 312L] and is order σ-continuous by [4, 314F] in both directions.

Observe that every element p of $\widehat{\mathcal{B}}(\widehat{P})$ can be represented as follows:

$$p = \bigsqcup_{\ell \in A} \widehat{b}_{2^k+\ell}, \quad \text{where } k \in \mathbb{N} \text{ and } A \subseteq \{0, \ldots, 2^k - 1\}. \tag{21.3}$$

For every $p \in \widehat{\mathcal{B}}(\widehat{P})$ of form (21.3), we set

$$J_1(p) = \bigsqcup_{\ell \in A} b_{2^k+\ell}, \quad \text{where } k \in \mathbb{N} \text{ and } A \subseteq \{0, \ldots, 2^k - 1\}. \tag{21.4}$$

We omit a routine exercise to prove that the value of $J_1(p)$ defined by (21.4) is actually independent of the expansion p given by (21.3). In particular, we have that $J_1(\widehat{b}_n) = b_n$ for all $n \in \mathbb{N}$. Denote by $\mathcal{B}(P)$ the smallest subalgebra of $\mathcal{B}$ containing b_n for every $n \in \mathbb{N}$, which equals the set of all disjoint unions of elements of (b_n). So, we have defined a bijection $J_1 : \widehat{\mathcal{B}}(\widehat{P}) \to \mathcal{B}(P)$, which is order preserving (here we omit another routine procedure to prove that $p \le q$ for any $p, q \in \widehat{\mathcal{B}}(\widehat{P})$ implies $J_1(p) \le J_1(q)$). By [4, 312L], J_1 is a Boolean isomorphism, and by [4, 314F], J_1 is order σ-continuous.

Now we extend J_1 from $\widehat{\mathcal{B}}(\widehat{P})$ to $\widehat{\mathcal{B}}$. Let $\widehat{\mathcal{G}}$ be the set of all equivalence classes of open subsets of $[0, 1)$. By the well-known property of open subsets of $\mathbb{R}$, for every $g \in \widehat{\mathcal{G}}$ one has

$$g = \sup\{p \in \widehat{\mathcal{B}}(\widehat{P}) : p \le g\}, \tag{21.5}$$

and so we set

$$J(g) = \sup\{J_1(p) : p \in \widehat{\mathcal{B}}(\widehat{P}), p \le g\}. \tag{21.6}$$

Finally, let $x \in \widehat{\mathcal{B}}$ be any element. Since every measurable subset of $[0, 1)$ could be approximated by open sets from above, one has

$$x = \inf\{g \in \widehat{\mathcal{G}} : x \le g\}. \tag{21.7}$$

Thus, we set

$$J(x) = \inf\{J(g) : g \in \widehat{\mathcal{G}}, x \le g\}. \tag{21.8}$$

Since J is obviously order preserving on $\widehat{\mathcal{G}}$, the new definition of $J(g)$ for any $g \in \widehat{\mathcal{G}}$, given by (21.8), coincides with the old one, given by (21.6).

By (21.5), (21.7), and σ-completeness of $\mathcal{B}$, $J : \widehat{\mathcal{B}} \to \mathcal{B}$ is well-defined by (21.4), (21.6), and (21.8). Now we prove that J possesses the desired properties using several claims, the first of which is clear from the definitions.

Claim 1. *J is order preserving, that is, for every $x, y \in \widehat{\mathcal{B}}$, if $x \le y$ then $J(x) \le J(y)$.*

Claim 2. *Let $p \in \widehat{\mathcal{B}}(\widehat{P})$, $g', g'' \in \widehat{\mathcal{G}}$, and $p \le g' \vee g''$. Then there are sequences (p'_n) and (p''_n) in $\widehat{\mathcal{B}}(\widehat{P})$ such that $p'_n \le g'$ and $p''_n \le g''$ for all n, and $p'_n \vee p''_n \uparrow p$.*

Proof of Claim 2. Observe that, for every $g \in \widehat{\mathcal{G}}$ there exists a sequence (q_n) in $\widehat{\mathcal{B}}(\widehat{P})$ such that $q_n \uparrow g$. Then we choose sequences (q'_n) and (q''_n) in $\widehat{\mathcal{B}}(\widehat{P})$ with $q'_n \uparrow g'$ and $q''_n \uparrow g''$. Now set $p'_n = p \wedge q'_n$ and $p''_n = p \wedge q''_n$. Then $p'_n, p''_n \in \widehat{\mathcal{B}}(\widehat{P})$ for all $n \in \mathbb{N}$ and

$$p'_n \vee p''_n = (p \wedge q'_n) \vee (p \wedge q''_n) = p \wedge (q'_n \vee q''_n) \uparrow p \wedge (g' \vee g'') = p. \qquad \square$$

Claim 3. *Given any $g', g'' \in \widehat{\mathcal{G}}$, one has $J(g' \vee g'') \leq J(g') \vee J(g'')$.*

Proof of Claim 3. Fix any $p \in \widehat{\mathcal{B}}(\widehat{P})$ with $p \leq g' \vee g''$. Choose by Claim 2 sequences (p'_n) and (p''_n) in $\widehat{\mathcal{B}}(\widehat{P})$ such that $p'_n \leq g'$ and $p''_n \leq g''$ for all n, and $p'_n \vee p''_n \uparrow p$. Then for every $n \in \mathbb{N}$ one has

$$J_1(p'_n \vee p''_n) = J_1(p'_n) \vee J_1(p''_n) \leq J_1(g') \vee J_1(g''). \tag{21.9}$$

Since J_1 is order preserving and order continuous, $J_1(p'_n \vee p''_n) \uparrow J_1(p)$. Hence, by (21.9), $J_1(p) \leq J_1(g') \vee J_1(g'')$. By the arbitrariness of $p \in \widehat{\mathcal{B}}(\widehat{P})$, $J(g' \vee g'') \leq J(g') \vee J(g'')$. □

Claim 4. *For any $x, y \in \widehat{\mathcal{B}}$, one has $J(x \vee y) \leq J(x) \vee J(y)$.*

Proof of Claim 4. Suppose $g', g'' \in \widehat{\mathcal{G}}$, $x \leq g'$, $y \leq g''$. Then $x \vee y \leq g' \vee g''$, $g' \vee g'' \in \widehat{\mathcal{G}}$, and hence

$$J(x \vee y) \overset{\text{Claim 1}}{\leq} J(g' \vee g'') \overset{\text{Claim 3}}{\leq} J(g') \vee J(g'').$$

Then for a fixed $g' \in \widehat{\mathcal{G}}$ with $x \leq g'$ and every $g'' \in \widehat{\mathcal{G}}$ with $y \leq g''$, one has

$$J(x \vee y) - (J(g') - J(g'')) \leq J(g') \vee J(g'') - (J(g') - J(g'')) = J(g''). \tag{21.10}$$

Since $J(y) \leq J(g'')$, one has

$$J(x \vee y) - (J(g') - J(y)) \leq J(x \vee y) - (J(g') - J(g'')). \tag{21.11}$$

Now (21.10) and (21.11) imply

$$J(x \vee y) - (J(g') - J(y)) \leq J(g'').$$

By the arbitrariness of $g'' \in \widehat{\mathcal{G}}$, we obtain

$$J(x \vee y) - (J(g') - J(y)) \leq J(y),$$

which in turn gives

$$J(x \vee y) \leq J(g') \vee J(y).$$

Doing a similar step as above for every $g' \in \widehat{\mathcal{G}}$ with $x \leq g'$, we obtain $J(x \vee y) \leq J(x) \vee J(y)$. □

Claim 5. *For every $x \in \widehat{\mathcal{B}}$, the condition $J(x) = \mathbf{0}$ implies $x = \mathbf{0}$.*

Proof of Claim 5. Fix $x \in \widehat{\mathcal{B}}$ with $J(x) = \mathbf{0}$. Let $f \in \widehat{\mathcal{B}}$ be an equivalence class containing a closed subset of $[0, 1)$ such that $f \leq x$. Then $-x \leq -f$ and, therefore, $J(-x) \leq J(-f)$. Hence

$$b = J(x \vee -x) \overset{\text{Claim 3}}{\leq} J(x) \vee J(-x) = J(-x) \leq J(-f) \leq b.$$

Thus, $J(-f) = b$. Observe that $-f \in \widehat{\mathcal{G}}$. So, by the injectivity of $(b_n)_{n=1}^{\infty}$, $-f = [0,1)$, that is, $f = \mathbf{0}$. Since every measurable subset of $[0,1)$ is the supremum of an increasing sequence of closed subsets, this yields $x = \mathbf{0}$ by the arbitrariness of f. □

Claim 6. *If $g', g'' \in \widehat{\mathcal{G}}$ and $g' \wedge g'' = \mathbf{0}$, then $J(g') \wedge J(g'') = \mathbf{0}$.*

Proof of Claim 6. Observe that, by the distributivity laws, if $u_n, u, v_n, v \in \mathcal{B}_0$ for $n \in \mathbb{N}$, $u_n \uparrow u$, $v_n \uparrow v$ and $u_i \wedge v_j = \mathbf{0}$ for all i, j, then $u \wedge v = \mathbf{0}$.

Choose sequences $p'_n, p''_n \in \widehat{\mathcal{B}}(\widehat{P})$ such that $p'_n \leq g'$, $p''_n \leq g''$ for all $n \in \mathbb{N}$, $J(p'_n) \uparrow J(g')$ and $J(p''_n) \uparrow J(g'')$. Then $p'_n \wedge p''_n \leq g' \wedge g'' = \mathbf{0}$, and hence $p'_n \wedge p''_n = \mathbf{0}$ for all $n \in \mathbb{N}$. Since J_1 is a Boolean isomorphism, $J(p'_n) \wedge J(p''_n) = \mathbf{0}$ for all $n \in \mathbb{N}$. By the above observation, $J(g') \wedge J(g'') = \mathbf{0}$. □

Claim 7. *For every $g \in \widehat{\mathcal{G}}$, one has $J(g) \wedge J(-g) = \mathbf{0}$.*

Proof of Claim 7. With some abuse of notation, we will write with the same letters elements of the measure algebra and the canonical representatives of their equivalence class. Assume that $g \subseteq [0,1)$ is an open set. Observe that for every $p \in \widehat{\mathcal{B}}(\widehat{P})$ one has $p = \sup\{c \in \widehat{\mathcal{B}}(\widehat{P}) : \overline{c} \subseteq p\}$. Then by (21.6), we obtain

$$J(g) = \sup\{J(c) : c \in \widehat{\mathcal{B}}(\widehat{P}), \overline{c} \subseteq p\}. \tag{21.12}$$

Given any $c \in \widehat{\mathcal{B}}(\widehat{P})$ with $\overline{c} \subseteq p$, we find, by the normality, open sets g'_c and g''_c such that $\overline{c} \subseteq g'_c$, $-g \subseteq g''_c$ and $g'_c \wedge g''_c = \mathbf{0}$. By Claim 6, $J(g'_c) \wedge J(g''_c) = 0$, and hence

$$J(g'_c) \wedge J(-g) = \mathbf{0}. \tag{21.13}$$

Then

$$\begin{aligned} J(g) \wedge J(-g) &\overset{\text{by (21.12)}}{=} \sup\{J(c) \wedge J(-g) : c \in \widehat{\mathcal{B}}(\widehat{P}), \overline{c} \subseteq p\} \\ &\leq \sup\{J(g'_c) \wedge J(-g) : c \in \widehat{\mathcal{B}}(\widehat{P}), \overline{c} \subseteq p\} \overset{\text{by (21.13)}}{=} \mathbf{0}. \end{aligned}$$ □

Claim 8. *Let $x, y \in \widehat{\mathcal{B}}$. If $J(x) \leq J(y)$, then $x \leq y$.*

Proof of Claim 8. First, we prove the claim for the case where $g := y \in \widehat{\mathcal{G}}$. We show that $x \wedge -g = \mathbf{0}$. By Claim 1, $J(x \wedge -g) \leq J(x) \leq J(g)$ and $J(x \wedge -g) \leq J(-g)$. Hence,

$$J(x \wedge -g) \leq J(x) \wedge J(-g) \leq J(g) \wedge J(-g) \overset{\text{Claim 7}}{=} \mathbf{0}.$$

By Claim 5, $x \wedge -g = \mathbf{0}$.

Now let $x, y \in \widehat{\mathcal{B}}$ be arbitrary. Then for every $g \in \widehat{\mathcal{G}}$, if $y \leq g$ then by Claim 1, $J(x) \leq J(y) \leq J(g)$. By the above case, $x \leq g$. Thus, x is a lower bound for $\{g \in \widehat{\mathcal{G}} : g \geq y\}$. Since $y = \inf\{g \in \widehat{\mathcal{G}} : g \geq y\}$, we obtain $x \leq y$. □

It is left to resume that property (1) for J is clear from the definition of J, (2) is stated in Claim 1 and (3) is stated in Claim 8. □

21.4 Rademacher families

In this section, we introduce and analyze some new notions. We obtain that an independent family in a σ-complete Boolean algebra is hereditarily σ-atomless if and only if it is vanishing. Another natural property of an independent family is to be irredundant. This property appears to be strictly weaker that the injectivity for a vanishing independent family.

Definition 21.24. An independent family $\mathcal{R}$ in a Boolean algebra $\mathcal{B}$ is called *τ-irredundant* if for any $r \in \mathcal{R}$ one has that $r \notin \mathcal{B}_\tau(\mathcal{R} \setminus \{r\})$.

Simple examples (like a disjoint family of nonzero elements) show that an irredundant family need not be independent. On the other hand, by Theorem 21.21, there is an independent family which is not τ-irredundant.

Definition 21.25. An infinite independent family $(r_i)_{i\in I}$ in a Boolean algebra $\mathcal{B}$ is called *hereditarily σ-atomless* if every of its infinite subfamily is σ-atomless.

The following statement is a consequence of Theorem 21.17.

Proposition 21.26. *Let $\mathcal{B}$ be a Boolean algebra. Then the following assertions hold:*
(1) *Every vanishing independent family in $\mathcal{B}$ is hereditarily σ-atomless.*
(2) *If, moreover, $\mathcal{B}$ is σ-complete then the converse also holds: an infinite independent family in $\mathcal{B}$ is hereditarily σ-atomless if and only if it is vanishing.*

As Example 21.19 shows, a hereditarily σ-atomless independent family need not be vanishing. So, the σ-completeness assumption in (2) of Proposition 21.26 is essential.

In the final section, we will show that every injective vanishing independent family is τ-irredundant.

Definition 21.27. Let $\mathcal{B}$ be a Boolean algebra.
- A vanishing injective independent family in a Boolean algebra $\mathcal{B}$ is called a *Rademacher family* in $\mathcal{B}$.
- A vanishing τ-irredundant independent family in a Boolean algebra $\mathcal{B}$ is called a *weak Rademacher family* in $\mathcal{B}$.

As a direct application of the definitions, we obtain the following fact.

Proposition 21.28. *A subfamily of a Rademacher (weak Rademacher) family is a Rademacher (weak Rademacher) family itself.*

To emphasize the importance of the vanishing property of a Rademacher family, we provide an example showing the variety of distinct (nonisomorphic) types of countable τ-irredundant independent families without this property.

Proposition 21.29. *Let $\mathcal{B}$ be a purely atomic τ-complete Boolean algebra with the set $\mathcal{A}_0$ of atoms of cardinality $\aleph_0 \le |\mathcal{A}_0| \le \mathfrak{c}$, where $\mathfrak{c}$ is the cardinality of continuum. Then there exists a countable τ-generating τ-irredundant independent family in $\mathcal{B}$.*

Proof. With no loss of generality, we assume that $\mathcal{A}_0 = \{\{a\} : a \in \mathcal{B}_0\}$, where $\mathcal{B}_0$ is a dense subset of $[0,1)$ and $\mathcal{B}$ is the power set of $\mathcal{B}_0$, that is, the set of all subsets of $\mathcal{B}_0$. Any number $x \in [0,1)$ we represent as $x = \sum_{n=1}^{\infty} a_n(x)\,2^{-n}$, where the dyadic digits $a_n(x) \in \{0,1\}$ are not eventually 1's. We set $r_n = \{x \in \mathcal{B}_0 : a_n(x) = 1\}$. Then for any finite collection of distinct numbers $n_1, \dots, n_k \in \mathbb{N}$ and signs $\theta_1, \dots, \theta_k = \pm 1$ one has

$$\theta_1 r_{n_1} \wedge \dots \wedge \theta_k r_{n_k} = \left\{x \in [0,1) : (\forall i \le k)\, a_{n_i}(x) = \frac{\theta_i + 1}{2}\right\} \wedge \mathcal{B}_0,$$

which is nonempty because $\mathcal{B}_0$ is dense in $[0,1)$. To show that $(r_n)_{n=1}^{\infty}$ is σ-generating, observe that for any $y \in [0,1)$ one has

$$\{y\} = \bigcap_{n=1}^{\infty} \{x \in [0,1) : a_n(x) = a_n(y)\} = \bigcap_{n=1}^{\infty} (2a_n(y) - 1)\{x \in [0,1) : a_n(x) = 1\}.$$

In particular, for any $y \in \mathcal{B}_0$ one has $\{y\} = \bigcap_{n=1}^{\infty} (2a_n(y) - 1)\, r_n$.

Finally, we show that $(r_n)_{n=1}^{\infty}$ is τ-irredundant. Fix any $n_0 \in \mathbb{N}$. We prove the following claim.

Claim. *For any $A \in \mathcal{B}_\tau(\{r_n : n \ne n_0\})$, one has*

$$(\forall x \in A) \quad x^* \stackrel{def}{=} \sum_{n \ne n_0} a_n(x) 2^{-n} + (1 - a_{n_0}(x))\, 2^{-n_0} \in A. \tag{21.14}$$

First, observe that (21.14) holds for $A = r_n$ with any $n \ne n_0$. Hence, by (1) of Proposition 21.8, (21.14) holds for all $A \in \mathcal{B}(\{r_n : n \ne n_0\})$. Now fix any $A' \in \mathcal{B}_\tau(\{r_n : n \ne n_0\})$. By (2) of Proposition 21.8, there exists a net (A_α) in $\mathcal{B}(\{r_n : n \ne n_0\})$ and a net (u_α) in $\mathcal{B}$ with the same index set such that $u_\alpha \downarrow \mathbf{0}$ and $A_\alpha \bigtriangleup A' \le u_\alpha$. Hence, given any $x \in A'$, there exists α_0 such that $x \notin u_\alpha$ and, therefore, $x \in A_\alpha$ for all $\alpha \ge \alpha_0$. Since (21.14) holds for $A = A_\alpha$, we have that $x^* \in A_\alpha$ for all $\alpha \ge \alpha_0$, and so $x^* \in A$. Thus, (21.14) holds for $A = A'$ and the claim is proved.

Since (21.14) does not hold for $A = r_{n_0}$, we deduce that $r_{n_0} \in \mathcal{B}_\tau(\mathcal{R}) \setminus \mathcal{B}_\tau(\{r_n : n \ne n_0\})$. □

We finish the section with an example which shows that a weak Rademacher family need not be Rademacher.

Example 21.30. There exists a σ-complete Boolean algebra with a countable weak Rademacher family which is not Rademacher.

Proof. Let $\mathcal{B} = \mathrm{Borel}\,\{-1,1\}^{\mathbb{N}}/M$ be the quotient Boolean algebra of the σ-algebra of Borel sets in the Cantor set $\{-1,1\}^{\mathbb{N}}$ modulo the σ-ideal M of meager sets. Then $\mathcal{B}$ is

σ-complete. Consider the sequence $\mathcal{R} = (r_n)_{n=1}^{\infty}$ defined by

$$r_n = \{x = (x_1, x_2, \dots) \in \{-1,1\}^{\mathbb{N}} : x_n = 1\}, \quad n \in \mathbb{N}$$

and denote by s_n the element of $\mathcal{B}$ containing r_n. We show that the family $\mathcal{S} = (s_n)_{s=1}^{\infty}$ possesses the desired properties. Obviously, $\mathcal{S}$ is an independent family. Since r_n are clopen sets, an infinite intersection of r_n or their complements is closed and, having empty interior, is therefore meager. Hence, any infinite meet of s_n or their complements is zero in $\mathcal{B}$. Thus, $\mathcal{S}$ is vanishing.

Now we prove that $\mathcal{S}$ is τ-irredundant. Assume, on the contrary, that there is $j \in \mathbb{N}$ and a sequence (because $\mathcal{B}$ is σ-complete) $\widehat{v}_n \in \mathcal{B}_\sigma(\mathcal{S} \setminus \{s_j\})$, $n \in \mathbb{N}$ such that $(\widehat{v}_n)_{n=1}^{\infty}$ order tends to s_j in $\mathcal{B}$. For every $n \in \mathbb{N}$ pick any $v_n \in \widehat{v}_n$. Then

$$t := \bigcap_{n=1}^{\infty} \bigcup_{m=n}^{\infty} (r_j \vartriangle v_m) \in M. \tag{21.15}$$

Define a function F : $\mathrm{Borel}\,\{-1,1\}^{\mathbb{N}} \to \mathrm{Borel}\,\{-1,1\}^{\mathbb{N}}$ by setting for all $A \in \mathrm{Borel}\,\{-1,1\}^{\mathbb{N}}$

$$F(A) = \{(x_1, \dots, x_{j-1}, -x_j, x_{j+1}, \dots) \in \{-1,1\}^{\mathbb{N}} : (x_1, \dots, x_{j-1}, x_j, x_{j+1}, \dots) \in A\}.$$

Then F is a Boolean σ-isomorphism which sends meager sets to meager sets. Observe that $F(v_m) = v_m$ for all $m \in \mathbb{N}$ and $F(r_j) = -r_j$. Hence, by (21.15),

$$F(t) := \bigcap_{n=1}^{\infty} \bigcup_{m=n}^{\infty} (-r_j \vartriangle v_m) \in M. \tag{21.16}$$

By (21.16), $(\widehat{v}_n)_{n=1}^{\infty}$ order tends to $-s_j$ in $\mathcal{B}$, a contradiction.

It remains to show that $\mathcal{S}$ is not injective. Consider a “fat Cantor set,” a compact set Z in $\mathrm{Borel}\,\{-1,1\}^{\mathbb{N}}$ with empty interior and measure 1/2. Then the sequence of dyadic intervals which have been thrown when constructing Z is contained inside $\{-1,1\}^{\mathbb{N}} \setminus Z$ and have supremum $\mathbf{1}$ in $\mathrm{Borel}\,\{-1,1\}^{\mathbb{N}}$, because their union is co-meager. But the sum of measures is not 1, it cannot exceed 1/2. □

21.5 Main results

The present section is devoted to several important implications of Theorem 21.23 (Theorem A is rewritten in an equivalent form; see Theorem A.1).

Theorem A.1. *Let $\mathcal{R} = (r_n)_{n=1}^{\infty}$ be a sequence of elements of a σ-complete Boolean algebra $\mathcal{B}$. Then the following assertions are equivalent:*

(1) *There is a positive σ-additive measure $\mu : \mathcal{B}_\sigma(\mathcal{R}) \to [0,1]$ possessing (21.1).*

(2) *$\mathcal{R}$ is σ-equivalent to the usual Rademacher family* $(\widehat{\boldsymbol{r}}_n)_{n=1}^{\infty}$.
(3) *$\mathcal{R}$ is a Rademacher family.*

Proof of Theorem A.1. (2) ⇒ (3). If $\mathcal{R}$ is σ-equivalent to $(\widehat{\boldsymbol{r}}_n)_{n=1}^{\infty}$, then obviously the independence, injectivity and the property to be vanishing for $\mathcal{R}$ follows from the same properties of $(\widehat{\boldsymbol{r}}_n)_{n=1}^{\infty}$.

(3) ⇒ (2). Now let $\mathcal{R}$ be a Rademacher family. We age going to define a regular tree $(b_n)_{n=1}^{\infty}$ in $\mathcal{B}$. For this purpose, given any $k \in \mathbb{N}$, $\ell \in \{0, 1, \dots, 2^k - 1\}$ and $j \in \{1, \dots, k\}$ by $\alpha_{k,\ell,j}$ we define the digit $\alpha_{k,\ell,j} \in \{0, 1\}$ such that $\ell = \sum_{j=1}^{k} \alpha_{k,\ell,j} 2^{j-1}$, and then define signs by setting $\theta_{k,\ell,j} = 1 - 2\alpha_{k,\ell,j} \in \{-1, 1\}$. Now set $b_1 = \mathbf{1}$, $b_2 = r_1$, $b_3 = -r_1$, and more generally

$$b_{2^k+\ell} = \bigwedge_{j=1}^{k} \theta_{k,\ell,j} r_j, \quad k \in \mathbb{N},\ \ell \in \{0, 1, \dots, 2^k - 1\}. \tag{21.17}$$

It is straightforward that $(b_n)_{n=1}^{\infty}$ is a regular tree, every element of which is a particle of $\mathcal{R}$. Show that $(b_n)_{n=1}^{\infty}$ is vanishing. Let $n_1 < n_2 < \cdots$. If $b_{n_i} \wedge b_{n_j} = \mathbf{0}$ for some indices $i \neq j$ then surely $\bigwedge_{k=1}^{\infty} b_{n_k} = \mathbf{0}$. Assume now that $b_{n_i} \wedge b_{n_j} > \mathbf{0}$ for all $i \neq j$. Observe that, if two distinct particles p', p'' are not disjoint then either $p' < p''$ of $p'' < p'$. By the ordering of b_n's, one has that $b_{n_k} > b_{n_{k+1}}$ for all $k \in \mathbb{N}$. This means that $b_{n_{k+1}} = b_{n_k} \wedge \bigwedge_{j \in J_k} \theta_j r_j$ with disjoint nonempty sets of indices J_k and some $\theta_j \in \{-1, 1\}$. Thus,

$$\bigwedge_{k=1}^{\infty} b_{n_k} = b_{n_1} \wedge \bigwedge_{k=1}^{\infty} \bigwedge_{j \in J_k} \theta_j r_j = \mathbf{0},$$

because $\mathcal{R}$ is vanishing. So, $(b_n)_{n=1}^{\infty}$ is vanishing. The injectivity of $(b_n)_{n=1}^{\infty}$ follows from the injectivity of $\mathcal{R}$ due to the observation that b_n is a particle for all n, and $\mu^*(b_n) = 2^{-[\log_2 n]}$ by (21.1). By Theorem 21.23, $(b_n)_{n=1}^{\infty}$ is σ-equivalent to the tree of dyadic intervals $(\widehat{b}_n)_{n=1}^{\infty}$. Let $\tau : \mathcal{B}_\sigma(\{b_n : n \in \mathbb{N}\}) \to \widehat{\mathcal{B}}$ be a Boolean σ-isomorphism such that $\tau(b_n) = \widehat{b}_n$ for all $n \in \mathbb{N}$. One can inductively show that

$$r_n = \bigvee_{j=0}^{2^{n-1}-1} b_{2^n+2j}, \quad n \in \mathbb{N}.$$

Hence, taking into account (21.17), we deduce that $\mathcal{B}_\sigma(\{b_n : n \in \mathbb{N}\}) = \mathcal{B}_\sigma(\mathcal{R})$ and, moreover,

$$\tau(r_n) = \bigvee_{j=0}^{2^{n-1}-1} \tau(b_{2^n+2j}) = \bigvee_{j=0}^{2^{n-1}-1} \widehat{b}_{2^n+2j} = \widehat{\boldsymbol{r}}_n, \quad n \in \mathbb{N}.$$

So, $\tau : \mathcal{B}_\sigma(\mathcal{R}) \to \widehat{\mathcal{B}}$ is a Boolean σ-isomorphism with $\tau(r_n) = \widehat{\boldsymbol{r}}_n$ for all $n \in \mathbb{N}$. So, the equivalence (2) ⇔ (3) is proved.

Implications (1) ⇒ (3) and (2) ⇒ (1) are easy to prove. □

Theorem 21.31. *A transfinite sequence $\mathcal{R} = (r_\alpha)_{\alpha<\omega_\delta}$ in a σ-complete Boolean algebra $\mathcal{B}$ is σ-equivalent to the generalized Rademacher family $(\overline{r}_\alpha)_{\alpha<\omega_\delta}$ of the same cardinality if and only if $\mathcal{R}$ is a Rademacher family.*

Proof. Let P be any of the properties: independent, vanishing, injective. Observe that a transfinite sequence $\mathcal{R} = (r_\alpha)_{\alpha<\omega_\delta}$ in a σ-complete Boolean algebra $\mathcal{B}$ possesses P if and only if every countable subsequence of $\mathcal{R}$ possesses P. Hence, $\mathcal{R}$ is a Rademacher family if and only if every of its countable subfamily is. Another observation, useful for the proof, is that the generalized Rademacher family $(\overline{r}_\alpha)_{\alpha<\omega_\delta}$ is a Rademacher family in $\widehat{\Sigma}_{\omega_\delta}$. Hence, if $\mathcal{R}$ is σ-equivalent to $(\overline{r}_\alpha)_{\alpha<\omega_\delta}$ then it is a Rademacher family.

Let $\mathcal{R}$ be a Rademacher family. Our goal is to construct a Boolean σ-isomorphism $S : \mathcal{B}_\sigma(\mathcal{R}) \to \widehat{\Sigma}_{\omega_\delta}$ with $S(r_\alpha) = \overline{r}_\alpha$ for all $\alpha < \omega_\delta$. Denote by $\mathfrak{M}$ the set of all Boolean σ-isomorphisms $S_A : \mathcal{B}_\sigma(\{r_\alpha : \alpha \in A\}) \to \widehat{\Sigma}_A$ with $S(r_\alpha) = \overline{r}_\alpha$ for all $\alpha \in A$, where A runs through all infinite subsets of ω_δ and $\widehat{\Sigma}_A$ denotes the minimal σ-complete subalgebra of $\widehat{\Sigma}_{\omega_\delta}$ including $\{\overline{r}_\alpha : \alpha \in A\}$. Observe that if $A \subseteq B$ then $S_A \subseteq S_B$, that is, the function S_B is an extension of S_A (see Remark 21.4). Hence, every chain $\mathfrak{L}$ in $\mathfrak{M}$ has an upper bound $\bigcup \mathfrak{L}$ in $\mathfrak{M}$. By Zorn's lemma, $\mathfrak{M}$ has a maximal element S which must be obviously S_{ω_δ}. □

Corollary 21.32. *Any two Rademacher families in σ-complete Boolean algebras of the same cardinality, that are arbitrarily well ordered, are σ-equivalent. In particular, a Rademacher family in a σ-complete Boolean algebra, which is arbitrarily well ordered, is σ-equivalent to every of its rearrangement, as well, as to every of its transfinite subsequence of the same cardinality.*

Corollary 21.33. *Let $\mathcal{R} = (r_\alpha)_{\alpha<\omega_\delta}$ be a σ-generating Rademacher family in a σ-complete Boolean algebra $\mathcal{B}$, where ω_δ is an infinite cardinal. Then the following assertions hold:*
(1) *There is a unique Boolean σ-isomorphism $S : \mathcal{B} \to \widehat{\Sigma}_{\omega_\delta}$ such that $S(r_\alpha) = \overline{r}_\alpha$ for all $\alpha < \omega_\delta$.*
(2) *There is a unique positive σ-additive measure $\mu : \mathcal{B} \to [0,1]$ which extends the dyadic measure μ^* with respect to $\mathcal{R}$, that is, (21.1) holds.*
(3) *$\mathcal{B}$ is a Maharam homogeneous measurable algebra.*

Proof. (1) follows from Theorem 21.31 and Remark 21.4.

(2) Let $S : \mathcal{B} \to \widehat{\Sigma}_{\omega_\delta}$ be the Boolean σ-isomorphism such that $S(r_\alpha) = \overline{r}_\alpha$ for all $\alpha < \omega_\delta$. Then the measure $\mu : \mathcal{B} \to [0,1]$ defined by setting $\mu(x) = \mu_{\omega_\delta}(S(x))$ for all $x \in \mathcal{B}$ is as desired.

(3) follows from (2). □

By the Maharam theorem (see [8] for the original paper, and [4], [7] for different proofs), we obtain the following new characterization of homogeneous measurable algebras.

Corollary 21.34. *A σ-complete Boolean algebra $\mathcal{B}$ admits a σ-generating Rademacher family if and only if $\mathcal{B}$ is a Maharam homogeneous measurable algebra.*

Since every Rademacher family is σ-equivalent to the generalized Rademacher system which is τ-irredundant, we have the following implication.

Corollary 21.35. *A Rademacher family in a σ-complete Boolean algebra is a weak Rademacher family.*

Bibliography

[1] S. V. Astashkin, *Rademacher functions in symmetric spaces*, J. Math. Sci. **169** (6) (2010), 725–886.
[2] V. I. Bogachev, *Measure Theory, vol. 1*, Springer, Berlin, 2007.
[3] D. H. Fremlin, *Measure Theory, vol. 1*, Torres Fremlin, Colchester, 2000.
[4] D. H. Fremlin, *Measure Theory, vol. 3. Measure Algebras*, Torres Fremlin, Colchester, 2004.
[5] S. Givant and P. Halmos, *Introduction to Boolean Algebras*, Springer, New York, 2009.
[6] S. Koppelberg, *Handbook of Boolean Algebras, vol. I*, J. Donald Monk and Robert Bonnet (eds.), North Holland, 1989.
[7] H. E. Lacey, *The Isometric Theory of Classical Banach Spaces*, Springer-Verlag, Berlin, Heidelberg, New York, 1974.
[8] D. Maharam, *On homogeneous measure algebras*, Proc. Nat. Acad. Sci. U.S.A. **28** (1942), 108–111.
[9] M. Popov and B. Randrianantoanina, *Narrow Operators on Function Spaces and Vector Lattices*, De Gruyter Studies in Mathematics, **45**, De Gruyter, 2013.
[10] W. Rudin, *Well-distributed measurable sets*, Amer. Math. Mon. **90** (1983), 41–42.

Index

Bei Fragen zur Produktsicherheit wenden Sie sich bitte an:
If you have any questions regarding product safety,
please contact:

Walter de Gruyter GmbH
Genthiner Straße 13
10785 Berlin
productsafety@degruyterbrill.com